XXII Fluid Mechanics Conference (KKMP 2016)

Journal of Physics: Conference Series Volume 760

Belchatow, Poland
11 – 14 September 2016

Editors:

Henryk Kudela
Katarzyna Strzelecka
Pawel Regucki

ISBN: 978-1-5108-3274-9
ISSN: 1742-6588

Printed from e-media with permission by:

Curran Associates, Inc.
57 Morehouse Lane
Red Hook, NY 12571

Some format issues inherent in the e-media version may also appear in this print version.

Printed by Curran Associates, Inc. (2016)

For permission requests, please contact the Institute of Physics
at the address below.

Institute of Physics
Dirac House, Temple Back
Bristol BS1 6BE UK

Phone: 44 1 17 929 7481
Fax: 44 1 17 920 0979

techtracking@iop.org

Additional copies of this publication are available from:

Curran Associates, Inc.
57 Morehouse Lane
Red Hook, NY 12571 USA
Phone: 845-758-0400
Fax: 845-758-2633
Email: curran@proceedings.com
Web: www.proceedings.com

Table of contents

Volume 760

XXII Fluid Mechanics Conference (KKMP2016)
11–14 September 2016, Słok near Bełchatów, Poland

Accepted papers received: 23 September 2016
Published online: 14 November 2016

Preface

Papers

Journal of Physics: Conference Series **760** (2016) 011001 doi:10.1088/1742-6596/760/1/011001

XXII FLUID MECHANICS CONFERENCE
Słok near Bełchatów city, Poland, 11–14 September 2016

Preface

This *Journal of Physics: Conference Series* contains of papers that have been presented at XXII Fluid Mechanics Conference (XXII FMC) held in Słok near Bełchatów in Poland during 11th–14th September of 2016. The Conference is organized by Wrocław University of Science and Technology, Polish Academy of Sciences – Committee of Mechanics and Foundation for Development of Wrocław University of Science and Technology.

Let us recall some historical facts: Fluid Mechanics Conferences have been taking place every two years since 1974, which makes a total of forty-two years. The goal of this conference is to provide a forum for exposure and exchange of ideas, methods and results in fluid mechanics.

We have already met in Bełchatów 10 years ago (XVII KKMP). It was a successful meeting. Since then National Conference on Fluid Mechanics has changed the title and started to be named as Fluid Mechanics Conference in the hopes that it will attract more participants from other countries. English became the Conference's first language and we started to invite world leading scientists – working in the field of fluid mechanics. At the 2006 conference we hosted for the first time prof. Keith Moffatt from the Cambridge University. In this year prof. Moffatt once again promised us to arrive to Bełchatów. The whole fluid mechanics community celebrates 9^2 anniversary of his birthday. So let us also wish happy anniversary to prof. Moffatt.

In the mean time we had to pay last respects to our collages. Prof. Prosnak who is regarded as a founder of Notational Conference on Fluid Mechanics and is well known through his books. Prof. Puzyrewski who was present at all conferences so far. He was providing by his discussions a special value to these conferences, and our colleague prof. Konrad Bajer who was intended to be the organizer and host of the present conference. Short memories to them will be given during the opening ceremony.

Conference topics include, but are not limited to aerodynamics, atmospheric science, bio-fluids, combustion and reacting flows, computational fluid dynamics, experimental fluid mechanics, flow machinery, general fluid dynamics, hydromechanics, heat and fluid flow, measurement techniques, micro- and nano-flow, multi-phase flow, non-Newtonian fluids, rotating and stratified flows and turbulence.

Within the general subjects of this conference, the Professor Janusz W. Elsner's Competition for the best fluid mechanics paper presented during the Conference is organized. Authors holding a M.Sc. or a Ph.D. degree and who are not older than 35 may enter the Competition. Authors with a Ph.D. degree must present individual papers; authors with a M.Sc. degree may present papers with their supervisors as coauthors, including original results of experimental, numerical or analytic research.

Six state-of-the-art keynote papers will be delivered by world leading experts. All contributed papers were peer reviewed. Recommendations were received from the Scientific Committee of the Conference; reviewers were from all Polish scientific-academic centres that are involved in fluid mechanics. Accordingly, of the 67 eligible extended abstracts submitted, after a review process by the Scientific Committee, all papers were selected for presentation at XXII Fluid Mechanics Conference.

We hope that this publication will be used not only as a guide through the conference's programme but also help in the future to access people and topics in fluid mechanics research.

XXII FLUID MECHANICS CONFERENCE

Słok near Bełchatów city, Poland, 11–14 September 2016

We would like to express grateful appreciation to our colleagues from the Polish Academy of Sciences – Committee of Mechanics, as well as to the Scientific Committee of this conference. Their advices and efforts have helped us to overcome the problems normally associated with organising international meeting. Special thanks goes to the reviewers for their work in encouraging the submission of papers. Their contribution cannot be overestimated.

Thank you very much.

Organizing Committee,
Faculty of Mechanical and Power Engineering,
Wrocław University of Science and Technology

Henryk Kudela
Katarzyna Strzelecka
Paweł Regucki

Journal of Physics: Conference Series **760** (2016) 011001 doi:10.1088/1742-6596/760/1/011001

SCIENTIFIC COMMITTEE

CHAIRMAN:

Tomasz Kowalewski,
Institute of Fundamental Technological Research,
Polish Academy of Sciences

MEMBERS

Janusz Badur, Szewalski Institute of Fluid-Flow Machinery Polish Academy of Sciences
Andrzej Bogusławski, Częstochowa University of Technology
Tadeusz Chmielniak, Silesian University of Technology
Michał Ciałkowski, Poznań University of Technology
Piotr Doerffer, Szewalski Institute of Fluid-Flow Machinery Polish Academy of Sciences
Stanisław Drobniak, Częstochowa University of Technology
Zbigniew Kosma, Kazimierz Pułaski University of Technology and Humanities in Radom
Anna Kucaba-Piętal, Rzeszów University of Technology
Henryk Kudela, Wrocław University of Science and Technology
Jan Kulczyk, Wrocław University of Science and Technology
Marek Morzyński, Poznań University of Technology
Zbigniew Peradzyński, University of Warsaw
Jacek Pozorski, Szewalski Institute of Fluid-Flow Machinery Polish Academy of Sciences
Romuald Puzyrewski, Gdańsk University of Technology
Jacek Rokicki, Warsaw University of Technology
Kazimierz Rup, Tadeusz Kościuszko Cracow University of Technology
Andrzej Styczek, Warsaw University of Technology
Janusz Szmyd, AGH University of Science and Technology
Jacek Szumbarski, Warsaw University of Technology
Ewa Tuliszka-Sznitko, Poznań University of Technology

Journal of Physics: Conference Series **760** (2016) 011001 doi:10.1088/1742-6596/760/1/011001

ORGANISERS

Polish Academy of Science
Committee of Mechanics

Wrocław University of Science
and Technology

Fundacja Rozwoju
Politechniki Wrocławskiej

ORGANIZING COMMITTEE

Henryk Kudela

Conference chairman,
Wrocław University of Science and Technology

Katarzyna Strzelecka

Conference secretary,
Wrocław University of Science and Technology

Paweł Regucki

Conference manager,
Wrocław University of Science and Technology

HONORARY PATRONAGE

RECTOR OF WROCŁAW UNIVERSITY OF SCIENCE AND TECHNOLOGY

Wrocław University of Science and Technology

PGE GÓRNICTWO I ENERGETYKA KONWENCJONALNA S.A.

PGE

PGE Górnictwo i Energetyka Konwencjonalna SA

SPONSORS

mpwik WROCŁAW

..unisens

SymKom

intel

casp system ADVANCED TECHNOLOGIES

LAVISION FOCUS ON IMAGING

DANTEC DYNAMICS

Eurotek International

SCIENTIFIC REVIEWERS

Janusz Badur, Szewalski Institute of Fluid-Flow Machinery Polish Academy of Sciences
Jerzy Banaszek, Warsaw University of Technology
Artur Bartosik, Kielce University of Technology
Andrzej Bogusławski, Częstochowa University of Technology
Michał Ciałkowski, Poznań University of Technology
Tadeusz Chmielniak, Silesian University of Technology
Piotr Doerffer, Szewalski Institute of Fluid-Flow Machinery Polish Academy of Sciences
Stanisław Drobniak, Częstochowa University of Technology
Sławomir Dykas, Silesian University of Technology
Witold Elsner, Częstochowa University of Technology
Paweł Flaszyński, Szewalski Institute of Fluid-Flow Machinery Polish Academy of Sciences
Elżbieta Fornalik-Wajs, AGH University of Science and Technology
Marek Jaszczur, AGH University of Science and Technology
Krzysztof Jóźwik, Lodz University of Technology
Zbigniew Kosma, Kazimierz Pułaski University of Technology and Humanities in Radom
Tomasz Kowalewski, Institute of Fundamental Technological Research, Polish Academy of Science
Krzysztof Kubryński, Air Force Institute of Technology, Warsaw
Anna Kucaba-Piętal, Rzeszów University of Technology
Henryk Kudela, Wrocław University of Science and Technology
Jan Kulczyk, Wrocław University of Science and Technology
Marek Morzyński, Poznań University of Technology
Romuald Mosdrof, Białystok University of Technology
Jacek Pozorski, Szewalski Institute of Fluid-Flow Machinery Polish Academy of Sciences
Jacek Rokicki, Warsaw University of Technology
Kazimierz Rup, Tadeusz Kościuszko Cracow University of Technology
Piotr Sawicki, Poznań University of Technology
Wojciech Sobieski, University of Warmia and Mazury in Olsztyn
Andrzej Styczek, Warsaw University of Technology
Janusz Szmyd, AGH University of Science and Technology
Jacek Szumbarski, Warsaw University of Technology
Ryszard Szwaba, Szewalski Institute of Fluid-Flow Machinery Polish Academy of Sciences
Krzysztof Tesch, Gdańsk University of Technology
Ewa Tuliszka-Sznitko, Poznań University of Technology
Artur Tyliszczak, Częstochowa University of Technology
Włodzimierz Wróblewski, Silesian University of Technology

XXII Fluid Mechanics Conference
11–14.09.2016, Słok near Bełchatów, Poland

Peer review statement

All papers published in this volume of *Journal of Physics: Conference Series* have been peer reviewed through processes administered by the proceedings Editors. Reviews were conducted by expert referees to the professional and scientific standards expected of a proceedings journal published by IOP Publishing.

Analysis of unsteady flow forces acting on the thermowell in a steam turbine control stage

J Badur[1,2], S Kornet[1,3], D Sławiński[1] and P Ziółkowski[1,4]
[1] Energy Conversion Department, The Szewalski Institute of Fluid-Flow Machinery PAS-ci, Fiszera 14, 80-231 Gdańsk, Poland

E-mail: [2]jb@imp.gda.pl, [3]skornet@imp.gda.pl, [4]pziolkowski@imp.gda.pl

Abstract. In the present paper the phenomenon of unsteady flow forces acting on the thermowell for measuring steam temperature in a steam turbine control stage has been presented. The non-stationarity of fluid acting on the thermowell such as: Strouhal frequency, pressure amplitude, pressure peaks, pressure field, velocity field etc. have been studied analytically and numerically. There have been examined two cases of flow with changing mass flow rate, pressure and temperature in the control stage chamber of a turbine high-pressure cylinder. The problem of entry into resonance by thermowell has been described in the ASME standard PTC19.3 TW-2010 with providing detailed guidelines for thermowell designs.

1. Introduction

Generally, from the scientific and engineering point of view, unsteady flow forces acting on a solid structure are a very important issue. Therefore, in the paper, in order to specify the similarity between periodical phenomena and to take into account the case of non-stationary flows forces acting on the thermowell for measuring steam temperature in a steam turbine control stage, a novel CFD+CSD approach has been used.

However, it should be underlined that the presence of a cylindrical element in the steam flow is a reason of occurrence of the vortex shedding phenomenon, where the flowing steam detaches from the cylinder surface creating downstream a vortex structure [1]. The accompanying pressure pulsations generate flow-induced vibrations which may lead to thermowell damage due to high-cycle fatigue [2,3]. The problem is described in the ASME standard PTC19.3 TW-2010 [4] providing detailed guidelines for thermowell calculations.

1.1. Resonance and Strouhal number Sh

In order to specify the similarity between periodical phenomena and non-stationary flows, the Strouhal number Sh is introduced, which is the similarity number defined as [5]:

$$\mathrm{Sh} = \frac{a}{v\,t} = \frac{a\,f}{v}\,, \tag{1}$$

where: a - characteristic dimension, v - steady ambient velocity of the uniform flow, t - period of tested phenomenon, f - frequency of the phenomenon.

It should be added, that the Strouhal number Sh is used in the engineering technology to analysis of e.g. flow around the vibrating body (aerofoil, rod, thermowell), flow through a blade row mounted in the rotor disc, flow around a rotating propeller or screw propeller, etc.

There is broad literature discussing the problem of the formation of vortices, changes in their structure, the impact of vortex structures on objects on which they form. The research on the vorticity and the formation of structures at flow has been carried out for hundreds years, what is evidenced by sketches of Leonardo da Vinci [6]. Great importance in the development of knowledge of periodical phenomena had works of von Kármán about vortices generation and works of Strouhal about frequency of their formation. Detailed description of these issues can be found in the works of the pioneers [7,8] and of contemporary authors [5,9]. Most of the published articles related to the non-stationary structures in flow, concern the experimental work and analysis [10-13], but the importance of numerical works is growing [14-16].

The fundamental work in this regard is the analysis of the formation and development of vortex structures in a flow around obstacles. A significant number of studies in this field concerns laminar-turbulent transition in the area, where unrecognized and not well recognized phenomena in single theory occurs. There are known works, in which the obstacle has the form of sphere [13] , rotating sphere [12], trapeze [10] or cube [11].

Especially important is the problem of vortices forcing vibrations, and vice versa - the vibration of devices affecting the vortex structure. In the last decade, a number of works was published [15] concerning interaction of fluid and structures arising in it, which resulted in creation of a new norm for ASME PTC 19.3 TW-2010. However, the main aim of this paper is analysis of interaction between steam and thermowell. Steam temperature measurements in thermal turbines are conducted using thermocouples placed in thermowell which is inserted into the steam flow. The thermowell protects the temperature sensor against detrimental action of steam, simultaneously introducing to the measurement system a thermal inertia resulting from the solid design of thermowell. The presence of a cylindrical element in the steam flow is a reason of occurrence of the vortex shedding phenomenon, where the flowing steam detaches from the cylinder surface creating downstream a vortex structure [17]. The accompanying pressure pulsations generate flow-induced vibrations which may lead to thermowell damage due to high-cycle fatigue [1,18-21].

The problem is well described in the ASME standard PTC19.3 TW-2010 [4] providing detailed guidelines for thermowell calculations. However, thermal and strength analyses of elements working in high-pressure were presented in [18-21]. Other studies [22,23] have modelled and predicted the stress rupture of alloys and the residual life of the most critical components while considering the effects of creep, thermo-mechanical fatigue, corrosion, and oxidation.

In the framework of root cause analysis, steam flow, dynamic and mechanical integrity calculations were performed aiming at determining steam excitations acting on the thermowell, its natural frequencies and safety margins. For this purpose, CFD+CSD (Computational Fluid-Solid Dynamic) approach has been adopted.

2. Model description

In analyses case, the similarity criterion is the frequency of the force which can be directly compared with the free vibration frequency of the device, so that it can be determined whether the resonance occurs. An important tool in the analysis of this phenomenon is the combination of CFD + CSD, often called FSI - Fluid Solid Interaction or Fluid Structure Interaction.

2.1. CFD

Traditionally, the five balance equations (that consist of one mass balance equation, three momentum balance equations and one energy balance equation) and two evolution equation for parameters defining turbulence (equation for turbulent kinetic energy evolution k and equation for turbulence dissipation evolution ε) have been described in [6,22]. To maximize the use of the easily implementable matrix calculus, the starting point for CFD computation is formulation of universal set of mass, momentum and energy balance equations for the fluid, supplemented with equations for turbulence evolution *k-ε* in the form of:

$$\frac{\partial}{\partial t}\begin{Bmatrix}\rho\\ \rho\mathbf{v}\\ \rho e\\ \rho k\\ \rho\varepsilon\end{Bmatrix}+\operatorname{div}\begin{Bmatrix}\rho\mathbf{v}\\ (\rho\mathbf{v}\otimes\mathbf{v})+p\mathbf{I}\\ (\rho e+p)\mathbf{v}\\ \rho\mathbf{v}k\\ \rho\mathbf{v}\varepsilon\end{Bmatrix}=\operatorname{div}\begin{Bmatrix}0\\ \mathbf{t}^{res}\\ \mathbf{t}^{res}\,\mathbf{v}+\mathbf{q}^{c}\\ \mathbf{J}_{k}\\ \mathbf{J}_{\varepsilon}\end{Bmatrix}+\begin{Bmatrix}0\\ \rho\mathbf{b}\\ \rho S_{e}\\ \rho S_{k}\\ \rho S_{\varepsilon}\end{Bmatrix} \tag{2}$$

where: $\rho = \rho(\mathbf{x},t)$ - fluid density, generally dependent on time t and location $\mathbf{x}$, $\mathbf{v} = v_i\mathbf{e}_i$ - fluid velocity, p - thermodynamic pressure, $\mathbf{I} = \delta_{ij}\mathbf{e}_i \otimes \mathbf{e}_j$ - unit tensor, $\mathbf{t}^{res} = \mathbf{t}^{lam} + \mathbf{t}^{tur}$ - stress flux components, laminar end turbulent respectively, $\mathbf{b}$ - mass force of earth gravity, $e = u + \frac{1}{2}\mathbf{v}^2$ - sum of internal and kinetic energy , $\mathbf{q}$ - heat flux, $\mathbf{J}_k$, $\mathbf{J}_\varepsilon$ - diffusive flux of k and diffusive flux of ε with sources S_k, S_ε (various definitions of different authors exist in literature). For each finite volume of the computational grid, seven equations are solved (one for mass, energy, k and ε transport balance equation and three momentum balance equations) [6].

2.2. CSD

The problem of natural vibration has been discussed in [2,6]. The problem of determination of non-stationary load coming from steam flow, amplitudes of excitation from non-stationary flow and partial load have recently been discussed in [2,6,21].

In analogy to CFD, the appropriate set of CSD governing equations is determined as follows [6]:

$$\frac{\partial}{\partial t}\begin{Bmatrix}1\\ \rho\mathbf{v}\\ \rho e\\ \rho\mathbf{e}^{pl}\\ \rho\boldsymbol{\alpha}\\ \rho r\end{Bmatrix}+\operatorname{div}\begin{Bmatrix}0\\ \rho\mathbf{v}\otimes\mathbf{v}\\ \rho e\mathbf{v}\\ \rho\mathbf{e}^{pl}\otimes\mathbf{v}\\ \rho\boldsymbol{\alpha}\otimes\mathbf{v}\\ \rho r\mathbf{v}\end{Bmatrix}=\operatorname{div}\begin{Bmatrix}0\\ \boldsymbol{\sigma}\\ \boldsymbol{\sigma}\mathbf{v}+\mathbf{q}\\ 0\\ 0\\ \mathbf{J}_{r}\end{Bmatrix}+\begin{Bmatrix}0\\ \rho\mathbf{b}\\ \rho S_{e}\\ \rho S_{pl}\\ \rho S_{\boldsymbol{\alpha}}\\ \rho S_{r}\end{Bmatrix} \tag{3}$$

where: $\boldsymbol{\sigma}$ - tensor of solid stresses; $\mathbf{e}^{pl}$ - tensor of plastic strain; $\boldsymbol{\alpha}$ - kinematic hardening; r - isotropic hardening; $\mathbf{J}_r$ - diffusive flux of r; S_{pl}, $S_{\boldsymbol{\alpha}}$, S_r sources of plasticity , kinematic and isotropic hardening typical for the Chaboche model. Our analysis and calculations in solid body are based on the 3D modal and harmonic analysis typical for the CSD [2]. CSD is point-blank analogy of CFD (Computational Fluid Dynamics). Both methods use the same balance equations (mass, momentum and energy). Discretization method for the CSD and CFD is arbitrary (FEM, FVM, etc) but the governing equations are identical. This architecture of solving equations greatly simplifies FSI (Fluid Solid Interaction) [2,6] and thermal-FSI [22] analyses.

3. Analysis

3.1. Geometry and boundary conditions

The thermowell analyzed here operated in the control stage chamber of a turbine high-pressure cylinder (figure 1). Figure 2 presents the geometry model discretized with finite volume mesh which was used to simulate the general steam flow conditions within the control stage. It shows four inlet pipelines, nozzle sectors, control stage rotating wheel and chamber, and first full-arc admission stage.

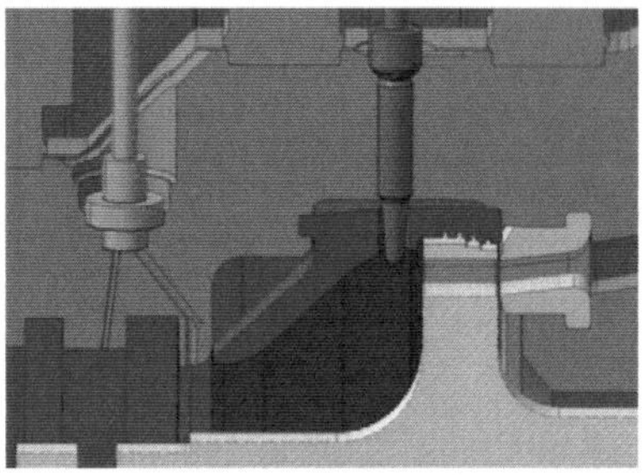

Figure 1. Location of the thermowell (marked in red) in the control stage chamber.

For appropriate description of the flow nature within the control stage chamber, the numerical model precisely describes dimensions of the control stage (64 blades), nozzle rings whose shape exactly describes the geometry of real stage. In order to better describe the flow conditions and obtain a real picture of the control stage flow, the computational domain was extended to the first not-regulated stage. The whole model was discretized using structured mesh providing the most reliable results with optimum density. The areas of expected big gradients were additionally refined, which resulted in the number of elements reaching 6 million.

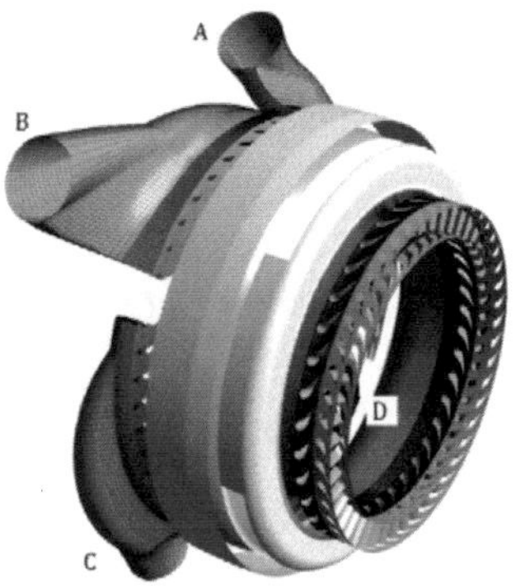

Figure 2. Geometry model with partially shown mesh for complete computations of entire control stage (near nozzle box B was located the thermowell; both nozzle boxes A and C includes 14 nozzles, next pair of nozzle boxes (B and D) include 9 nozzles).

Boundary conditions for CFD calculations are presented in table 1. There were considered two cases of work of the control stage at reduced pressure and mass flow rate of steam - night work with 180 MWe and daily work with 380 MWe at nominal parameters. Previously described model was used. However the ideal gas was considered so viscosity effect was omitted.

Table 1. Boundary conditions such as ($\dot{m}, p, T$) assumed to calculations for two cases of work (N).

Parameter	Unit	Daily work – full load	Night work – partial load
$\dot{m}$	kg/s	101	76
p	bar	153.7	119.6
T	°C	535	520.7
N	MWe	380	180

3.2. Calibration of the CFD model

In order to confirm the correctness of the results, some resultant points from [2] were marked on the diagram presenting dependence of Reynolds number on Strouhal number (figure 3), taken from the ASME norm [4]. The position of points in figure 3 should be considered as satisfactory because they are within the area defined experimentally. It is worth mentioning that the diagram from refers to the obstacles, immersed in the whole height of the channel with a uniform cross-section, and in the case of analyses thermowell the flow was only partially disturbed, and additionally the thermowell has the shape of a truncated cone.

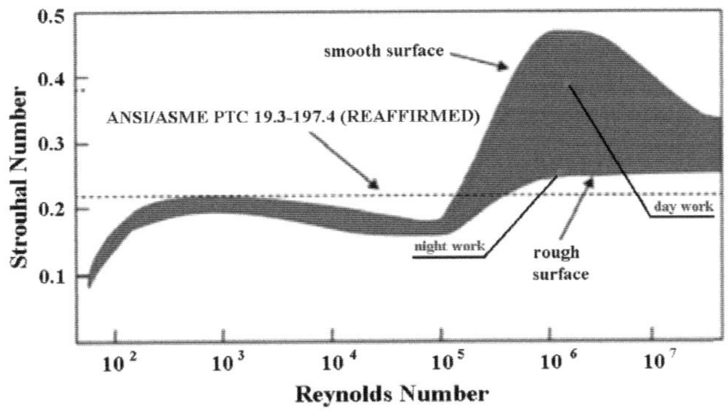

Figure 3. Comparison of St(Re) from ASME PTC 19.3 TW-2010 with numerical results.

3.3. Boundary condition of the CSD model

Fig. 4a) shows the geometry of the thermowell after making discretization which was used to calculate the frequency of free vibrations of the housing of steam temperature sensor. To mount a structure, 6 degrees of freedom was taken away from the area marked by bold lines. As the material of the thermowell was taken steel ST12. In the Fig. 4a), there are marked places (*p*) of application the aerodynamic loads, which values were determined on the basis of pressure field on the surface presented on Fig.4 b).

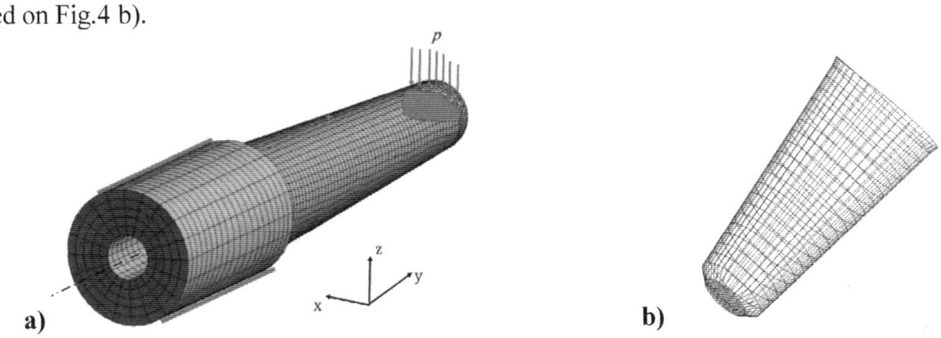

Figure 4.a) The housing of the thermowell after making discretization and with marked place (*p*) of application of the boundary conditions. **b)** surface used to determine pressure field acting on solid.

4. Results and discussion

Numerical simulations were carried out with boundary conditions describing turbine operation with nominal and part load. Figure 5a) presents spatial fields of static pressure computed for one full rotation of the rotor at nominal load. The results were presented for selected time step mainly t=0.0036 s. The time step was equal to Δt=0.000312 [s] which corresponds to 10 steps per each inter-blade channel for 64 rotor blades. Local pressure variations do not exhibit significant fluctuations. At this phase, after examining distribution of the flow parameters and its character, it can be stated that the flow at the control stage chamber is stable and does not exhibit symptoms of big oscillations.

The subsequent figures show the local variations of flow parameters. Figure 5b),c) presents the static pressure fields calculated at initial phase of rotor rotation at the control stage chamber: Fig.5b) – after the control stage at the inlet to control stage chamber and; Fig.5c) – outlet of the control stage chamber before the first not-regulated stage. The results from Figure 5b) and 5c) correspond to the turbine operating with partial load. This operation mode differs from the nominal load by opening of two nozzle boxes only. The remaining two are fully closed by the stop valves and steam is flowing only through the boxes with a smaller number of nozzles (compare with Figure 5a - nominal load). Consequently, higher non-uniformity of pressure Fig.5b) and velocity distribution Fig.6 is observed for the steam leaving the rotor blades. Figure 5c) presents static pressure variations at the exit plane of the control stage chamber. The exit is located close to the inlet to stationary blades of the first non-regulated stage. Clear difference between open and closed nozzle sectors is seen from the pressure distributions. This distinction is visible even though the steams flow through the entire control stage chamber.

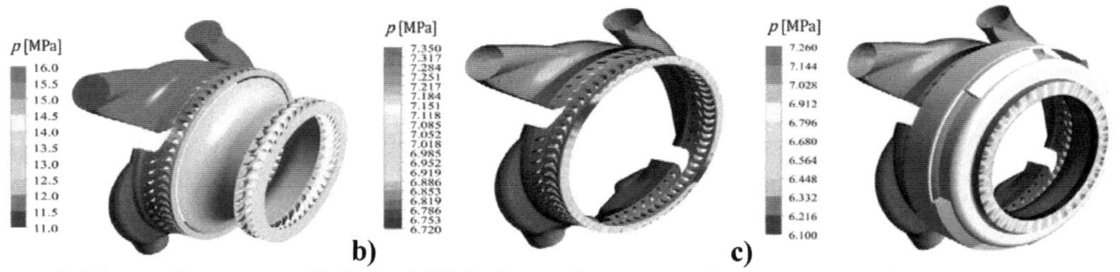

Figure 5. The static pressure fields at initial phase of rotor rotation: **a)** spatial distributions; **b)** cross section after the control stage; **c)** cross section before the first not-regulated stage.

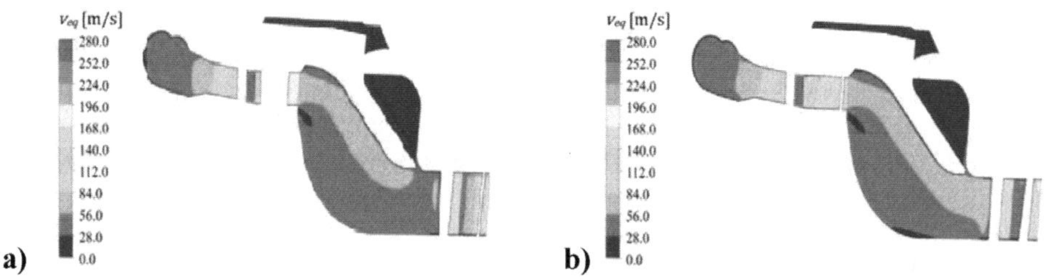

Figure 6. Velocity distribution at plane cutting the thermowell at subsequent time steps: **a)** 0.0036 [s] ; **b)** 0.007 [s].

Figure 6 presents the velocity fields at selected time steps. It can be noticed that the flow forms the main core of higher velocity beginning at the exit from the control stage and ending at the inlet to the stator blades of the first non-regulated stage. At the bottom part of the control stage chamber, a stagnation region persists due to high divergence of the chamber and reversed flow can be expected there. This flow can generate vortices and continuous circulation of steam inside the chamber.

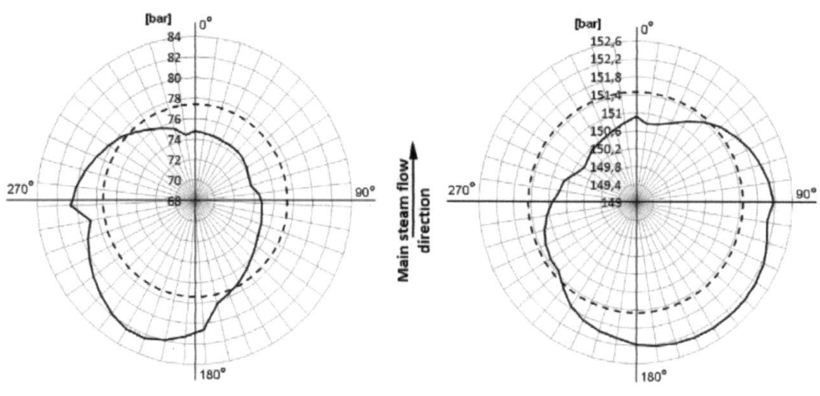

Figure 7. Plots of total pressure distribution around thermowell: left–180MW; right–380 MW.

Figure 7 presents distributions of the total pressure around the thermowell at 180MW (partial load) and 380 MW (full load). The pressure distribution changes both its values and shapes with changing load, which means that both pressure amplitude and resultant force direction vary with turbine load. It can be thus concluded that direction of the pressure force acting on the thermowell depends on the operation regime (figure 7). In the case of analyses thermowell the flow has different character of

Journal of Physics: Conference Series **760** (2016) 012001 doi:10.1088/1742-6596/760/1/012001

turbulence and pressure deviation in comparison to the steam main flow direction which is behaviour as axial.

These loads expressed as difference of total pressure for the thermowell section inserted into the flow are presented in figure 8. It is seen that the aerodynamic loads are one-sided which means that steam pushes the thermowell in line with the steam flow direction but with different time-dependent force. The key role in variable loading of the thermowell is played by the total pressure. The difference in steam pressure varies between 0.5 - 4.5 bar for part load and 0 - 0.9 bar for full load.

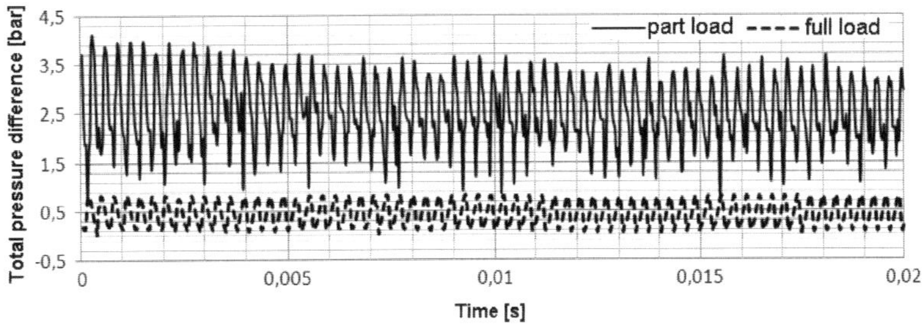

Figure 8. Plot of total pressure difference variation between upstream and downstream side of the thermowell section inserted into the flow (part load - solid line, full load – dotted line).

4.1. CSD Analysis

Figure 9 presents a variation of natural frequencies with temperature and comparison of the excitation frequencies with forbidden ranges of natural frequencies. As the main excitation the trailing edge wakes from 64 rotating blades were assumed and based on this the excitation frequency n_{zw} = 3200 Hz was computed. The excitations due to non-uniform steam supply to nozzle boxes were not analyzed here, as the frequencies equal to 50 Hz and 100 Hz for part and full load conditions are very small comparing with the thermowell natural frequencies.

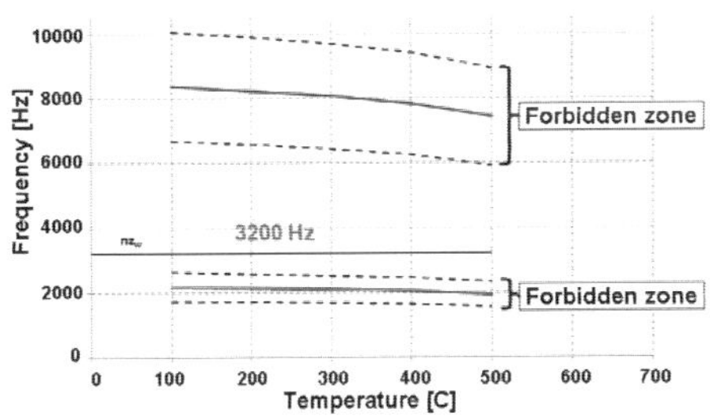

Figure 9. Variation of natural frequencies with temperature.

According to the ASME standard PTC 19.3 TW-2010, the required safety margin for natural frequency is ±20%. As it is seen from figure 9 this condition is satisfied with a considerable margin for the analyzed natural frequencies of the thermowell. Taking into account the above results it can be concluded that the thermowell natural frequencies are within a safe zone as compared with the excitation frequencies.

5. Conclusions

Summarizing the calculation results it can be said that, operation with partial load is characterized by larger amplitudes of pressure acting on the thermowell. The total pressure amplitude for, part load reaches 4.5 bar, while at full load is equal maximum 0.9 bar. Pressure distribution curve around the thermowell changes both its shape and value bringing about a variation of not only the pressure amplitude but also the load direction.Two major excitations from the steam flow in the control stage chamber can be distinguished, namely due to non-uniform steam supply to the nozzle boxes and due to aerodynamic wakes of rotor blades trailing edges. The excitation frequency due to non-uniform steam supply to the nozzle boxes is 100 Hz for 180 MW load and 50 Hz for 380 MW load. The excitation frequency due to aerodynamic wakes is 3200 Hz in the entire load range. At steady state operations the criteria of tuning the natural frequencies against steam forces basic frequencies are met and the thermowell is free of resonance.

References

[1] El-Batahgy A and Fathy G 2013 Fatigue failure of thermowells in feed gas supply downstream pipeline at a natural gas production plant *Case Stud Eng Fail Anal* **1**(2) 79–84
[2] Kornet S, Sławiński D, Ziółkowski P and Badur J 2015 Analysis of unsteady flow forces on the thermowell of steam temperature *Trans. IFFM* **129** 25-49
[3] Haslinger K H 2003 Flow-induced vibration testing of replacement thermos-well designs *Journal of Fluids Structures* **18** 425-440
[4] ASME PTC19.3 TW-2010: performance test codes: thermowells. ASME 2010
[5] Shyy W 1994 *Computational modeling for fluid flow and interfacial transport* Dover Pub. (NY)
[6] Badur J 2009 *Development of energy concept Wyd. IMP PAN* (Gdańsk)
[7] Fuchs R, Hopff L and Seewald F 1934 *Aerodynamik* Springer (Berlin)
[8] Prandtl L 1956 *Flow dynamics* Tłum. J. Wysocki PWN in Polish (Warszawa)
[9] Prosnak W J 1970 *Fluid mechanics* PWN in Polish (Warszawa)
[10] Goujon-Durand S, Jenffer P and Wesfreid J E 1994 Downstream evolution of the Benard-von Kármán instability *Physical Review E* **50** no 1 308-313
[11] Klotz L, Goujon-Durand S, Rokicki J and Wesfreid J E 2014 Experimental investigation of flow behind a cube for moderate Reynolds numbers *J. Fluid Mech.* **750** 73–98
[12] Skarysz M, Pryśko J, Goujon-Durand S and Wesfreid J E 2014 The wake behind a rotating sphere *Journal of Physics: Conference Series* **530** 012043
[13] Gumowski K, Miedzik J, Goujon-Duran S, Jenffer P and Wesfreid J E 2008 Transition to a time-dependent state of fluid flow in the wake of a sphere *Physical Review* E77 055308(R)
[14] Saha A K 2006 Three-dimensional numerical study of flow and heat transfer from a cube placed in a uniform flow *International Journal of Heat and Fluid Flow* **26** 80–94
[15] Sarpkaya T 2004 A critical review of the intrinsic nature of vortex-induced vibrations *Journal of Fluid and Structures* **19** 389-447
[16] Lee A H, Campbell R L and Hambric S A 2014 Coupled delayed-detached-eddy simulation and structural vibration of a self-oscillating cylinder due to vortex-shedding *Journal of Fluids and Structures* **48** 216–234
[17] Megahed M M and Attia M S 2015 Failure analysis of thermowell weldment cracking *Eng Fail Anal* **50** 51-61
[18] Zarruk G A, Cowen E A, Wu T –R and Liu P L-F 2015 Vortex shedding and evolution induced by a solitary wave propagating over a submerged cylindrical structure *Journal of Fluid and Structures* **52** 181-198
[19] Morishita M and Dozaki K 1998 History of flow-induced vibration incident occurred in Monju *ASME Trans PVP* **363** 103–8
[20] Odahara S, Murakami Y, Inoue M and Sueoka M 2005 Fatigue failure by in-line flow-induced vibration and fatigue life evaluation *JSME Int J, SerA* **48**(2) 109-17
[21] Banaszkiewicz M 2015 Multilevel approach to lifetime assessment of steam turbines *International Journal of Fatigue* **73** 39–47
[22] Badur J, Ziółkowski P, Sławiński D and Kornet S 2015 An approach for estimation of water wall degradation within pulverized-coal boilers *Energy* **92** 142-152
[23] Banaszkiewicz M and Rehmus-Forc A 2015 Stress corrosion cracking of a 60 MW steam turbine rotor *Engineering Failure Analysis* **51** 55–68

Simulation of Reynolds number influence on heat exchange in turbulent flow of medium slurry

A Bartosik[1]

[1] Kielce University of Technology, Faculty of Management and Computer Modelling
Al. Tysiąclecia P.P. 7, 25-314 Kielce, Poland, artur.bartosik@tu.kielce.pl

Abstract. The paper deals with the numerical simulation of mass and heat exchange in turbulent flow of solid-liquid mixture in the range of averaged solid particle diameter from 0.10mm to 0.80mm, named further as the medium slurry. Physical model assumes that dispersed phase is fully suspended and a turbulent flow is hydro-dynamically, and thermally developed in a straight horizontal pipeline. Taking into account the aforementioned assumptions the slurry is treated as a single-phase flow with increased density, while viscosity is equals to a carrier liquid viscosity. The mathematical model constitutes time averaged momentum equation in which the turbulent stress tensor was designated using a two-equation turbulence model, which makes use of the Boussinesq eddy-viscosity hypothesis. Turbulence damping function in the turbulence model was especially designed for the medium slurry. In addition, an energy equation has been used in which a convective term was determined from the energy balance acting on a unit pipe length, assuming linear changes of temperature in main flow direction. Finally, the mathematical model of non-isothermal medium slurry flow comprises four partial differential equations, namely momentum and energy equations, equations of kinetic energy of turbulence and its dissipation rate. Four partial differential equations were solved by a finite difference scheme using own computer code. The objective of the paper is to examine the influence of Reynolds number on temperature profiles and Nusselt number in turbulent flow of medium slurry in the range of solids concentration from 0% to 30% by volume. The effect of influential factors on heat transfer between the pipe and slurry is analysed. The paper demonstrates substantial impact of Reynolds number and solids volume fraction on the Nusselt number. The results of numerical simulation are reviewed.

1. Introduction

Hydraulic transport of solids is available in a wide range of industrial applications. Designing of slurry pipeline transportation with heat exchange requires a better understanding of flow mechanism, which includes pressure drop, velocity and solids concentration profiles, deposition velocity, and as a consequence the temperature profile. Therefore a solid-liquid transport belongs to main challenges of Computational Fluid Dynamics [1-3]. Various approaches have been investigated to predict a solid–liquid mechanism during turbulent flow, however, mixture theory models are the most promising and are based on a rigorous fluid mechanics framework [3-5].

Solid-liquid flow is recognised as stationary or moving bed, heterogeneous, and pseudo–homogeneous, or as settling or non-settling [6]. Settling slurry is formed mainly by coarse particles, however, it can exist as well if fine or medium solid particles have sufficiently low velocities.

If simulation of coarse dispersive slurry flow is considered one should mention basic researches of Bagnold [7] and recent mathematical models [8-10], which include solid-solid and/or solid-liquid interactions. In order to simulate frictional head loss in slurry flow, which contains coarse or medium

Journal of Physics: Conference Series **760** (2016) 012002 doi:10.1088/1742-6596/760/1/012002

solid particles, it is reasonable to assume the Newtonian model, as no one can measure rheology in such slurries [11].

If non-settling slurry is considered averaged solid particle diameters are sufficiently fine to form stable homogeneous mixture exhibiting increased apparent viscosity. Such slurry usually exhibits yield stress and require proper rheological model built into the momentum equation. Slurry with fine solid particles demonstrates increased viscous sublayer, which means that damping of turbulence appears in a near-wall region. In this case, the apparent viscosity concept, and suitable rheological model, and properly defined wall damping function is required [12-15].

Slurry with medium solid particles of averaged diameter from d_S=0.10 □m to d_S=0.80 □m is usually assumed as Newtonian flow or as flow of two separated phases [16]. However, if slurry velocity is sufficiently high such slurry can be considered to be non-settling and pseudo-homogeneous. It is important to emphasize that slurry with medium solid particles exhibits enhanced damping of turbulence since frictional head loss is almost equal or below that for carrier liquid flow [17,18]. Damping of turbulence depends mainly on particle size, solids concentration and flow condition. The damping of turbulence could be a consequence of thickening of a viscous sublayer, which tends to increase throughput velocity and thus promotes a drag reduction [15]. Simulation of turbulent flow of medium slurry requires a proper approach, which includes damping of turbulence, in order to predict the pressure drop and velocity distribution. The velocity distribution and the viscous sublayer are closely related to temperature distribution, which is a special interest in mining industry, pipeline erosion, food industry, medicine, etc.

Sundaresan et al. [4], outlined a number of scientific challenges, which represent the building blocks for the comprehensive understanding of disperse flows encountered in a variety of technologies and in nature. Researchers concluded that new experiments and/or analyses are needed to cast light on the vital phenomena that cause turbulence damping or generation. The authors suggested that the experiments should be made in simple turbulent flows, such as: fully developed pipe or channel flow, or axially symmetrical flow. Regardless of geometry, experiments must include a wide range of particle parameters in a single fixed facility. Their conclusions are still outstanding in mathematical modelling of solid-liquid flow.

Silva et al. and Cotas et al. [19-21] used commercial CFD software to simulate fully developed turbulent flow of eucalyptus and pine pulps as such flows exhibit drag reduction. They applied a pseudo-homogenous approach by considering the viscosity of pulp as a function of shear rate and solids concentration. They assumed there exists lubrication layers responsible for the drag reduction, which consists of pure water with a few fibres. They adopted into account the Chang-Hsieh-Chen turbulence model [22], in which they used damping function, proposed by Bartosik [23], to study the flow of solid particles. They validated the mathematical model by comparing measured and predicted frictional head loss obtaining satisfying agreement.

Transport of solid particles is often strongly influenced by heat exchange between the transported materials and the surrounding. Researchers are developing methods and techniques to enhance or attenuate of heat exchange. Enhancement of heat exchange contributes to energy saving and preserve the environment. Those methods are recognised as active or passive [24]. The passive ones like inserted wire coils or mechanically deformed pipes, have been studied for the last several years, and have become commercial solutions. The active techniques can produce very high increases in heat exchange and there are still main challenges of fluid mechanics of two-phase flows [25].

Influence of solid particles submerged in a carrier liquid on heat exchange in turbulent flow has been experimentally investigated be several researchers [26-30]. Rozenblit et al. [30] made an experimental study on heat transfer coefficient associated with solid-liquid mixture transport in horizontal pipe using electro-resistance sensor and infrared imaging. They considered flow of acetyl-water mixture with moving bed for solids volume fraction from 5% to 15% by volume. The experimental results included the temperature profile on the heated wall and the vertical solids volume fraction profile. They observed that the local heat transfer coefficient changes from its lowest value at the bottom of the pipe, to highest values above carrier liquid at the upper heterogeneous layer.

It clearly means that the heat transfer coefficient is strongly influenced by the cross sectional profile of the solid phase in the pipe. In order to analyse the near wall structures they used algebraic three layer model to predict pressure drop and solids volume fraction distribution across the pipe [30].

Wang and Xu [31] examined experimentally thermal conductivity of nanoparticles of Al_2O_3 and CuO mixed with water, vacuum pump liquid, engine oil, and ethylene glycol. The averaged particle diameter of the Al_2O_3 was 28 nm, and the averaged particle diameter of the CuO was 23 nm. All particles were loosely agglomerated in chosen liquids. Experiments proved that the thermal conductivity of the mixture increases with the volume fraction of the Al_2O_3 particles. Increase of the mixture thermal conductivity depends on liquid and solid phase properties. They proved that increase of the mixture thermal conductivity in ethylene glycol and engine oil are the highest, whereas that in the pump liquid is the lowest. The effective thermal conductivity of ethylene glycol increases by 40% when solids concentration of Al_2O_3 was 8% by volume [31].

Summarising the literature review one can say there are no available experiments and simulations presenting the influence of Reynolds number on heat exchange in turbulent flow of medium slurry with high solids concentration.

The objective of the paper is to examine influence of the Reynolds number on the Nusselt number in turbulent flow of medium slurry in the range of solids concentration from 0% to 30% by volume.

2. The physical and the mathematical approach

Physical model assumes that slurry comprises medium solid particles with averaged particle diameter from 0.10 mm to 0.80 mm and water as a carrier liquid. All solid particles are rounded and narrowly sized with density equals to ρ_S=2440 kg/m^3 and ρ_S=2650 kg/m^3. The solids concentration (C) varies from 0% to 30% by volume. The slurry flow takes place in a horizontal pipe with sufficiently high bulk velocity, so the flow can be considered as pseudo-homogeneous. The slurry flow is stationary, turbulent, fully developed and axially symmetrical. It is assumed that slurry viscosity is equal to the carrier liquid viscosity as no one can measure rheology for such particle sizes [7]. The slurry density (ρ_m), specific heat (c_P), and thermal conductivity (λ) are dependent on temperature and volume fraction of solids (C) and were calculated, as follows:

$$\Phi_m = \Phi_L(1-C) + \Phi_S C \tag{1}$$

where Φ_m is general variable for slurry (mixture of solid and liquid phase), $\Phi = (\rho,\ c_P,\ \lambda)$, and subscript L - means liquid phase, while S - means solid phase.

The final form of the mathematical model is, as follows:

- Momentum equation:

$$\frac{1}{r}\frac{\partial}{\partial r}\left[r\left(\mu\frac{\partial\bar{U}}{\partial r} - \bar{\rho}_m\overline{u'v'}\right)\right] = \frac{\partial\bar{p}}{\partial x} \tag{2}$$

where **U** - velocity component in main flow direction (m/s); ρ_m - slurry density (kg/m^3); μ - molecular viscosity of carrier liquid phase (Pa s).

The component of the turbulent stress tensor, which appears in equation (2), was designated by the Boussinesq hypothesis, as follows:

$$-\bar{\rho}_m\overline{u'v'} = \mu_t\frac{\partial\bar{U}}{\partial r} \tag{3}$$

where ρ_m is slurry density and subscript t - means turbulent.

Launder and Sharma developed turbulent viscosity (μ_t), stated in equation (3), on the basis of dimensionless analysis, as follows [32]:

$$\mu_t = f_\mu \frac{\bar{\rho}_m k^2}{\varepsilon} \tag{4}$$

It was proved that the Launder and Sharma turbulence model has great potential to predict pseudo-homogeneous solid-liquid flows [33]. Therefore the Launder and Sharma turbulence model was chosen into the mathematical model in order to predict pressure drop, velocity and temperature profiles, and Nusselt number for medium slurry flow. The kinetic energy of turbulence (k) and its dissipation rate (ε), which exists in equation (4), were delivered from the Navier-Stokes equations [35].

For the aforementioned assumptions the final form of k and ε equations, proposed by Launder and Sharma [32], are following:

- Equation of kinetic energy of turbulence:

$$\frac{1}{r}\left[r\left(\mu + \frac{\mu_t}{\sigma_k}\right)\frac{\partial k}{\partial r}\right] + \mu_t\left(\frac{\partial \bar{U}}{\partial r}\right)^2 = \bar{\rho}_m \varepsilon + 2\mu\left(\frac{\partial k^{1/2}}{\partial r}\right)^2 \tag{5}$$

- Equation of dissipation rate of the kinetic energy of turbulence:

$$\frac{1}{r}\left[r\left(\mu_{ap} + \frac{\mu_t}{\sigma_\varepsilon}\right)\frac{\partial \varepsilon}{\partial r}\right] + C_1 \frac{\varepsilon}{k}\mu_t\left(\frac{\partial \bar{U}}{\partial r}\right)^2 = C_2\left[1 - 0.3\exp\left(-Re_t^2\right)\right]\frac{\bar{\rho}\varepsilon^2}{k} - 2\frac{\mu_{ap}}{\bar{\rho}}\mu_t\left(\frac{\partial^2 \bar{U}}{\partial r^2}\right)^2 \tag{6}$$

On the basis of dimensionless analysis the turbulent Reynolds number in equation (6) can be expressed, as follows [35]:

$$Re_t = \frac{\bar{\rho}_m k^2}{\mu \varepsilon} \tag{7}$$

The crucial point in the above turbulence model is proper expression of turbulence damping function (f_μ), stated in equation (4). The function is known as the wall damping or turbulence damping function. This is an empirical function, which causes decrease of turbulent viscosity in the vicinity of solid wall and in consequence decrease of the turbulent stress tensor component described by equation (3). As it is an empirical function it is possible to redesign the function (f_μ) for certain application, like medium slurry flows for instance. Recently, Ruffin and Lee [34] successfully used the standard k-ε model of Launder and Spalding [35] and a new wall damping function for the unstructured Cartesian grid solver.

Taking into account medium slurry flows, which exhibits enhanced damping of turbulence; author developed a new turbulence damping function [23]. The new turbulence damping function includes averaged particles diameter (d), and solids concentration by volume (C). The new turbulence damping function proposed to medium slurry flow is described, as follows:

$$f_\mu = 0.09\, EXP\left[\frac{-3.4\left[1 + A^3 d^2 (8 - 88\, A\, d) C^{0.5}\right]}{\left(1 + \frac{Re_t}{50}\right)^2}\right] \tag{8}$$

where A is an empirical constant, and Re_t is turbulent Reynolds number described by equation (7).

If particle diameter and/or solids concentration goes to zero the new turbulence damping function, described by equation (8), approaches the standard turbulence damping function of a single-phase flows proposed by Launder and Sharma [32]. This is in accordance with our expectation as the mathematical model should be suitable for single-phase flow if d_S=0 and/or C=0. The new turbulence damping function works well if slurries with averaged solid particle diameters from d_S=0.10 mm to d_S=0.80 mm are considered. The new turbulence damping function proposed for the medium slurry flows, described by equation (8), demonstrates that averaged solid particle diameter plays primarily role, while the solids concentration plays secondary role in damping of turbulence near a pipe wall. This is in accordance with previous observations made by several researchers [15, 17-18]. The new turbulence damping function, described by (8), was developed on the basis of comparison between numerical predictions and measured global parameters in comprehensive range of averaged solid particle diameter equals to d_S=(0.125; 0.240; 0.471; 0.780) mm and solids concentration C=(10%; 20%; 30%; 40%).

In order to simulate heat exchange in turbulent flow of medium slurry the energy equation, written in form of the temperature equation, is needed. Taking into account aforementioned assumptions one can write final form of the temperature equation, as follows:

- Temperature equation:

$$\bar{\rho}_m \bar{U} \frac{\partial \bar{T}}{\partial x} = \frac{1}{r}\frac{\partial}{\partial r}\left[r\left(\frac{\mu}{Pr} + \frac{\mu_t}{Pr_t}\right)\right]\frac{\partial \bar{T}}{\partial r} \tag{9}$$

where Pr is the Prandtl number.

In reference to the turbulent Prandtl number, researches of Blom [36] confirmed that in the flow on a plate the turbulent Prandtl number is about Pr_t=0.5, while for the boundary layer is about Pr_t=0.9.

The convective term, which appears in the equation (9), was determined from the energy balance. Taking into account the steady heat flux acting on the unit pipe length and assuming that the temperature in main flow direction (**ox**) varies linearly, the final form of the convective term can be described, as follows:

- Convective equation:

$$\frac{\partial \bar{T}}{\partial x} = \frac{2\,q}{\rho_b\, U_b\, (c_p)_b R^2} \tag{10}$$

where **q** is an input power of heat per unit pipe length (W/m), c_p is specific heat at constant pressure (J/(kg K) and R is the pipe radius (m), while subscript b - means cross sectional averaged value.

The final form of the mathematical model of non-isothermal medium slurry flow comprises four partial differential equations, namely momentum equation (2), temperature equation (9), equation for kinetic energy of turbulence (5), and its dissipation rate (6), together with the wall turbulence damping function (8), and complementary equations (3), (4), (7) and (10).

All constants in the turbulence model are the same as those proposed by Launder and Sharma [35] and are equal to: C_1=1.44; C_2=1.92; σ_k=1.0; σ_e=1.3, Pr_t=0.9. The constant in the turbulence damping function was found as equal to A=100.

The mathematical model assumes non-slip velocity at the pipe wall, i.e. U=0, and k=0, ε=0. Axially symmetrical conditions were applied at the pipe centre, therefore: dU/dr=0, dT/dr=0, dk/dr=0 and dε/dr=0. Taking into account boundary conditions the equations set have been solved by finite difference scheme using own computer code. After determining boundary and convergence conditions the mathematical model was solved for 80 nodal points distributed on the radius of the pipe. The majority of the nodal points were localized in a close vicinity of the pipe wall to ensure the convergence process. The number of nodal points was set up experimentally to ensure nodally independent computations. Numerical simulations were performed for *a priori* set of dp/dx.

The mathematical model is able to predict frictional head loss, velocity and temperature profiles, and Nusselt number in turbulent flow of slurries with averaged solid particle diameters from d_S=0.10 mm to d_S=0.80 mm, known as medium slurries.

3. Simulation results

The mathematical model of slurry flow with solid particles of averaged diameter from d_S=0.10 to d_S=0.80 mm, presented in chapter 2, was validated for isothermal turbulent flow only [23]. The predictions and measurements of slurry frictional head loss were matched for two different types of solid particles, named as Canasphere and Sand. Solid particle density of Canasphere and Sand were ρ_S=2440 kg/m^3 and ρ_S=2650 kg/m^3, respectively. The measurements and simulations were performed for solids concentration from 10% to 40% by volume. Measurements and simulations demonstrate that frictional head loss significantly depends on particle size and solids concentration [23]. Experiments and simulations confirmed that in slurry flow with averaged solid particles diameter about d_S=0.50mm the frictional head loss is equal or below that for carrier liquid flow even for solids concentration equal up to 30% by volume. It is worth to emphasise that for averaged solid particle diameters, which are below or above d_S=0.50 mm, the frictional head loss was higher than that for the carrier liquid flow, however, was still below values we expected.

The mathematical model, presented in chapter 2, was not validated for non-isothermal slurry flow, as there are not any available experimental data in literature. Limited validation has been made for carrier liquid flow only. Figure 1 presents comparison of simulated and empirically calculated Nusselt number expressed by equation (11) for the carrier liquid flow (water) in the range of Reynolds number from Re=10,000 to Re=140,000. Simulations confirmed that the relative error of the predicted Nusselt number vary from +2.5% to -3,3%.

$$Nu = 0.023\, Re^{0.8}\, Pr^{1/3} \tag{11}$$

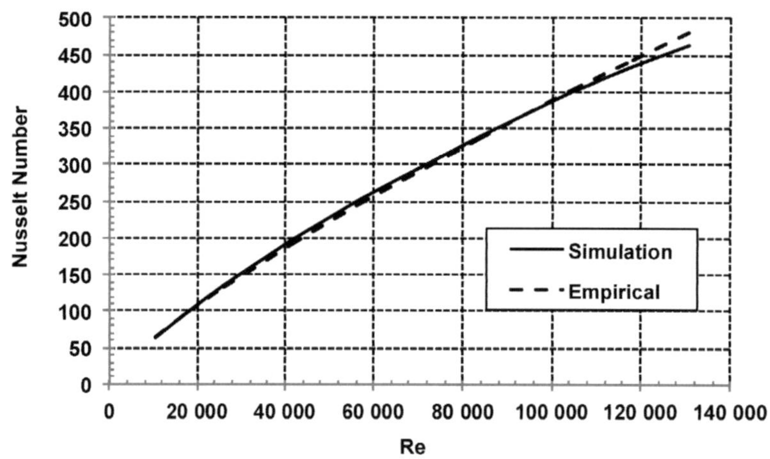

Figure 1. Comparison of simulations with empirically predicted Nusselt number for carrier liquid flow (water): q= 200 W/m, T_w=303.15 K, D=0.0127 m.

All predictions presented in the paper have been made for slurries with averaged particle diameter equals to 0.240 mm. Simulations of heat exchange in turbulent flow of medium slurry indicate that Reynolds number and solids concentration have significant influence on the process. Figure 2 presents temperature profiles across the pipe radius in the carrier liquid and in the medium slurry flows at solids concentration from 10% to 30% by volume, where y - is the distance from the pipe wall. If solids concentration arises from 10% to 30% the temperature difference between the pipe wall and the centre of the pipe increases. If temperature difference increases the coefficient of heat transfer decreases because the heat flux per unit pipe length **q** and area perpendicular to the heat flux are constant. If solids concentration increases, the Nusselt number decreases as the coefficient of heat transfer decreases.

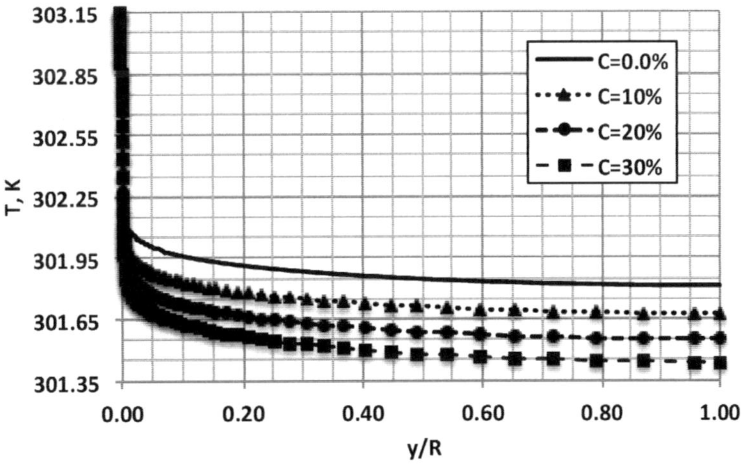

Figure 2. Temperature profiles in carrier liquid and slurry flows: d_S=0.24 mm; ρ_S=2440 kg/m^3 C=0%; 10%; 20%; 30%; T_w=303.15 K, q= 200 W/m, D=0.0127 m.

Figure 3 presents influences of Reynolds number on the Nusselt number in the flow without solid particles (C=0%) and in flow with solids concentration equals to 10%, 20% and 30% by volume.

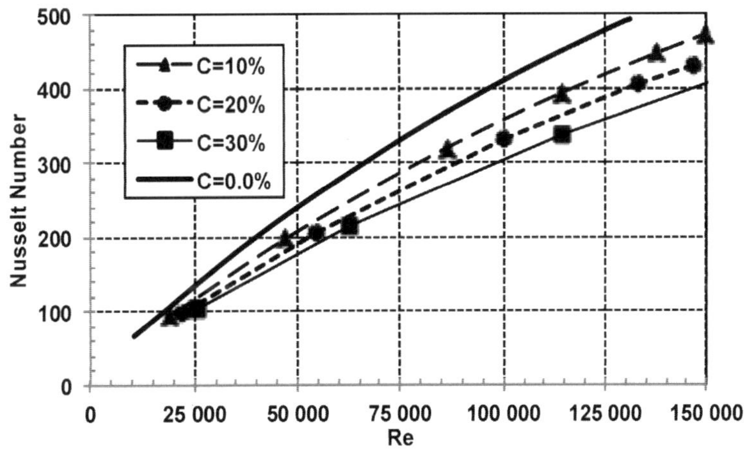

Figure 3. Influence of Reynolds number on the Nusselt number for different solids concentration d_S=0.24 mm; ρ_S=2650 kg/m^3 C=0%; 10%; 20%; 30%; T_w=303.15 K, q= 200 W/m, D=0.0127 m.

Simulations made for turbulent flow of medium slurry emphasized substantial influence of Reynolds number and of solids concentration on the temperature profile and as a consequence on the Nusselt number.

4. Discussion

Simulations of heat exchange in turbulent flow of medium slurry requires mathematical model, which is suitable to predict increased damping of turbulence. The mathematical model presented in the paper has been successfully validated for isothermal slurry flow in a comprehensive range of solid particle diameters, solids concentration and Reynolds number [23]. The crucial point of the mathematical model is a properly designed turbulence damping function incorporated into the turbulence model, which has been described by equation (8).

The mathematical model of heat exchange in turbulent flow of medium slurry was not validated, as there are no such experimental data available in literature. Such experiments are extremely tough for performance. Measurements at high solids concentration of the dispersed phase are prevented because of the risk of damage or contamination of intrusive hot-film or hot wire probes. With optical methods, attenuation of the beams by particles occurs. As a result of these difficulties, most of the measurements in the literature concern gas-liquid or gas-solid flows.

There are evidences in literature that some slurries exhibit increased damping of turbulence. Slurries with averaged solid particle diameters in the range from d_S=0.10 mm to d_S=0.80 mm exhibit increased damping of turbulence; however, the highest damping of turbulence takes place if averaged solid particle diameter is about d_S=0.50 mm [17,18]. Schreck and Kleis [37] proved that swirls with sizes lower than solid particle diameter, drastically reduce particles shade causing that the level of turbulence decreases. Furthermore, there are evidences in literature that solids concentration decreases towards a pipe wall, which depends on diameter of solid particles [11,17-18, 33]. This could be mainly caused by lift forces, which act from the pipe wall towards symmetry axis. Such forces causes that the contact of solid particles with a pipe wall is less intensive. This could be responsible for thickening of viscous sublayer as larger solid particles are pushed away from the pipe wall and are replaced by very fine particles. Very fine particles are responsible for enhancement of viscous forces in vicinity of a pipe wall. If the viscous forces are increasing at the pipe wall the *laminarisation* of the flow takes place [38]. However, it should be noted that the complexity of turbulent slurry flow causes that many parameters influence the flow phenomena and some of them could act oppositely to each other. Therefore, slurry flow still belongs to main challenges of Computation Fluid Dynamics. In literature, there is still lack of a simple parameter, which can resolve whether there is an enhancement or attenuation of turbulence in slurry flow.

Numerical simulations presented in the paper, which have been made for slurry flow with averaged solid particle diameter equals to d_s=0.240 mm, demonstrate the influence of Reynolds number and volume fraction of solids on the heat exchange process. Both mentioned factors affect the velocity profile across the pipe and as a result the temperature profile too. Changes of temperature profile are due to changes in heat transfer coefficient, which affects the Nusselt number.

References

[1] Agarwal P K 1988 Transport phenomena in multi-particles systems-II. Particle-fluid heat and mass transfer *Chem. Eng. Sci.* **43** pp 2501-2510
[2] Govier, G.W., and Aziz, K. (1972). The Flow of Complex Mixtures in Pipes, Robert E. Krieger Publishing Company, 1st ed., 326-327, New York
[3] Soo S L 1990 *Multiphase fluid dynamics* Science Press Beijing
[4] Sundaresan S, Eaton J, Koch D L and Ottino J M 2003 Appendix 2: Report of study group on disperse flow *Int. J. Multiphase Flow* **29** pp 1069–87
[5] Ravikumar S K, Ziazi R M, Chambers F W, McNally M E and Hoffman R M 2013 Proc. ASME 2013, Fluids Engineering Division, Summer Meeting FEDSM2013 July 7-11 2013 pp 1-10 Incline Village Nevada USA
[6] Doron P and Barnea D 1996 Flow pattern maps for solid-liquid flow in pipes *Int. J. Multiphase Flow* **22** pp 273–83
[7] Bagnold R A 1954 Experiments on a gravity–free dispersion of large solids spheres in a Newtonian fluid under shear *Proc. Roy. Soc.* **225** Ser A pp 49–63

[8] Wijk J M 2016 *Vertical hydraulic transport for deep sea mining - a study into flow assurance* (Drukkerij Haveka Alblasserdam Nederland)
[9] Talmon A M 2013 Analytical model for pipe wall friction of pseudo-homogenous sand slurries *Particulate Science and Technology: An International Journal* 31:**3** pp 264-70
[10] Bartosik A 2010 Influence of coarse-dispersive solid phase on the 'particles–wall' shear stress in turbulent slurry flow with high solid concentration *The Archive of Mech. Eng.* **LVII** No 1 pp 45–68
[11] Shook C A and Roco M C 1991 *Slurry flow: Principles and Practice* (Butterworth-Heinemann Boston)
[12] Bartosik A 2009 *Simulation and experiments of axially-symmetrical flow of fine- and coarse-dispersive slurry in delivery pipelines* Monograph M-11 (Kielce University of Technology PL)
[13] Kelessidis V C, Dalamarinis P and Maglione R 2011 Experimental study and predictions of pressure losses of fluids modeled as Herschel–Bulkley in concentric and eccentric annuli in laminar, transitional and turbulent flows *J. Petroleum Sc. and Engineering* **77** pp 305–12
[14] Bartosik A 2010 Application of rheological models in prediction of turbulent slurry flow, *Flow, Turbulence and Combustion* **84** pp 277-93
[15] Wilson K C and Thomas A D 1985 A new analysis of the turbulent flow of non-Newtonian fluids *Can. J. Chem. Eng.* **63** pp 539-46
[16] Messa G V and Malavasi S 2015 Improvements in the numerical prediction of fully-suspended slurry flow in horizontal pipes *Powder Technology* **270** pp 358–67
[17] Sumner R J, McKibben M and Shook C A 1991 Concentration and velocity distribution in turbulent vertical slurry flow *J. Solid Liquid Flow* **2** No 2 pp 33–42
[18] Sumner R J 1992 *Concentration variations and their effects in flowing slurries and emulsions* PhD Thesis (University of Saskatchewan Saskatoon Canada)
[19] Silva R, Garcia F A P, Faia P M G M and Rasteiro M G 2015 Settling suspensions flow modelling: A Review *KONA Powder and Particle Journal* No 32 pp 41–56
[20] Cotas C, Asendrych D, Garcia F, Faia P and Rasteiro M G 2015 *COST Action FP1005 Final Conf. EUROMECH Colloquium 566* Trondheim Norway June 9-11 2015 pp 31-3
[21] Cotas C, Asendrych D and Rasterio M G 2015 Proc. *8th Int. Conference for Conveying and Handling of Particulate Solids* Tel-Aviv Israel May 2015 pp 1-18
[22] Chang K C, Hsieh W D and Chen C S 1995 A modified low-Reynolds-number turbulence model applicable to recirculating flow in pipe expansion *Journal of Fluids Engineering* **117** pp 417-23
[23] Bartosik A 2011 Mathematical modelling of slurry flow with medium solid particles, *Mathematical Models and Methods in Modern Science* ed Mastorakis N, Mladenov V, Travieso-Gonzalez C M, Kohler M pp 124-9
[24] Hagiwara Y 2011 Effects of bubbles, droplets or particles on heat transfer in turbulent channel flows *Flow, Turbulence and Combustion* **86** No 3-4 pp 343-67
[25] Crespí-Llorens D, Martínez P, Vicente P and Viedma A 2013 Effect of the axial scraping velocity on enhanced heat exchangers *Int. J. Heat and Fluid Flow* **44** pp 265–75
[26] Wilkinson G T and Norman J R 1967 Heat transfer to a suspension of solids in a gas *Trans. Int. Chem. Eng.* **45** pp 314-8
[27] Harada E, Toda M, Kuriyama M and Konno H 1985 Heat transfer between wall and solid-water suspension flow in horizontal pipes *J. Chem. Eng. Japan* **18** pp 33-8
[28] Harada E, Kuriyama M and Konno H 1989 Heat transfer with a solid liquid suspension flowing through a horizontal rectangular duct *Heat Transfer - Japanese Research* **18** pp 79-94
[29] Hetsroni G and Rozenblit R 1994 Heat transfer to a solid-liquid mixture in a flume *Int. J. Multiphase Flow* **20** pp 671-89
[30] Rozenblit R, Simkhis M, Hetsroni G, Taitel Y 2000 Heat transfer in horizontal solid-liquid pipe flow *Int. J. Multiphase Flow* **26** pp 1235-46

Journal of Physics: Conference Series **760** (2016) 012002 doi:10.1088/1742-6596/760/1/012002

[31] Wang X, Xu X and Choi S U S 1999 Thermal conductivity of nanoparticle - fluid mixture *J. Thermophysics and Heat Transfer* **13** pp 474-80

[32] Launder B E and Sharma B I 1974 Application of the energy-dissipation model of turbulence to the calculation of flow near a spinning disc *Letters in Heat and Mass Transfer* No 1 pp 131-8

[33] Bartosik A and Shook C 1991 *Proc. 7th Int. Conf. on Numerical Methods in Laminar and Turbulent Flow*, ed Taylor C, Chin J H, Homsy G M, Pineridge Press 7 Part 1 pp 277-87

[34] Ruffin S M and Lee J D 2011 Adaptation of a k-epsilon model to a Cartesian grid based methodology *Int. J. Mathematical Models and Methods in Applied Sciences* Issue 3 **3** pp 238-45

[35] Launder B E and Spalding D B 1974 The numerical computation of turbulent flows *Comp. Meth. in Appl. Mech and Eng.* **3** pp 269-89

[36] Blom J 1970 Proc. 4th Int. Conf. on Heat Transfer VII Paper FC 2.2 (Elsevier Amsterdam)

[37] Schreck S and Kleis S J 1993 Modification of grid–generated turbulence by solid particles J. Fluid Mech. 249 pp 665–88

[38] Bartosik A 2008 Laminarisation effect in fine-dispersive slurry flow *Archives of Thermodynamics* **29** No 3 pp 69-82

Study of Reynolds number effect on turbulent boundary layer near the separation

A Dróżdż[1] **and W Elsner**[1]

[1] Częstochowa University of Technology, Faculty of Mechanical Engineering and Computer Science, Armii Krajowej 21, 42-200 Czestochowa, Poland

E-mail: welsner@imc.pcz.czest.pl

Abstract. The paper deals with the experimental analysis of strong decelerated turbulent boundary layer developed on a flat plate. The special design of the test section equipped with perforated, movable upper wall allow to generate on the bottom wall turbulent boundary layer, which is at the verge of separation. The objective of the work is to examine the effect of Reynolds number on non-equilibrium boundary layer. A momentum thickness Reynolds number at the inlet to a test section was in the range from 6300 to 10150 what was achieved by varying wind tunnel speed. The emphasis is on the analysis of the streamwise Reynolds stress and mean velocity profiles and on related scaling issues problem. It was found that for the same external conditions defined by the pressure gradient coefficient Cp the effect of Reynolds number close the wall is seen even for such a narrow range of Reynolds number. In particular, the shape factor revealed more flow resistance on separation with increase in Reynolds number. The difference where observed mainly at the beginning of incipient detachment, where there is a drop of turbulence activity near the wall while it reaches a maximum value in the outer region.

1. Introduction

Although a significant amount of research has been devoted to understanding canonical flat plate zero pressure gradient boundary layer, it is not the case for adverse pressure gradient boundary layers and even more when the flow is approaching separation. The most common problem in the similarity analysis is the Reynolds number effect [1]. For zero pressure gradient boundary layer it is known that scaling parameters of classical inner or outer scaling are a function of Reynolds number. It is not the case for more complex system, where additional length or time scales are introduced by the geometry or system dynamics. These scales do not develop in the same manner with Reynolds number as the traditional ones [1]. It is the reason why the non-linear effect is observed, what is seen by the rapid decrease of near wall viscous length scale and week variation of the outer length scale with Reynolds number. This is particularly important when trying to analyze adverse pressure gradient (APG) flows and to predict the separation location on curved surface, where due to considerable change of the wall shear in the streamwise direction, the turbulence cannot be characterized solely in terms of local parameters [2]. In the case when a boundary layer approaches separation the mean velocity inflectional profiles produce conditions for emergence of large eddies which interacts with those already present in the flow [3]. A different process of decay of those eddies may be responsible for the lack of similarity of those non-equilibrium flows.

Difficulties in scaling of APG wall-bounded flows may also emerge from amplitude modulation effect of small scales by large scale structures, which was elucidated by skewness changes of filtrated velocity small and large scale signals [4]. The physical phenomenon of amplitude modulation is still unknown, but it seems that it influences the wall-normal momentum transport by the changing the convection velocity of small scales [5]. If so, this may postpone separation with rise of the Reynolds number.

In the most available experimental research with APG flows there is a lack of canonical setup [6]. The paper is an attempt to fill this gap. The main motivation of the present study is to verify the effect of Reynolds number at matched conditions of pressure gradient leading to detachment and to analyze how it influence the scaling of mean and fluctuating velocity profiles.

2. Experimental facility and measurement technique

The experiment was performed in an open-circuit wind tunnel, where the turbulent boundary layer was developed along the flat plate, which was 6870 mm long. The wind tunnel is designed with large dimension settling chamber and three contraction sections, which allows to achieve free stream turbulence intensity below 1% at the inlet plane. In order to bypass laminar-turbulent transition, the tripping of boundary layer after the leading edge was used. The inlet rectangular channel with a length of 5.035 m located upstream the proper test section has two pairs of suction gaps aimed to control the two-dimensionality of the flow by minimizing of boundary layers on the side walls. In order to reduce the effect of secondary vortices developing along rectangular channels the triangular corner inserts were used in the whole inlet channel. A slight inclination of the upper wall helped to keep zero pressure gradient (ZPG) conditions at the at the entire length of this section.

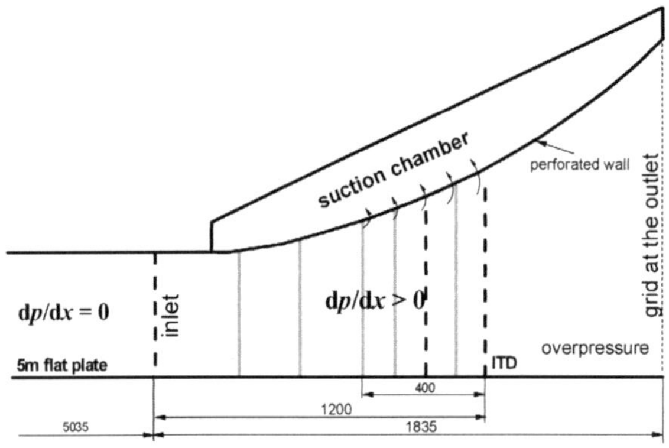

Figure 1. Test section geometry

The specially design test section located at the end of the wind-tunnel (see figure 1) is equipped with perforated, movable upper wall. Computer-controlled suction system equipped with a low power axial compressor allows for smooth adjustment of the amount of the exhausted air from the top of the wall. By playing with the shape and position of the upper wall as well with the suction flux it is possible to generate wide range of pressure gradient conditions, while at the inlet channel the zero pressure gradient conditions were secured. The static pressure in the test section is increased by the throttling at the outlet of the test section. For certain pressure conditions it is possible to generate on the bottom wall the turbulent boundary layer, which is at the verge of separation.

Static pressure measurements were done through pressure holes of 0.5 mm dia, drilled along the side line along the streamwise direction, from -700 to 1750 mm of the x coordinate where zero is set at the inlet plane (see Fig. 1). The spacing of pressure taps was equal 50 mm. Pressure differences were measured by means of DATA INSTRUMENTS DCXL01DN pressure transducer connected to KULITE D486 amplifier. The mean relative error of pressure measurements equals 2%.

The velocity measurements were performed with hot-wire anemometry CCC developed by Polish Academy of Science in Krakow. A single hot-wire probe of a diameter $d = 3$ µm and length $l = 0.4$ mm was used. In the experiment for each case the wire length was always below 20 in inner variables as it was recommended by Ligrani and Bradshaw [15]. The hot-wire bridge was connected to a 16 bit

PC card. Acquisition was maintained at frequency 25 kHz with 30 s sampling records. The ambient conditions were carefully controlled during the measurements. In the course of a single profile measurement the scatter of ambient temperature at the end of the test section did not exceed ±0.2°. In case when the measured temperature was different from temperature during calibration, the temperature correction of CTA voltage was used by Jorgensen [7]. At the same time the free-stream velocity was monitored by the means of a Prandtl's tube.

The facility is equipped with the computer-controlled 2D traversing system (in streamwise and wall-normal direction). The traverse carriage was driven over the maximum displacement of 180 mm by a servo motor. Uncertainty of the drive step in wall normal direction was 0.001 mm with the smallest step equal 0.01 mm, while in the streamwise direction the uncertainty was equal drive step of 0.37 mm.

To have the verified reference friction velocity u_τ along the flow except the Clauser plot method, the fringe skin friction (FSF) technique was also applied [8]. This method, in contrast to hot-wire technique, does not require any assumptions about the form of velocity profile and is very simple and reliable. It is based on the relationship between the thinning of an oil film deposited on the surface exposed to the flow and the local shear stress. In the experiment described in [9] SOX Whitecroft Lighting sodium lamp, emitting the monochromatic light of the wavelength λ = 0.5893 μm, for illuminating the oil-film was used. The relative error was in the range of 1% for maximal of skin friction value estimated by FSF technique. The shear stresses were compared with the ones obtained by modified Clauser plot method with satisfactory results with standard parameters of logarithmic law of the wall $\kappa = 0.38$ and $B = 4.1$ taken from hot-wire measurements up to x = 700 mm [10]. The u_τ values used in the paper were obtained based on FSF method above location of x = 700 mm apart from the case of 15 m/s, which was obtained by scaling using the inlet value for each Reynolds number.

3. Test cases description

Flow parameters determined in core flow at the inlet plane to test section (i.e. 5035 mm downstream the flat plate leading edge), located in the zero pressure gradient area, are the mean velocity U ≈ 10, 15 and 20 m/s and turbulence intensity Tu < 1%. The inlet Reynolds number based on friction velocity and boundary layer thickness was equal 1900, 2600 and 3300 respectively. To activate the side suction gaps the overpressure 15 Pa was set at the inlet plane. Basic inlet parameters have been summarized in Table 1, where Tu is turbulence intensity, U_{in} mean velocity outside of TBL, u_τ friction velocity $u_\tau = \sqrt{\frac{\tau_w}{\rho}}$, θ momentum loss thickness and $Re_\theta = \frac{U_{in}\theta}{\nu}$ is Reynolds number, where ν is the kinematic viscosity.

Table 1. Inlet conditions of ZPG turbulent boundary layer.

	Tu [%]	U_{in} [m/s]	u_τ [m/s]	θ [mm]	Re_θ [-]
□	<1%	10	0.72	8.26	6300
○	<1%	15	0.55	8.91	8200
Δ	<1%	20	0.37	10.3	10150

The grey vertical lines in Fig. 1 represent the positions where the measurements of velocity profiles were performed, while detailed comparison was done at three locations marked as dark dashed vertical lines. In the last position the boundary layer has a features typical for Intermittent Transitory Detachment (ITD) defined by Simpson [11] as 20% of reversed flow. Measurements were performed for three different Reynolds number and perfectly matched distributions of pressure coefficient:

$$C_p = 1 - \left(\frac{U_\infty}{U_{\infty 0}}\right)^2, \quad (1)$$

shown in figure 2a, where $U_{\infty 0}$ is the mean inlet velocity. In figure 2b the Clauser-Rotta pressure gradient parameter β defined as:

$$\beta = \frac{\Delta}{u_\tau}\frac{dU_\infty}{dx}, \tag{2}$$

where $\Delta = \delta^* \frac{U_\infty}{u_\tau}$ and $\delta^*$is displacement thickness, is presented.

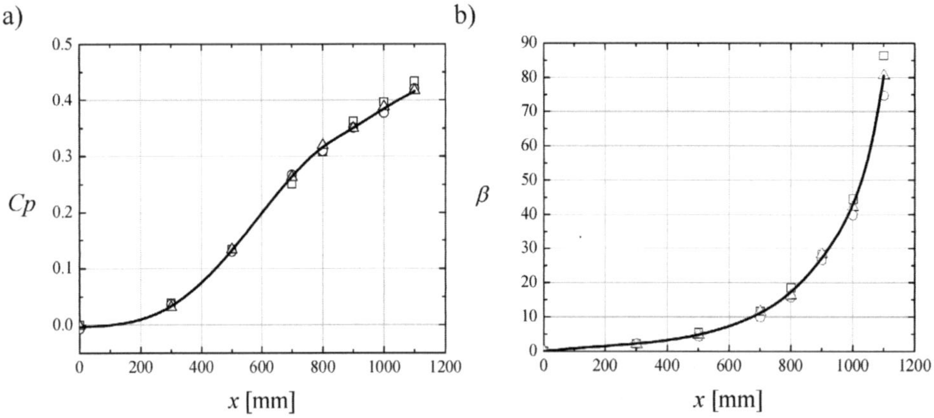

Figure 2. Distributions of the flow parameters: Pressure gradient parameter C_p a) and Clauser-Rotta pressure gradient parameter β b)

It is seen that, irrespective of different Reynolds number both C_p and β distributions matches well. This confirms that from the point of view of pressure gradient it was possible to get similar external conditions. For the completeness of data figure 3 presents the pressure and the pressure gradient distributions.

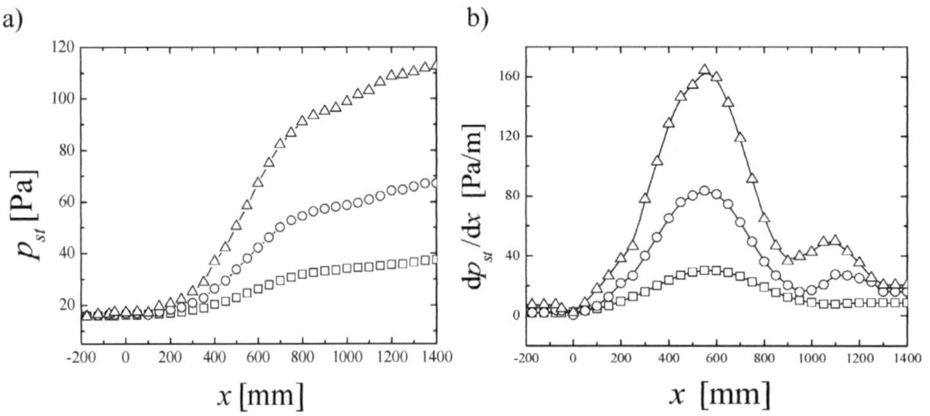

Figure 3. Pressure a) and pressure gradient b) distributions for analyzed range of *Re*.

It is seen that the shape of pressure gradient distribution at the wall varies slightly across the cases. In particular, the second maximum of pressure gradient (figure 3b) is placed at different position. This may be caused by interfering impact of separation on pressure measurements and must be clarified in the future.

4. Results

Data analysis starts with the inlet plane. Figure 4 shows the mean streamwise velocity profiles and streamwise Reynolds stresses presented in outer scaling (a) and 9b) and inner scaling (c) and (d). In the figure U_o is the outer scale velocity defined as $U_o = 2(U_\infty - U_{y=0.5\delta})$, which is similar to Zagarola-Smith scale [12].

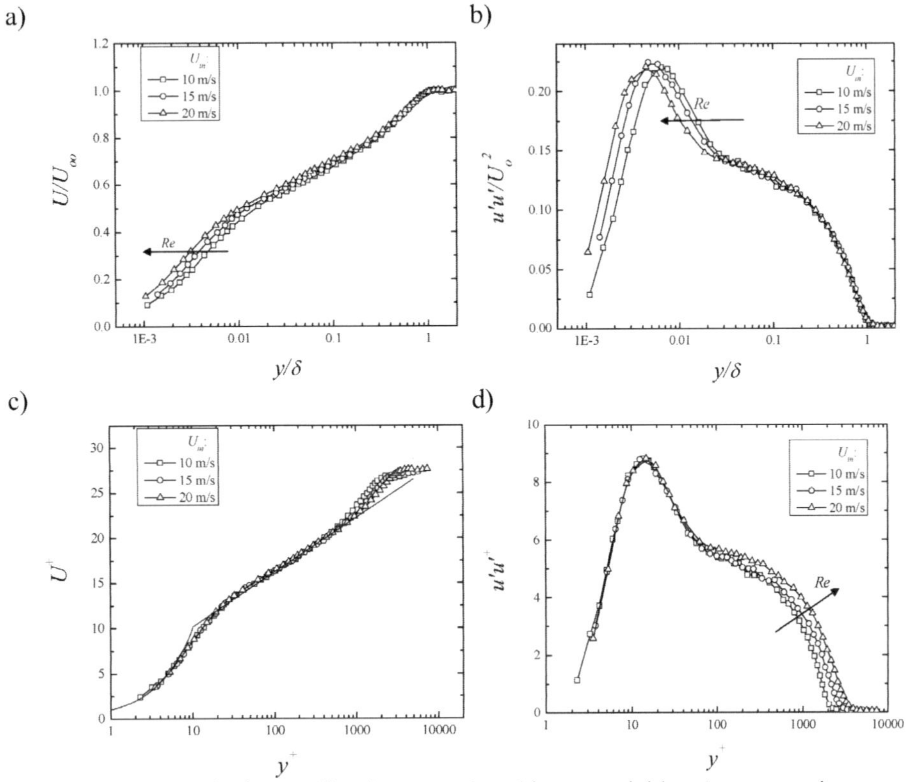

Figure 4. Mean velocity profiles in outer a) and inner variables c); streamwise Reynolds stress distributions in outer b) and inner variables d) at inlet location for analysed Reynolds numbers.

Scaling with outer variables aims to show the difference near the wall, while inner scaling reveal changes in the outer layer. In the figure the typical effect of Reynolds number is observed. Particularly there is an increase of log-layer range for mean profiles and increase of streamwise normal Reynolds stress in the outer region, which is related to the rise of a large scales energy in this region. For such a narrow range of Reynolds number the near wall peak reveal only minor increase of maximum what was also indicated by Marusic et al. [14].

Figure 5 presents the downstream evolution of the boundary layer thickness δ and shape factor $H=\delta^*/\theta$. While the inlet values behave as expected with increase of Reynolds number it is not the case for APG region. It is seen that in the direction of the flow and especially at the verge of separation the difference between δ values decreases. On the other hand, the reaction of shape parameter on *Re* is weaker. Nevertheless, in each case a strong almost linear increase of *H* from x = 700mm is seen, while the beginning of this growth delays with Reynolds number. Such a tendency in *H* is due to strong drop of mean momentum defect.

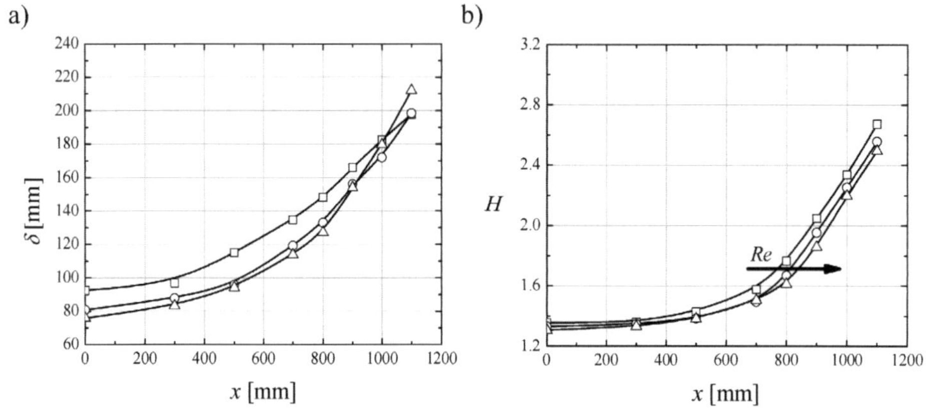

Figure 5. Distributions of the flow parameters: boundary layer thickness δ (a) and shape factor H (b)

Figure 6 presents the scaling of mean velocity and streamwise Reynolds stress profiles along the APG region for the largest *Re* number (Re_θ = 10150). Because the flow dynamics in separation region is dominated by outer layer large scale motions the scaling analysis were performed using outer variables introduced already in figure 4a-b.

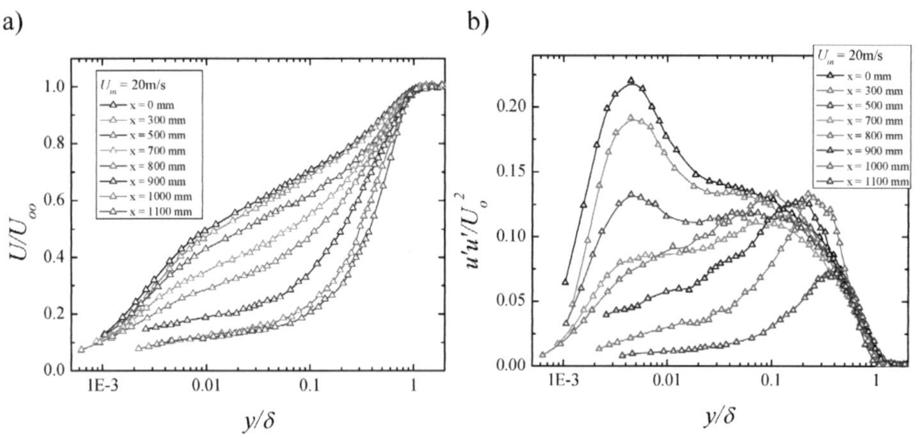

Figure 6. Mean velocity profiles (a) and streamwise Reynolds stress distributions (b) for 20 m/s case

In the downstream direction a strong deformation of velocity profile and the change of the shape of streamwise Reynolds stress can be noticed. The biggest difference between particular profiles appear near the wall. Approaching separation the near-wall peak decreases rapidly and vanishes near the separation. On the other hand, the rise of the outer peak is observed, which coincides well with the increase of d*U*/d*y* observed in figure 6a. In figure 6b the sharp increase in the value of *u'u'* in the outer region close to separation is seen, what is due to the strong activity of large scale structures. It is clear from the above graphs that outer scaling fails in the adverse pressure gradient. This also indicates that the constant scaling factor throughout the boundary layer thickness is not appropriate for Reynolds stress profiles.

To compare the Reynolds number effect on APG boundary layer two different locations were chosen. First one is x = 900 mm with $\beta \approx 28$, where there is the greatest difference in the shape factor (see figure 5). The second one is the last measuring position x = 1100 mm with the highest pressure parameter $\beta \approx 80$. The comparison is presented in figure 7.

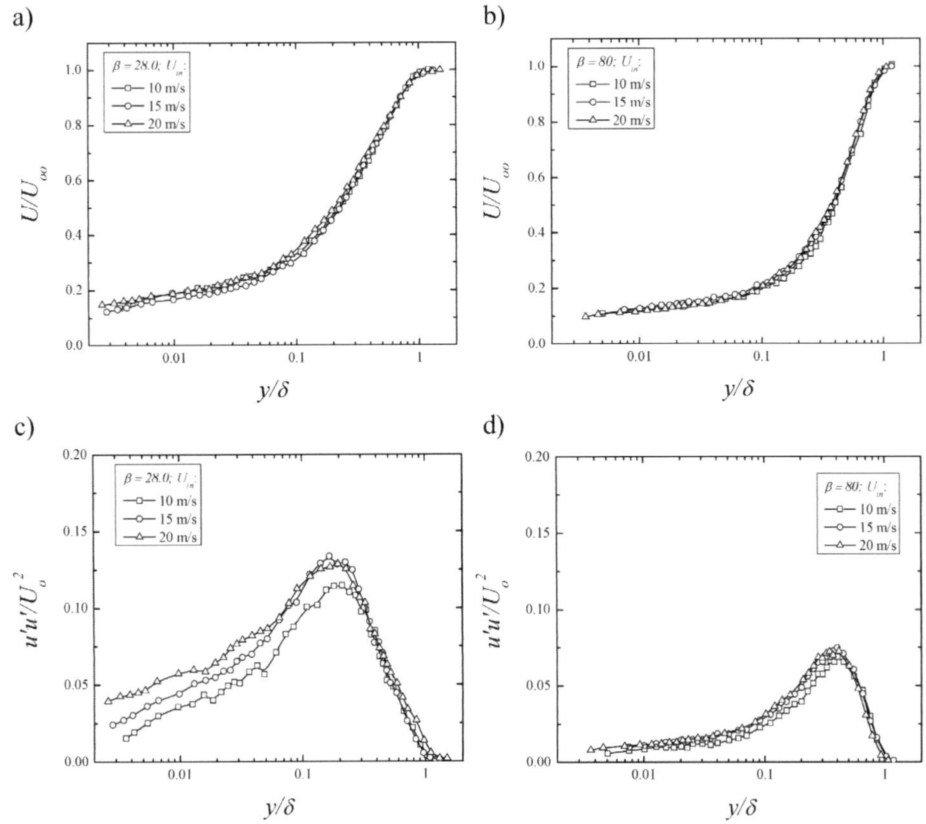

Figure 7. Mean velocity profiles a) and b) and streamwise Reynolds stress distributions b) and d) respectively for $\beta \approx 28$ and for $\beta \approx 80$

Despite that mean velocity profiles well overlaps (see figures 7a and 7b the difference is seen for $u'u'$ especially close to the wall. This is more evident for the first location (x = 900 mm), where strong enhancement of the outer peak with the rise of Reynolds number is observed. It confirms the dominant role of the outer length scales and it is different to the observed behaviour for canonical zero pressure gradient flows visible in figure 4b. Surprisingly, the effect of Reynolds number decays at the last location. It is probably due to vanishing of small-scale turbulence and only large scales left to drive the flow. The growth of the outer peak, seen in figure 7c, may be also related to mentioned in the introduction amplitude modulation effect and the resulting changes of convection velocity of small scales which influences wall-normal momentum transport. The symptom of increased momentum transport is visible as the elevation of Reynolds stress near the wall (see figure 7c). It is not so clearly visible on mean profiles probably due to narrow range of Reynolds number.

5. Conclusions

The influence of Reynolds number on turbulent strong decelerated turbulent boundary layer developed on a flat plate was analysed. A momentum thickness Reynolds number at the inlet to a test section was in the range from 6300 to 10150.

It was found that for the same external conditions defined by the pressure gradient coefficient Cp the effect of Reynolds number close the wall is seen even for such a narrow range of Reynolds number. In particular, the shape factor revealed more flow resistance on separation with increase in Reynolds number. The difference where observed mainly at the beginning of incipient detachment, where there is a drop of turbulence activity near the wall while it reaches a maximum value in the outer region. The sharp increase in the value of $u'u'$ in the outer region is probably due to the strong activity of large scale structures. The growth of the outer peak may be also related to mentioned in the introduction amplitude modulation effect and the resulting changes of convection velocity of small scales which influences wall-normal momentum transport. Nevertheless the primary reason this phenomenon causes the failing of outer scaling methods. Surprisingly, the effect of Reynolds number decays at the last location, where the boundary layer has a features typical for Intermittent Transitory Detachment having 20% of reversed flow. It is probably due to vanishing of small-scale turbulence and only large scales left to drive the flow.

Acknowledgment
The investigations presented in this paper have been obtained with funding from Polish National Centre of Science within the grant DEC 2011/03/B/ST8/06401 as well as statutory fund BS 1-103-301/2015

References

[1] Degraaff D B, Webster D R and Eaton J K 1999 The effect of Reynolds number on boundary layer turbulence *Exp. Therm. Fluid Sci.* **18** 341–6
[2] Dróżdż A, Elsner W and Drobniak S 2015 Scaling of streamwise Reynolds stress for turbulent boundary layers with pressure gradient *Eur. J. Mech. B/Fluids* **49** 137–45
[3] Song S and Eaton J K 2004 Reynolds number effects on a turbulent boundary layer with separation, reattachment, and recovery *Exp. Fluids* **36** 246–58
[4] Mathis R, Marusic I, Hutchins N and Sreenivasan K R 2011 The relationship between the velocity skewness and the amplitude modulation of the small scale by the large scale in turbulent boundary layers *Phys. Fluids* **23** 121702
[5] Dróżdż A 2014 Influence of pressure gradient on streamwise skewness factor in turbulent boundary layer *J. Phys. Conf. Ser.* **530** 012061
[6] Gungor A G, Maciel Y, Simens M P and Soria J 2014 Analysis of a Turbulent Boundary Layer Subjected to a Strong Adverse Pressure Gradient *J. Phys. Conf. Ser.* **506** 012007
[7] Jorgensen F E 2002 *How to measure turbulence with hot-wire anemometers - a practical guide* ed Dantec Dynamics (Dantec Dynamics A/S, P.O. Box 121, Tonsbakken 16-18, DK-2740 Skovlunde, Denmark)
[8] Tanner L H and Blows L G 1976 A study of the motion of oil films on surfaces in air flow, with application to the measurement of skin friction *J. Phys. E.* **9** 194–202
[9] Dróżdż A, Elsner W and Drobniak S 2008 Application of oil-fringe interferometry for measurements of wall shear stress *Turbomachinery* **133** 103–10
[10] Dróżdż A, Elsner W and Sikorski D 2016 Skin Friction Measurements by Thin-Oil Film Interferometry in Turbulent Boundary Layer with Pressure Gradient *12th International Symposium On Compressor & Turbine Flow Systems Theory & Application Areas SYMKOM,* pp 18–9
[11] Simpson R L 1989 Turbulent Boundary-Layer Separation *Annu. Rev. Fluid Mech* **21** 205–34
[12] Zagarola M V. and Smits A J 1998 Mean-flow scaling of turbulent pipe flow *J. Fluid Mech.* **373** 33–79
[13] George W K and Castillo L 1997 Zero-pressure-gradient turbulent boundary layer *Appl. Mech. Rev.* **50** 689–730
[14] Marusic I, Mathis R and Hutchins N 2010 High Reynolds number effects in wall turbulence *Int. J. Heat Fluid Flow* **31** 418–28
[15] Ligrani P M and Bradshaw P 1987 Spatial resolution and measurement of turbulence in the viscous sublayer using subminiature hot-wire probes *Exp. Fluids* **5** 407–17

Validation of numerical models for flow simulation in labyrinth seals

D Frączek[1] and W Wróblewski[1]

[1] Silesian University of Technology, Institute of Power Engineering and Turbomachinery, Gliwice Konarskiego Street 18, Poland

E-mail: daniel.fraczek@polsl.pl

Abstract. CFD results were compared with the results of experiments for the flow through the labyrinth seal. RANS turbulence models (k-epsilon, k-omega, SST and SST-SAS) were selected for the study. Steady and transient results were analyzed. ANSYS CFX was used for numerical computation. The analysis included flow through sealing section with the honeycomb land. Leakage flows and velocity profiles in the seal were compared. In addition to the comparison of computational models, the divergence of modeling and experimental results has been determined. Tips for modeling these problems were formulated.

1. Introduction

Labyrinth seals are widely used in the rotating machinery. In spite of leakages this type of sealing has many operational advantages. In design process the possible reduction in leakages is nowadays one of the main aims and therefore detailed analysis of flow phenomena in the seal is of great importance to the design of high efficient and reliable turbomachines. The flow structures in the labyrinth seal are very complex. The flow is unsteady with strong jets, flow detachment, and vortex structures. The structures modelling are very important for the evaluation of the seal effectiveness, which is directly related to the possibility of kinetic energy dissipation in seals chambers. The process of seal selection requires conducting an optimization and the numerical model of the flow to identify a solution with the greatest possible precision. CFD calculations are used to resolve the flow structures and provide the optimization of various flow channels. Numerical modelling deserves special appreciation for the design of labyrinth seals in turbines and compressors due to the high costs of the test rigs. The study [1] shows that the use of RANS turbulence models can give solutions which are incompatible with the experimental data.

Numerical calculations are used for flow simulation through the seal for many years. Some authors suggest that the seal flow can be simulated practically only with hybrid LES models [1] although, LES models require higher computational cost. LES errors presented in [1] were below approximately 5 % and predict better the flow physics, errors and scatter between RANS solutions reached levels of 23 %. Steady RANS solvers are still commonly used for seal flow simulation. Comparisons of CFD and the experiment were described in the paper [2] for seal with three fins and step. Authors obtained the relative difference of about 1.7% between the results in case of mass leakage. Straight four fins model with a honeycomb land was analyzed in [3]. The difference between CFD and experiment was at the level of 11%. Authors of the paper [4] compared results for a smooth land seal. Fidelity of the CFD results increased with the number of fins. They also obtained better results for seal land with steps.

CFD results of leakage, in case of straight seal with two fins, were inflated of about 20% comparing to the experiment results.
Complex flow with detachments or heat flow can be modeled using SST-SAS turbulence model. Comparisons [5] show that by introducing von Karman length scale, this model can be an alternative to hybrid RANS/LES in case of complex flow simulations.
The analyzed problem concerns straight-through labyrinth seals model with honeycomb land. The geometrical data were selected on base of the literature sources (occurrence in citied papers) and experimental data availability. In the paper comparison between the CFD results and results of the experiment described in [6], [7] is presented. The RANS turbulence models (k-epsilon, k omega, SST, SST-SAS), commonly used for the industrial calculations, were selected for the analysis. Depending on the selected turbulence model and parameters the independent results may occur. Therefore, the grid independence study was performed for each of the three turbulence models in case of steady-state flow. The comparison of steady and transient solutions for selected grid was presented. The results averaged over time and instantaneous distributions of the velocity field depending on the turbulence model were analyzed.

2. Test cases

For the numerical study two configurations of the straight-through labyrinth seals were chosen, which were well experimentally investigated and described in the literature. The first configuration called Test A has a honeycomb land of diameter 3.2 mm and clearance of 0.51 mm. The labyrinth seal with four tapered fins was investigated by Stocker et al. [6] among many other cases on the static test rig. The definition of geometrical parameters of the seal is presented in Figure 1 and Table 1.

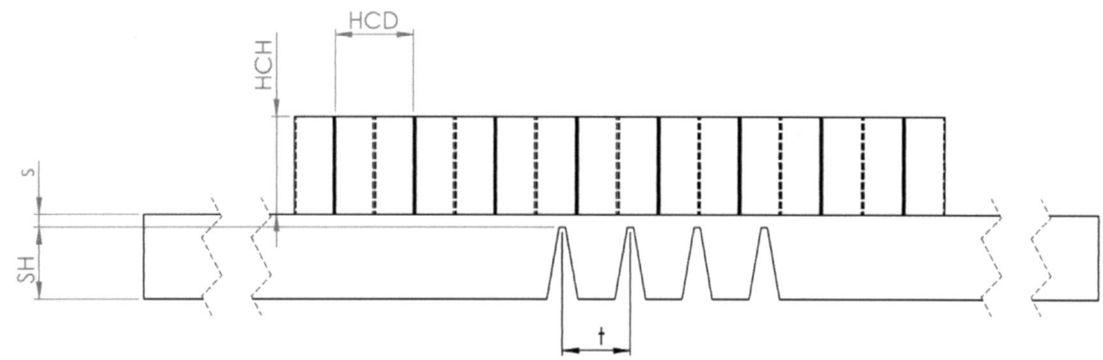

Figure 1. Geometry of the labyrinth seal Test A

Table 1. Dimensions of the seal Test A

s	0.51 mm
t	2.79 mm
SH	2.79 mm
HCD	3.18 mm
HCH	3.81 mm

The computational domain was simplified and the pitch consist two honeycomb cells in the circumferential direction. The study performed with different pitch dimensions showed the negligible influence on the results. The domain width was equal to 5.5 mm. The computational domain has been extended at the inlet and outlet. Inlet extension is 30 mm long. It is sufficient to form a velocity profile and to avoid the influence of the inlet vortex on the inlet boundary. The outlet channel (120 mm long) is shifted downstream to reduce the influence of the outlet jet and large recirculation zone on the outlet boundary. Boundary condition was set according to the experimental data. The inlet pressure is calculated from the pressure ratio and outlet static pressure value equal 100 kPa. The inlet turbulence intensity level was set to 5% (medium intensity with $\mu_t/\mu=10$). Epsilon, omega and k values at the boundary for turbulence equations were calculated from turbulence intensity and viscosity ratio [8]. The inlet temperature is 293 K. The air ideal gas model was assumed. The honeycomb walls and channel wall were adiabatic.

Translational periodicity was set on the left and right domain surfaces. The computational domain in the seal region consists of two parts (honeycomb part and main flow part) which are connected by the general grid interface (GGI). The O-grid was used inside the honeycomb structure and the hexahedral mesh was used in the main flow domain. The GGI connection allows the use independent meshes inside the honeycomb and main channel. The GGI connection was successfully implemented in [3], [9], [10], [11]. The mesh independence study was performed for steady state conditions on the four meshes (0,45M, 3.6M, 7.5M, and 15M). The grid was refined in the areas of honeycomb cells and near the fins. The grid cell size was changed proportionally in all directions. The basic parameter compared in the calculations is the discharge coefficient C_D, which is defined as:

$$C_D = \frac{\dot{m}}{\dot{m}_{ideal}}, \tag{1}$$

$$\dot{m}_{ideal} = \frac{p_0 A}{\sqrt{T_0}} \sqrt{\frac{2\kappa}{R(\kappa-1)} \left[\left(\frac{1}{\pi}\right)^{2/\kappa} - \left(\frac{1}{\pi}\right)^{(\kappa+1)/\kappa} \right]} \tag{2}$$

C_D describes the effectiveness of energy dissipation in the seal by comparing a mass flow rate in the seal to the mass flow rate in the isentropic flow through a single ideal nozzle (2) at the same conditions.

The mesh independence study was performed only for one pressure ratio equals 1.5. The stopping criteria for the solution were: discharge coefficient C_D variation less than ±1% and the solution residuals level less than 10E-4. Figure 2 indicates that for meshes with a number of nodes higher than 3.6M there is no significant variation in the C_D with increasing number of nodes. There are slight differences between results depending on turbulence models. The C_D values obtained for the k-epsilon turbulence model are the highest and values for SST and k-omega turbulence model are almost the same. Values obtained for steady state solution are at least higher by about 5.5% from the results of the experiment. The mesh with 3.8M elements which will be used in further calculations is presented in Figure 3.

The geometrical data and experimental results derived from [7] were used for the second test case (Test B). Two fins straight through seal with a honeycomb land was chosen (Figure 4). Because the geometrical data in [7] were given relatively to the fin pitch it was necessary to assume the fin pitch value. It was set to 10 mm. Final seal dimensions are presented in Table 1.

Boundary conditions were set according to the conditions of experiment with assumption of adiabatic conditions. At the inlet the total pressure of 200 kPa and total temperature of 400K were assumed. The inlet turbulence intensity level was set to 5% (medium intensity). The outlet pressure was set from the pressure ratio.

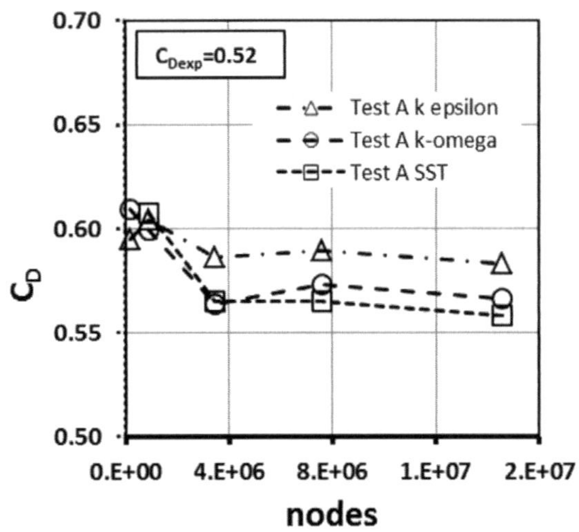

Figure 2. Discharge coefficient C_D in function of number of nodes Test A

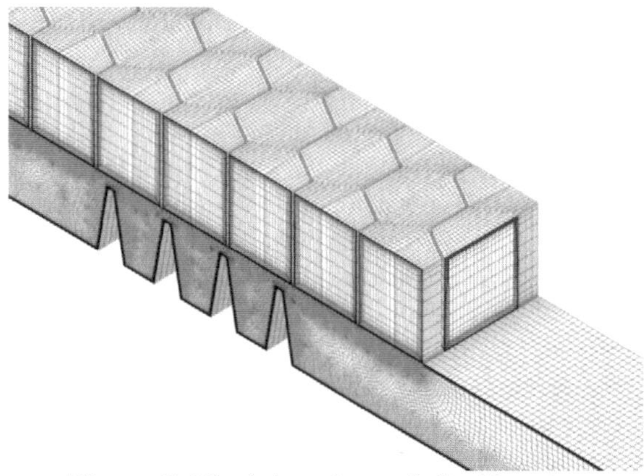

Figure 3. Final domain mesh for Test A

Table 2. Dimensions of the seal Test B

s	0.75 mm
t	10 mm (set)
b	0.5 mm
HCD	1.6 mm
HCH	8 mm
SH	3.5 mm

The extensions of the computational domain at the inlet and outlet were similar to the Test A. Sutherland's formula was used to derive the dynamic viscosity and heat transfer viscosity of an ideal gas as a function of the temperature.

The mesh independent study was performed for 6 different meshes with nodes number from about 0.4M to 15M. Mesh independence study was performed for the pressure ratio of π=1.2. The calculations were performed with the same turbulence models as in the previous case: k-epsilon, k-omega and SST. Discharge coefficient values obtained for the different meshes and turbulence models are presented in Figure 5. The C_D values for meshes with node number greater than 2M are changing very slowly. The difference of about 1.5% is between results for the mesh with 2M and 15M nodes. Meshes with the smallest number of nodes got results only about 5% higher than mesh with the highest number of nodes. The highest value of C_D was obtained for SST turbulence model, the lowest for k-epsilon turbulence model. The mesh with 2M nodes was chosen for further calculations and is presented in Figure 6. The difference between experimental data and calculation results in the case with SST turbulence model was the highest and amounted about 11%.

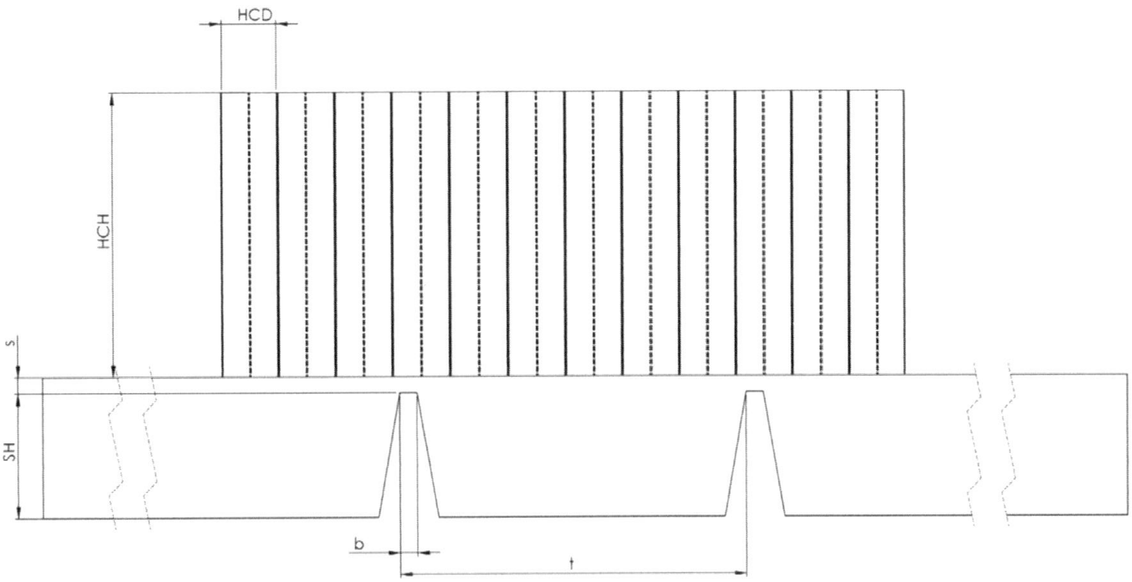

Figure 4. Geometry of the labyrinth seal for the Test B

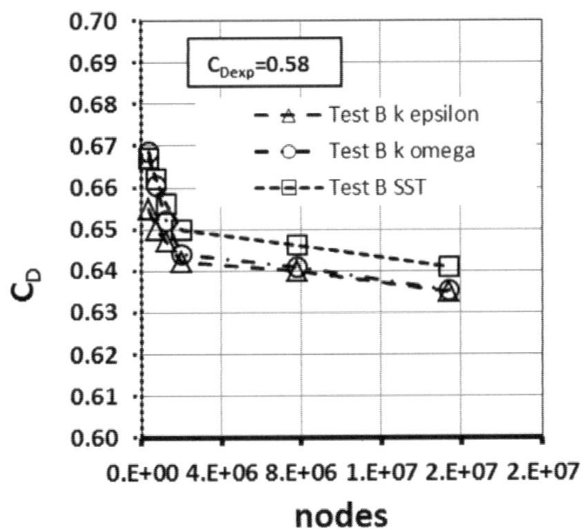

Figure 5. Discharge coefficient C_D in function of number of nodes Test B

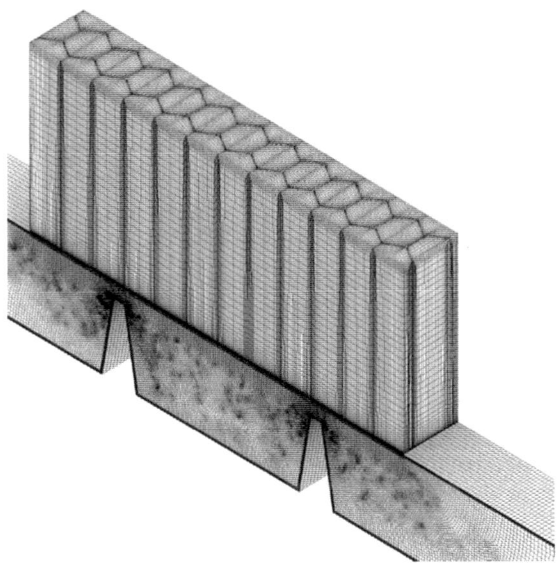

Figure 6. Final domain mesh for Test B

3. Results

3.1 Test case A

Discharge coefficient depending on pressure ratio for steady calculation of Test A is presented in Figure 7. The difference between CFD and experiment results increase with increasing pressure ratio. Values from the calculation increase along one trend line. There is still some difference between turbulence models for each pressure ratio. Highest C_D values were calculated with k-epsilon turbulence model.

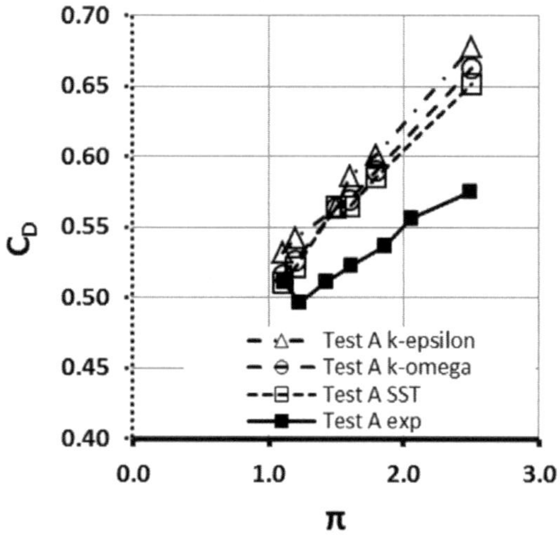

Figure 7. $C_D(\pi)$ Test A

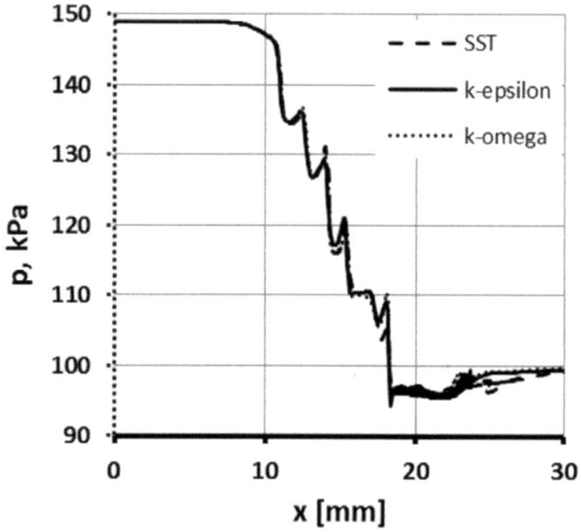

Figure 8. Pressure distribution along streamline Test A, $\pi = 1.5$

There were no available results of pressure or velocity distribution in the seal from experiment so only results between various turbulence models were selected for comparison. Pressure distribution along streamline from inlet to outlet looks similar for all turbulence models (Figure 8). The streamline is not the same in all cases, but it passes through the same seal regions clearances and chamber between fins. There are five clearly visible regions with pressure decrease connected with fins position and honeycomb meridional direction change. There are also similar pressure recovery processes after acceleration. Honeycomb meridional direction change and difference between the average values of effective clearance above each fin caused differences in pressure decrease along a streamline. An example of velocity distribution in the middle of the seal width model cross section is showed in Figure 9. Velocity profile for others turbulence models with steady calculation looks similar. Maximum velocity in cross-section occurs above third fin. The CFD models reproduce single vortices filling chamber between the fins and the honeycomb cells above fins.

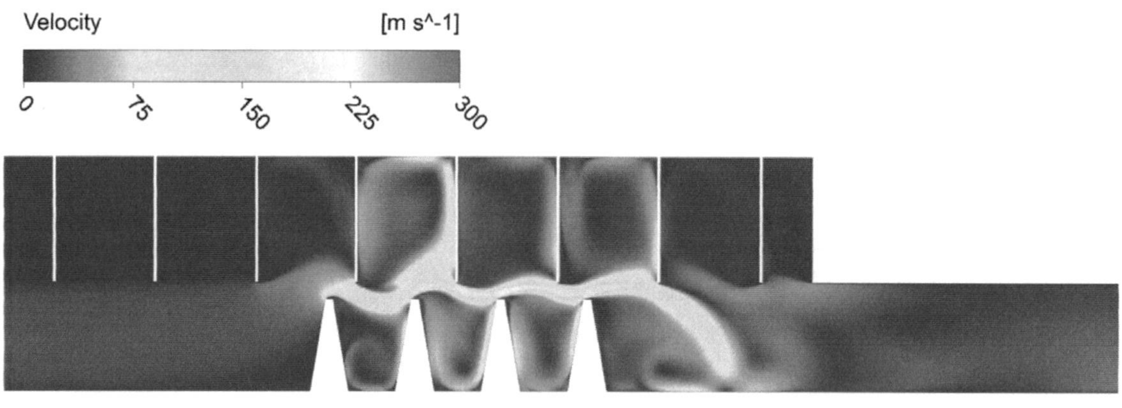

Figure 9. Velocity contour in the mid-section of the domain (SST, steady solution, $\pi = 1.5$)

Transient calculations were initialized from steady state solution. Time step was set as a function of Courant number, in the case of k-epsilon, k-omega and SST turbulence model with a max Co<5. Final time step values take a value close to 1E-7 s. Using short time step can help to show the smallest disruption in the flow field but because it extends the time of calculation values of Courant was limited to 5. An SST-SAS turbulence model were used for calculation with a Co<1 as a control case. Simulation requires calculating over 20 000 time steps. Calculation for transient Test A CFD models were limited to the case with π=1.5.

Discharge coefficient from transient results (Figure 10) for k-omega SST and SST-SAS turbulence models decrease below steady value and start to oscillate around experiment value. Average values for k-omega, SST and SAS-SST were almost the same. Results with k-epsilon turbulence model fail to simulate unsteady flow field inside seal land, and remains equal to the steady results discharge coefficient. Flow oscillations periods are a few ranges larger than selected time step which suggests that unsteady flow condition can be simulated with larger time step.

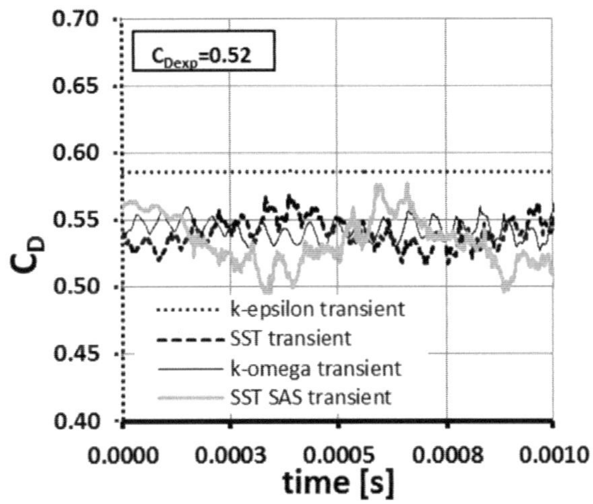

Figure 10. C_D(time) fluctuation Test A

3.2 Test case B

Calculations for the Test B were made with a similar manner. Discharge coefficient C_D apart from pressure ratio is above results from experiment (Figure 11). Relative error $(C_{DCFD}-C_{Dexp})$ $/C_{Dexp}$ between steady calculation and experiment results decrease with increase pressure ratio which is tendency reverse to Test A. Error level is on the similar level to the Test A case. The result with SST turbulence model got the highest C_D values, but a globally difference between each C_D results was small.

Pressure distribution along streamline (Figure 12) and velocity contours (Figure 13 and Figure 14) in Test B were compared for π=1.2 and various turbulence models. The pressure reduction between fins was clearly higher for k-epsilon turbulence model despite comparable values of C_D. In this case, the pressure difference is connected with a difference in velocity flow field. Maximum velocity occurs above first fin for SST (and also k-omega) turbulence model and energy dissipation takes places with different manner. Maximum value of velocity for SST turbulence model reaches about 200 m/s while for k-epsilon turbulence model maximum velocity reaches maximum about 150 m/s above second fin. Since the pressure distribution for SST and k-omega are comparable, velocity distributions were similar and figure showing velocity contour for k-omega turbulence model was omitted. The SST turbulence model is a combination of k-epsilon and k-omega turbulence models [8] solving in the region close to the wall equation of k-omega turbulence model. This indicate a strong influence of the modeling the boundary layer on the global nature of the seal flow.

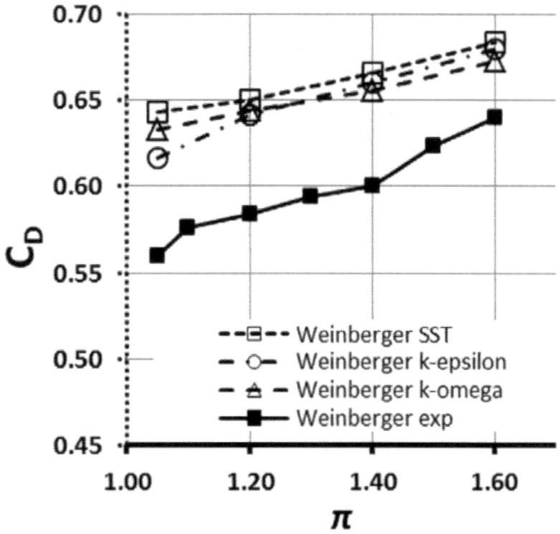

Figure 11. $C_D(\pi)$ Test B

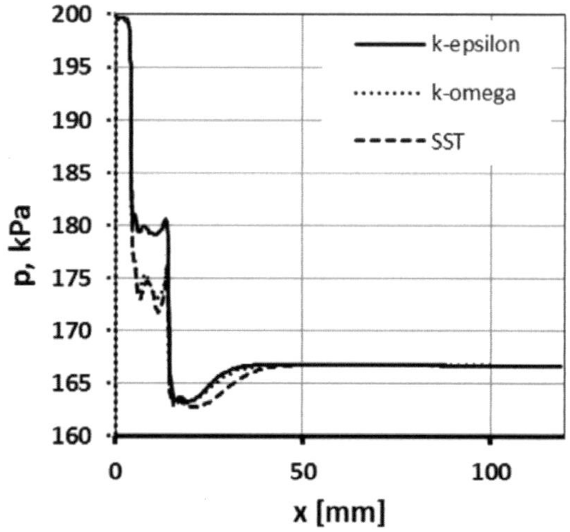

Figure 12. Pressure distribution along streamline Test B, $\pi = 1.2$

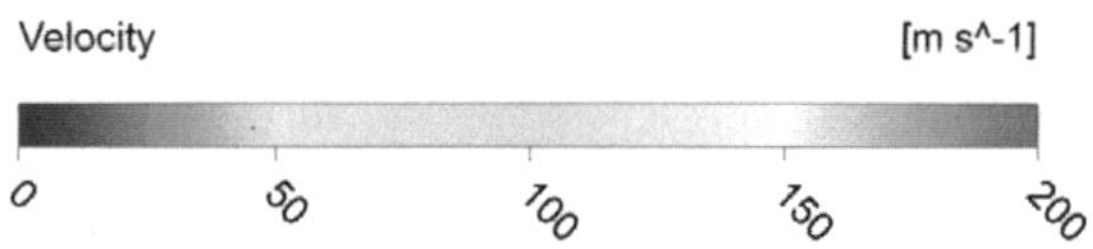

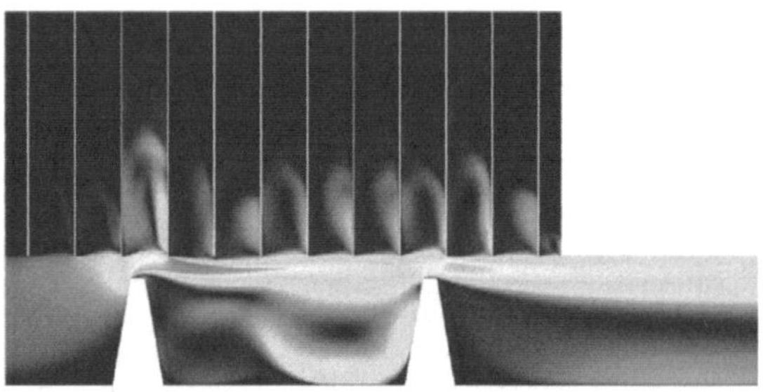

Figure 13. Velocity contour in middle of domain for SST Test B steady $\pi = 1.2$

Figure 14. Velocity contour in middle of domain for k-epsilon Test B steady $\pi = 1.2$

There are more differences in flow distribution inside seal channel. The region with the highest value of velocity is more visible in case of SST turbulence model; also vortices inside honeycomb are clearer. These differences in flow can potentially help to show energy dissipation processes related with flow deceleration and reduction of the leakage [12].

Transient results for Test B allowed showing specific and instant flow field inside seal land. Transient calculation were performed with constraints like in Test A, k-epsilon, k-omega and SST with Co<5 and for SST-SAS Co<1. Chart from Figure 15 illustrated C_D(time). SST, k-omega and SST-SAS turbulence models showed similar tendency to Test A transient C_D fluctuation. CFD C_D values decreased and oscillate around experiment C_D results. The standard k-epsilon model did not predict flow fluctuation. Flow oscillations have a frequency with the clearly lower period than selected time step. Flow fluctuation are higher than in Test A because for a solution with fewer fins the flow is less stable and unsteady flow and pressure fluctuation are less suppressed. Maximum amplitudes were obtained for the models SST and SST-SAS.

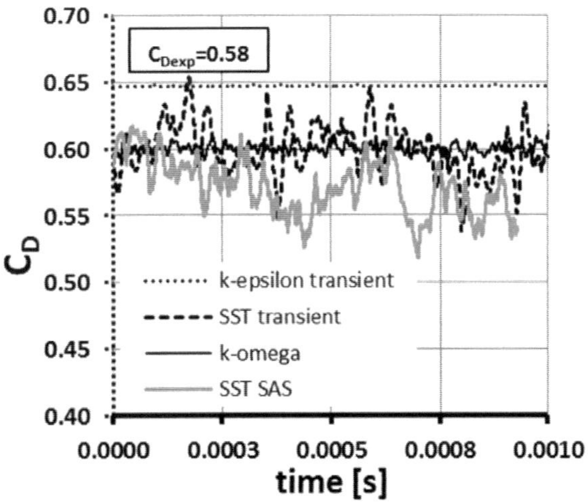

Figure 15. C_D(time) fluctuation Test B

Leakage oscillations in Figure 15 and Figure 10 are also connected with velocity and pressure field disruption. Results for Test B with SST-SAS turbulence model allowed showing very complex flow compared to the steady solution. An example of the instantaneous velocity profile (showed in Figure 16) inside seal mid-section and pressure distribution along streamline (in Figure 17) from inlet showed complex fluctuation started above fins and developed in seal chambers.

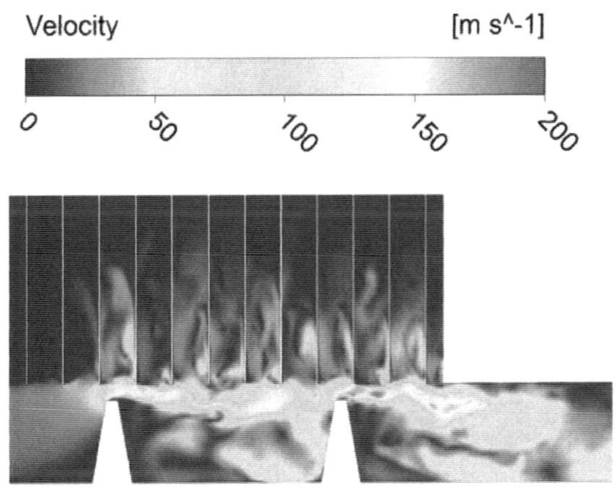

Figure 16. Instantaneous plot in middle of domain for SST SAS Test B transient $\pi = 1.2$

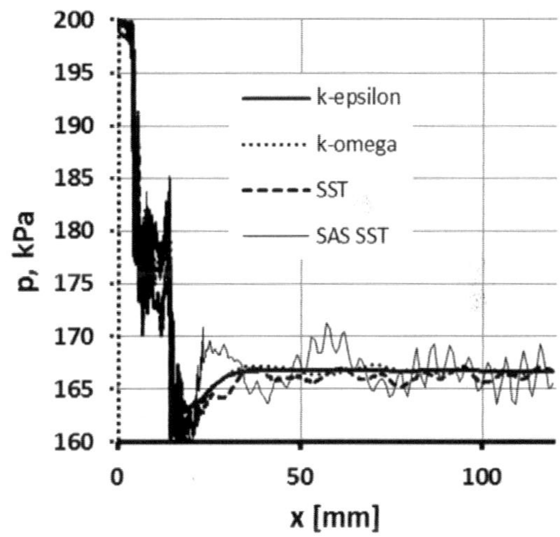

Figure 17. Instant pressure distribution along streamline Test B transient $\pi = 1.2$

4. Summary

CFD modelling of flow through the seal can provide results which are not in full compliance with the results of the experiment. Discrepancies amounting to more than few percent could be too high to predict the seal performance. Differences between solutions can also be observed. However, it is possible to predict the global trends. The analysis can be useful in justifying the decision on the use of numerical simulation of flow through the channels.

Analyses indicate that SST and k-omega turbulence models with transient solvers are suitable for modelling compressible flows with numerous flow separation and large velocity gradients. Results with more precise flow projection can be obtained using SST-SAS turbulence model but with higher computational cost.

References

[1] J. Tyacke, R. Jefferson-Loveday and P. G. Tucker, "On the Application of LES to Seal Geometries," *Flow, Turbulence and Combustion,* vol. 91, no. 4, pp. 827--848, 2013.

[2] I. S. Bambang, C. K. Johan, M. J. d. C. Koen, B. K. Arjen, A. K. Gerrit and F. A. V. Joris, "GT2007-27905 Performance Evaluation of Gas Turbine Labyrinth Seals Using Computational Fluid Dynamics," in *ASME Turbo Expo 2007: Power for Land, Sea, and Air*, Montreal, 2007.

[3] H. H. Chougule, D. Ramerth and D. Ramachandran, "Low Leakage Designs for Rotor Teeth and Honeycomb Lands in Labyrinth Seals," *ASME Turbo Expo 2008: Power for Land, Sea, and Air,* no. GT2008-51024, pp. 1613-1620, 9–13 June 2008.

[4] S. Wittig, U. Schelling, S. Kim and K. Jacobsen, "Numerical Predictions and Measurements of Discharge. Coefficients in Labyrinth Seals," *87-GT-188,* 1987.

[5] Y. Egorov, F. R. Menter, R. Lechner and D. Cokljat, "The Scale-Adaptive Simulation Method for Unsteady Turbulent Flow Predictions. Part 2: Application to Complex Flows," *Flow, Turbulence and Combustion,* vol. 85, no. 1, pp. 139--165, 2010.

[6] H. L. Stocker, D. M. Cox and G. F. Holle, "Aerodynamic performance of conventional and advanced design labyrinth seals with solid-smooth abradable, and honeycomb lands," NASA-CR-135307, 1977.

[7] T. Weinberger, Einfluss geometrischer Labyrinth- und Honigwabenparameter auf das Durchfluss- und Wärmeübergangsverhalten von Labyrinthdichtungen, Stutensee, 2014.

[8] *ANSYS® Academic Research, Release 15.0, Help System, CFX Theory Guide, ANSYS, Inc.*

[9] W. Wróblewski, S. Dykas, K. Bochon and S. Rulik, "Optimization Of Tip Seal With Honeycomb Land In Lp Counter Rotating Gas Turbine Engine," *TASK Quarterly : scientific bulletin of Academic Computer Centre in Gdansk,* pp. 189-207, 2010.

[10] J. Gao, K. Zheng and Z. Wang, "Effect of Honeycomb Seals on Loss Characteristics in Shroud Cavities of an Axial Turbine," *Chinese Journal of Mechanical Engineering,* vol. 26, no. 1, pp. 69-77, 1 02 2013.

[11] I. Mahle, "Improving the interaction between leakage flows and main flow in a low pressure turbine," *ASME Turbo Expo 2010: Power for Land, Sea, and Air,* vol. 7, pp. 1177-1186, 14–18 June 2010.

[12] J. Denecke, V. Schramm, S. Kim and S. Wittig, "Influence of Rub-Grooves on Labyrinth Seal Leakage," *Journal of Turbomachinery,* vol. 123, pp. 387-393, April 2003.

Symbols and notes

A	– seal clearance area, m^2
b	– fin tip thickness, m
C_D	– discharge coefficient
SH	– fin height, m
HCD	– honeycomb diameter, m
HCH	– honeycomb cell height, m
$\dot{m}$	– mass flow rate, kg/s
p	– pressure, Pa
s	– clearance, m
t	– fin pitch, m
R	– specific gas constant, kJ/kgK
μ	– dynamic viscosity
π	– pressure ratio (p_0/p_{out})
κ	– ratio of specific heats

Index:

nom	– nominal
0	– total inlet
out	– outlet
t	– turbulence

LBM estimation of thermal conductivity in meso-scale modelling

A Grucelski

Institute of Fluid Flow Machinery, Fiszera 14, 80-231 Gdańsk, Poland

E-mail: agrucelski@imp.gda.pl

Abstract. Recently, there is a growing engineering interest in more rigorous prediction of effective transport coefficients for multicomponent, geometrically complex materials. We present main assumptions and constituents of the meso-scale model for the simulation of the coal or biomass devolatilisation with the Lattice Boltzmann method. For the results, the estimated values of the thermal conductivity coefficient of coal (solids), pyrolytic gases and air matrix are presented for a non-steady state with account for chemical reactions in fluid flow and heat transfer.

1. Introduction

Modelling of coupled phenomena that occur in complicated and random geometry has recently become a subject of intensive research in fields of fluid mechanics and chemical engineering. To obtain a detailed description for the number of technological processes one has to undertake the interdisciplinary research of the physico-chemical phenomena occurring in the multi-level geometry. Such approach gives the opportunity to bring information obtained through the micro-scale description to the macro-level as a closure relationship. For example: effective values of physico-chemical properties describing transport processes occurring in the representative element of volume (REV) along with modelling of coupled phenomena.

In the coal-processing industry, coking plants are widely used to obtain chemically cleaner coal (coke) with side products like benzene, ammonium hydroxide and other chemical species (generally known as coal gas) as well as tar. From the fluid thermomechanics point of view, coking occurs in a bed of granular material (coal grains) that upon heating becomes a homogeneous porous medium. During the coking process, complex phenomena occur at the level of a single grain (pore), including fluid flow through the deposit, thermal conduction and convection, chemical reactions (heterogeneous as well as homogeneous), plastic deformation and others. For accurate modelling of the coupled subprocesses above the micro level, detailed description of effective transport coefficients is essential. For this purpose more traditional CFD methods become far too expensive because of the random geometry and phenomena occurring at multiple scales. All this has to be modelled at the pore level, with additional change of the pore geometry (due to change of size and shape of coal grains) [1]. The common way for devolatilisation modelling is to provide simulation at the averaged level, usually solving the most important phenomena. A number of recent publications review the numerical tools used for simulation of the coking or pyrolysis process at the averaged level; a good example of such a description is [2]. Averaged simulation results are also used for general process description [3].

Journal of Physics: Conference Series **760** (2016) 012005 doi:10.1088/1742-6596/760/1/012005

For the purpose of modelling of fluid thermomechanics we utilize the lattice Boltzmann method, recently enjoying an increasing interest as a numerical tool, especially for modelling the fluid thermo-mechanics in complex geometry. It is based on the Boltzmann equation with subsequent discretisation [4, 5]. The method has proven to be suitable for simulation of viscous and nearly incompressible fluid flow in a simple [6] and complex geometry [7, 8], as well as heat transfer [9, 10], with addition of chemical reactions [11]. In the context of practical application, Asinari *et al* [12] report the LBM modelling of complex phenomena occurring in fuel cells at single pore level for prediction of the overall system behaviour. Modelled process is treated here at the level of meso-scale and complex phenomena occurring at the micro-level (like transport processes inside the coal particle and related chemical reactions) are simplified. The LBM was adopted for determination of the effective heat conductivity for heat transfer in complex geometry. An example of such work is [7]; the authors use LBM to compute effective thermal conductivity of randomly created porous foams.

In this work, we present results for estimation of effective heat conductivity obtained for 2D bed of coal grains, where coal is modelled by uniform cylinders. Fluid thermomechanics as well as chemical specie evolution is modelled by LBM and a brief description of the numerical tool is provided in section 2. In the same section, simplified one specie chemical model of devolatilisation along with temperature dependence of the material properties is described. Before conclusion, results obtained for random geometry are presented.

2. Numerical model details

Chosen geometry, for this particular case, is the representative element of volume (REV) for the packed bed; this simulates coal particles in coking plants. The REV geometry requires that the size of computational domain be much smaller than the length scale present in typical devices in industry but also much larger than the smallest length in modelled phenomena (here, the diameter of the coal particle).

2.1. Lattice Boltzmann method

The Lattice Boltzmann equation, discretised in time, space (by lattice), and advection velocity ($\mathbf{e}_i$) on a regular square lattice, describes the evolution of a relevant physical field in terms of its distribution function. Description of LBM in this section is valid for D2Q9 discretisation of advection velocity; so in presented simulation approach we consider 2-dimensional lattice with 9 velocities of advection of distribution function. Evolution of density distribution function f_i is described with (see [10]):

$$f_i(\mathbf{r}+\mathbf{e}_i, t+\delta t) - f_i(\mathbf{r}, t) = -\tau_\nu^{-1}[f_i(\mathbf{r}, t) - f_i^{eq}(\mathbf{r}, t)] + Q_i^f, \tag{1}$$

where Q_i^{f} is a source term, f_i^{eq} represents the equilibrium state of distribution function corresponding to the averaged fields at $(\mathbf{r}, t)$; τ_ν is a non-dimensional relaxation time for fluid flow, it takes the form:

$$\tau_\nu = 0.5 + \frac{1}{2}\frac{\nu}{c_s^2 \delta t}. \tag{2}$$

here ν and δt are viscosity and time step; $c_s = c/\sqrt{3}$ is the speed of sound in the simulation and c is the lattice speed of advection of distribution functions. In a similar fashion, we track the evolution of temperature with internal energy density distribution function (IEDDF) and the distribution function for chemical species concentration. Respectively for IEDDF we use g_i to describe distribution function and the source term is Q_i^{g} (in this work, it is accounted only in solid phase as change of enthalpy by chemical reactions occurs only in solid phase). Relaxation time $\tau_{\alpha,m}$ written both for fluid f (gas) and solid s (coal grains) in the computational domain

($m \in \{\mathrm{s}, \mathrm{f}\}$) takes a form:

$$\tau_{\alpha,m} = 0.5 + \frac{3}{2}\frac{\lambda_m}{\rho_m C_{\mathrm{p,m}} c^2 \delta t} \tag{3}$$

C_{pm} and λ_m are the heat capacity and heat conductivity of m, respectively. The evolution of chemical specie (being a mixture of gaseous products of devolatilisation) is modelled with use of LBM with additional distribution function. Because of numerical instabilities occurring close to large gradients of macroscopic fields typical of Eulerian methods, Di Rienzo *et al* [11] proposed a numerical scheme for chemical species evolution. In presented approach, we track a single specie with distribution function prescribed with ϕ (only in fluid domain) and non-dimensional relaxation time:

$$\tau_D = 0.5 + 3\frac{D_\phi}{\psi c^2 \delta t}.$$

Modifications proposed by DiRienzo *et al* [11] account for varied density field in the volume of fluid by introducing $\psi = \rho^*/\rho$ a ratio of the minimum density of fluid in the entire domain and the density in a given lattice node; D_ϕ is a diffusivity coefficient of the mixture and the source term is prescribed with Q_i^ϕ.

Equilibrium distribution function for every accounted evolution equation has similar form (f_i^{eq} is usually used regarding fluid flow; here, for presentation purpose, a generic form is presented):

$$f_i^{eq}(\mathbf{r}, t) = f_i^{eq}(\beta, \mathbf{u}) = \beta\Omega_i\left(A_i + \frac{B_i}{c^2}\mathbf{e}_i\cdot\mathbf{u} + \frac{C_i}{c^4}(\mathbf{e}_i\cdot\mathbf{u})^2 - \frac{D_i}{c^2}\mathbf{u}\cdot\mathbf{u}\right), \tag{4}$$

where $\mathbf{u}$ is the local fluid velocity and Ω_i are weight coefficients; for D2Q9 $\Omega_{i=0} = 4/9$, $\Omega_{i\in\{1,2,3,4\}} = 1/9$, $\Omega_{i\in\{5,..,8\}} = 1/36$. In Eq. (4), coefficients A_i to D_i are not constant and their values are different for various modelled phenomena; furthermore, they can depend on discretisation scheme and directions of advection of distribution function. Full description of equations (with exact arrays of coefficients for Eqs. (1)-(4) for fluid flow and heat transfer) in 2D and 3D can be found in [10] and references therein. Modelling of the fluid mechanics (where subscript i of coefficients in Eq. 4 can be removed) is done with coefficients $A = 1$, $B = 3$, $C = 4.5$, $D = -1.5$ and $\beta = \rho$. Heat transfer modelling requires that $\beta = \theta$ in Eq. (4) and for IEDDF the coefficients, generally, are not constant; namely for D2Q9 discretisation scheme:

$$\{A_i, B_i, C_i, D_i\} = \begin{cases} \{0, 0, 0, -3/2\}, & i = 0 \\ \{3/2, 3/2, 9/2, 1\}, & i = 1-4 \\ \{3, 6, 9/2, 1\}, & i = 5-9 \end{cases} \tag{5}$$

Utilized equilibrium function for chemical specie transport was first presented and validated in [11] for D2Q5; here we present results for D2Q9 where the equilibrium functions are:

$$\phi_0^{eq} = \frac{\rho Y}{9}(9 - 5\psi), \qquad \phi_{i=1,..,4}^{eq} = \frac{\rho Y}{9}(\psi + 3\mathbf{e}_i\cdot\mathbf{u}), \qquad \phi_{i=5,..,9}^{eq} = \frac{\rho Y}{36}(\psi + 3\mathbf{e}_i\cdot\mathbf{u}),$$

here $\phi_{i,\mathrm{k}}^{eq}$ is the distribution function corresponding to modelled specie evolution.

Macroscopic values (density ρ, velocity $\mathbf{u}$, temperature θ, heat flux $\mathbf{q}$ and concentration Y) are obtained by integration of respective distribution function [7]:

$$\rho = \sum_i f_i, \qquad \mathbf{u} = \rho^{-1}\sum_i f_i\mathbf{e}_i, \qquad \theta = \sum_i g_i, \qquad \mathbf{q} = \sum_i \mathbf{e}_i g_i, \qquad Y = \sum_i \phi_i \tag{6}$$

2.2. Convection

Model of free convection by Guo *et al* presented in [6] is used here to account for density variation in fluid caused by local temperature gradient in a field of gravity force **g**. Coupling of the fluid flow and heat transport by convection is done by introducing the source term:

$$Q_i^f = 3\Omega_i \alpha \delta t \rho (\theta - \langle \theta \rangle) \frac{\mathbf{e}_i \cdot \mathbf{g}}{c} ,$$

where α is heat expansion coefficient for air; θ and $\langle \theta \rangle$ are local and averaged (in the whole domain) temperature, respectively.

2.3. Pyrolysis gas release

In presented approach, a mixture of gases is modelled; in recently presented works detailed investigation has been taken into account with additional evolution of tar (see [14]). This type of analysis, at the moment, is far too expensive for physically sound simulation at the meso-scale.

For the purpose of simplified modelling, we use an empirical correlation known from literature, in details presented in [15]; for a grain S the total amount of released gas is described by equation:

$$Q_S = \varepsilon \delta t V_S [\langle \rho \rangle - \rho_0 g(\theta)] \, f(\varepsilon_S^{\text{vol}}, \theta) , \tag{7}$$

where ε is the porosity (here the fluid volume fraction), V_S and $\varepsilon_S^{\text{vol}}$ for grain S are the lattice volume and the volume fraction of volatiles, respectively; $\langle \rho \rangle$ is the density averaged over the volume of fluid. The source term in Eq. (1) $Q_i^{\phi}(\mathbf{r}) = \sum_{S|_\mathbf{r}} \Omega_i Q_S$.

The reaction rate of degasification of mixed gases in Eq. (7) is represented by the Arrhenius type equation:

$$f(\varepsilon_S^{\text{vol}}, \theta) = a(\varepsilon_S^{\text{vol}}) \, \exp[-b(\varepsilon_S^{\text{vol}})/\theta];$$

where the activation energy $b(\varepsilon_S^{\text{vol}})$, as well as the frequency factor $a(\varepsilon_S^{\text{vol}})$, known from the Arrhenius equation, have a simple low-order polynomial form and correspond to different gases which are released at the given temperature range.

The empirical function in Eq. (7) has been proposed in [15]:

$$g(\theta) = 1 - 0.1 a_Y - 0.1 \begin{cases} a_Y \, \text{erf}(b_Y \theta - c_Y), & \text{if } \theta \le 573K \\ (d_Y \sqrt{\theta} - e_Y)[1 + \text{erf}(f_Y \theta - g_Y)], & \text{otherwise.} \end{cases}$$

It corresponds to change of density due to heating and coupled phenomena like gas release.

2.4. Change of material properties

During the pyrolysis solid and fluid properties change rapidly; change of temperature along with elemental composition has a significant impact on the solid phase. Modelling of technological processes like devolatilisation requires introducing simplification for density and heat capacity for solid ($\rho_\text{S}, C_\text{p,S}$) that depends on distribution of elements in coal grain. In presented study, for the solid phase two properties are functions of temperature (namely, density and heat capacity) and for fluid the viscosity. Figure 1 presents change of these properties in function of temperature along with its influence on the relaxation time in Eq. (3,2).

2.5. Further details of numerical procedure

Arithmetic mean of grain diameter is $d = 4.8\ mm$ (Sauter mean diameter is $4.9\ mm$). Domain is constructed as the REV element, with the diagonal of about 80 medium-sized obstacles. Obstacles (coal grains) are chosen by selecting a random position and diameter, from the uniform and Gaussian distributions, respectively. Domain is discretized by 200×400 nodes.

Journal of Physics: Conference Series **760** (2016) 012005 doi:10.1088/1742-6596/760/1/012005

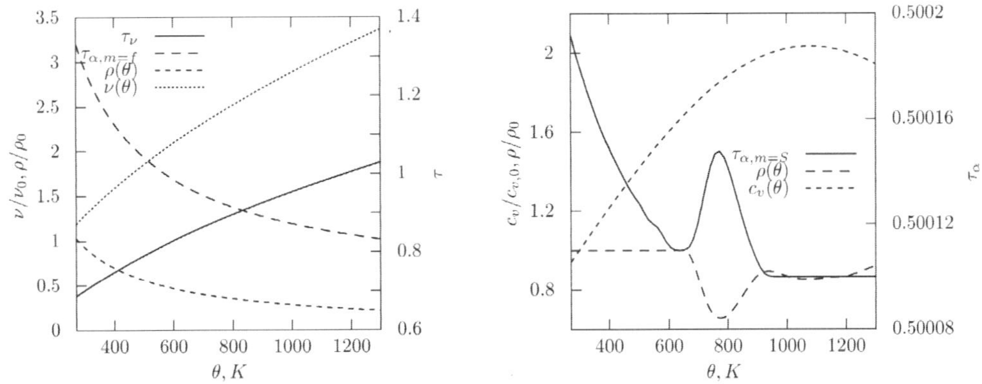

Figure 1. Fluid and solid properties change in function of temperature and its influence on relaxation time for distribution functions f_i and g_i, respectively.

In the preliminary examination, we choose that left boundary is a heated wall and the bottom wall is adiabatic. This is done by simply introducing no-slip wall of constant temperature (for heated wall) and zero heat flux (for adiabatic wall). At right boundary of the domain symmetry plane is imposed. In LBM, this condition is achieved by assigning to the unknown distribution functions a value of populations leaving the numerical domain, with the symmetry taken into account. For the top boundary-free outlet is implemented by Gradd boundary scheme, see [13]. In case of the heat transfer and specie evolution, zero flux boundary is implemented.

For simplicity fluid-solid interface is treated as no-slip wall with LBM schemes known from literature: the bounce-back (see [13]) and He scheme for heat transfer (see [9]). In case of the fluid flow, additional source term at the boundary is implemented to account for mass increase due to gas release. As the initial condition we impose zero concentration of the chemical species in the volume of the fluid; during heating of the obstacle, chemical compounds are released and transported to the surface of grain. Solid grains are simplified by 0D model, that is: release of gaseous product and change of internal energy are uniformly distributed along surface nodes (for fluid and chemical specie evolution) and nodes of solid grain (for heat transfer).

3. Results and effective thermal conductivity estimation

The purpose of this work is to calculate effective transport coefficient for geometry of packed grains of coal accounting for phenomena occurring during the coking process. A number of accompanying phenomena play an important role in heat transfer through the modelled geometry. important ones are not implemented in proposed model; namely solid grain Some dilatation occurring during heating as well as radiation are still under implementation.

Resulting pressure in the numerical domain along with the velocity map are presented in Fig. 2 (left and right, respectively). Due to heating, release of gaseous products along with convection plays a major role in fluid flow through the deposit. Decrease of solid density (due to release of gases and dilatation) results in the increase of the fluid pressure (which linearly influence the density in the LBM). Increase of the fluid pressure is slightly altered by convection. Although the gas release has a local effect on the fluid flow, the mass transfer is mainly influenced by convection phenomena; see right plot of Fig. 2, where the source term from chemical reactions in Eq. (1) is neglected.

Figure 3 presents an instant for heat transfer. Change of density lets the cold fluid to penetrate the deposit of coal grains, what (in the modelled geometry) slows down the heat transfer; on the other hand, although gas release improves the heat transfer (by release of hot gas from the grain volume), the chemical reaction is treated as the endothermic process which

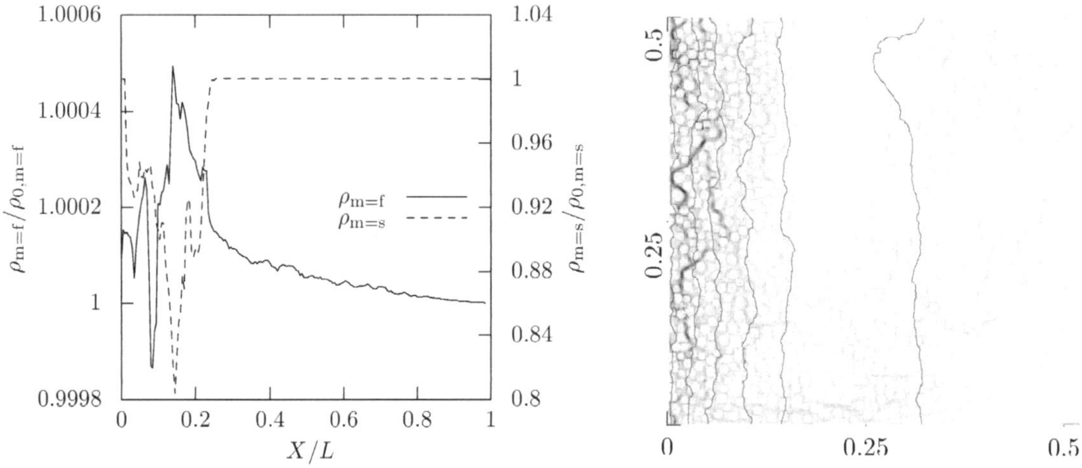

Figure 2. Density and velocity profiles averaged over fluid nodes in slices parallel to the heated wall (left plot); map of velocity profile with a few isolines of temperature in the trimmed computational area.

(uniformly in the volume of grain) reduces the temperature through a negative source term in the evolution equation for IEDDF. This phenomenon has its influence on heat flux what appears as the local maximum of mean heat flux value. This intensification of heat transfer appears only locally in a range of temperatures where chemical reactions occur.

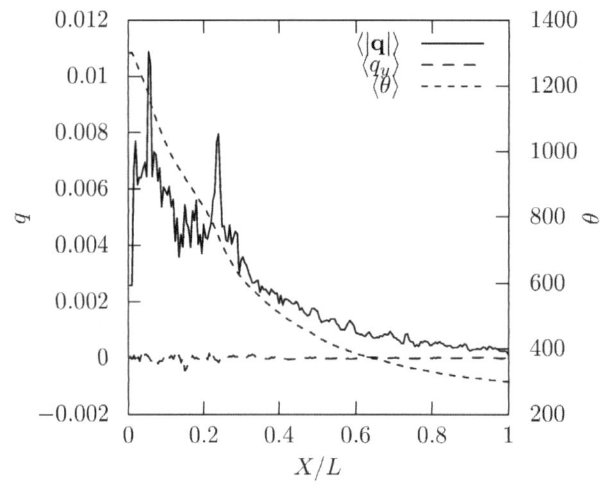

Figure 3. Averaged heat flux and temperature profiles. For comparison, the heat flux component parallel to the heated wall is also plotted. Averaging is done over fluid and solid nodes in the slices parallel to the heated wall.

The effective thermal conductivity coefficient is estimated by [7]:

$$k_{\text{eff}} = \frac{L \int q_x dY}{\Delta\theta \int dY}$$

where $\mathbf{q}|_x$ is heat flux projected on direction perpendicular to heated wall and Y is the area of the slice; $\Delta\theta$ is the temperature difference between adjacent slices of domain where heat flux is calculated. Resulting k_{eff} is presented in Fig. 4. Although presented results refer to unsteady state of the simulation, calculated heat conductivity is nearly constant (with negligible deviations). At this stage, it is difficult to distinguish areas of numerical domain where modelled phenomena: convection, chemical reactions, etc. play a crucial role in heat transfer and estimation of the effective heat conductivity.

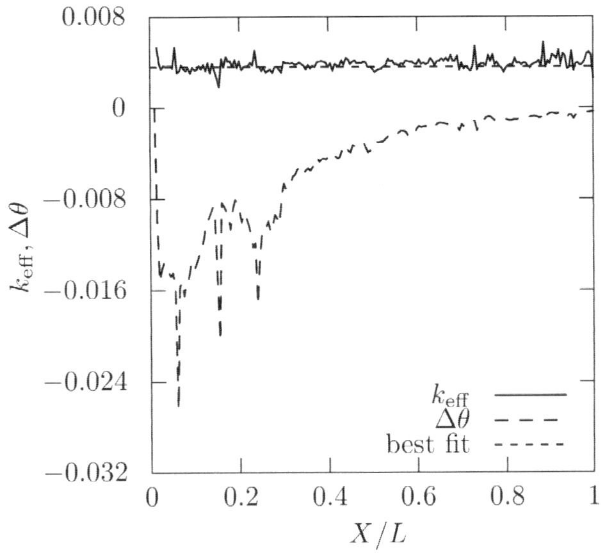

Figure 4. Estimated value of effective thermal conductivity and temperature difference along the numerical domain.

4. Conclusion and future work

Presented results show good agreement with the qualitative predictions concerning the influence of the subsequent phenomena in the coking process. Implemented phenomena allows to predict influence of the fluid flow, release of gases on heat transport in random, two dimensional geometry of the bed of coal particles. Described investigation, to the best knowledge of the author, is a first approach to meso-scale modelling of coking process in REV geometry.

The paper presents detailed analysis and first results; in order to achieve a good agreement some additional phenomena have to be considered; the most important seems to be a thermal dilatation, because it plays a major role for fluid flow and heat transfer phenomena. Also radiation is under considered to implementation to check its influence in heat transfer for this modelling approach. The future work will be oriented toward quantitative comparison of results with other known from literature simulations.

Acknowledgments

The author would like to thank Dariusz Kardaś and Sylwia Polesek-Karczewska for valuable comments regarding necessary simplification to the model and Jacek Pozorski for discussions regarding the model and the final manuscript.

References

[1] Miura S and Silveston P L 1980 Change of pore properties during carbonization of coking coal *Carbon* **18** 93–108
[2] Di Blasi C 2008 Modeling chemical and physical processes of wood and biomass pyrolysis *Prog. Energ. Combust. Sci.* **34** 47–90
[3] Polesek-Karczewska S, Kardaś D, Wardach-Święcicka I *et al* 2013 Transient one-dimensional model of coal carbonization in a stagnant packed bed *Arch. Thermodyn.* **34** 39–52
[4] Aidun C K and Clausen J R 2010 Lattice-Boltzmann method for complex flows *Annual Review of Fluid Mechanics* **42** 439–472
[5] Succi S *The Lattice Boltzmann Method for Fluid Dynamics and Beyond*, Clarendon Press, Oxford
[6] Guo Z, Shi B and Zheng C 2002 A coupled lattice BGK model for the Boussinesq equations *Int. J. Numer. Meth. Fluids* **39** 325 - 342
[7] Wang M and Ning P 2008 Modeling and prediction of the effective thermal conductivity of random open-cell porous foams *Int. J. Heat Mass Transf.* **51** 1325 - 1331
[8] Chai Z, Shi B, Lu J and Guo Z 2010 Non-Darcy flow in disordered porous media: A lattice Boltzmann study *Comp. Fluid* **39** 2069–2077
[9] He X, Chen S and Doolen G D 1998 A novel thermal model for the lattice Boltzmann method in incompressible limit *J. Comp. Phys.* **146** 282–300
[10] Wang J, Wang M and Li Z 2007 A lattice Boltzmann algorithm for fluid–solid conjugate heat transfer *Int. J. Therm. Sci.* **46** 228–234
[11] Di Rienzo A F, Asinari P, Chiavazzo E *et al* 2012 Lattice Boltzmann model for reactive flow simulations *Eur. Phys. Lett.* **98** 34001
[12] Asinari P, Calì Quaglia M *et al* 2007 Direct numerical calculation of the kinematic tortuosity of reactive mixture flow in the anode layer of solid oxide fuel cells by the lattice Boltzmann method *J. Power Sources* **170** 359–375
[13] Grucelski A and Pozorski J 2015 Lattice Boltzmann simulations of heat transfer in flow past a cylinder and in simple porous media, *Int. J. Heat Mass Transf.* **86** 139–148
[14] Grucelski A and Pozorski J 2013 The application of the Lattice Boltzmann method to thermo-mechanics of granular media flow and modelling of devolatilization *Proc. of the Colloquium Fluid Dynamics* **ed.** Jonáš P and Uruba V
[15] Postrzednik S 1994 Solid fuel carbonization method of determination, basic relations. *Karbo, Energochemia, Ekologia* **39** 220-228 (in Polish)

Numerical research on unsteady cavitating flow over a hydrofoil

D Homa[1, 2] **and W Wróblewski**[1]

[1]Institute of Power Engineering and Turbomachinery, Silesian University of Technology, Konarskiego Street 18, Gliwice, Poland
[2]To whom any correspondence should be addressed

Abstract. Cavitation is a widely known phenomenon in pumps and water turbines installations. It can lead to significant damage of blades and walls of the rotor therefore it is crucial during pump designing and exploitation to avoid working in flow conditions, that enabled cavitation to occur. Nowadays numerical simulations of flow can provide valuable information concerning pressure and velocity distribution and can indicate if there is a risk of cavitating flow appearance. There are a few mathematical models which describe cavitating flow. In the paper Schnerr & Sauer model was chosen for simulation. Aim of the paper is to verify its utility in case of different cavitating flow regimes over Clark-Y hydrofoil. After performing the grid independence study four different cavitation regimes were investigated. The vapour areas appearance, their shapes and changes in time were observed. The assumption of isothermal, two – phase flow was made. The calculations were performed using OpenFOAM and were compared to the available measurements data. The presented results concerned sheet and cloud cavitation regimes.

1. Introduction

Cavitation phenomenon was first observed and described by Osborne Reynolds in 1894. It is a complex process of forming and collapsing vapour bubbles in liquid flow. The formation and growth of bubbles is possible under low pressure conditions. The bubbles next flow into higher pressure region and collapse suddenly. It leads to great pressure and temperature augmentation at the bubble center. The pressure wave is generated and it propagates through the flow. It causes high frequency noise which is one of first cavitation symptoms. The others effects of cavitating flow occurrence are vibrations and erosion of solid surfaces near cavitation structures. The detailed descriptions of bubble physics and cavitation types can be found in work of Brennen [1] and Franc and Michel [2]. Numerical modelling of cavitating flow has to deal with instabilities on liquid – vapour interface, at which large variations in structures size and velocity over short periods appear. The scaling of cavitation phenomenon is challenging, taking into consideration that collapse of a single bubble takes about one millisecond, but collapse of a whole structure – order of microseconds. According to the [2], the velocities of an interface vary between meters per second to hundreds of meters per second. The gauge pressure generated by the implosion of cavitation structures can reach several thousand bars. Regarding to all these facts the numerical modelling of cavitation make use of some simplifications of the process, such as negligence of the surface tension.

The topic of following investigation was flow over a foil. For this type of flow, which is typical for pumps' rotors, the characteristic dimensionless variable is called cavitation number σ, which is defined as [1, 2]:

$$\sigma = \frac{p - p_s}{0.5\rho_l u_\infty^2} \tag{1}$$

where: p – pressure, Pa; p_s – saturation pressure, Pa; ρ_l – liquid density, kg·m^{-3}; u_∞ - free stream velocity, m·s^{-1}.

With decrease of cavitation number the following cavitation regimes can be observed: incipient, sheet, cloud and supercavitation [3]. Wang et al. [3] performed an experiment including flow over Clark-Y foil and provided wide range of information about cavitation dynamics and structures for these four cavitation regimes. The aim of the following study was to verify results obtained by means of Schnerr & Sauer model in case of two cavitation regimes: sheet and cloud cavitation. The sheet type cavitation is the first type when the developed cavitation structures can be observed. Sheet cavitation occurs when cavitation number σ reaches 1.4 [3]. According to [3] it is characterized by finger-like structure of vapour-water interface on the leading edge of the blade. Such shape is due to the combination of different vortex pairs and detachment of the head of each pair. However, the rear region of the sheet is unsteady and rolls up into a series of bubbly eddy that are shed intermittently. The frequency of changes of the whole structures is about 200 Hz (period 5 of ms). The next cavitation regime is called cloud cavitation. It occurs when cavitation number drops to 0.8 [3]. The frequency of cloud cavitation for these parameters is equal to 20 Hz (50 ms for one cycle). It is characteristic of attached front portion and unsteady rear region. In the beginning of the cycle, the cavity is attached near the leading edge. It grows and travels downstream with packet of bubbles moving with a clockwise rotation till about the midpoint of the cycle. Then the massive vortex shedding appears. Due to large-scale vortex dynamics the higher pressure near wall region occurs and cavitating flow is pushed away from the wall. A re-entrant flow in the wall region is induced toward the upstream. As it reaches the vicinity of leading edge of the cavity, the existing cavitating flow detaches from the wall and a new cavitating flow structure forms there.

2. Cavitation model

Cavitation modelling methods can be gathered into two main groups: one-fluid and two-fluid models [4]. In one-fluid models group the flow is assumed to be a mixture of two phases and the conservation equations (mass and momentum) for the mixture are to be solved. In this simulation no slip between the phases was assumed. To calculate fraction of gaseous phase mass conservation equation of vapour is solved [4]:

$$\frac{\partial \alpha\rho_v}{\partial t} + \nabla(\alpha\rho_v u) = \begin{cases} R_e & p \le p_s \\ -R_c & p > p_s \end{cases} \tag{2}$$

where: α – vapour volume fraction, -; ρ_v – vapour density, kg·m^{-3}; u – velocity, m·s^{-1}; R_e, R_c – source terms, kg·m^{-3}s^{-1}.

In Schnerr & Sauer model the source terms are derived from Rayleigh-Plesset (RP) equation, which describes the dynamics of vapour bubbles. The RP equation is simplified, no surface tension is assumed, as in formula below [1]:

$$\frac{\mathrm{D}r_B}{\mathrm{D}t} = \sqrt{\frac{2}{3}\frac{p_B - p}{\rho_l}} \tag{3}$$

where: r_B – radius of bubble, m; p_B – pressure in the bubble, Pa.
The final formulas for source terms are as follows [5]:

$$R_e = \frac{\rho_l \rho_v}{\rho}\alpha(1-\alpha)\frac{3}{r_B}\sqrt{\frac{2}{3}\frac{p_s - p}{\rho_l}} \tag{4}$$

$$R_c = \frac{\rho_l \rho_v}{\rho}\alpha(1-\alpha)\frac{3}{r_B}\sqrt{\frac{2}{3}\frac{p - p_s}{\rho_l}} \tag{5}$$

Where: ρ – mixture density, derived from volume fractions and densities of the phases, kg·m^{-3}.

The radius of bubble r_B is derived from vapour volume fraction and number of bubbles per volume of liquid n_B, using the following relation [5]:

$$r_B = \left(\frac{\alpha}{1-\alpha} \frac{3}{4\pi} \frac{1}{n_B} \right)^{\frac{1}{3}} \tag{6}$$

Number of bubbles per volume of liquid is to be set as the model constant.

3. Numerical model description

3.1. Geometry

The simulation concerned flow over a Clark-Y hydrofoil. It is a typical foil used in water machinery. The overview of the geometry is shown in figure 1.

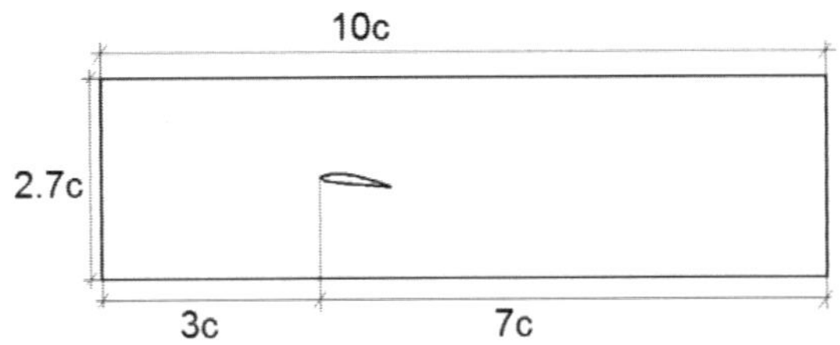

Figure 1. Overview of the geometry.

The foil was placed in distance of 3c from the inlet and 7c from the outlet, as the cavitation structures can reach much further than the chord length. The height of the channel was set to 2.7c and foil was placed symmetrically from the upper and lower wall of the channel. Chord of the foil was equal to 70 mm.

3.2. Grid parameters

The 3D grid was made by means of ICEM CFD software. The geometry has been divided into 8 blocks and the O-grid was generated around the blade. The blade profile was split into 4 edges: near leading edge, upper side, lower side and trailing edge. On lower and upper side edges 85 grid nodes were set, on the leading and trailing edges – 30 nodes. The surface block mesh was used at first, next it was extruded in the perpendicular direction by 10 layers of elements of 1 mm thickness each. The whole mesh consisted of 216k hexahedra elements. The overview of the mesh is shown in figure 2.

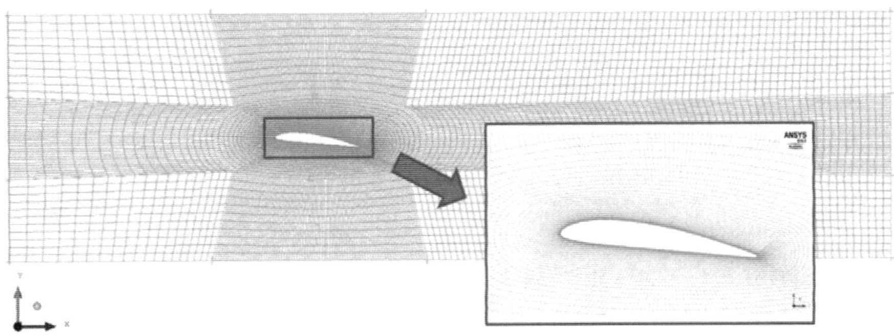

Figure 2. Overview of the mesh, zoom of the O-grid region.

3.3. Calculations set up

The simulation was performed with assumption of constant inlet velocity. The pressure at the outlet was consequently lowered. For sheet cavitation the outlet pressure was equal to 72 000 Pa, for cloud cavitation: 42 260 Pa. The calculations were performed in OpenFOAM open source code with the solver interPhaseChangeFoam, which captures well the dynamics of the cavitating flow [6]. The simulation set up was based on other investigations [7-9], the turbulence model was k -ω SST, chosen after literature study [10-12]. The adjustable time step was used, with the condition that Courant number must remain under 1. The transient, first order numerical scheme was chosen. The simulation set up is shown in table 1.

Table 1. Calculation set up (bc – boundary condition).

Chord length	70 mm
Angle of attack	8°
Heat transfer model	Isothermal
Flow temperature	20°C
Turbulence model	k-ω SST
Side walls bc	Symmetry
Upper/lower wall bc	Wall
Outlet bc	Static pressure (case dependent)
Inlet velocity	10 m/s
Turbulence intensity at inlet	5%
Reynolds number	700 000
n_B in Schnerr&Sauer model	1.6×10^{13} [5]
r_B in Schnerr&Sauer model	10^{-6} m [5]

4. Results

The end of the calculation was determined by the end time. The final residuals dropped to the level of 10^{-6}. The maximum values of y+ on the foil were equal to 21.1 on the upper and lower side and 32.4 in the elements near leading and trailing edge.

4.1. Structures description

The calculations were performed for two different cavitation regimes: sheet and cloud cavitation. In figure 3. one period of changes for each cavitation model is shown. For both regimes the periodic changes of structures growth and collapse were observed. The chosen parameters of changes are compared in table 2.

Table 2. Comparison of calculations results.

Parameter	Sheet cavitation	Cloud cavitation
Shedding frequency	32 Hz	30 Hz
Period	31 ms	33 ms
Beginning of the structure coordinate	X=0.1038 mm	X=3.31mm
(measured from beginning of the foil)	Y=0.1886 mm	Y=2.6mm
Maximum height of a structure	12 mm	13 mm
Maximum length of a structure	81 mm	104 mm

The frequency of shedding for sheet cavitation is much lower compared to the available measurement data [3], which states that the frequency is equal to 200 Hz. For cloud cavitation the frequency is lower comparing to the sheet cavitation regime, which is compatible to the observations [3]. The frequency observed during calculations is similar to the one described in [3], equal to 30 Hz. For cloud cavitation regime the cavitation structures beginning was located further that for sheet cavitation regime, which is shown in figure 4. The maximum height and length of the cavitation structures grew as the cavitation number decreased. For both regimes they occurred after 2/3 of the period. The dynamics of cavitation structures was different for the two investigated flows. First the

Journal of Physics: Conference Series **760** (2016) 012006 doi:10.1088/1742-6596/760/1/012006

bubble cluster was formed, starting from leading edge and covered whole of the upper side of the foil for sheet cavitation after 5/12 of the period, for cloud cavitation after 1/3 of the period.

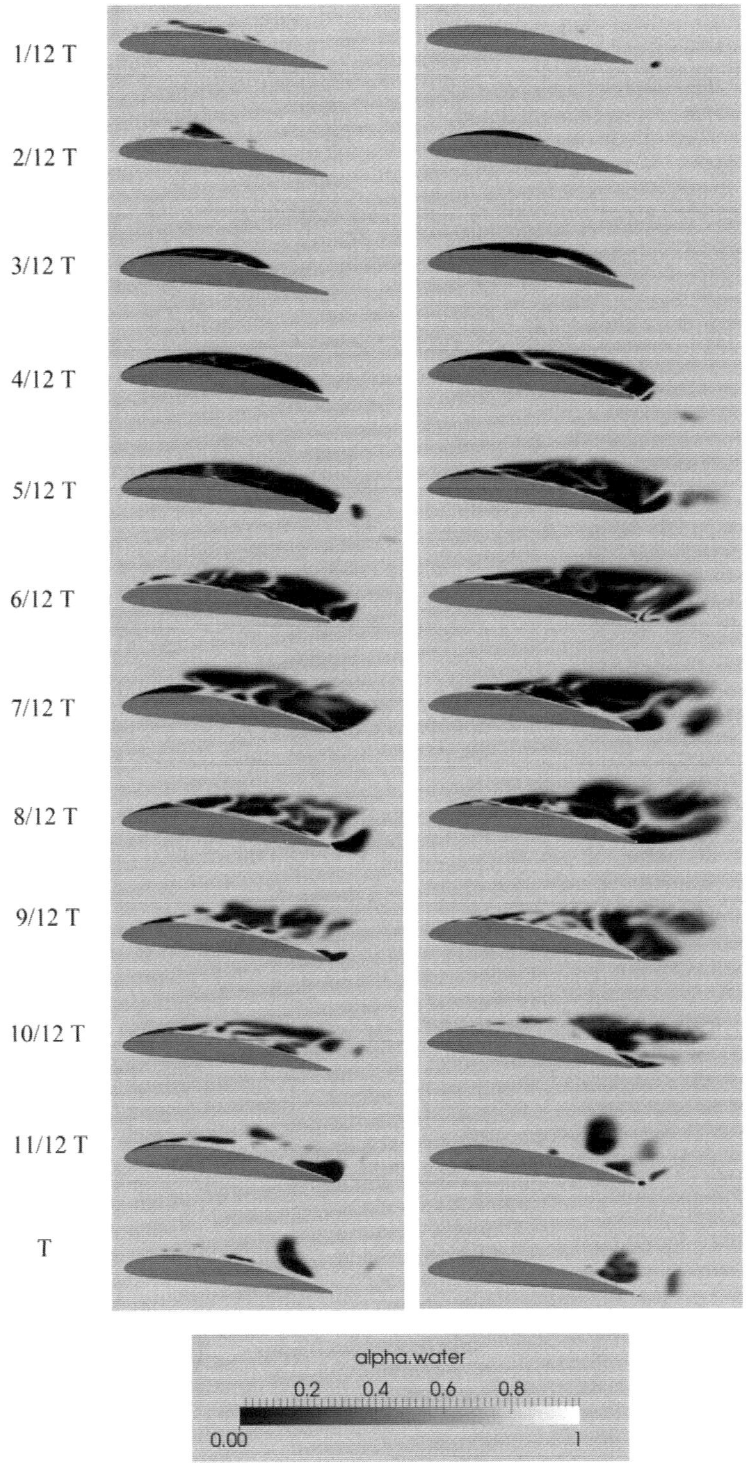

Figure 3. One period of changes for sheet cavitation – left, cloud cavitation – right.

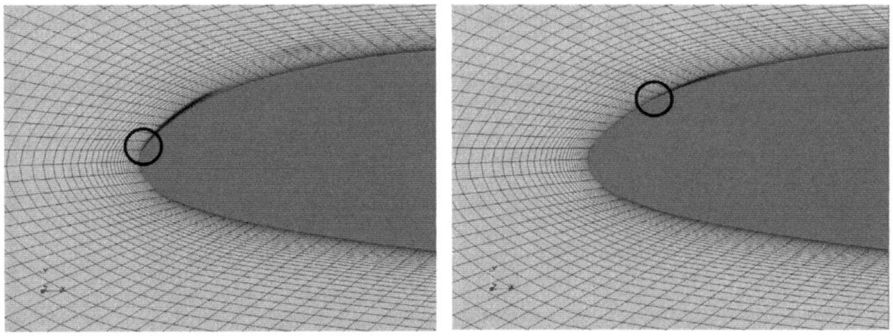

Figure 4. The beginning of the cavitation structure for sheet cavitation – left, cloud cavitation – right.

The re-entrant jet starting from trailing edge can be observed for both regimes. It travelled upstream the foil and caused the structures to detach from the foil wall and collapse. For sheet cavitation it was observed that for the whole period the cavitation bubbles are attached to the leading edge, for cloud cavitation the bubbles are separated from leading edge after half of the period, but then the small structure appeared after about 75% of the period. For both regimes the flow near trailing edge was highly unsteady, which was also observed during the experiment.

4.2. Pressure and vapour volume fraction distributions

During the calculations the pressure and vapour volume fraction values were recorded in monitor points located along the foil. In figure 5 the time-averaged (averaged in one period) pressure and vapour volume fraction distributions are shown. For cloud cavitation the pressure on the upper and lower side of the foil was significantly lower, but near the trailing edge, for x/c coordinate 0.7 to 1, the averaged pressure values were close to each other, which can be caused by the strong unsteady flow. The distribution of the vapour volume fraction shows that for sheet cavitation regime that parameter reached the maximum equal to 0.87 near 18% of x/c coordinate, which corresponds to the attached bubble cluster near trailing edge. For cloud cavitation the structure was located further from leading edge and the vapour volume fraction was near 0.7 at the same x/c coordinate. For both cavitation types near the trailing edge the vapour volume fraction increased, which is probably connected to the process of detaching the structures from the foil end.

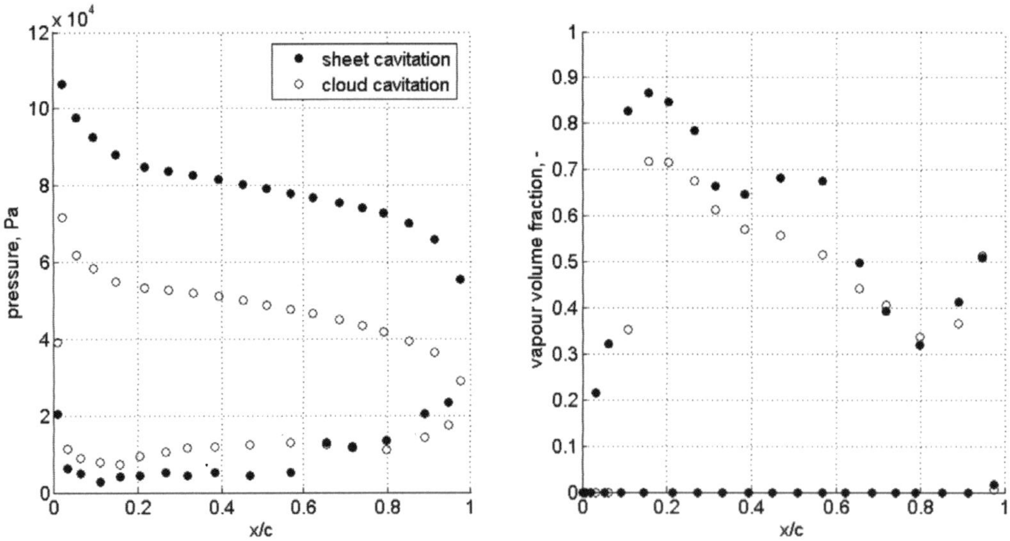

Figure 5. Pressure and vapour volume fraction distributions for sheet and cloud cavitation.

4.3. Pressure fluctuations

The changes of pressure in time at different monitor points were also studied and are shown in figure 6. The chosen points are located on upper side of the blade, at *x/c* coordinate: point 1 – 0, point 2 – 0.2, point 3 – 0.55, point 4 – 0.95.

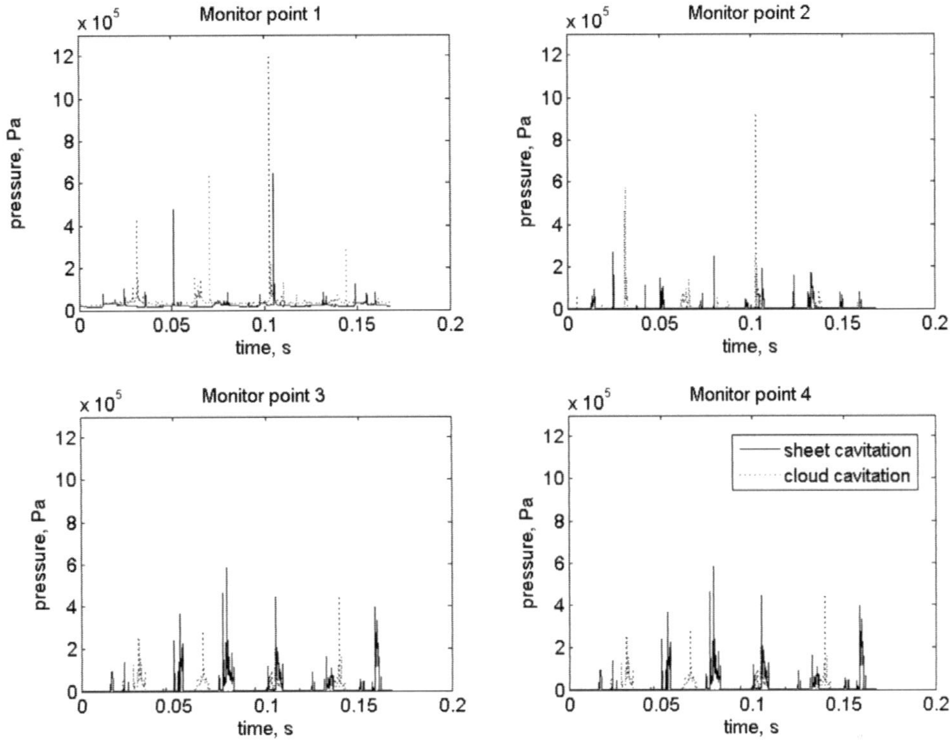

Figure 6. Pressure changes in time at monitor points.

For each cavitation regime the peaks of pressure were observed. For cloud cavitation they were much higher, especially for two first monitor points. The pressure changes in these points reached up to 10^6 Pa, but the pressure course was in general softer than in case of the other two monitor points. Pressure course in monitor point 4, located closest to the trailing edge, is highly irregular, but without such great peaks of pressure. This confirms the highly unsteady character of cavitation structures near the end of the foil.

The sudden changes of pressure are described in the [1, 2] as one of the characteristic feature of the cavitating flow connected with the process of forming and collapsing of vapour bubbles and need to be further investigated in next numerical simulations.

5. Summary

In the paper the results of numerical simulation of two cavitating flow types are presented. The sheet and cloud cavitation regimes were investigated in case of flow over a Clark-Y foil with use of Schnerr & Sauer cavitation model. For both cavitation types the beginning of cavitation structure and the maximum dimensions of cavitation cloud were described and compared to each other. For cloud cavitation the structures were longer and higher, but the beginning of the cavitation cloud was located further from the leading edge of the foil. The shedding frequency for cloud cavitation was in good agreement with the experimental results (30 Hz compared to 20 Hz obtained during the experiment), but for sheet cavitation the frequency of changes was significantly lower comparing to the experimental one (32 Hz in numerical simulation compared to 200 Hz in the experiment). This need to be solved in the future numerical calculation. The dynamics of parameter changes was generally similar to the one described in the [3] for both regimes. The highly unsteady region near trailing edge as well as re-entrant jet was observed. Also the time-averaged pressure and vapour volume fraction distributions for both cavitation types were compared. The time-averaged vapour volume fraction distributions vary for both regimes, for sheet cavitation the higher values were observed in about 0.2 of normalized chord x/c, for cloud cavitation they were slightly moved downstream the foil. Near the trailing edge vapour volume fractions increased for both cavitation regimes. The other important aspect was pressure changes in time, as cavitating flow is characterized by great pressure differences occurring in short period of time. During the calculations the pressure was monitored and the sudden peaks of that parameter were observed. The pressure reached up to 10^6 Pa. The peaks occurred mainly in the first half of the chord length and they were very rapid. For the second half of the chord, especially for monitor point located near the trailing edge, the changes were not so high but the course of the pressure was strongly irregular.

The results of performed simulations enable to state that the used cavitation model and the chosen OpenFOAM solver capture the dynamics of sheet and cloud cavitation sufficiently. The main features of these flows such as periodic character of changes, forming and collapsing of vapour structures, occurrence of re-entrant jet and sudden pressure fluctuations were observed. However, there are still aspects to improve, such as the obtained shedding frequency, which was far from the experimental data for sheet cavitation regime. That will be the object of interest in the future investigations.

Acknowledgments
The presented work was supported by the Polish National Science Centre founds within the project UMO-2014/15/B/ST8/00203 and within the project BK/267/RIE5/2015/502.

References
[1] Brennen C E 1995 *Cavitation and bubble dynamics* (Oxford: Oxford University Press)
[2] Franc J P and Michel J M 2004 *Fundamentals of cavitation* (Dodrecht: Kluwer Academic Publishers)
[3] Wan G, Senocak I, Shyy W, Ikohago T and Cao S 2001 Dynamics of attached turbulent cavitating flows *Progress in Aerospace Sciences* vol 37 pp 551-81
[4] Puffary B 2006 Numerical modelling of cavitation, design and analysis of high speed pumps *Educational Notes RTO-EN-AVT-143* paper 3, pp 1–54
[5] Schnerr G H and Sauer J 2001 Physical and numerical modelling of unsteady cavitation dynamics *Fourth International Conference on Multiphase Flow (New Orleans)*
[6] Gosset A, Diaz Casas V and Lopez Pena F 2010 Evaluation of the CavitatingFoam solver for low Mach number flow around a 2D hydrofoil *Fifth OpenFOAM Workshop (Gothenburg)*
[7] Wang G, Zhang B, Huang B and Zhang M 2009 Unsteady dynamics of cloud cavitation flows around a hydrofoil *Proceedings of the Seventh International Symposium on Cavitation CAV2009 (Ann Arbor)*
[8] Roohi E, Zahiri A P and Pasandideh-Fard M 2012 Numerical simulation of cavitation around a two-dimensional hydrofoil using VOF method and LES turbulence *Proceedings of the Eight International Symposium on Cavitation CAV2012*
[9] Bensow R E and Bark G 2010 Simulating cavitating flows with LES in OpenFOAM *5th European Conference On Computational Fluid Dynamics*
[10] Li Z, Pourquie M and Van Terwisga T 2010 A numerical study of steady and unsteady cavitation on a 2D hydrofoil *Journal of Hydrodynamics* vol 22(5) supplement 770-777
[11] Li D Q, Grekula M and Lindell P 2009 A modified SST k-ω turbulence model to predict the steady and unsteady sheet cavitation on 2D and 3D hydrofoils *Proceedings of the Seventh International Symposium on Cavitation CAV2009 (Ann Arbor)*
[12] Huang B and Wang G 2011 Partially averaged Navier-Stokes methods for time dependent turbulent cavitating flows *Journal of Hydrodynamics* vol 23(1) pp 26-33

Influence of cooling intensity on shock wave boundary layer interaction region in turbine cascade

P Kaczyński[1] **and R Szwaba**[1]
[1] Institute of Fluid-Flow Machinery, Polish Academy of Sciences (IMP PAN), Fiszera 14, 80-231 Gdansk, Poland

E-mail: Piotr.Kaczynski@imp.gda.pl

Abstract. The shock wave boundary layer interaction on the suction side of a transonic turbine blade was one of the main objectives of the TFAST project. For this purpose a model of a turbine passage was designed, manufactured and assembled in a transonic wind tunnel. The paper presents the experimental investigations concerning the flow structure on the transonic turbine blade. A clean case (without a cooling system) with a normal shock wave interacting with a laminar boundary layer and also the influence of the blade cooling system with three different coolant blowing intensities on the laminar interaction region were investigated.

1. Introduction

The research of which some parts are presented in this paper was carried out within the European Commission 7th Frame Work project under the acronym TFAST dedicated to transition location effects on the shock wave boundary layer interaction. A more restrictive regulation concerned fuel gas emissions and increasing demands of transport system effectiveness enforcing modification in aircraft engines. It is usually realized by reduction of mass and size of newly designed engines. This can be achieved by a reduction of the number of blades in specific turbine cascades. This leads to an increase in the blade load, and as a consequence, to the appearance of a local supersonic region terminated by the shock wave in the blade passages. In turbine passages the strong acceleration favours maintaining a laminar boundary layer in a long distance and very often in these cases the laminar shock wave boundary layer interaction (SWBLI) takes place which could be problematic for the engine performance in case that separation appeared. The process of separation usually becomes unsteady and is connected with shock oscillations in the internal flows. Shock oscillation causes the pulsation of pressure, thus resulting in blade load fluctuation. From that point of view, a more desirable type of interaction is a turbulent one [1] which can be induced by introducing disturbances upstream of the interaction area.
The main objective of the TFAST project was to study the transition location effects on the interaction structure of the shock wave with the boundary layer (BL). Steps, roughness elements, and air-jet vortex generators [2] [3] were used for the transition control. Moreover, the film cooling introduced into the flow by rows of holes has influence on the boundary layer development and furthermore on the SWBLI region. Therefore an objective of TFAST was also to investigate the boundary layer development on the gas turbine blade suction side and the effects on the shock wave boundary layer interaction [4]. Despite the cooling introduced into the boundary layer small disturbances, the strong acceleration of the flow is leading to relaminarisation of the boundary layer and there is a possibility that still the laminar boundary layer will be exposed to interaction with the shock wave.
In order to investigate the flow structure on the suction side of the turbine blade in a rectilinear test section of the transonic wind tunnel, a model of a turbine passage was designed, manufactured and

assembled in the wind tunnel. The model can reproduce the flow structure, pressure distribution and boundary layer development similar to the obtained in a reference turbine cascade profile. The blade is a Nozzle Guide Vanes (NGV) with cooling in the gas turbine. The blade model was equipped with a cooling system which is a crucial part of measurements. The thermal efficiency of the turbine stage requires operating on a high turbine entry temperature. Such heat loads of blades are intended to work in temperatures between 800÷1000 °C which are typical for aircraft and heavy duty gas turbines [5]. To avoid heat-up of blades a thermal barrier coating is used and coolant air is injected to create a film cooling layer on the cooled surface [6]. The air for film-cooling in the turbine cascade is delivered from the compressor stage, thus the increase in the coolant mass flow reduces the overall engine performance. This approach imposes that the film-cooling has to be as efficient as possible. The film cooling effectiveness and its interaction with the main flow has been studied for many decades in various configurations [7], [8], [9].

The paper presents the experimental investigations concerning the flow structure on the transonic turbine blade. A clean case (without a cooling system) with a normal shock wave interacting with a laminar boundary layer and also the influence of a blade cooling system with three different coolant blowing intensities on a laminar interaction region were investigated. A cooling design process at most takes into account the cooling effectiveness and the coolant is introduced to the flow smoothly, at a small angle not to induce too many mixing losses. The conducted investigations will help to estimate how the increase in the cooling intensity upstream of the interaction region influences the losses downstream of the passage. Such study can also cast more light on some phenomena which are not always taken into account during the very complex turbine cascade design process.

2. Experimental arrangement

The measurement campaign was carried out in the Transonic Wind Tunnel of IMP PAN. Figure 1 shows the measurement system which consists of an inlet system with a drying unit, a test section, a vacuum tank, vacuum pumps and an additional pressure tank for dedicated tasks.

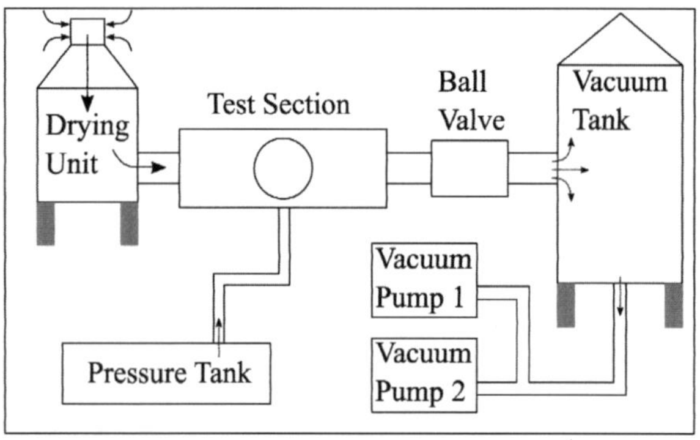

Figure 1. Measurement system scheme.

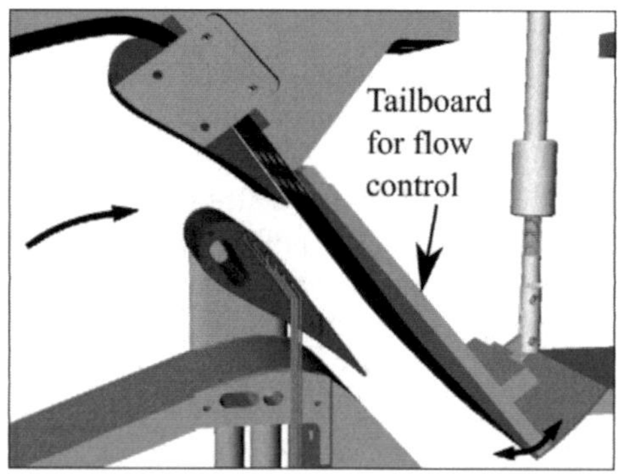

Figure 2. Sketch of turbine passage model.

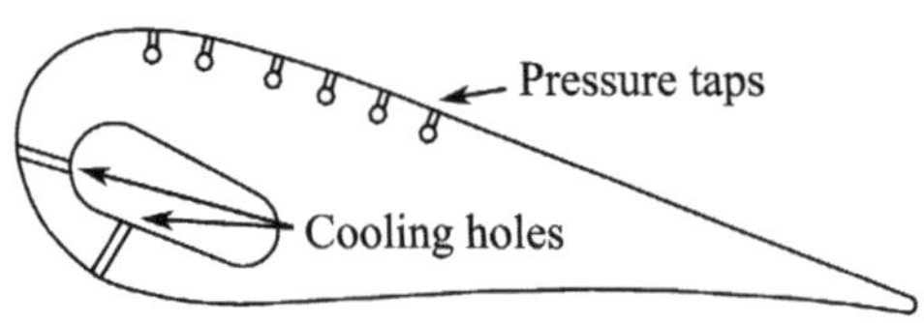

Figure 3. Cross section of lower profile.

In order to investigate the flow structure on the suction side of the blade, a turbine passage model was designed and assembled in the wind tunnel. The model can reproduce the flow structure, the pressure distribution and the boundary layer development similar to the one obtained in a reference turbine cascade profile [10].

The single passage turbine test section (Figure 2) is a model of the reference linear turbine cascade supplied by Rolls-Royce Deutschland (RRD). The measurement configuration was also adjusted according to the reference cascade and was set up by RRD. The test section consists of two profiles and limiting upper and lower walls. The shape of the limiting walls and also the position of blades have significant influence on the Mach number distribution in the whole test section and in particular in the passage and the provided specific inflow conditions. The lower (Figure 3) and upper blade profiles are located in a similar configuration as in a linear cascade in order to keep a similar flow structure as in the reference passage. Both profiles are mounted in the channel at the required inlet Mach number M=0.13. The Reynolds number at the inlet of the passage based on the blade cord length is $Re_C=3\times10^5$, and the cord length c is 112.5 mm. The axial cord length Cax =52 mm.

In order to achieve the required flow structure the upper limiting wall was equipped with a movable part, the so-called tailboard. This part starts just above the upper blade and can influence the flow structure in the investigated passage. The investigation was focused on the flow analysis in the turbine passage, in particular on the suction side of the blade. The blade profile was equipped with two rows of cooling holes located near the leading edge. The cooling holes were 1.35 mm in diameter. The hole angle of the first row to the stream direction (tangential) was 90°, whereas it was 16.25° in second row. The measured profile was also equipped with pressure taps located along the blade at the middle span. The cooling holes were connected to the main cavity where pressurised air was delivered from the pressure tank. Three different cooling intensity blowing ratios (cavity pressure in relation to the main flow stagnation pressure) were tested in the experiments. The investigation concerned the influence of the cooling intensity on the shock wave boundary layer interaction at the suction side of the turbine blade, therefore four configurations were tested:

- Clean case, uncooled.
- Film-cooling with blowing ratio Pc/P0=1.05.
- Film-cooling with blowing ratio Pc/P0=1.10.
- Film-cooling with blowing ratio Pc/P0=1.15.

3. Results of Experimental Investigation

The results are presented in form of charts obtained from the measured data and in form of photographs taken from the schlieren and oil visualizations. To investigate some aspects of the cooling intensity effects on the SWBLI region the following measurements were carried out:

- Static pressure along the lower wall and blade.
- Schlieren visualization.
- Oil visualization.
- PSP (Pressure Sensitive Paint) measurements.
- Shock oscillations measurements.
- Velocity traverse measurements by means of LDA.

3.1. Pressure distribution on blade

Film cooling introduces some disturbances into the flow which have influence on the SWBLI region, and consequently on the shock wave system on the suction side of the profile. Figure 4 shows normalized static pressure distribution on the suction side along the profile at middle span versus normalized cord length. Static pressure was measured by means of PSI scanner with accuracy of 0.05% of full scale (1 bar), i.e. ±0.5hPa. For all measured configuration the inlet stagnation temperature was the ambient one. The plot shows the results for flow without cooling (clean case) and with cooling of various blowing intensity. One can observe that in the uncooled configuration (the green line with triangles) the pressure jump is spread over a longer distance opposite to the film-cooling case. For cooling configurations the pressure lines mostly coincide, except for the main shock wave region. The differences in pressure distributions result from the shock system distinction for specific cases.

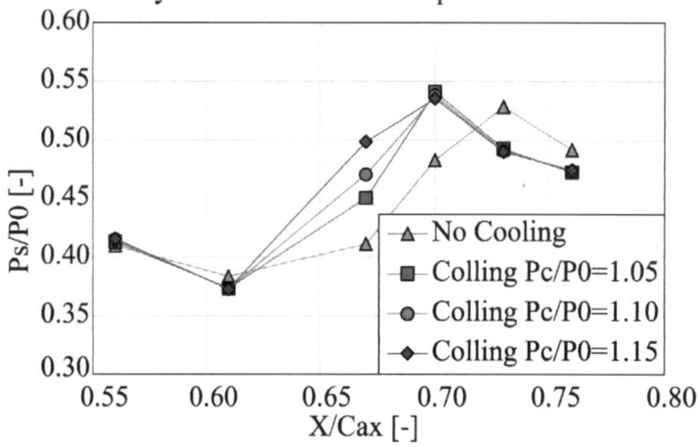

Figure 4. Pressure distribution on turbine blade suction side.

3.2. Schlieren visualization

Figures 5-7 show the schlieren visualization for a clean case and for cooling cases with pressure ratios of Pc/P0=1.05 and 1.15. In every single measurement in specific set-up, pressure and velocity distribution were repeatable. For all configurations one can observe a shock wave interaction with the laminar boundary layer. The laminar type of interaction indicates that a very large λ-foot (as it's visible on schlieren pictures) appears in the flow at such a relative small Mach number upstream of the shock (M=1.26). Moreover, the shock pressure jump is realized by means of two shock waves in the clean case. Film cooling just downstream of the jets locally turbulises the boundary layer, nevertheless, when the flow accelerates one can observe relaminarization of the boundary layer upstream of the interaction area. Downstream of the interaction region the flow accelerates again to a local supersonic region which is closed by the shock at the blade trailing edge.

Figure 5. Schlieren visualization, clean case.

Figure 6. Schlieren visualization, Pc/P0=1.05.

Increasing the cooling intensity modifies the shock wave system topology where the second shock gradually vanishes with the increasing blowing ratio. The case with a maximal cooling intensity produces only one main shock wave. In general one can notice that the cases with cooling are characterized by smaller λ-foot in comparison to the clean case. The film cooling introduces some energy into the boundary layer, and as a result, the shock pressure jump can be realised at a shorter distance, thus the λ-foot height decreases.

Figure 7. Schlieren visualization, Pc/P0=1.15.

3.3. Oil visualization

The flow structure on the suction side of the turbine profile from oil flow visualization in Figures 8-9 is shown. The view onto the blade is through the sidewall window of the test section. The flow is from left to right. One can observe a two dimensional flow with a separation region induced by the SWBLI. The separation onset starts at the front λ-foot of the first normal shock wave (see Figure 5). It is worth noticing that the interaction at the sidewall between the main flow and the corner vortex is very weak, thus the separation region spreads almost from one lateral wall to the opposite one. For the clean case, the separation size length is about 9 mm (Figure 8).

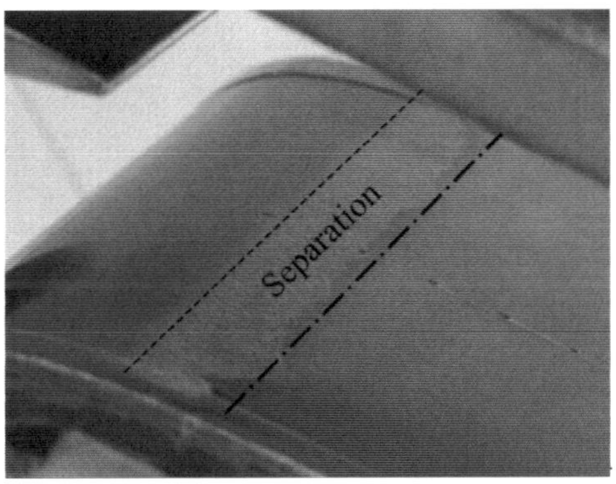

Figure 8. Oil visualization, clean case.

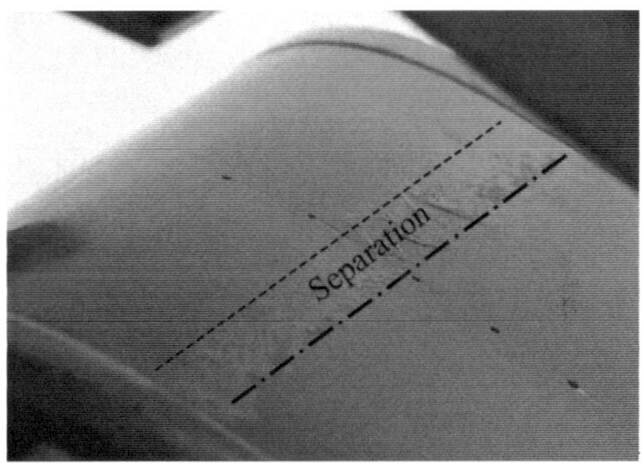

Figure 9. Oil visualization, cooling Pc/P0=1.15.

The film cooling introduced to the flow (Figure 9) reduces the separation region in relation to the clean case (Figure 8). For this case the separation length is about 6mm. The increasing of cooling intensity has no effect on the separation length. The cooling increases the fluid momentum in the boundary layer and therefore the separation region decreases. Nevertheless, the differences in the fluid momentum for various cooling intensities are probably too small to influence the separation size.

3.4. Pressure distribution on blade by means of PSP

In the PSP (Pressure Sensitive Paint) measurements the 'in situ' calibration was used, what means that the paint light intensity response was determined based on the pressure measured by the taps (see Figure 3) located along the blade. Figures 10 and 11 present normalized static pressure distribution on the blade suction side measured by means of PSP. The results are presented only for the flow with cooling (Pc/P0=1.05 and 1.15). The results for the clean case are not presented because the PSP in this flow modifies the shock system on the blade, hence the pressure distribution of the same flow with and without paint is not comparable. As in the foregoing oil visualization analysis one can clearly notice in

PSP measurements a two dimensional flow structure on the suction side of the turbine profile. Furthermore, very weak corner effects are also observed. Figures 10 and 11 show a very well displayed pressure jump (dashed lines) on the shock wave following a strong acceleration region (blue area). Downstream of the shock wave the flow accelerates again to the supersonic region which is terminated by another shock at the trailing edge.

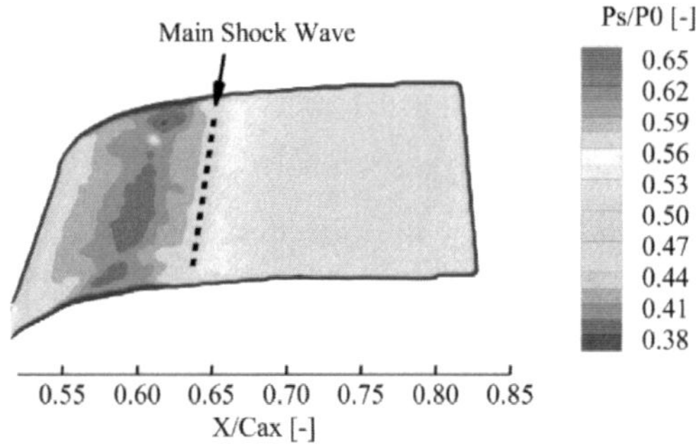

Figure 10. Blade pressure distribution for cooling of blowing ratio Pc/P0=1.05.

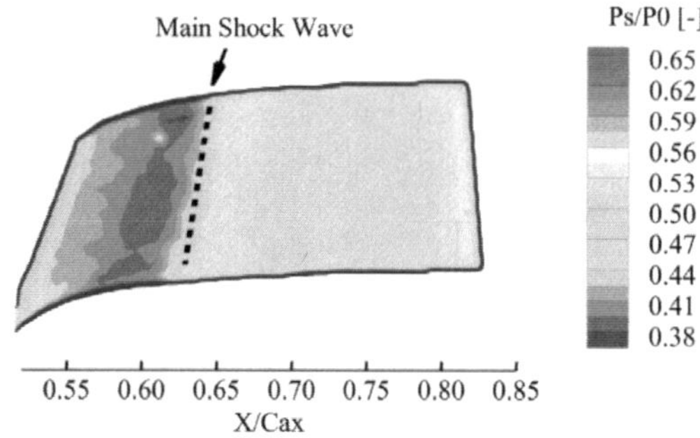

Figure 11. Blade pressure distribution for cooling of blowing ratio Pc/P0=1.15.

The plot in Figure 12 shows normalized static pressure distribution at the blade middle span obtained by PSP measurements. In the region just upstream of the shock wave the flow slightly decelerates (the pressure increases) showing that the onset of the interaction starts upstream of the shock wave what is a typical phenomenon for a laminar type of interaction [11] [12] and [13]. This observation is in agreement with schlieren visualization where one can also observe that a boundary layer upstream of the shock system is laminar. It is also visible that increasing the cooling intensity shifts the shock pressure jump upstream of the flow (see Figures 10 and 11).

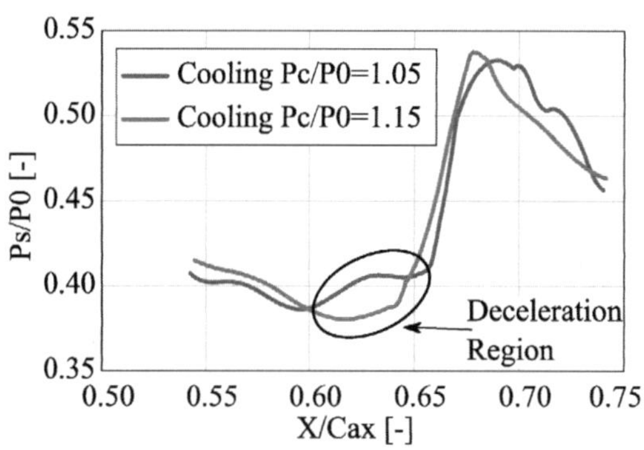

Figure 12. Pressure distribution on suction side of lower profile from PSP measurements.

3.5. Velocity profiles

The location of traverses for velocity measurements by means of LDA (Laser Doppler Anemometry) is presented in Figure 13. The first traverse is placed 10 mm upstream of the main shock wave position at X/Cax=0.55 and the second is located 20 mm upstream of the trailing edge at X/Cax=0.94. All traverses were measured in the channel centre. In every traverse the Y coordinate (distance from the wall) is perpendicular to the wall. Spatial resolution of LDA measurements in perpendicular direction towards the wall was approximately 80 μm.

Journal of Physics: Conference Series **760** (2016) 012007 doi:10.1088/1742-6596/760/1/012007

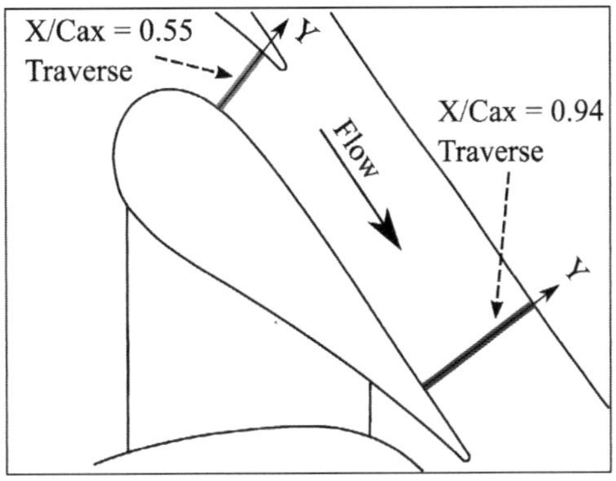

Figure 13. Locations of traverses for velocity measurements by means of LDA.

Figure 14 shows the velocity profiles in the passage upstream of the shock wave. The curves for all configurations nearly coincide. One can notice a supersonic region (velocity ratio above U/Umax=0.87) in the vicinity of the lower profile suction side. At some distance from the lower wall towards the upper blade the flow decelerates to subsonic values. The distribution of seeding in the laminar boundary layer was poor (due to the weak mixing), thus the reliability of the results permits only to establish the BL thickness which reaches approximately 0.5 mm. The film cooling introduced into the flow has no impact on boundary layer thickness in that region.

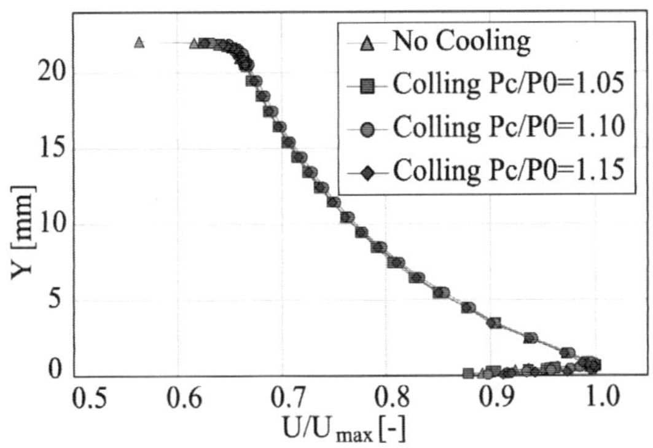

Figure 14. Velocity profile upstream of interaction, X/Cax=0.55.

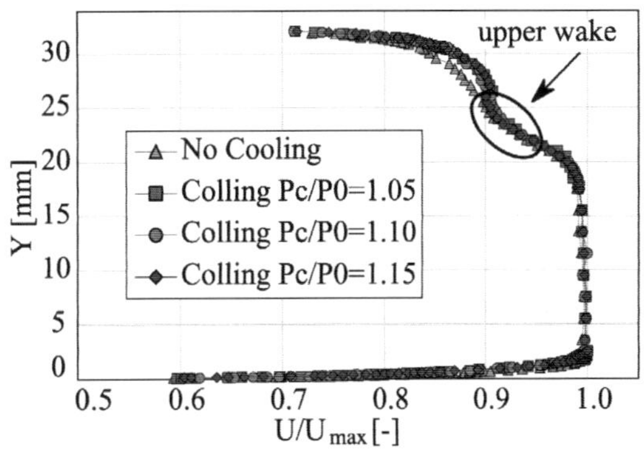

Figure 15. Velocity profiles downstream of interaction, X/Cax=0.94.

The results of measurements in traverse upstream of the trailing edge are shown in Figure 15. Velocity distribution for all cases are very similar except the boundary layers. In the plot one can observe a barely visible wake from the upper blade where the deceleration connected with the wake has almost dissipated due to a strong mixing process. The detailed velocity distributions in the boundary layer at X/Cax=0.94 are shown in Figure 16. The curves are mostly overlapped with one exception for the clean case. The BL in the clean configuration is not disturbed by any film cooling, therefore the boundary layer is thinner in relation to the cooling case.

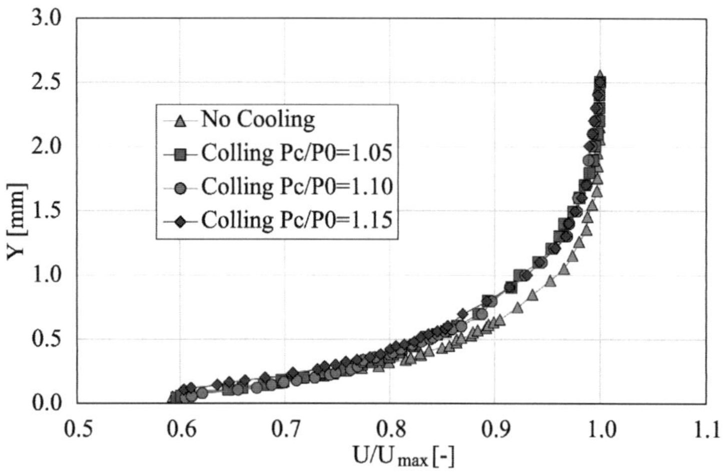

Figure 16. Velocity profiles in boundary layer downstream of interaction, X/Cax=0.94.

In order to calculate the boundary layer parameters the following formulas were used:

$$\delta^* = \int_0^\infty \left(1 - \frac{u(y)}{u_0}\right) dy \text{ - displacement thickness} \quad (1)$$

$$\delta^{**} = \int_0^\infty \frac{u(y)}{u_0}\left(1 - \frac{u(y)}{u_0}\right) dy \text{ - momentum thickness} \quad (2)$$

$$H_{12} = \frac{\delta^*}{\delta^{**}} \text{ - shape factor,} \quad (3)$$

the δ is determined as 0.99 of local free stream velocity.

Cooling increases the boundary layer thickness by about 30% in relation to the uncooled flow in the passage. The cooling intensity has very limited effect on the boundary layer size. The integral parameters like displacement and momentum thickness increase about 20% with introducing film cooling into the flow. Moreover, the shape parameter indicates that the boundary layer is turbulent downstream of the interaction region.

Table 1. Boundary layers integral parameters downstream of interaction.

	No Cooling	Cooling Pc/P0=1.05	Cooling Pc/P0=1.10	Cooling Pc/P0=1.15
δ [mm]	1.48	1.80	2.05	1.98
δ^* [mm]	0.16	0.20	0.21	0.19
δ^{**} [mm]	0.13	0.16	0.17	0.15
H_{12}	1.27	1.25	1.24	1.24

4. Conclusions

The influence of the cooling intensity on the interaction region between the shock wave and the boundary layer was analysed in these investigations. The following conclusions can be drawn from the above presented results:

- A turbine passage model was manufactured and assembled in the rectilinear test section of a transonic wind tunnel. The flow field contains characteristic features of the flow in the reference NGV turbine cascade.
- Both in the reference flow (without cooling) and in the flow with cooling the laminar shock wave boundary layer interaction is observed; the transition is induced in the interaction region, downstream of the separation the boundary layer is turbulent.

- The blade surface cooling intensity changes influence to a certain extent the pressure distribution on the blade suction side only in the interaction region. Increasing the blowing intensity also causes gradual shifts of the shock position towards the leading edge.
- Film cooling alters the shock system in the interaction area, from a double normal shock wave in the clean case to the gradual modification to a single normal shock wave for the highest cooling intensity.
- The cooling flow causes a reduction of the separation size induced by the shock wave, nevertheless, the cooling intensity increase does not affect the separation region size.
- Film cooling energizes the boundary layer a little (the separation size decreases), however the boundary layer velocity deficit downstream of the interaction increases.
- The boundary layer upstream of the interaction area has the same thickness for all cases, approximately 0.5 mm. Downstream of the interaction the cooling system causes the boundary layer to grow by about 30%, also the displacement and momentum thickness increase.

Acknowledgments
The research leading to these results received funding from the European Union Seventh Framework Programme (FP7/2007-2013) under Grant Agreement No. 265455 (TFAST).

References

[1] Babinsky H, Harvey J K 2011 Shock Wave-Boundary-Layer Interactions*Cambridge University Press vol 1*
[2] Szwaba R 2013 Influence of Air-Jet Vortex Generator Diameter on Separation Region *Journal of Thermal Science* vol 22 no 4 p 294-303
[3] Szwaba R 2011 *Comparison of the Influence of Different Air-Jet Vortex Generators on the Separation Region* Aerospace Science and Technology vol 15 p 45-52
[4] Petersen A 2014 Numerical Transition Prediction in a Straight Turbine Cascade *11th World Congress on Computational Mechanics* (WCCM XI)
[5] Naik S, Krueckels J, Grisch M, Schnieder M 2014 Multirow Film Cooling Performances of High Lift Blade and Vane *Journal of Turbomachinery* vol 136 / 051003-1
[6] Narzary D P, Liu K-C, Rallabandi A P, Han J-C 2012 Influence of Coolant Density on Turbine Blade Film Cooling Using Pressure Sensitive Paint Technique *Journal of Turbomachinery* vol 134 / 031006-1
[7] Goldstein R J 1971 Film Cooling Advances in Heat Transfer *New York: Academic Press* vol 7 p 321-379
[8] Bogard D G, Thole K A 2006 Gas Turbine Film Cooling *Journal of Propulsion and Power* vol 22, no 2
[9] Mhetras S, Han J-C, Rudolph R 2012 Effects of Flow Parameter Variations on Full Coverage Film-Cooling Effectiveness for a Gas Turbine *Blade Journal of Turbomachinery* vol 134/ 011004-1
[10] Doerffer P, Flaszynski P, Szwaba R 2014 Design of the test section *TFAST Deliverable D4.1.1* TFAST Webside - www.tfast.eu
[11] Flaszyński P, Doerffer P, Szwaba R, Kaczyński P, Piotrowicz M 2015 Shock Wave Boundary Layer Interaction on Suction Side of Compressor Profile in Single Passage Test Section *Journal of Thermal Science* vol 24 vol 6 p 510-515
[12] Szwaba R, Flaszyński P, Szumski J A, Doerffer P 2015 Influence of Air Cooling and Air- Jet Vortex Generator on Flow Structure in Turbine Passage *TASK Quarterly* vol 19 no 2 p 153-166
[13] Gomes R A, Niehuis R 2012 Aerothermodynamics of a High-Pressure Turbine Blade With Very High Loading and Vortex Generators *Journal of Turbomachinery* Vol. 134 pp 011020 1-9

Combined Differential Interferometry and Long-range μPIV measurements of a temperature driven boundary layer flow

S Kordel[1], T Nowak[1], R Skoda[1] and J Hussong[1]

[1]Ruhr-Universität Bochum, Chair of Fluid Machinery, Universitätsstraße 150, 44801 Bochum, Germany

E-mail: `stephan.kordel@ruhr-uni-bochum.de`

Abstract. In the present study Differential Interferometry and Long-range μPIV are successfully combined for a benchmark experiment of an accelerating temperature driven boundary layer flow. Spatial resolutions of 405 μm for the interference and 101 μm for the μPIV measurements could be achieved. The results of the combined measurements are compared with results of numerical simulations. Temperature and velocity profiles are compared with theoretical values and a relation between Nusselt number and Grashof number could be worked out.

1. Introduction

Several optical measurement techniques are available nowadays that have been used for simultaneous measurements of velocity and temperature or concentration distributions. Praisner et al. [1] for example combined a liquid crystal sensing system and PIV to determine the convective heat transfer and the flow field of a turbulent water channel flow. In flow situations where transient scalar fields such as density or pressure distributions have to be determined, interference methods have been the first choice for many decades. The most common interference techniques for flow studies are Mach-Zehnder Interferometry (MZI) and Differential Interferometry (DI) [2]. Goldstein and Eckert [3] used a Mach-Zehnder Interferometer to the study the steady state and transient development of a free convection thermal boundary layer about a uniformly heated vertical plate. Early studies on simultaneous density and velocity field measurements were utilized by Skarman et al. [4]. They combined holographic interferometry with 3D-PTV measurements to capture temperature and flow fields. While interference techniques are often used to measure density fields in gas flows, applications to weakly compressible liquid flows are scarce [5]. Iben et al. [6] performed pressure field measurements in cavitating liquid flows by means of Mach-Zehnder Interferometry.

In the present work simultaneous velocity and temperature field measurements were successfully performed, combining for the first time Long-range PIV and Differential Interferometry with a standard double-pulsed Nd:YAG laser. In the present study we present results on the time dependent temperature and velocity fields in a temperature driven circulation flow in a cuvette that was created by a heated side wall.

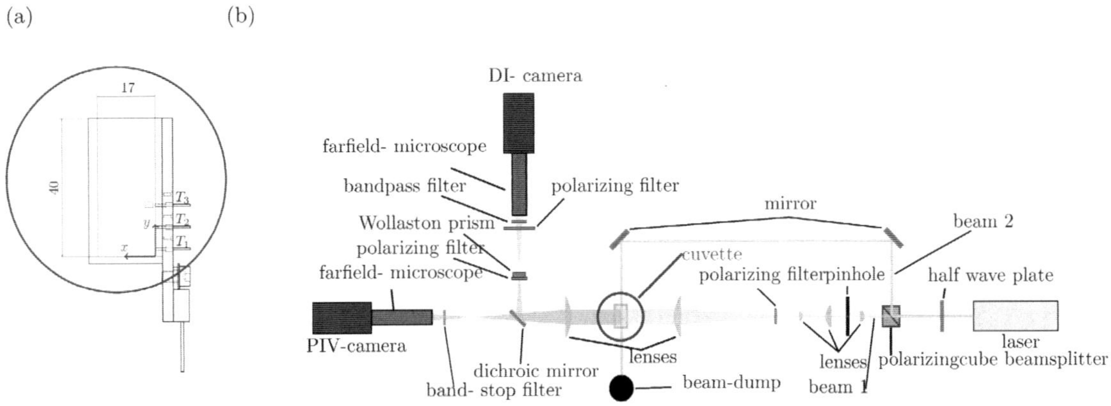

Figure 1: (a) Side view of cuvette; thermocouples are indicated with T_1 to T_3; (b) Optical set-up of the combined measurement technique

2. Set-up

2.1. Optical System

The experimental set-up of the combined measurement technique of Long-range µPIV and Differential Interferometry (DI) as shown in Fig.1b was established and tested. Details can be found in Kordel et al [7]. A double pulsed dual cavity Nd:YAG laser (Nano S 65-15 PIV, Nitron Lasers, $\lambda = 532\,\mathrm{nm}$) is used at a frequency of 5 Hz. Using a combination of a half wave plate and a polarizing beam splitter cube the pulsed beam is split into two separate parts of approximately 5 % (beam 1) and 95 % (beam 2) laser intensity. After passing a spatial filter system beam 1 is widened to a bundle of parallel light which passes the test section in positive z-direction. The second beam (beam 2) is deflected unchanged in such a way that it enters the test section in negative x-direction. Suspended Rhodamine B coated fluid tracers (Microparticles GmbH) of $25.07 \pm 0.2\,\mu\mathrm{m}$ nominal diameter in the middle plain of the test section (z=0) are excited in means of a volume illumination by the laser beam. Both, the laser light and the fluorescence signal are passing along z-axis, are then focused by a lens before separated again by a dichroic mirror. While the particle light transmits the dichroic mirror and is then recorded directly by a 12 bit dual frame CCD camera (Imager Pro SX, LaVision) the laser light is reflected. The reflected laser light passes a Wollaston prism and a polarizer which creates interference fringes in the image plane. Two far-field microscopes (Infinity K2, DistaMax) with 2.18 x magnification yield a field of view (FOV) of 3.246 mm x 3.874 mm for the full camera sensor of 2456 x 2058 pixel.

2.2. Measurement Section

An optically accessible cuvette (Hellma GmbH Co. KG) is used for all measurements (see Fig. 1a). The cuvette has a thickness of $b = 10\,\mathrm{mm}$ and one side wall was replaced by a brass plate. It is closed with a plastic plug to avoid a free surface. Heat is supplied to the plate by a $10\,\Omega$ resistor and a 4 W power supply. Along the y-axis the near wall water temperature is measured by three thermocouples (T_1 - T_3) which reach approximately 400 µm into the fluid and have a spacing of 2 mm, 8 mm and 14 mm to the bottom wall of the cuvette.

3. Measurement procedure

For all measurements a total time series of 60 s at a frequency of 5 Hz was recorded. The heating process was started with 2 s delay to measure the initial fluid temperature (ambient temperature) and eventual background flows in the cuvette. At each time step both, DI and µPIV camera, record a double image pair. This leads to a total amount of 600 double images for one measurement sequence. The interframing time was set to $dt = 5\,\mathrm{ms}$. Hence, the particle image displacement is ~ 10 pixel in the final stage of the measurement sequence. Due to refraction effects in the DI image the particle image density for the used 25.07 µm Rhodamine B particles was limited to 3.9×10^{-4} particle pixel^{-1} within the FOV. The orientation of the fringe pattern can be adopted by the rotation Wollaston prism. In this evaluation all measurements are done with an fringe orientation angle of $\alpha = \pm 45°$.

4. Evaluation

4.1. Differential Interferometry

An in-house matlab code is used for the evaluation of the recorded interferograms. It evaluates the density gradient based on the relative fringe displacement $\Delta S/S$ as follows [8]:

$$\frac{\partial \overline{\rho}(x,y)}{\partial \zeta} = \frac{\lambda}{Kbd}\frac{\Delta S(x,y)}{S} \tag{1}$$

Here, ζ denotes the direction normal to the fringe orientation. K is the Gladstone-Dale constant which was evaluated for a temperature range of $20\,°\mathrm{C} \leq T \leq 40\,°\mathrm{C}$. Wavelength λ, measurement volume thickness d and the distance between neighboring light rays b [9] are characteristic quantities of the set-up and constant throughout all measurements.

To determine $\Delta S/S$ a fourth order polynomial was fitted to each fringe of the pattern and was then compared to a linear equation extracted from a initial not displaced reference fringe. The local gradient evaluation by the fringe pattern limits the spatial resolution to the fringe spacing S. In order to receive a continuous fringe displacement field a 2D surface polynomial was fitted to the displacement values. From this the density gradient field in ζ-direction was derived. Hence, this is not necessary the gradients main direction, measurements are repeated with 90° rotated Wollaston prism. Again the density gradient field, now in η direction is determined. A final coordinate transformation gives the gradient in x-and y-direction.

With a reference temperature from the thermocouples the temperature field can be derived by integration [10].

To compensate for refraction effects, which may lead to an overestimation of the density gradient, an a priory calibration procedure was implemented in the set-up. With the help of a calibration plate and a blue LED the refraction induced image distortion is measured by means of an Background Oriented Schlieren (BOS) evaluation. A mapping function for each time step is applied for fringe image.

4.2. Long-range µPIV

The evaluation of the recorded particle images is done with commercial software DaVis 8.2.2 (LaVision GmbH). Before the evaluation of the particle image shift several preprocessing steps are performed to improve the signal-to-noise ratio reaching a mean background signal of about 0.16 counts. The PIV evaluation uses a 2D multi-pass cross-correlation algorithm. To achieve reliable cross-correlation results of more than 95 % probability the particle density has to be > 5 per interrogation window [11]. Therefore, a final window size of 128 x 128 pixel with 5 % overlapping was chosen, leading to a vector spacing of 101 µm.

5. Results

5.1. Temperature and Velocity fields

Figs. 2a and 2b show the temperature and a velocity field within a Region of Interest (ROI) of $0.82\,\mathrm{mm} \leq \Delta x \leq 3\,\mathrm{mm}$ spacing to the heated side wall and $12.8\,\mathrm{mm} \leq \Delta y \leq 15.5\,\mathrm{mm}$ spacing to the bottom wall of the combined measurement technique at $t = 30\,\mathrm{s}$. Since a fully transient

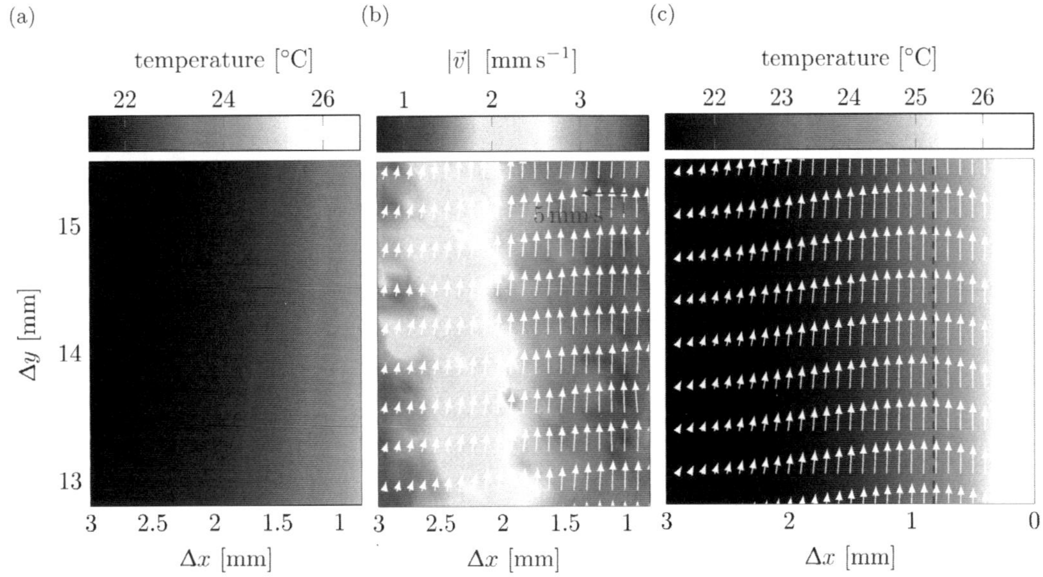

Figure 2: (a) Measured temperature field; (b) Measured velocity field; (c) Temperature field (contour plot) and velocity field (vectors) derived from simulations; Experimental and numerical results at $t = 30\,\mathrm{s}$

flow is investigated, the velocity field (see Fig. 2b) is computed by cross-correlation of one double image pair. As mentioned in section 4.1 the temperature field (Fig. 2a) is a result of two separate measurements. Out of three independent repetitive measurements the local maximum standard deviation of measured density gradient fields in ζ direction was determined to be less than $\delta_{max} = 4\,\%$. Hence, the computation of the main density gradient out of two measurements is a fair procedure. Nevertheless, our aim in future studies is to record the density gradients in ζ and η direction simultaneously. Therefore a second Wollaston prism is needed, which will be placed in a separated beam path before recombined again. The result is a grid pattern as discussed in section 6.

Fig. 2c shows results of numerical simulations at the same time step. The three dimensional flow problem has been simulated using ANSYS CFX. For simulations both, the brass plate and the fluid have been modeled with cuboid elements on a Cartesian grid. The first 3 mm spacing to the heated side wall were resolved with 300 elements in x-direction with 4.6 million cells in the domain. The heat source is represented by a point source supplying 4 W. For the brass plate a specific heat capacity of $376\,\mathrm{J\,kg^{-1}\,K^{-1}}$ and a heat transfer coefficient of $454\,\mathrm{W\,m^{-2}\,K^{-1}}$ for the heat transfer between brass and fluid have been estimated. Glass walls are modeled with no slip wall boundary conditions and a heat transfer coefficient of $5.1\,\mathrm{W\,m^{-2}\,K^{-1}}$.

Both, experimental and numerical results are in good agreement. Local temperature deviations of up to 1.2 °C in the near wall region can be detected. Our assumption is that these differences originate from deviations in the actual and numerically assumed heat transfer coefficient and power input, which lead to different wall temperatures in the simulation and the experiment. As the actual alloy composition for the used brass wall is unknown the heat transfer coefficient was assumed from standard material tables [12].

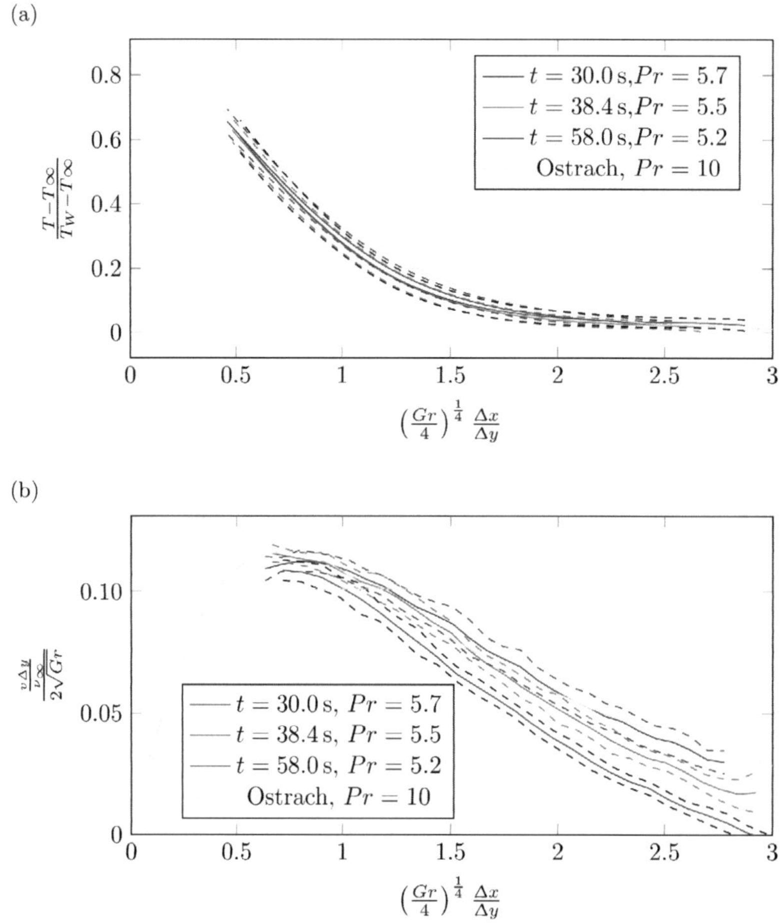

Figure 3: (a) Dimensionless temperature profiles and (b) Dimensionless velocity magnitude profiles in the dimensionless normal wall direction Δx for height of $\Delta y = 14\,\mathrm{mm}$ from the bottom wall at different time steps; yellow indicated theoretical computations of Ostrach [13]

5.2. Temperature and Velocity profiles of the accelerating flow

Fig. 3 shows the temperature and velocity profiles of the wall normal direction Δx averaged over $12.8\,\mathrm{mm} \leq \Delta y \leq 15.5\,\mathrm{mm}$ and the corresponding standard deviation as a function of time and therefore of the Prandtl number. The profiles are shown for $0.82\,\mathrm{mm} \leq \Delta x \leq 3$ which corresponds to the depicted ROI. Because of the steady growth of the temperature profile a time independent and dimensionless scale was chosen. For comparison theoretical profiles for a laminar convection flow of a vertical plate with homogeneous temperature distribution according to Ostrach [13] are plotted in yellow for a Prandtl number of 10. The Prandtl number decreases with higher temperature and in this case with higher velocities. The profiles are in good qualitative agreement with theoretical predictions. Differences emanate from different inflow conditions of the present experimental compared to those depicted by Ostrach.

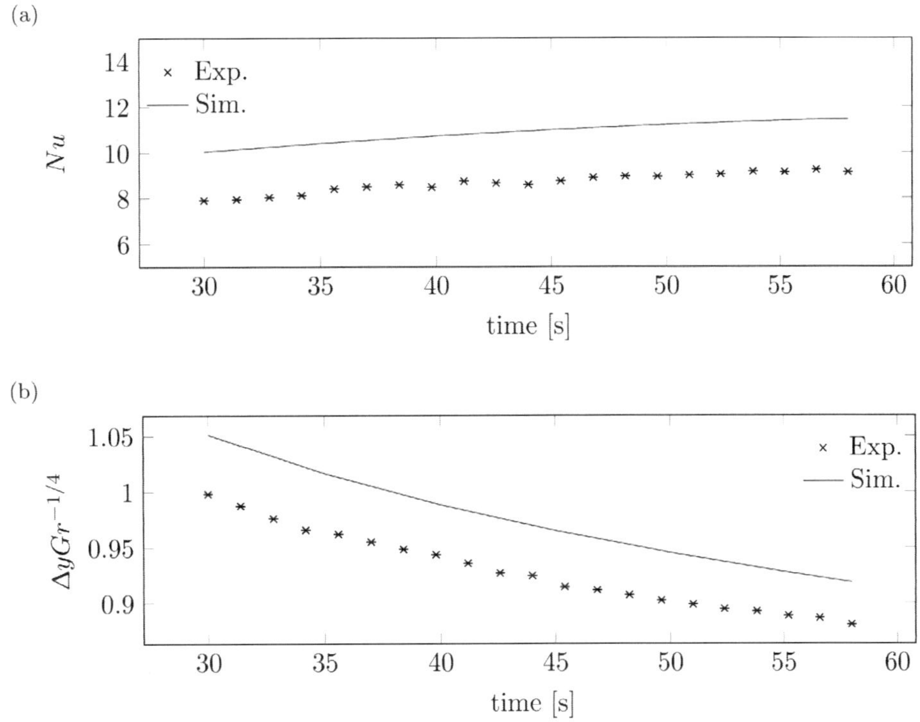

Figure 4: (a) Nusselt numbers averaged over for $12.8\,\text{mm} \leq \Delta y \leq 15.5\,\text{mm}$ according to (2) at different time steps; (b) averaged scaled Grashof number $\Delta y Gr^{-1/4}$ for $12.5\,\text{mm} \leq \Delta y \leq 15.5\,\text{mm}$

Additionally Nusselt and Grashof numbers were determined according to:

$$Nu = \frac{\Delta y \cdot \frac{\partial T}{\partial x}|_w}{(T_w - T_\infty)} \tag{2}$$

$$Gr = \frac{g \Delta y^3 \beta_{inf} (T_w - T_\infty)}{\nu_{inf}^2} \tag{3}$$

$$\bar{\delta_h} \propto \Delta y Gr^{-1/4} \tag{4}$$

Here, $\Delta y = 14\,\text{mm}$ denotes the distance from the bottom wall. $\partial T / \partial x$ is the temperature gradient in normal wall direction and T_w the wall temperature. Both values were extracted from a first order polynomial fit to the temperature profiles. $T_\infty = 21.2\,°\text{C}$ denotes the starting temperature of the fluid measured by the thermocouples, g denotes the acceleration of gravity, β is the isobaric cubic expansion coefficient and ν is the kinematic viscosity for water at $T = 21.2\,°\text{C}$. $\bar{\delta_h}$ denotes the hydrodynamic boundary layer thickness.

Since there is a proportionality between Grashof number and hydrodynamic boundary layer thickness according to (4), Nu and $\Delta y Gr^{-1/4}$ are shown in Fig.4a and 4b.

Fig. 4 shows the Nusselt numbers and the hydraulic boundary layer thickness as a function of time for both, numerical and experimental results. While in both cases the slopes are similar, there is an offset between the absolute values. It may be noted that both, Nusselt number and Grashof number are mainly a function of wall temperature and as mentioned in section 5.1, the near wall temperature in the simulation differs from that in the actual experiment.

Early studies of Goldstein and Eckert [3] showed that the unsteady free convection boundary layer grows with time before shrinking again to approach a steady state value. Obviously the time series shown here, covers the phase of hydrodynamical boundary layer shrinking. According to a relation obtained by Sparrow and Gregg [14] for this phase of flow problem following relation can be expressed:

$$Nu = k \cdot \left(\frac{Gr Pr}{4}\right)^{1/4} \tag{5}$$

It could be found that the factor $k \approx 0.5$ here for $Pr = 5.7$.

6. Conclusion and Outlook

Combined Differential Interferometry and Long-range µPIV measurements could be successfully applied to a temperature driven boundary layer flow. Spatial resolutions of 405 µm for Differential Interferometry and 101 µm for Long-range µPIV could be achieved. The comparison between measurement and simulations results showed local temperature deviations of less than 1.2 °C, which lead to deviations in the temporal evolution of the Nusselt number and the thickness of the hydrodynamic boundary layer at the same time. While the Nusselt numbers increased for all observed time steps, the hydrodynamics boundary layer reduces simultaneously. Theoretical results of a steady laminar convective flow with a vertical heated plate are in good qualitative agreement to the observed flow problem.

In future studies the measurement technique will be improved by implementing a second Wollaston Prism into the set-up. For this the separated laser light after the dichroic mirror (see Fig 1b) has to be split into two parallel beams each passing one Wollaston prism. Before entering the camera the signals have to be recombined again. The result on the camera chip is a grid pattern. Furthermore, to investigate all characteristics of the boundary layer development, long time scale measurement are planned.

References

[1] Praisner T J, Sabatino D R and Smith C R 2001 *Experiments in Fluids* **30** 1–10

[2] Egbers C, Brasch W, Sitte B, Immohr J and Schmidt J R 1999 *Measurement Science and Technology* **10** 866–877

[3] Goldstein R and Eckert E 1960 *International Journal of Heat and Mass Transfer* **1**

[4] Skarman B, Becker J and Wozniak K 1996 *Flow Measurement and Instrumentation* **7** 1–6

[5] Woisetschläger J, Pretzler G, Jericha H, Mayrhofer N and Pirker H P 1998 *Experiments in Fluids* **24** 102–109

[6] Iben U, Morozov A, Winklhofer E and Wolf F 2011 *Experiments in Fluids* **50** 597–611

[7] Kordel S, Nowak T, Skoda R and Hussong J (accepted) 2016 *Experiments in Fluids*

[8] Merzkirch W 1987 *Flow visualization* 2nd ed (Orlando: Acad. Press)

[9] Small R D, Sernas V A and Page R H 1972 *Applied Optics* **11** 858

[10] Agrawal A, Raskar R and Chellappa R 2006 *Computer Vision ECCV 2006* vol 3951 (Berlin, Heidelberg: Springer Berlin Heidelberg) pp 578–591

[11] Raffel M (ed) 2007 *Particle image velocimetry: a practical guide* 2nd ed (Heidelberg ; New York: Springer)

[12] Verein Deutscher Ingenieure and Gesellschaft Verfahrenstechnik und Chemieingenieurwesen (eds) 2013 *VDI-Wärmeatlas* 11th ed (Berlin: Springer Vieweg)

[13] Ostrach S 1953 *Trans. Am. Soc. Mec. Engrs* **75** 1287–1290

[14] Sparrow E and Gregg J 1956 *Trans. ASME* **78** 435–440

Evaporation level of the condensate droplets on a shock wave in the IMP PAN nozzle depending on the inlet conditions

S Kornet[1] and J Badur[1]

[1] The Szewalski Institute of Fluid-Flow Machinery PAS-ci, Energy Conversion Department, Fiszera 14, 18-231 Gdansk, Poland

E-mail: skornet@imp.gda.pl, jb@imp.gda.pl

Abstract. In the present paper we have focused on the phenomena of condensate re-evaporation in the shock wave zone. Having observed the finishing of a foggy flow within the shock wave, according to Puzyrewski's observations, we would like to analyse the critical inlet conditions which cause total evaporation of condensate droplets on the shock in the IMP PAN nozzle. In the paper some original mechanistic model of droplet evaporation is involved, numerically implemented and compared with the IMP PAN experiment. The single continuum model of wet steam with a special microstructure growing up during phase transitions was validated on IMP PAN experiment performed for inlet conditions close to the saturation line. The present work includes simulations results of total and partial evaporation liquid phase on the shock wave for different boundary conditions.

1. Introduction

In the space between the turbine blades, which resembles the shape of the de Laval nozzle, the shock wave can appear, which has a negative impact on the flow of the steam [1]. If a supersonic flow passes through a duct of certain length, under certain conditions, successive shocks can appear downstream of the first shock. This series of shocks is a so-called shock train [2, 3]. The shock train in the IMP PAN nozzle obtained from CFD simulations is shown in figure 1.

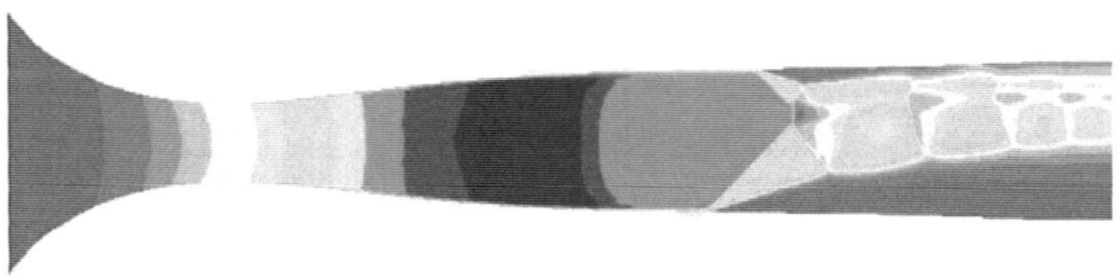

Figure 1. Typical shock train in supersonic part of the IMP PAN nozzle.

Shock wave is the result of interaction between a normal shock wave and the boundary layer, which produces a λ-foot structure. It is well-know and experimentally confirmed that for high flow velocities (Ma > 1.4) the λ-feet generated on the side walls of the symmetric nozzle become different in size. It has also been observed that the tendency towards asymmetry depends on the nozzle divergence angle and the walls roughnesses [7, 8, 9, 10, 13, 14]. This asymmetry does not flip during an experiment but may change sides from one experiment to the next one [11, 12].

According to Puzyrewski's observations, partial or total evaporation of the condensate droplets in the shock wave zone is possible and depends on the boundary conditions and the thermodynamic conditions in divergent part of the nozzle [1, 4, 5].

In this paper we have focused on the precise prediction of the critical inlet temperature above which occurs total evaporation of condensate on the shock. In our model, the evaporation is governed not only by mass transport but also by internal structure energy that is based on balance of heat energy transported into a droplet. Numerical analysis was performed on the IMP PAN nozzle. The present work includes simulations results of the critical inlet temperatures which determine the border between total and partial evaporation on the shock wave. Calculations were performed for set of various inlet pressures with keeping the same value of pressure ratio for each case.

2. Single continuum model of wet steam

A wet steam model include governing equations which are based on balance of liquid-vapour mixture. For a consistent non-equilibrium condensation model a set of nine transport equations can be written in general form [1, 5, 6, 15]:

$$\partial_t(\rho\phi) + \mathrm{div}(\rho\phi\,\mathbf{v}) = \mathrm{div}(\mathbf{J}_\phi) + \rho S_\phi \tag{1}$$

where: $\phi = \{\, 1, \mathbf{v}, e, k, \varepsilon, x, a \,\}$ represents the relevant conserved variable. First three variables in ϕ set come from well know equations: balance of mass ($\phi = 1$), balance of momentum ($\phi = \mathbf{v}$ - three equations), balance of energy ($\phi = e$). The rest of variables include in ϕ set are explained in paragraph 2.1. Process of growth of individual droplet is governed by mass, momentum and energy transport mechanism between gas and liquid phases. Droplet growth can be described by an evolution of radius of droplet that moves in the wet steam field. Evolution equation of dryness fraction is given by:

$$\partial_t(\rho x) + \mathrm{div}(\rho x\,\mathbf{v}) = \mathrm{div}(\mathbf{J}_x) + \rho S_x \tag{2}$$

The dryness fraction sources S_x (in equation (2)) can be divided into homo- and heterogeneous sources of the mass generation rate due to condensation and evaporation. Homogenous source includes part responsible for inception of droplets and part responsible for growth and evaporation of droplets:

$$S_x = \frac{4}{3}\Pi\rho_l \mathrm{I}\mathrm{r}_*^3 + 4\Pi\rho_l\,\alpha\,\bar{\mathrm{r}}^2\,\frac{\partial\bar{\mathrm{r}}}{\partial t} \tag{3}$$

where: $\bar{\mathrm{r}}$ - an average radius of droplet [m], ρ_l - density of condensed phase [kg/m^3], r_* - the critical radius of droplet [m], I - a volumetric rate of nucleation [number of droplets/m^3], α - volume fraction of condensate [-][1, 5, 6, 15].

2.1. Interaction of turbulent and phase microstructure

In our approach the phase microstructure has similar size like the turbulent microstructure which evaluate under following concept: according to possible splitting of modes of momentum transport only two scalar parameters can describe a evolution of turbulent microstructure, e.i. the turbulent energy k and the turbulent dissipation ε. The evolution equation for turbulent microstructure has a following form:

$$\partial_t(\rho k) + \mathrm{div}(\rho k\,\mathbf{v}) = \mathrm{div}(\mathbf{J}_k) + \rho S_k \tag{4}$$

$$\partial_t(\rho\varepsilon) + \mathrm{div}(\rho\varepsilon\,\mathbf{v}) = \mathrm{div}(\mathbf{J}_\varepsilon) + \rho S_\varepsilon \tag{5}$$

Transport of non-equilibrium properties between the phase and turbulent microstructure can be made directly on the microstructure level, without exploring the total momentum equation. It may be done by introducing the crossing effects in phase and turbulence fluxes [6, 16, 17]:

- Turbulent kinetic energy k [m^2/s^2]:

$$\mathbf{J}_k = (D_{kk})\mathrm{grad}\, k + (D_{kxo} + D_{kxr})\mathrm{grad}\, x \tag{6}$$

- Dryness fraction x [-]:

$$\mathbf{J}_x = (D_{kxo} + D_{kxr})\mathrm{grad}\, k + (D_{xxo} + D_{xxr})\mathrm{grad}\, x \tag{7}$$

- Turbulent energy dissipation rate ε [m^2/s^3]:

$$\mathbf{J}_\varepsilon = (D_{\varepsilon\varepsilon})\mathrm{grad}\, \varepsilon + (D_{\varepsilon ao} + D_{\varepsilon ar})\mathrm{grad}\, a \tag{8}$$

- Interphase surface density a [m^2/m^3]:

$$\mathbf{J}_a = (D_{a\varepsilon})\mathrm{grad}\, \varepsilon + (D_{aao} + D_{aar})\mathrm{grad}\, a \tag{9}$$

Diffusion coefficients D, which are related with homogeneous diffusion known as the Ostwald mode (subscript o) and heterogeneous condensation mode (subscript r) need estimations and calibrations [1, 6, 17].

3. Validation of wet steam model by comparison with the experiment data

The model of a single continuum with a special microstructure growing up during phase transitions, proposed by Bilicki and Badur [6], was validated on the IMP PAN experiment carried out by Puzyrewski . The experiment was carried out on a symmetric nozzle of "rectangular" cross-section. The nozzle shape and 3D FVM model are shown in figure 2.

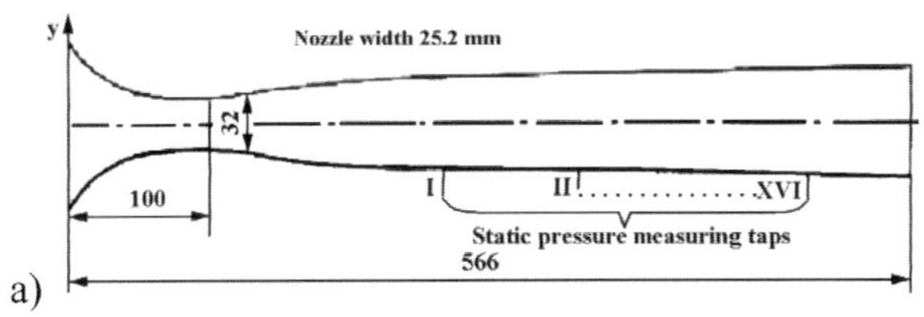

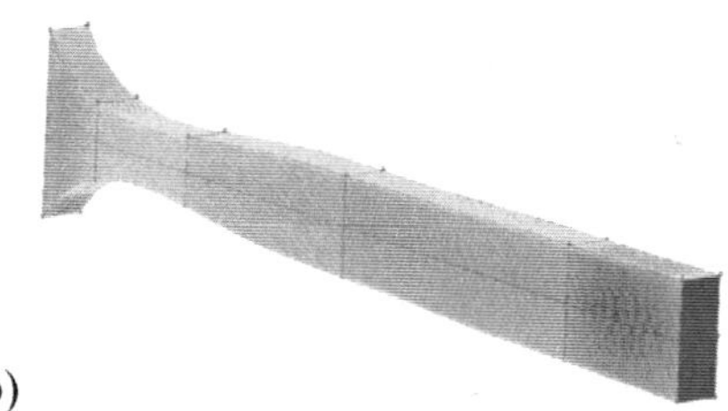

Figure 2. IMP PAN nozzle: a) shape and dimension, b) FVM discretization.

3.1. Boundary conditions of the IMP PAN experiment

To validation of CFD model selected boundary conditions which correspond to the V-th set of experimental conditions. The inlet pressure was p_{inlet} = 2.26 bar, the inlet temperature T_{inlet} = 502 K. Pressure at the point of intersection of the isentropic expansion line and saturation line was p_{sat} = 0.55 bar. These input conditions were close to the saturation line in order to obtain a condensation [4]. The static pressure during experiment was measurement on both side walls in sixteen points on each from wall (see figure 2a).

3.2. CFD results and comparison with the test data

For a given the de Laval nozzle (IMP PAN nozzle) geometry authors have conducted calculations for the single continuum model of wet steam. Figure 3 shows a comparison of the static pressure distribution obtained from numerical simulation with IMP PAN experimental data (for V-th set of boundary conditions[4]). For this case we obtained total evaporation of the condensate during passage through the strong shock what agree with experimental results: *For V-th set of experimental conditions, there were only very slight traces of mist ahead of the shock wave, disappearing in the shock.* The wet steam model of a single continuum, describing both condensation and re-vaporization, applied to the de Laval nozzle gives satisfactory results that agree well with experimental data (very well describes static pressure distribution and prediction of the shock wave).

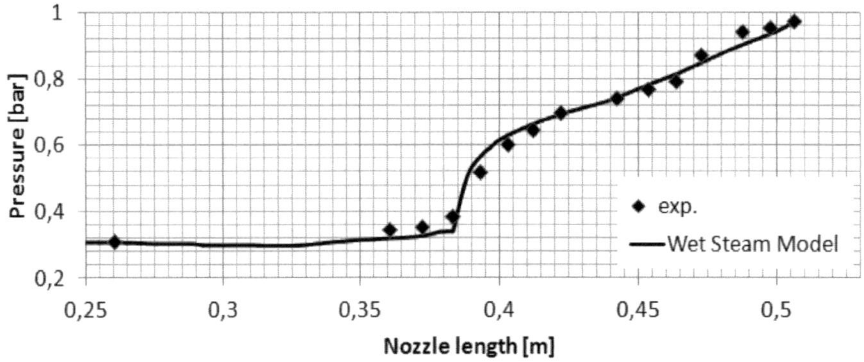

Figure 3. Comparison of the static pressure distributions along nozzle wall obtained from CFD calculations with experimental data.

4. Numerical analysis of the evaporation level of condensate droplets on the shock wave in the IMP PAN nozzle

Having observed full evaporation phenomena on the strong shock wave, according to Puzyrewski's observations (see figure 4 – more details can be find in [4]), it was decided carried out numerical analysis of the evaporation level on the shock in the IMP PAN nozzle depending on the inlet conditions. All geometry of the nozzle was splitted on nine subdomains what contributed to created the higher quality mesh (see figure 2b). For first step of simulation it was decided to decrease the grid spacing only in zones close to the inlet and close to the outlet of the channel. After having determined the position of the shock wave appearing in the flow and the oscillation zone in next steps increased the grid resolution in this places. Additionally decreased the size of the finite volume upstream and downstream of the oscillation zone. When the results in next step were the same (or approximately the same) as in previous step the modify of the mesh was ceased. These steps were used for each considered case of the boundary conditions.

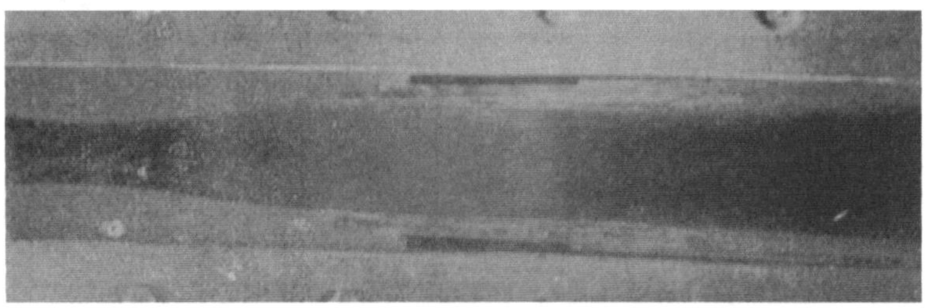

Figure 4. Evaporation of condensate droplets during passage through the shock wave [4].

4.1. Boundary conditions adopted to the numerical analysis

To numerical analysis adopted four different values of the inlet pressure: $p_{inlet} = 2.5$, $p_{inlet} = 3$, $p_{inlet} = 3.5$ and $p_{inlet} = 4$ bar. For each case of the inlet pressure carried out simulations with five different values of the inlet temperature: $T_{inlet} = 450;\ 440;\ 430;\ 420;\ 410\,\mathrm{K}$. The pressure ratio, in each considered case, was equal to $p_{inlet} / p_{outlet} = (p_{tot.} + p_{atm.})/(p_{stat.} + p_{atm.}) = 1.7$, where $p_{atm.} = 101325$ Pa. In the calculation it was assumed that steam is chemically very pure.

4.2. Results and discussion

The position of the nucleation zone mostly depended on the inlet temperature and the purity of steam. The effect of the inlet temperature on moving of the nucleation zone was analyzed by using CFD model in [18] based on the test data. For each case of the inlet pressure the nucleation zone is shifted to the nozzle entry with decreasing value of the inlet temperature. Calculated wetness fractions for different inlet temperatures and for the inlet pressure 2.5 and 4 bar are shown in figures 5 and 6 respectively. In each case of the inlet conditions the condensate mostly evaporate during passage through the Mach disc (in main flow zone). The evaporation level on the oblique shocks is lower than the evaporation level on the Mach disc. For the inlet pressure equal to 4 bar the value of wetness fraction upstream the shock is bigger than for the inlet pressure equal to 2.5 bar. The expansion line for the case with lower value of the inlet pressure is shifted to the right on i-s diagram. This means that for this case for bigger pressure expansion the saturation line will be exceeded. Therefore position of the nucleation zone for smaller inlet pressure is shifted to the nozzle exit and wetness fraction is lower. For cases of the inlet conditions which relate to conditions of wet steam on i-s diagram adopted that steam for these cases is sub-cooled and in metastable state. This assumption and assumption that steam is very pure means that on the nozzle inlet there doesn't exist two-phase flow.

In all cases smaller λ -feet is generated on a upper wall of the nozzle because all simulations were carried out by using one initialization. Of course during calculations with new initializations the asymmetry may change sides (as during new experiment [11, 12]), but for better comparisons it was decided show results for the same type of asymmetry.

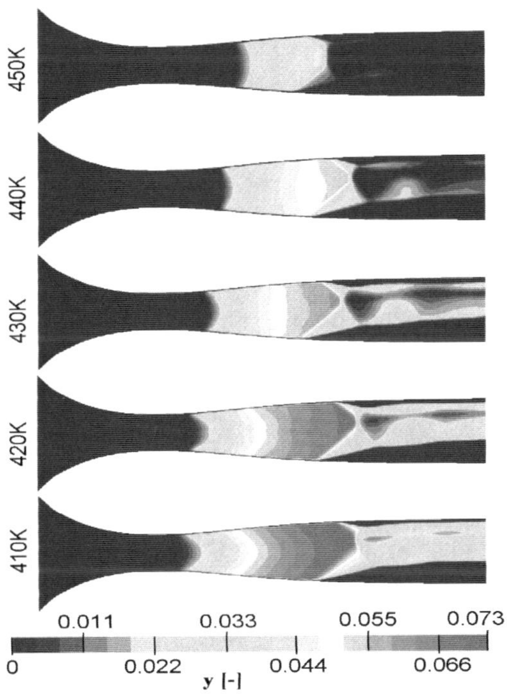

Figure 5. Calculated wetness fraction (y = 1- x) for the inlet pressure 2.5 bar and different values of the inlet temperature.

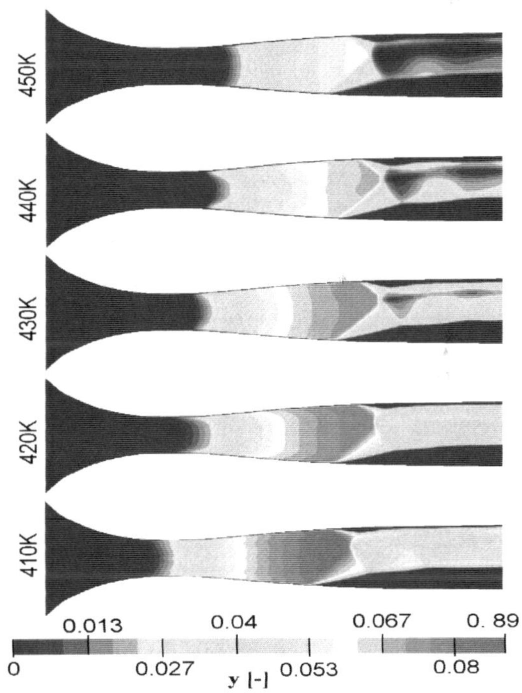

Figure 6. Calculated wetness fraction (y = 1- x) for the inlet pressure 4 bar and different values of the inlet temperature.

Nozzle "axis" in each case of boundary conditions crosses bigger λ -feet, therefore presentation of the steam expansion along this axis is hampered. In order to better visualisation of expansion were created the virtual straight lines for each case. Each from created line begins from the central point of the inlet, crosses the Mach disc in centre and ends few millimetres downstream the shock wave. The steam expansion (obtained from the numerical calculations) on "i-s" diagram along lines formed in this way are shown in figures 7-10. Red part of the solid line in figure 7 correspond to very short part of the virtual line (about 3-3.5 mm) which crosses the normal shock. Very strong steam compression on the shock and increase of steam temperature on very short distance can explain the shape of this part of curve (and shapes of curves in the places passage through the shock wave for other cases). For each case of the inlet pressure the expansion line on "i-s" diagram is shifted to the left with decreasing inlet temperature. For each case there is a jump of steam conditions caused by the shock wave which appears in supersonic part of the nozzle. In cases with bigger value of the inlet temperature there is a small peak in the wet steam zone (before reaching the lowest position) which determines start condensation of the supercooled water vapour. During condensation the thermal energy is transported from liquid droplets to the gas phase what can explain the direction and shapes of observed peaks.

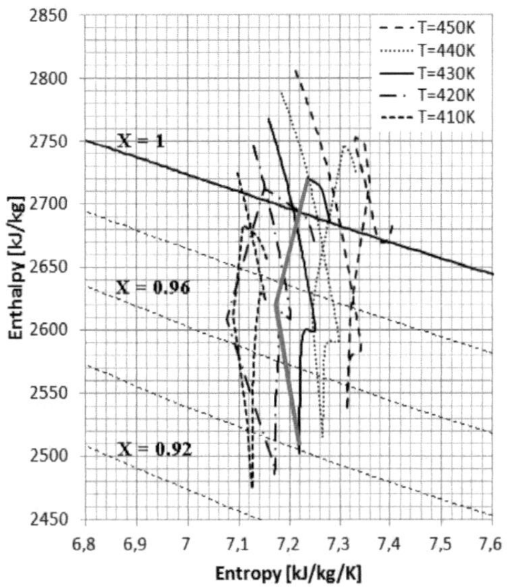

Figure 7. Change of steam entropy during nozzle expansion for case p_{inlet} = 2.5 bar.

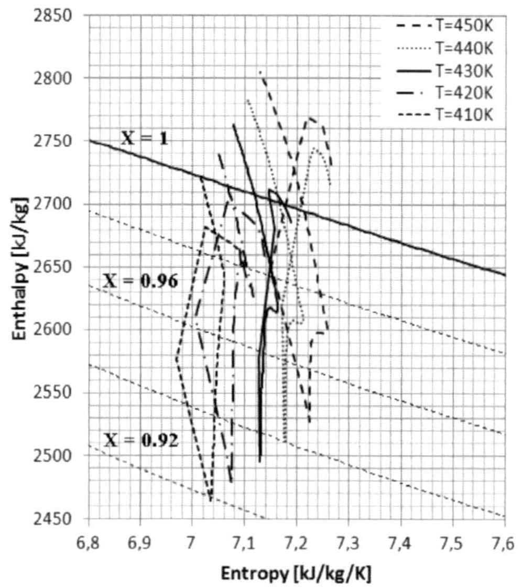

Figure 8. Change of steam entropy during nozzle expansion for case p_{inlet} = 3 bar.

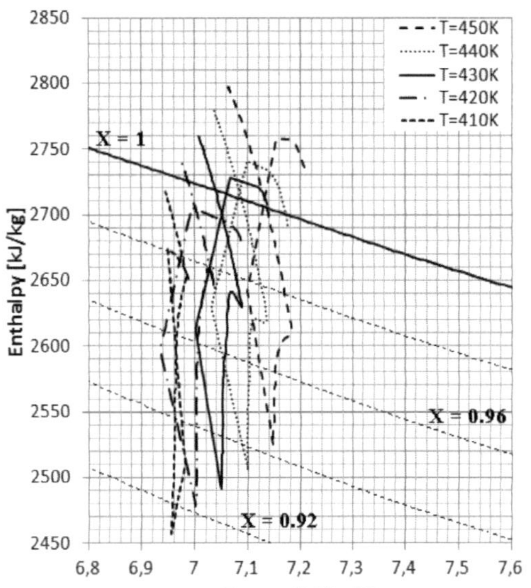

Figure 9. Change of steam entropy during nozzle expansion for case p_{inlet} = 3.5 bar.

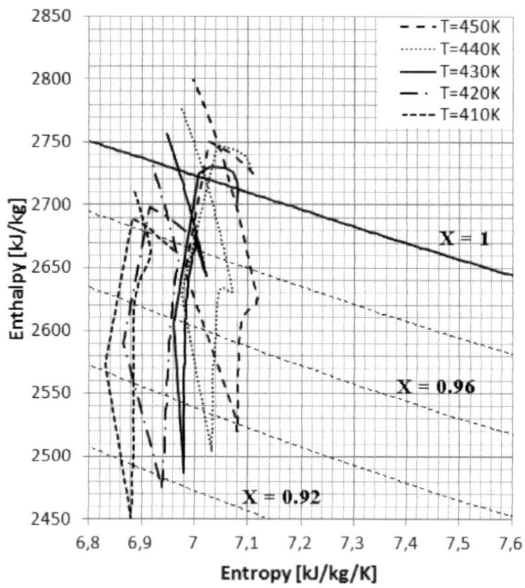

Figure 10. Change of steam entropy during nozzle expansion for case p_{inlet} = 4 bar.

In this work authors assumed that the evaporation level will be measured during passage through the Mach disc. For this purpose, for each considered case, appointed the values of the wetness fraction upstream and downstream the normal shock. In table 1 are presented the results of the evaporation level for all considered cases of the boundary conditions. In our approach the level of condensate evaporation on the shock wave is described under following concept: to show how big part of exist liquid phase located upstream the shock evaporate during passage through the Mach disc or the normal shock. Therefore results are presented in %. As expected the evaporation level increases with increase the value of the inlet temperature. During comparison results from "i-s" diagrams with results from table 1 we can note some differences. This differences results from definition of wetness fraction (or dryness fraction). On diagrams we have shown values of dryness fraction calculated as equilibrium while the evaporation level was computed by using non-equilibrium definition of the wetness fraction. In each case of the steam flow in the IMP PAN nozzle occurs phenomena called spontaneous condensation and phenomena "spontaneous evaporation". Differences between non-equilibrium and equilibrium values of the dryness fraction for case p_{inlet} = 4 bar and T_{inlet} = 410 K along virtual line are shown in figure 11.

Table 1. Evaporation level on the Mach disc for all considered boundary conditions

T_{inlet} [K]	p_{inlet} [bar]	Evaporation level [%]	p_{inlet} [bar]	Evaporation level [%]	p_{inlet} [bar]	Evaporation level [%]	p_{inlet} [bar]	Evaporation level [%]
450	2.5	100	3	100	3,5	100	4	99,00
440	2.5	100	3	100	3,5	99	4	97,00
430	2.5	97,00	3	91,00	3,5	92,93	4	92,00
420	2.5	90,00	3	88,00	3,5	78,79	4	68,04
410	2.5	81,63	3	66,33	3,5	56,00	4	64,00

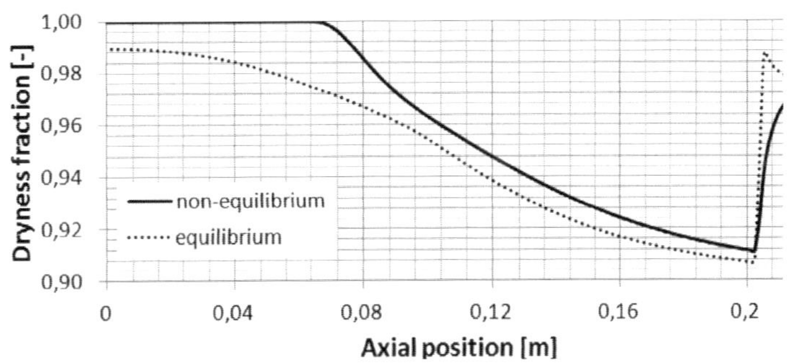

Figure 11. Comparison between non- and equilibrium dryness fraction along virtual line (p_{inlet} = 4 bar and T_{inlet} = 410 K).

The value of the critical inlet temperature which determine the border between total and partial evaporation on the shock wave depends on the pressure conditions, the purity of steam, position and shape of the shock wave, oscillation of the shock, level of the steam subcooling etc.. Additionally the calculation time for one set of the boundary conditions is very long. This causes that determination of this temperature is very time consuming.

5. Conclusions

The model of a single continuum with a special microstructure growing up during phase transitions was validated on a experiment carried out by Puzyrewski [4] on the planar symmetrical IMP PAN nozzle. The CFD results obtained by using our model agree very well with experiment data (pressure distribution along nozzle wall and full evaporation on the shock wave for V-th set of boundary conditions). Our numerical investigation has shown that:

- with decreasing value of the inlet temperature the evaporation level also decreases,
- the level of condensate evaporation on the shock wave depends on the boundary conditions, the purity of steam, localization and shape of the shock wave, oscillation of the shock, level of the steam subcooling, number of shocks in supersonic part of the nozzle, etc.,
- the evaporation level on normal shock is different with comparison with the evaporation level on oblique shocks,
- the position of the nucleation zone mostly depended on the inlet temperature and the purity of steam,
- precisely determination value of inlet temperature which determine the border between total and partial evaporation on the shock wave is very difficult.

References

[1] Kornet S and Badur J 2015 *Trans. of IFFM* **128** 119-130
[2] Grzona A and Olivier H 2012 *28th International Symposium on Shock Waves* **2** 141-146 (Springer Berlin Heidelberg)
[3] Weiss A and Olivier H 2014 *Shock Waves* **24** 11–19
[4] Puzyrewski R, Gardzilewicz A and Bagińska M 1973 *Archives of Mechanics* **25** 393-409
[5] Kornet S and Badur J 2015 *3rd Polish Congress of Mechanics and 21st Computer Methods in Mechanics* **2** 523-524 (ed M Kleiber et al.)
[6] Bilicki Z and Badur J 2003 *Journal of Non-Equilibrium Thermodynamics* **28** 145-172
[7] Namieśnik K and Doerffer P 2004 *TASK Quarterly* **9** 53-63
[8] Bourgoing A and Reijasse Ph 2006 *Shock Waves* **24** 251-258
[9] Sellam M *et al* 2014 *Shock Wave* **24** 33-39
[10] Verma S B and Manisanka C 2014 *Shock Wave* **24** 191-209
[11] Papamoschou D and Johnson A 2006 *36th AIAA Fluid Dynamics Conference and Exhibit* (San Francisco, California)
[12] Gawehn T *et al* 2010 *Shock Waves* **20** 297–306
[13] Bourgoing A and Reijasse Ph 2005 *Shock Waves* **14(4)** 251–258
[14] Papamoschou D, Zill A and Johnson A 2009 *Shock Waves* **19** 171–183
[15] Badur J *et al* 2011 *Proc. of microCAD "International Scientific Conference"* (Miscolc)
[16] Banaszkiewicz M and Badur J 2000 *TASK Quarterly* **4** 213-290
[17] Zakrzewski W, Karcz M and Kornet S 2012 *Trans. of IFFM* **124** 111-124
[18] Chandler K, Melas M and Jorge T 2015 *ASME Turbo Expo 2015: Turbine Technical Conference and Exposition* (Montreal, Canada)

Generation of the vorticity field by the flapping profile

T Kozłowski and H Kudela

Wrocław University of Technology, Faculty of Mechanical and Power Engineering, Wybrzeże Wyspiańskiego 27, 50-370 Wrocław, Poland

E-mail: `tomasz.kozlowski@pwr.edu.pl`

Abstract. Birds and insects move due to flapping their wings. In the paper we presented unsteady effects that led to the lift and thrust force generation on the flapping foil. It was assumed that the flow was two-dimensional. We demonstrated that flapping motion with low Reynolds produced well-ordered vorticity around the profile. The well-ordered vorticity atmosphere generates predictable aerodynamic force distribution on the profile. We demonstrated that high enough parameters of flapping motion caused the disordered vorticity distribution around the profiles that led to the loss of the lift force and limited the possibility of a flight.

1. Introduction

In contrast to classical airplanes, most insects and birds operate in low Reynolds number regime, below, 10^4 [1]. Such a flow is dominated by unsteady effects caused by vorticity dynamics. For low Reynolds number flows around the flapping foil one can observe the ordered fluid structures [2,3]. The ordered fluid structures behind the flapping profile create vortex atmosphere which can be divided into several types of vortex streets [6]. The type of vortex wake is related to aerodynamic force distribution on the profile [7]. Proper vortex wake generation is extremely important, especially for small insects which fly in very low Reynolds number flows (up to 500) [1,5]. In the paper we demonstrated three different flapping kinematics. First, we presented the formation of the vortex street and its transitions (into different types of vortex wake) for a simple plunging profile. We observed the whole family of vortex streets and created diagrams which linked the type of the observed vorticity field with the Reynolds, the Strouhal number and the plunging amplitude. The main purpose of our simulations was to investigate the flow phenomena related to transitions of the vortex wake and how this transition depends on the Reynolds number. The second part of our investigation was related to the simulation of flapping motion which is appropriate for small insects. It corresponds to the plunging and pitching profile [8]. We presented the dependence of aerodynamic forces calculated on the profile on the Reynolds number and different types of vorticity field. In the end, we simulated the free flapping motion. We demonstrated various profile trajectories for different flapping kinematics and different types of vortex wake behind the moving profile. We limited our research to two-dimensional flows. Although a flight in nature is three-dimensional, a two-dimensional simplification is widely used and seems to be appropriate to capture the essence of flapping flight aerodynamics [1,4,5,6]. For the study of a moving object in fluid we chose the vortex particle method – Vortex-in-Cell method. The importance of the VIC method lies in the possibility to analyze more directly the vorticity field since, in computations, particles that carry information about the vorticity field are used. The method was carefully tested for problems with known analytical solutions and well-documented experimental data.

2. Vortex particle method

The Navier-Stokes equations of the incompressible viscous fluid in primitive variables ($\boldsymbol{u}$,p) can be transformed to the Helmholtz equations in vorticity-stream function formulation. The Helmholtz equations in the non-inertial reference frame have the form

$$\frac{\partial(\omega+2\Omega)}{\partial t}+\boldsymbol{u}\cdot\nabla\omega=\upsilon\Delta\omega,\qquad \Delta\psi=-\omega \tag{1}$$

$$\boldsymbol{u}=\nabla\times(0,0,\psi). \tag{2}$$

where $\mathbf{u}=(u,v)$ means vector velocity, ψ stream function and υ viscosity. For our purposes, it was convenient to change the reference frame. The motion of the profile is made by the boundary condition for the stream function far from the body with specified angular Ω and translational U_0 velocity.

$$\psi(x,y)=U_0(ycos(\alpha)-xsin(\alpha))-\frac{\Omega}{2}(x^2+y^2) \tag{3}$$

where α means instantaneous angle of attack of the moving object in fluid and $U_0=|\boldsymbol{u}|$. A detailed description of the solution of the Helmholtz equations in the moving reference frame can be found in [9].

Numerical calculations of the moving profile were carried out by the vortex in cell method (VIC). In the calculation we used the numerical grid. To better fit the numerical grid to the solid boundary we used the conformal transformation. Through the conformal mapping the physical non-uniform flow region *(x,y)* is replaced by the rectangular one (ξ,η), where the calculations were performed very efficiently. The following conformal transformation was applied (Fig. 1)

$$x+iy=\cosh(\xi+i\eta). \tag{4}$$

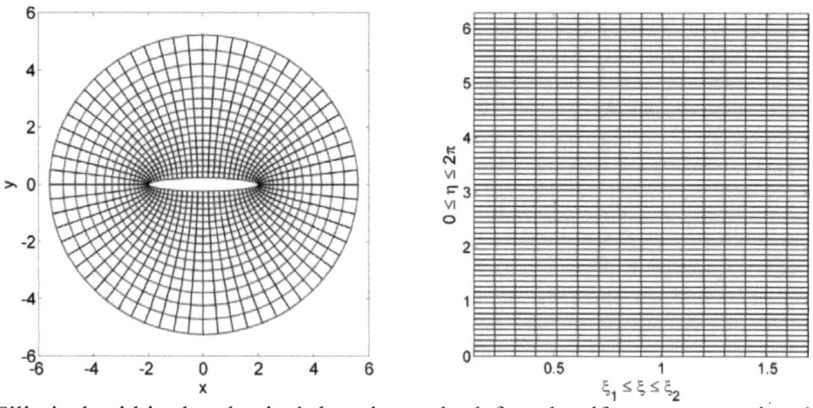

Figure 1. Elliptical grid in the physical domain on the left and uniform computational domain on the right.

In new variables the equations of motion have the form

$$\frac{\partial\omega}{\partial t}+\boldsymbol{u}\cdot\nabla\omega=\frac{\upsilon}{J}\Delta\omega,\qquad \Delta\psi=-J\omega, \tag{5}$$

where J denotes the Jacobian of the transformation.

The velocity field is obtained according to the formulas

$$u = \frac{1}{J}\frac{\partial \psi}{\partial \eta}, \qquad v = -\frac{1}{J}\frac{\partial \psi}{\partial \xi}, \tag{6}$$

In the vortex in cell method the continuos vorticity field is approximated with the discrete particle distribution. The calculation area is covered with the uniform grid $h = \Delta\eta = \Delta\zeta$ and in every grid node the particle with circulation: $\Gamma = \int_A \omega d\xi d\eta$ is placed, where $A = h^2$ and

$$\omega(\xi, \eta) = \sum_p \Gamma_p \, \delta(\xi - \xi_p)\delta(\eta - \eta_p) \tag{7}$$

The equations of motion were solved by means of viscous splitting algorithm [10]. First, the inviscid fluid motion equation was solved

$$\frac{\partial \omega}{\partial t} + \boldsymbol{u} \cdot \nabla\omega = 0. \tag{8}$$

The equation (8) proves that the vorticity is constant along the trajectory of the fluid particles. According to the Helmholtz theorem, vortex particles move in the same manner as material fluid particles [7,10], partial differential equation is replaced by the set of ordinary differential equations which describe the trajectory of the particles

$$u = \frac{\partial \xi}{\partial t}, \quad v = \frac{\partial \eta}{\partial t}, \quad \xi(0) = \alpha_1, \quad \eta(0) = \alpha_2 \tag{6}$$

where $\alpha = (\alpha_1, \alpha_2)$ denotes the Lagrangian coordinate of fluid particles. The number of particles equals the number of grid nodes. The set of ordinary equations (6) was solved by the four order Runge-Kutta method. The velocity field was obtained through the solution of Poisson's equation for the stream function and vorticity on the numerical grid (5). The velocities of the particles between the grid nodes were calculated by the two-dimensional bilinear interpolation.

In the second stage the viscosity was taken into account by the solution of the diffusion equation

$$\frac{\partial \omega}{\partial t} = \frac{v}{J}\Delta\omega \tag{7}$$

$$\omega(\xi, \eta, t^n) = \omega_E, \quad \omega|_{wall} = \omega_s \tag{8}$$

where ω_s denotes the vorticity at the solid boundary, ω_E means the Euler inviscid distribution of vorticity in time step t^n. The motion of the body in the fluid and the vorticity in the flow region generate non-zero tangent velocity component at the wall. The no slip condition is realized by putting a proper amount of vorticity at the wall. We used the Briley formula of the fourth order [11]. Nullification of the normal velocity component at the wall was performed by the constant stream function at the wall ψ = const. The diffusion equation was solved by the alternating direction implicit method (ADI).

In calculations we used the numerical grid, therefore, it was necessary to transfer the information from the vortex particle to the mesh nodes, Fig. 2. The redistribution of the vorticity from fluid particles to the mesh nodes was performed according to the formula

$$\omega_{ij} = \frac{1}{h^2}\sum_p \Gamma_p \varphi_h(\xi)\, \varphi_h(\eta) \tag{14}$$

where $\varphi_h(\cdot)$ denotes the interpolation kernel. The redistribution process plays a fundamental role in the accuracy of the vortex particle methods. In the current study we used the Z-splines interpolation functions [12], which, when applied to vortex particle methods have very good interpolation properties.

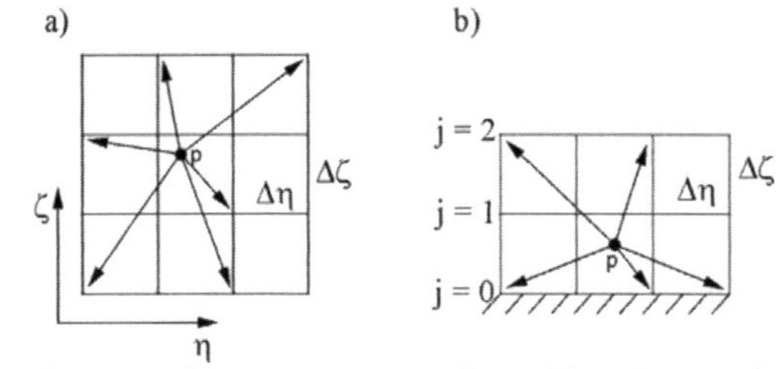

Figure 2. Redistribution of the particle masses onto the grid nodes, a) for particles lying inside the computational domain, b) for the particles in the vicinity of the wall.

3. Plunging simulation

We performed a numerical calculation for the simple plunging profile which moves according to the equation

$$y(t) = \frac{A_0}{2}\cos(2\pi f t), \tag{15}$$

where $y(t)$ denotes the instantaneous position of the profile center, A_0 is the amplitude and f denotes the frequency of the plunging. The profile's vertical velocity was calculated from the formula: $v = y'(t)$. We assumed that velocity of the fluid U_0 is constant far from the body, see figure 3. The flow over the plunging body can be characterized by the following set of non-dimensional parameters: The Reynolds number Re, the Strouhal number St and the plunging amplitude A_C, [1,2]:

$$Re = \frac{U_0 c}{\upsilon}, \quad St = \frac{fc}{U_0}, \quad A_C = \frac{A_0}{c}, \quad St_A = \frac{A_0 f}{U_0}, \tag{16}$$

where c is a chord of the profile and ν denotes a kinematic coefficient of fluid viscosity.

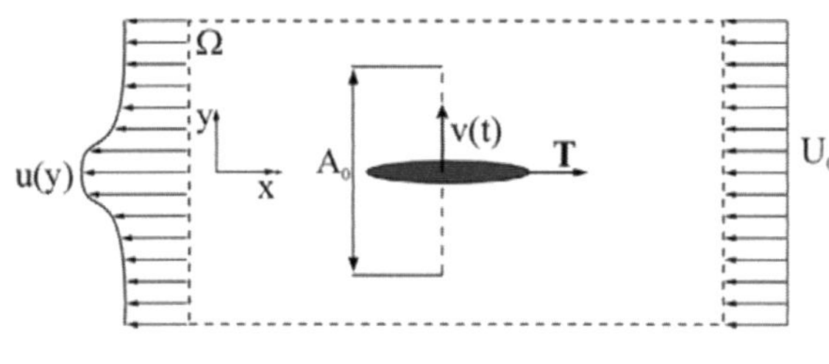

Figure 3. Scheme of the plunging ellipse immersed in the fluid.

The profile chord was set to $c = 2$, we assumed the thickness of the profile: $e = b/c = 0.2$. We performed numerical calculations for the Reynolds number $Re = 100$, $Re = 250$ and $Re = 500$, with the density: $\rho = 1$. The plunging frequency was fixed to $f = 0.5$ and the Strouhal number (St) was controlled by changing the free stream velocity U_0. The calculations were performed for dimensionless time $T = ft$, in the range of $T = (0, 10)$, which corresponds to ten periods in the equation (15). We used the elliptical grid shown in figure 1, with 256 grid nodes in a radial direction and 256 grid nodes in an azimuthal direction. In figure 4, the visualization of vortex wake around the profile was presented. If the amplitude and the Strouhal number were low, we observed a vortex Karman street (figure 1a).

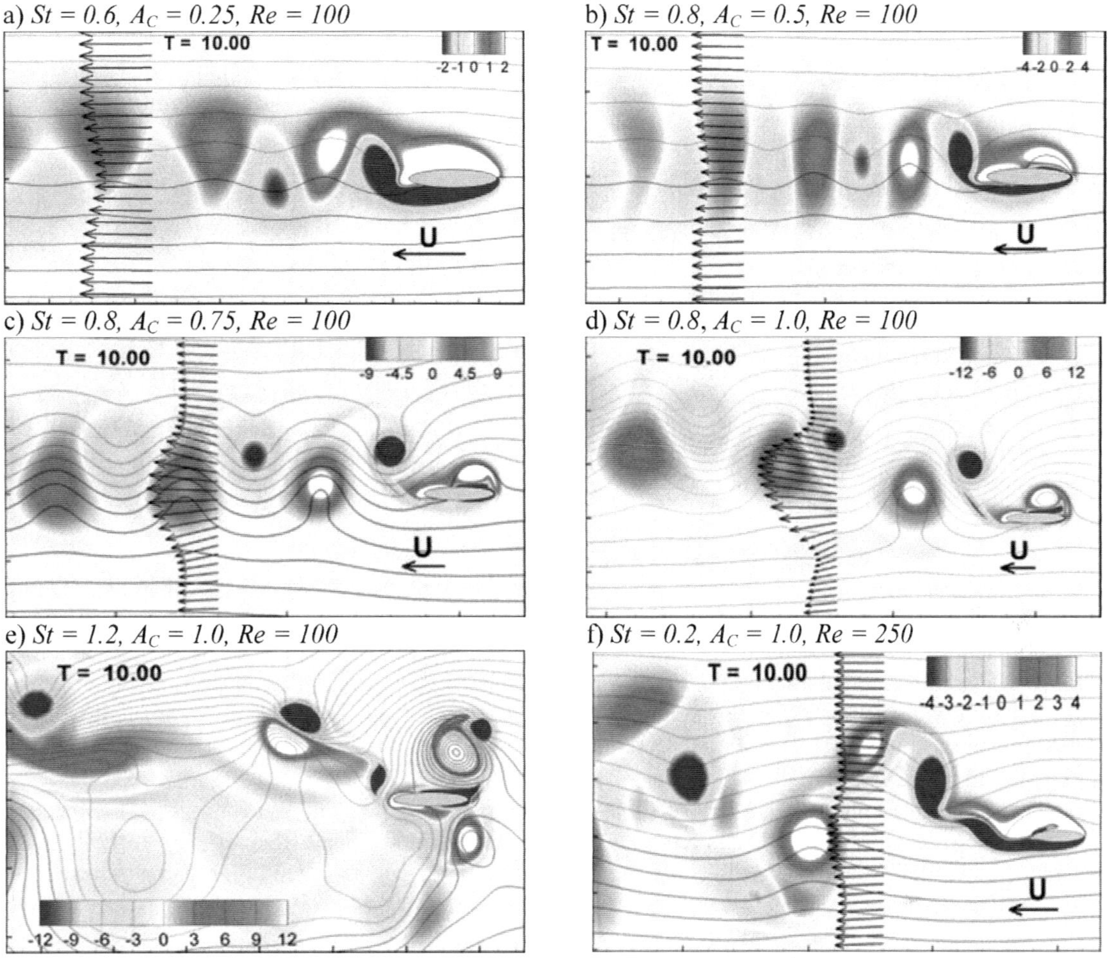

Figure 4. Vorticity field around the plunging profile. The arrows behind the profile denote the averaged velocity field.

A slight increase of the plunging parameters caused a change in the topology of the vortex street (figure 1b). This type of the vortex wake is called the aligned vortices. For proper plunging parameters we observed that vortices changed their positions in the vortex wake (figure 1c). This type of vorticity field increased momentum in the wake and generated thrust force on the profile. Further increase of the amplitude and the Strouhal number caused a deflection of the vortex street (figure 1d). Such a vorticity field generates thrust and lift force on the profile. The direction of the deflection depends on the first flap (upward or downward) and determines the lift force sign. If the flapping parameters are too high we observe the disordered vorticity field around the profile (figure 1e). The disordered vorticity field caused a random hydrodynamic force that acts on the profile. The last figure presents the vortex wake called 2P. Symbol 2P means that during a single foil motion (upward or downward) two vortex structures with the opposite circulation were created.

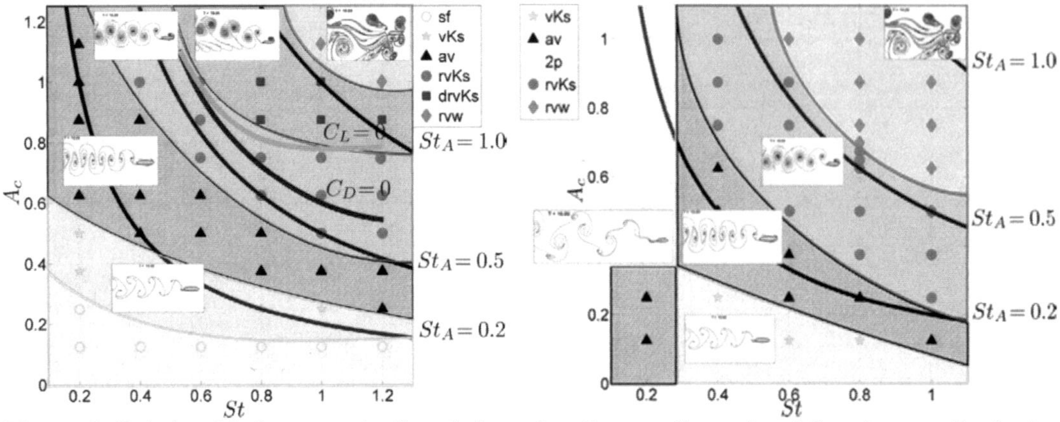

Figure 5. Relationship between the Strouhal number St, non-dimensional flapping amplitude Ac, and vortex topology in the wing wake, for the Reynolds number *Re = 100* on the left and for the Reynolds number *Re = 500* on the right.

We can summarize the above results in a phase transition diagram. In figure 5 we presented the relation between the vortex wake type and the plunging parameters. The phase diagram for the Reynolds number *Re = 500* can be found on the right. The increase in the plunging parameters caused vanishing of the stable deflected Karman vortex street. For these flapping parameters the deflection of the vortex wake was constantly changing. A similar effect for flapping with high Strouhal number: *St ~ 1* was observed in [13]. We showed in [6] that the effect which is responsible for changing the deflection is the boundary layer eruption phenomenon. A description of the discussed phenomenon can be found in [14].

The first part of our research was related to simple plunging motion. We found that for the low Reynolds number flows, flapping kinematics is very important to create the ordered vorticity field around the profile. The increase of the Reynolds number or improper flapping parameters caused the disordered vorticity field around the profile.

4. Pitching and plunging simulation

One of the most popular two-dimensional flapping kinematics is known as the incline hovering [1, 5, 15]. It may be described by the equations:

$$A(t) = \frac{A_0}{2}\cos(2\pi f t) \tag{17}$$

$$[x(t), y(t)] = A(t)[\cos(\beta), \sin(\beta)] \tag{18}$$

$$\alpha(t) = \alpha_0 + \alpha_A \cos(2\pi f t) \tag{19}$$

where $x(t)$ denotes the instantaneous position of the wing center, A_0: the plunging amplitude, f: the frequency, $\alpha(t)$ is the angle of attack measured relative to the horizontal line, α_0 denotes initial angle of attack, and α_A denotes pitching amplitude. The inclination of the stroke plane was set to: $\beta = 60^0$. The translation and angular velocities were given by the relation: $U_0 = \frac{dA(t)}{dt}$, $\Omega = \frac{d\alpha(t)}{dt}$. The Reynolds number (16) was calculated on the basis of maximal velocity: U_{max}. We assumed flapping parameters appropriate for small insects: $f = 0.5$, $\alpha_0 = \pi/4$, $\alpha_A = -\pi/4$, $A_C = 2.5$, $St = 0.12$. The time t was non-dimensionalised by the frequency of the motion: $T = ft$, therefore, $T = 1$ corresponds to one full stroke (upward and downward motion of the profile).

The hydrodynamic forces acting on the wing were calculated by means of the vorticity distribution on the surface of the profile

$$F_p = \upsilon\rho b \int_C \frac{\partial\omega}{\partial n} dC + \rho b A_b \frac{dU_0}{dt} \tag{20}$$

$$F_v = \upsilon\rho b \int_C \omega \mathbf{s^0}\, dC \tag{21}$$

where b denotes the width of the profile in the direction of perpendicular to the plane of flow, ρ – density of the fluid, **n** and $\mathbf{s^0}$ are normal and tangential unit vectors respectively, A_b denotes the area of the body. The last term on the right side of the equation (20) represents the inertial force of the fluid displaced by the profile. The forces obtained from the equations (20), (21) were decomposed into horizontal F_D and vertical F_L components, which in turn, correspond to the drag and lift forces. The force coefficients were calculated according to the relations

$$C_D = \frac{F_D}{\frac{\rho}{2}U_0^2 cb}, \quad C_L = \frac{F_L}{\frac{\rho}{2}U_0^2 cb}. \tag{22}$$

The visualization of one stroke is presented in figure 6. After the first four strokes the vorticity field was ordered and generated periodically. It can be noticed that the role of the flapping kinematics is to create a vortex pair structure and get rid of it [15]. The vortex pair structure moves due to self-induction. In the range of the Reynolds number below 125, it causes the removal of vorticity generated in the previous stroke far from the profile. Further, such a vortex pair creates the vortex wake. To check the generation of the vortex pair structures we performed numerical calculations in the wide range of the Reynolds numbers.

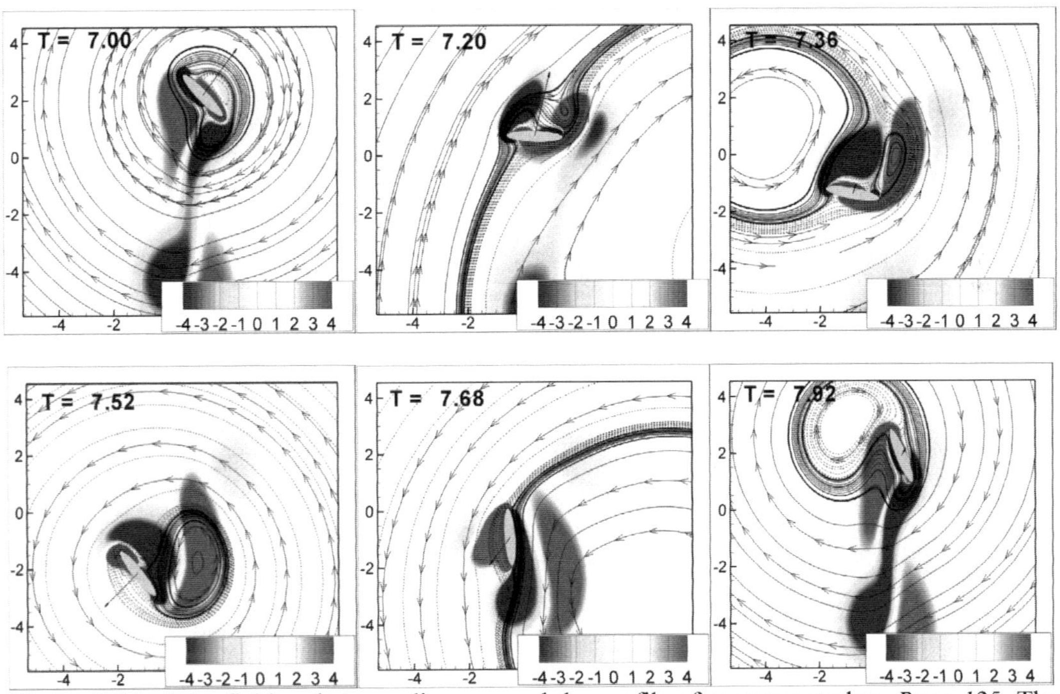

Figure 6. Vorticity field and stream lines around the profile after seven strokes, *Re = 125*. The stream lines are plotted in coordinates fixed to the moving profile.

In figure 7 the comparison of the vorticity field generated around the profile for different Reynolds numbers was presented. We noticed that the creation of the vortex pair did not occur for the Reynolds number below *Re = 40* (in the flow over the cylinder the Karman vortex wake created from a vortex pair occurred for Re > 49). With the increase of the Reynolds number, the vorticity field became disordered due to the interaction of the profile with random vortex structures.

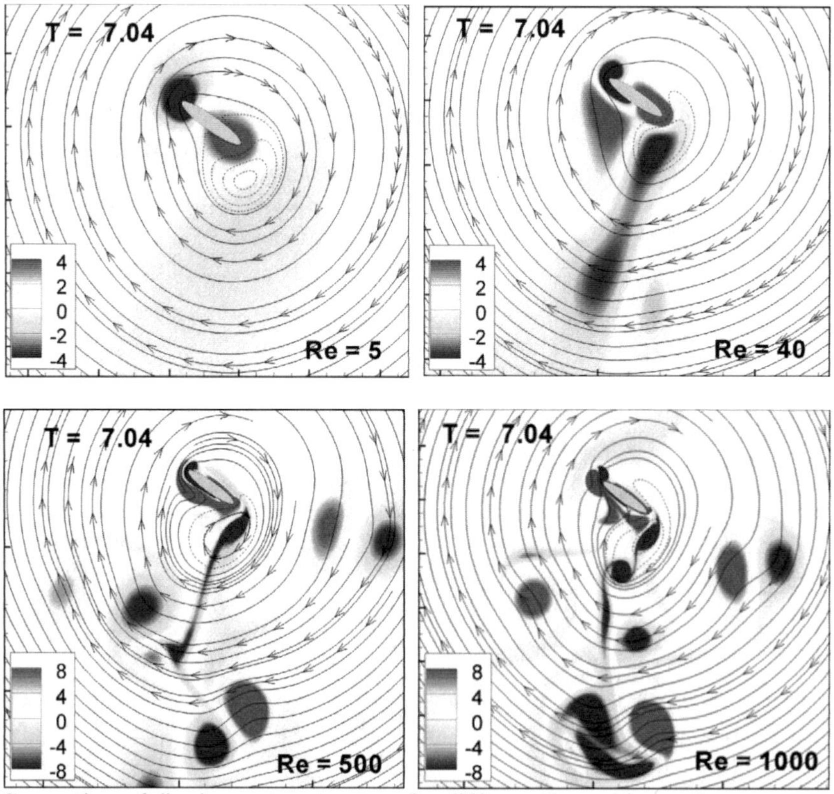

Figure 7. Generation of dipolar vortex structure for different Reynolds number.

To check the influence of the vorticity field around the profile for aerodynamic forces we calculated the lift force on the profile. In figure 8 we presented the lift force coefficient (22) for different Reynolds numbers. One can observe that for the Reynolds number: $Re = 125$ the generation of the aerodynamic forces was periodic. If the Reynolds number increased the characteristics of the lift force coefficient were disordered. It is difficult to estimate the difference in aerodynamic force creation from the lift force characteristic. To be able to compare the lift force generation for different Reynolds numbers we calculated the average lift force coefficient

$$C_{Lsr} = \frac{1}{T}\int_0^T C_L(t)dt. \tag{23}$$

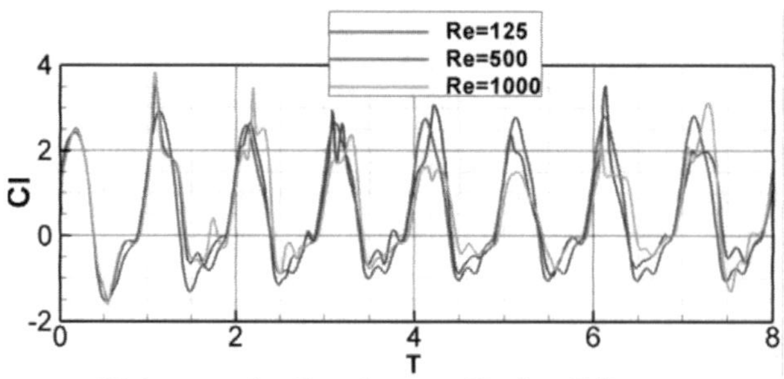

Figure 8. The lift force coefficient on the flapping profile for different Reynolds numbers.

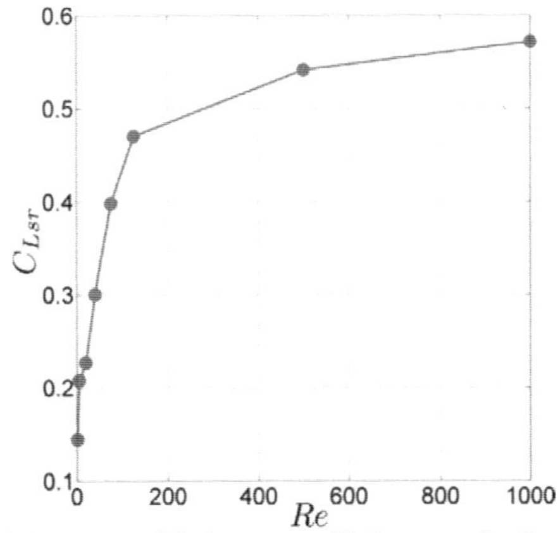

Figure 9. The dependence of the mean lift force coefficient on the Reynolds number.

The average lift force characteristic is presented in figure 9. Our calculations prove that the highest growth of the average lift force coefficient was in the range of the Reynolds number: $5 < Re < 125$. The increase of the Reynolds number from *125* up to $Re = 1000$ caused only 18% gain of the average lift force coefficient. We believe this is related to the disordered vorticity around the profile with higher Reynolds number flows.

5. Free flapping flight

To check in what way the disordered vorticity around the profile can affect the trajectory of the real flight we simulated a free flapping flight. The profile was replaced by the mass point which can move according to Newton's second law

$$m[\ddot{x}, \ddot{y}] = [0, -g] + \boldsymbol{F_{fluid}} + \boldsymbol{F}, \tag{24}$$

where m denotes the mass of the profile, $\boldsymbol{F_{fluid}}$ was calculated from the equations (20) and (21), and $\boldsymbol{F}$ denotes the driving force which is necessary to move the profile. The driving force was calculated on the basis of the equations (17), (18). We calculated the mass of the profile assuming the density of the profile: $\rho = 18 kg/m^3$. The equation (20) was solved by the second order Runge-Kutta method. Parameters of the flapping kinematics were the same as in section 4, however, the frequency of the motion was modified. We noticed that we can simply control the flight by a small modification of the angle α_0 in the equation (19).

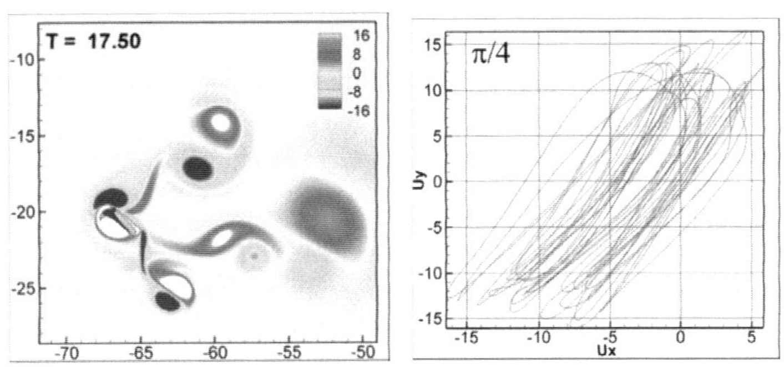

Figure 10. Vorticity field generation in a free flapping flight on the left and a phase diagram for the instantaneous profile velocity on the right, frequency: $f = 1.2$. Profile began from the origin.

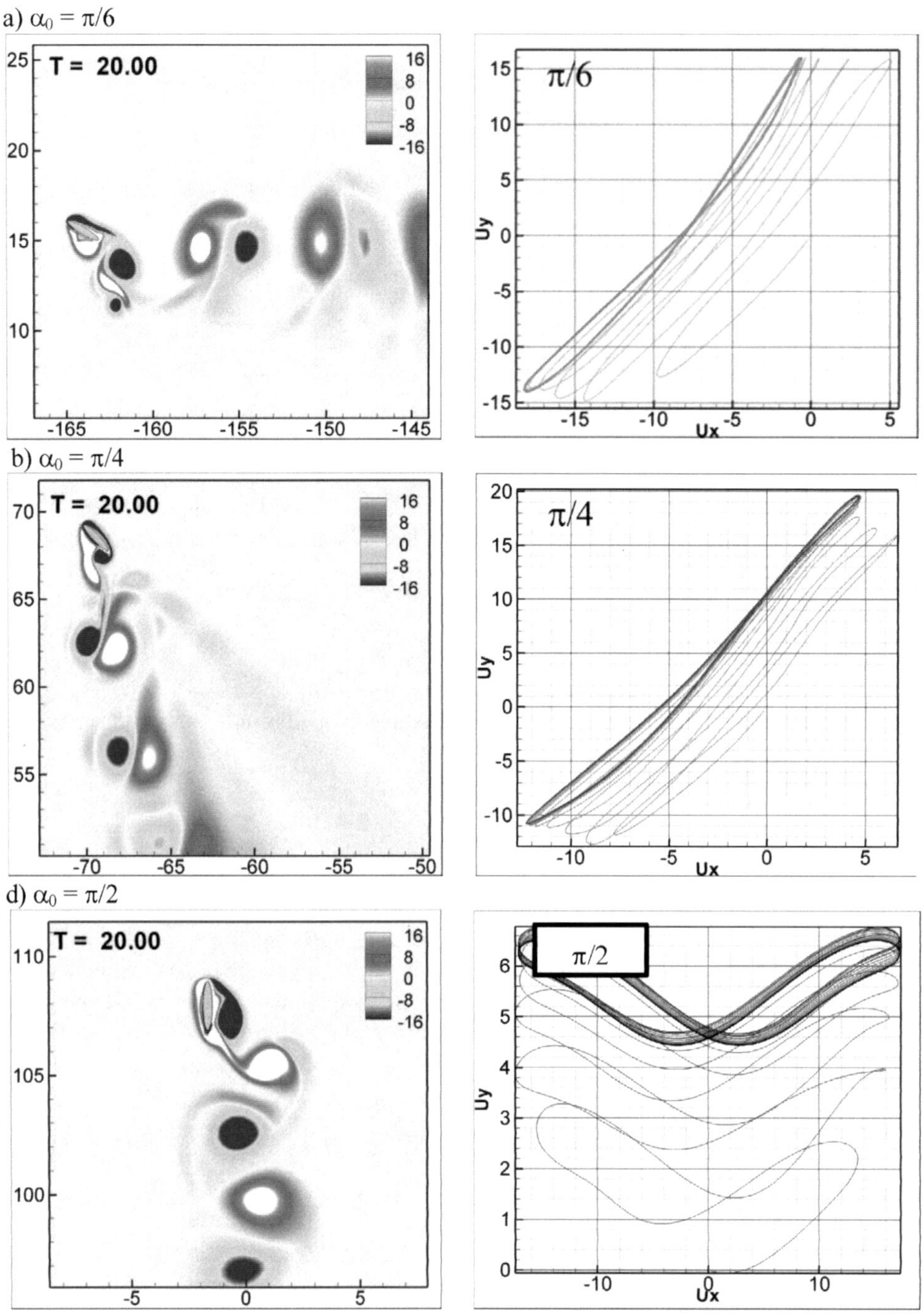

Figure 11. Vorticity field generation in a free flapping flight on the left and phase diagram for instantaneous profile velocity on the right, frequency: $f = 1.4$. Profile began from the origin.

In figures 10 and 11 the vorticity field generated by the free flapping profile and the instantaneous velocity of the profile were presented. It is well known that all birds and insects have some minimal frequency of flapping motion which is necessary to maintain in the air [16]. If this frequency is too low, we observed the disordered vorticity field and noticed that in such conditions a stable flight is not possible (see figure 10). For proper parameters of flapping motion we can observe that after a few strokes the vorticity field is generated periodically (figure 11). In figure 12 the trajectory of the flapping profile was presented. We can check how the angle of climb changes with a parameter α_0. Moreover, one can notice that the direction of the vortex wake deflection altered with the increase of the angle of climb (see figure 12a and b). It is worth noting that, although we set the plunging amplitude $A_0 = 5$, the flapping amplitude was lower due to the resistance of the motion, see figure 12.

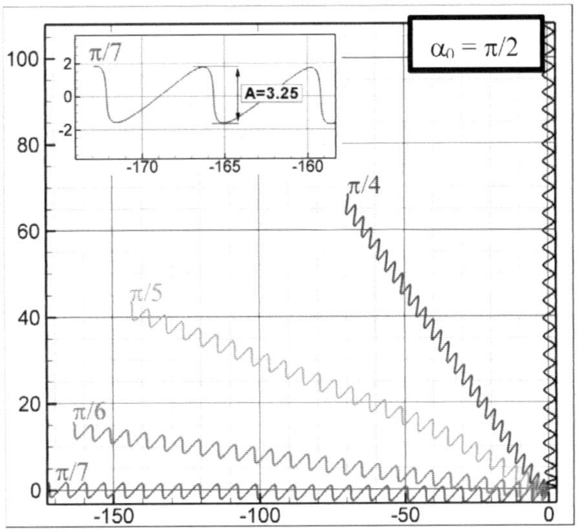

Figure 12. Trajectory of the free flapping profile.

6. Conclusion

In the paper the simulations of the flapping motion for low Reynolds number were presented. Due to a large number of fluid mechanics phenomena related to moving objects in fluid, we limited our research to two-dimensional flows. Despite the two-dimensional simplification the aerodynamic of flapping motion is very rich and full of vortex structures creation, and their interaction. All fluid mechanics phenomena depend strongly on the dynamic of vorticity. We showed how the vortex pair structures generated by the flapping profile form the vortex wake around the profile. For plunging motion of the profile we observed the whole family of different types of vortex wake. We built transition diagrams parameterised with the Strouhal number and the dimensionless plunging amplitude. For appropriate parameters of plunging motion we observed the ordered vorticity field around the profile. We noticed that for higher Reynolds number flows the generation of the stable lift force might be difficult. This being caused by the disordered vorticity field. Additionally, we discovered that the disordered vorticity field around the flapping profile led to the disordered, unpredictable aerodynamic forces. Through the simulation of a free flapping flight we proved that in such circumstances the stable flight may not be possible.

References

[1] Shyy W., Lian Y., Tang J., Viieru D., Liu H. Aerodynamics of Low Reynolds Number Flyers *Dover Publications*, *Cambridge University Press*, 2008
[2] Godoy-Diana R., Aider J-L., Wesfreid J. E. Transitions in the wake of a flapping foil, *Physical Review* E, 77, 2008
[3] Lentink D., Muijres F.T., Donker-Duyvis F.J., van Leeuwen J.L.,Vortex-wake interactions of a flapping foil that models animal swimming and flight, *J. Exp. Biol.* 211, 267-73, 2008

Journal of Physics: Conference Series **760** (2016) 012010 doi:10.1088/1742-6596/760/1/012010

[4] Wang, Z. J., Birch, J. M., Dickinson, M. H.: Unsteady forces and flows in low Reynolds number hovering flight: two-dimensional computations vs robotic wing experiments, *J. Exp. Biol.*, **207**, 449-460, 2007
[5] Wang Z. J., The role of drag in insect hovering, *J. Exp. Biol.*, **207**, 4147-4155, 2004
[6] Kozłowski T., Kudela H., Transitions in the vortex wake behind the plunging profile, *Fluid Dyn. Res.*, **46**, 2014, 061406
[7] Wu J. C., Theory for aerodynamics force and moment in viscous flows, *AIAA*, **19**, 432-441, 1981
[8] Wang. Z. J., Two dimensional mechanism for insect hovering. *Phys. Rev. Lett.*, **85** (10) 2000
[9] Gustafson K. E., Leben R., Freymuth P., *Vortex methods and vortex motion*, Society for Industrial and Applied Mathematics, 143-169, 1991
[10] Cottet G.-H., Koumoutsakos P. D., *Vortex Methods: Theory and Practice*, Cambridge University Press, 2000
[11] Koumoutsakos P., Leonard A., Pepin F., Viscous boundary conditions for vortex methods, *J. Comput. Phys.* **113**, 52-56, 1994
[12] Becerra Sagredo, Moment conserving Cardinal Splines Interpolation of Compact Support for Arbitrarily Spaced Data, *Research Report No. 10*, Zurich, Switzerland, 2003
[13] Lewin G. C., Haj-Hariri H., Modelling thrust generation of a two-dimensional heaving airfoil in a viscous flow, *J. Fluid. Mech.* **492**, 339- 362, 2003
[14] Kudela H., Malecha Z. M., Eruption of a boundary layer induced by a 2D vortex patch, Fluid Dyn. Res., **41**, 2009, 055502
[15] Wang Z. J., Two dimensional mechanism for insect hovering, *Physical Review Letters*, **85** (10), 2000
[16] Ellington C. P., The novel aerodynamics of insect flight: Application to micro-air vehicles. *J. Exp. Biol.*, **202**, 3439-48, 1999

Influence of a strong magnetic field on paramagnetic fluid's flow in cubical enclosure

A Kraszewska[1], **L Pyrda**[1] **and J Donizak**[1]

[1]AGH University of Science and Technology, Faculty of Energy and Fuels, Al. Mickiewicza 30, 30-059 Kraków, Poland

E-mail: kraszka@agh.edu.pl

Abstract. The fluid behaviour in thermo-magnetic convection of paramagnetic fluid in a strong magnetic field was studied. The fluid was 50% volume aqueous solution of glycerol with an addition of gadolinium nitrate hexahydrate ($Gd(NO_3)_3{\cdot}6H_2O$). Experimental enclosure – a vessel with aspect ratio (AR=height/width) equal to 1.0 – was heated from the bottom, and cooled from the top. Temperature difference between top and bottom walls was kept constant at ΔT = 5 and 11 [°C]. The magnetic induction was increased stepwise from 1 to 10 [T] and thermocouples placed inside the enclosures measured temperature changes of the fluid. On the basis of temperature measurements, analysis of the fluid flow was performed.

1. Introduction

Thermo-magnetic convection of a paramagnetic fluid – convective motion of a magnetic fluid with an external magnetic field present – is in scientists field of interest since superconducting magnets could be produced with relatively low costs, that is since superconductors working above boiling point of liquid nitrogen were invented. Braithwaite [1] studied influence of a non-uniform magnetic field on a paramagnetic fluid convection and presented mathematical description of forces acting in such a system. Huang and Edwards [2] presented research about intensification and suppression of convective motion in thermo-magnetic convection and Fornalik [3] published thermodynamic description of phenomena occurring in thermo-magnetic convection. Another field of interest, besides intensification and suppression of heat transfer in weakly magnetic fluids, is analysis of the flow in thermo-magnetic convection [4].

One of the ways to "look inside" a convection phenomenon is by a spectral analysis. An interest in a distribution of passive scalar quantities, such as temperature, was born with a development of metrology and a need to better understanding of turbulence mechanisms. One of the first works published in this field was by Obuchow [5], who applied Kolmogorow's local isotropy concept to describe temperature distribution in atmosphere. A decade later Batchelor [6] conducted an analysis of two extreme cases – one where momentum diffusion coefficient was much lower than heat diffusion confident and the opposite one. For the first instance he concluded that spectral function of temperature is located in inertial-convective subregime. In the second case convection subregime was more extensive and it depended more on fluid viscosity. He derived a spectrum correlation for this subregion which is fitted to a reverse wave number.

In presented work Authors conducted an experimental analysis of a magnetic field influence on a paramagnetic's fluid convection and performed spectral analysis on temperature signal changes during measurements.

2. Experimental stand

2.1. Experimental enclosure

For the purpose of analysing the magnetic field influence on a paramagnetic fluid flow, experimental research was conducted. As a geometry, cubical enclosure of dimensions 0.032x0x032x0.032 [m], shown in Figure 1, was chosen. It consisted of two thermally active walls – the bottom one built from a copper plate heated with a nichrome wire connected to a DC supply and the top one with built in cooling chamber. Six thermocouples were inserted at all times into the thermally active walls to measure the temperature. The enclosure was made from Plexiglas and in a front vertical wall six small holes were made and thermocouples were inserted through the wall into the fluid to measure temperature changes during experimental procedure. The experimental setup worked in Rayleigh-Benard configuration, and two temperature differences were applied – 5 and 11 [°C] with the top cooled wall temperature set to [18 °C] which was the ambient temperature in the magnet working section.

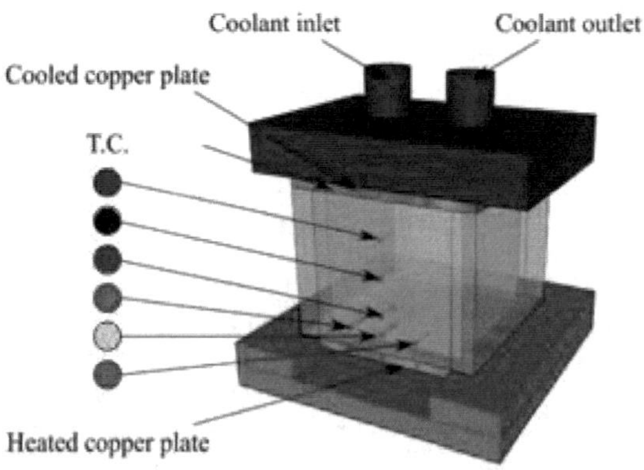

Figure 1. Schematic view of experimental enclosure with positions of the thermocouples.

2.2. Position in the magnet

In natural convection fluid motion is a result of gravitational force $\mathbf{f}^{gr}$ acting on a fluid particles:

$$\mathbf{f}^{gr} = \mathbf{g}\rho \quad (1)$$

Because density ρ of the fluid depends on a temperature T, gravitational force acts with different force on a particles with various temperature creating a gravitational buoyancy force $\mathbf{f}^{G}$:

$$\mathbf{f}^{G} = \mathbf{g}(\rho - \rho_0) \quad (2)$$

where $\mathbf{g}$ is gravity acceleration and ρ_0 is a reference density. Noting the Boussinesq model, which assumes that the density of the fluid is constant except the equation for buoyancy force, and that it can be applied only to small temperature differences, fluid density can be expressed as:

$$\rho \simeq \rho_0 \left(1 - \beta (T - T_0)\right) \quad (3)$$

where β is thermal expansion coefficient and T_0 is a reference temperature. Using the above relation to eq. (2), gravitational buoyancy force can be stated as:

$$\mathbf{f}^{G} = -\mathbf{g}\rho_0 \beta (T - T_0) \quad (4)$$

The effect of a gravitational buoyancy force is a downward cold fluid motion, while the hot fluid particles are lifted.

A force, acting on a paramagnetic fluid in a magnetic field can be described as [7]:

$$\mathbf{f}^{mg} = \mu_0 (\mathbf{M} \cdot \nabla)\mathbf{H} \quad (5)$$

where M is magnetization, H is the intensity of the magnetic field and μ_0 is a permeability of the vacuum. Noting that a physical value speaking about magnetic properties of the fluid is magnetic susceptibility χ:

$$\chi = \frac{\mathbf{M}}{\mathbf{H}} \tag{6}$$

Equation (5) can be rearranged into:

$$\mathbf{f}^{mg} = \mu_0 \chi (\mathbf{H} \cdot \nabla)\mathbf{H} = \frac{\mu_0 \chi}{2} \nabla \mathbf{b}^2 \tag{7}$$

Also knowing that magnetic induction **b** depends on an intensity of magnetic field and on a magnetic permeability of the material and that for a paramagnetic fluid $\chi << 1$ eq. (7) can be transformed into:

$$\mathbf{f}^{mg} = \frac{\chi}{2\mu_0} \nabla \mathbf{b}^2 \tag{8}$$

Using above relations, Curie's law and Boussinesq approximation, Tagawa [8] developed an equation describing magnetic buoyancy force:

$$\mathbf{f}^{M} = -\left(1 + \frac{1}{\beta T_0}\right) \frac{\chi_{m0} \rho_0 (T - T_0)}{2\mu_0} \nabla \mathbf{b}^2 \tag{9}$$

where χ_{m0} is mass magnetic susceptibility at reference temperature.
Equation (5) concludes that fluid, which temperature T is lower than the reference temperature T_0, is attracted to a square of magnetic field gradient $\nabla \mathbf{b}^2$. Taking into account both the gravitational and magnetic buoyancy forces, experimental enclosure was placed in the superconducting helium free magnet (HF10-100VHT-B *Sumito Heavy Industries Ltd.*) in an upper half position where the square of the magnetic field gradient is highest and intensification of fluid motion could be obtained.

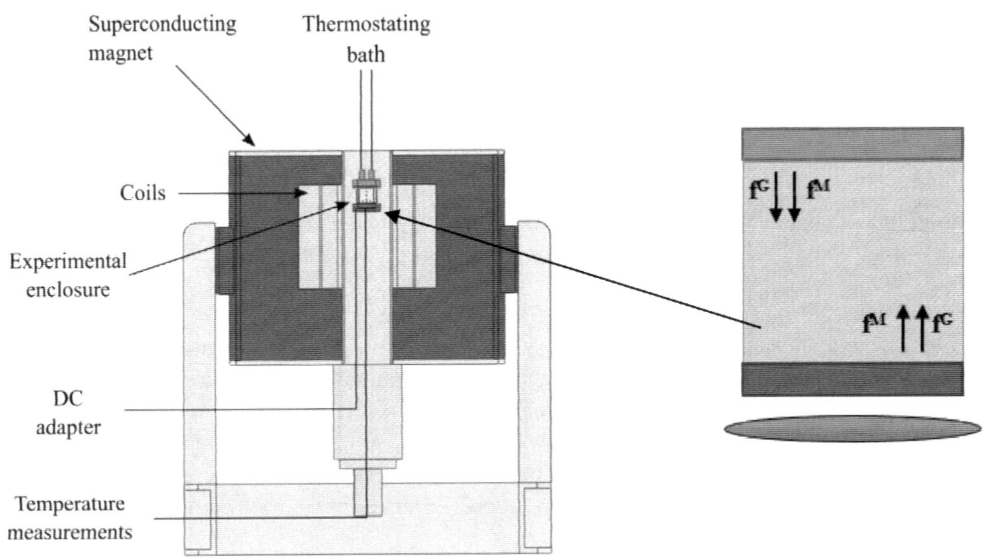

Figure 2. Schematic view of experimental setup (left) and configuration of the gravitational and magnetic buoyancy forces acting on a fluid (right).

2.3. Working fluid
A fluid chosen for the experiments was a non-electrically conducting 50% volume solution of distilled water and glycerol with an addiction of 0.8 [mol/kg_of_solution] concentration of gadolinium(III) nitrate hexahydrate ($Gd(NO_3)_3 \cdot 6H_2O$) to make it paramagnetic. Properties of the fluid are listed in Table 1 and were measured experimentally.

Table 1. Properties of the working fluid at 25 [°C].

Property	Symbol	Value	Unit
Heat capacity	c_p	$2.92 \cdot 10^{3}$	[J/kg·K]
Thermal diffusivity	α	$9.13 \cdot 10^{-8}$	[m²/s]
Thermal expansion coefficient	β	$4.78 \cdot 10^{-4}$	[1/K]
Dynamic viscosity	μ	$1.30 \cdot 10^{-2}$	[kg/m·s]
Thermal conductivity	λ	0.376	[W/m·K]
Kinematic viscosity	ν	$9.25 \cdot 10^{-6}$	[m²/s]
Density	ρ	1411	[kg/m³]
Mass magnetic susceptibility	χ_{m}	$2.39 \cdot 10^{-7}$	[m³/kg]

2.4. Experimental procedure

Because of the way a Nusselt number was calculated in this study, the first step in experimental analysis was to measure the heat losses of the experimental enclosure. To do that, experimental enclosure was filled with water, rotated 180° from the Rayleigh-Benard configuration, so the bottom wall was cooled and the top one heated, and placed in the designed position in the magnet's working section. This allowed, after the chosen temperature difference was set, to obtain a temperature stratification in the fluid and then calculation of the heat losses of the system, which is described in section 3.1

Then the main experiment took place. The experimental enclosure was rotated back to Rayleigh-Benard configuration and temperature difference was set and the setup was left to stabilize, which took about two hours. Then the first measurement took place – the neutral case – natural convection in the system, without external magnetic field present. The measurement was a 15-min record of temperature changes of the fluid. Next step was connected with thermo-magnetic convection - magnetic field was applied - and because magnetic buoyancy force was acting toward intensification of fluid motion, the voltage and current supplied to the heater had to be corrected to maintain chosen temperature difference between horizontal walls. After stabilization of the setup, measurement was made. The same procedure took place for magnetic inductions from 1 [T] to 10 [T] with s step of 1 [T].

3. Analysis of the thermocouples signals

Temperature signals obtained during experimental part can be analysed two ways – one to acquire information about heat transfer in the system and two – to calculate spectral functions.

3.1. Heat transfer analysis

To analyse an influence of a thermo-magnetic convection on a heat transfer in the system, a method proposed by Ozoe and Churchill [9] was utilized to calculate a Nusselt number, which is ratio of net convection heat flux $Q_{\mathrm{net\_conv}}$ and net conduction heat flux $Q_{\mathrm{net\_cond}}$:

$$\mathrm{Nu} = \frac{Q_{\mathrm{net\_conv}}}{Q_{\mathrm{net\_cond}}} \tag{10}$$

Where the net conduction heat flux is:

$$Q_{\mathrm{net\_cond}} = Q_{\mathrm{cond}} - Q_{\mathrm{loss}} \tag{11}$$

and the net convection heat flux is:

$$Q_{\mathrm{net\_conv}} = Q_{\mathrm{conv}} - Q_{\mathrm{loss}} \tag{12}$$

In proposed research it was assumed that the heat losses depend only on the temperature of the heated wall – hence the conduction state measurements. The heat losses were estimated from:

$$Q_{\mathrm{loss}} = Q_{\mathrm{cond}} - Q_{\mathrm{Fourier's\_law}} \tag{13}$$

where theoretical conduction through the fluid was calculated from Fourier's law:

$$Q_{\text{theor\_cond}} = \frac{a^2 \lambda \Delta T}{d} \tag{14}$$

And since a=d=0.032 [m] (14) takes form:

$$Q_{\text{theor\_cond}} = a\lambda\Delta T \tag{15}$$

Heat flux was calculated for conduction area of 0.032 [m] x 0.032 [m]. The estimated heat loss was approximated linearly:

$$Q_{loss} = 0.08 \cdot \Delta T \tag{16}$$

Applying equations 11, 13 and 14 to equation 10, Nusselt number can be expressed as:

$$\text{Nu} = \frac{Q_{\text{conv}} - Q_{\text{loss}}}{a\lambda\Delta T} \tag{17}$$

where:

$$Q_{conv} = UI \tag{18}$$

where U is electric voltage and I is current.

3.2. Spectral analysis

Temperature signals, which were recorded during experiments, allowed examination of the flow structure. This was obtained through Fast Fourier Transform (FFT), which calculates discrete Fourier transform (DFT) of a sequence:

$$F_n = \sum_{i=0}^{N-1} x_i e^{-\frac{2\pi i}{N} ni} \tag{19}$$

where x_i is a sequence of length N, and F_n is its discrete Fourier transform.

Calculated results, in the form of amplitude versus frequency, were used to analyse the fluid behaviour in thermo-magnetic convection.

Spectral functions of a scalar field (i.e. temperature), in general, are very useful tool to analyze a turbulent transport mechanisms. So, with an assumption that the turbulence is homogenous, power density P_{xx} or spectrum can be calculated with the utilization of FFT:

$$P_{xx}\left(e^{j\omega}\right) = \sum_{m=-\infty}^{\infty} r_{xx}(m) e^{-j\omega m} \tag{20}$$

and with the use of Peridogram method, which estimates the power as time-integral square amplitude from amplitude obtained with FFT:

$$\text{Power density (TISA)} = \frac{\Delta t\left(\text{Re}^2 + \text{Im}^2\right)}{n} \tag{21}$$

In general, temperature spectral functions depend on the energy dissipation, thermal diffusivity, kinematic viscosity and temperature. But in some ranges, spectral function does not depend on the diffusion processes, and therefore does not rely on kinematic viscosity and thermal diffusivity. This subrange is called intertial-convective [10] and the spectral function has an inclination of wave number k with -5/3 exponent. When thermal diffusivity becomes more significant, the spectral function has an inclination of reverse wave number, and this subrange is called viscous-diffusive [10].

4. Results

Figure 3 presents results of spectral analysis for ΔT=5 [°C] as a amplitude versus frequency and power spectrum diagrams. For natural convection case ($|b_0|_{max}$=0 [T]) power spectrum shows a partial match to -5/3 slope for low frequencies 0.01-0.05 [Hz] and a flat power spectrum for frequencies higher than 0.05 [Hz] which is specific to regular and stable flows. For this case two characteristic frequencies can be observed 0.00706 [Hz] and 0.00823 [Hz]. Applying magnetic induction to the system causes significant changes in flow character. For $|b_0|_{max}$=3 [T] only one characteristic frequency appears – 0.07411 [Hz] and power spectrum shows a good fit to viscous-diffusive subrange for small frequencies. Further increase of magnetic field causes rapid change in fluid behaviour. For magnetic induction of 10 [T] signals from thermocouples show high and irregular oscillations, indicating that magnetic field results in flow destabilization. For those cases power spectrum points towards intertial-convective regime of the flow.

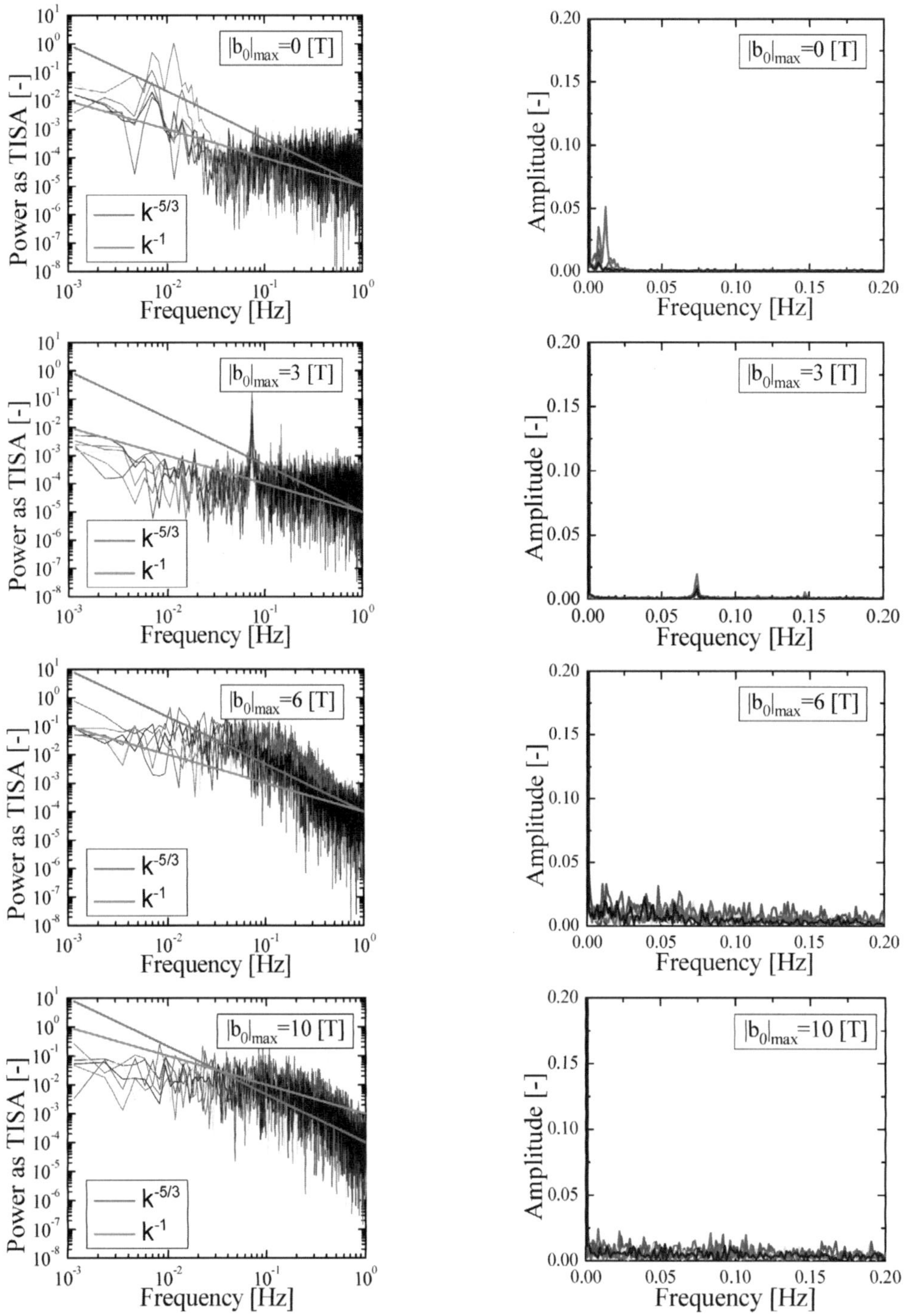

Figure 3. Power spectrum (left) and amplitude versus frequency (right) diagrams for ΔT=5 [°C].

Figure 4 presents results of spectral analysis for ΔT=11 [°C]. Natural convection case exhibits flat power spectrum indicating that the flow is stable and regular. Frequency analysis shows that only one characteristic frequency with small amplitude appears for that case – 0.00335 [Hz].

Journal of Physics: Conference Series **760** (2016) 012011 doi:10.1088/1742-6596/760/1/012011

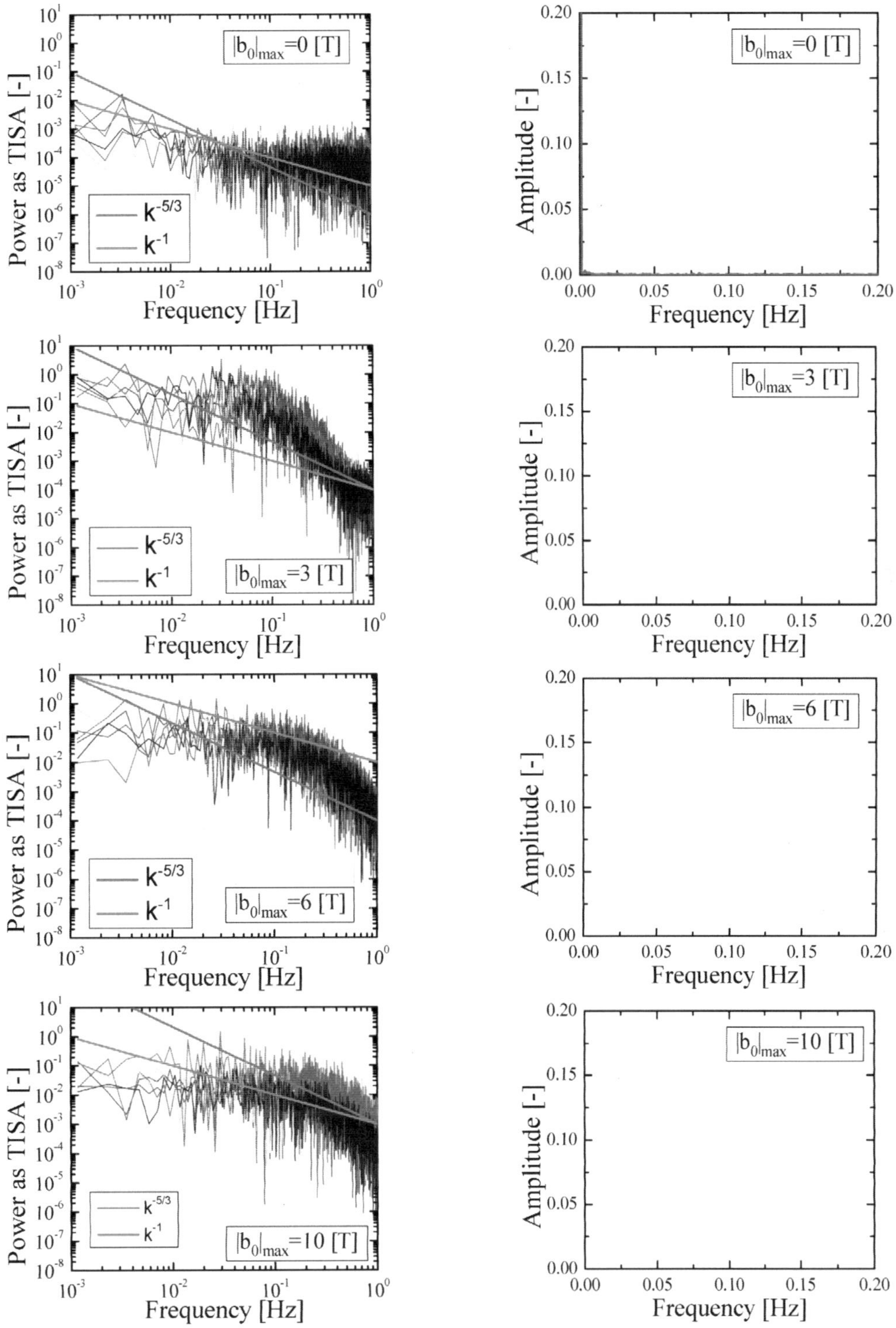

Figure 4. Power spectrum (left) and amplitude versus frequency (right) diagrams for ΔT=11 [°C].

A rapid change in flow behaviour appears for smaller magnetic induction that in the previous case. For magnetic induction 3 [T] in the centre of the magnet flow changes significantly. Power spectrum displays sharp inclination with steeper slope than -5/3. Many characteristic frequencies with relatively high amplitudes for this case point toward chaotic flow and transitioning regime. Further increase of magnetic induction indicates that flow is in viscous-diffusive subregime for small frequencies and in inertial-convective subregime for frequencies up to 1 [Hz].

Figure 5 presents results of heat transfer analysis as a Nusselt number ratio Nu/Nu_0 where Nu_0 is a Nusselt number value for a neutral case – without magnetic induction applied to the system. Intensification of heat exchange in the system is clearly seen. For small values of magnetic induction ($|b_0|_{max}=1$ [T]) Nusselt number value is about 3-5% higher than for natural convection. For maximal values of magnetic inductions, intensification of heat transfer reaches 290% for $\Delta T=11$ [°C] and over 310% for $\Delta T=5$ [°C]. This is caused by intensification of fluid motion with the magnetic induction increase, as was explained in section 2.2.

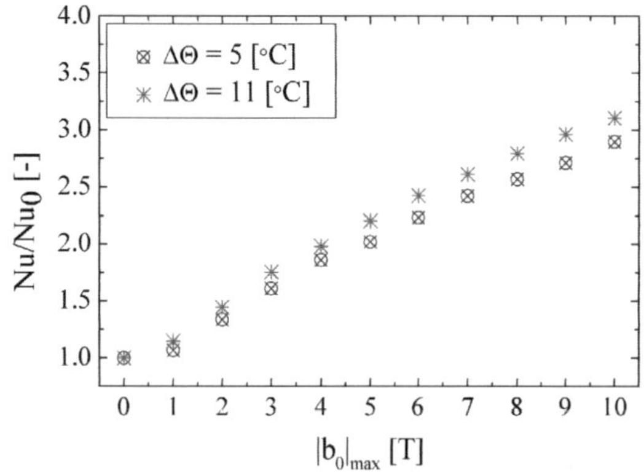

Figure 5. Ratio of Nu/Nu_0 versus magnetic induction in the centre of the magnet.

5. Summary

Spectral analysis of magnetic field, with different magnetic inductions, influence on natural convection was conducted for two temperature differences between thermally active walls. Presented research allowed to draw following conclusions: a) for natural convection in both temperature differences temperature field is stable state – a flat power spectrum can be observed; b) applying magnetic induction to the system significantly changes character of the flow; c) for temperature difference of 5 [°C] changes in flow behaviour appear for magnetic inductions higher than 3 [T] and for high magnetic inductions flow is in viscous-diffusive subregime; d) for higher temperature differences destabilization of the flow field appears for smaller magnetic induction values – 3 [T] and power spectrum functions have inclination sharper than -5/3; e) magnetic induction applied to convection of paramagnetic field, where magnetic field source is below experimental setup, causes intensification of fluid motion and heat transfer in the system; f) for values of $|b_0|_{max} > 5$ [T] intensification of heat transfer is more than 200%; g) Fast Fourier Transform is a good tool in analyzing temperature signals from the flows and h) spectral analysis allows an insight into phenomena occurring during thermo-magnetic convection processes.

Acknowledgment

This work was supported by the National Science Centre (Project No. 12/07/B/ST8/03109).

References

[1] Braithwaite D, Beaugnon E and Tournier R 1991 *Nature* **354** 134–6
[2] Huang J and Edwards B 1998 *Phys. Rev. E* **57** 29–31
[3] Fornalik E 2009 *Magnetic convection of paramagnetic fluid in an enclosure.* (Kraków: AGH Uczelniane wydawnictwo naukowo-dydaktyczne)
[4] Pyrda L 2014 *J. Phys. Conf. Ser.* **530** 012060
[5] Obukhov A U 1949 *Geogr. i Goofiz* **II** 58–69
[6] Batchelor G K 1959 *J. Fluid Mech.* **5** 113
[7] Bai B, Yabe A, Qi J and Wakayama N I 1999 *AIAA Journal* **37** 1538-1543.
[8] Tagawa T, Shigemitsu R and Ozoe H 2002 *Int. J. Heat Mass Transf.* **45** 267–77
[9] Ozoe H and Churchill S W 1973 *AIChE Symp. Ser. Heat Transf.* **69 (131)** 126–33
[10] Elsner J W 1989 *Turbulencja przepływów (Turbulence of Flows) PWN*

LES - IB analysis of a flow in channel with an adverse pressure gradient.

M Księżyk and A Tyliszczak

Czestochowa University of Technology, Faculty of Mechanical Engineering and Computer Science, Al. Armii Krajowej 21, 42-201 Czestochowa, Poland

E-mail: ksiezykm@imc.pcz.pl

Abstract. This paper presents results of Large Eddy Simulation (LES) of a flow in a channel with an adverse pressure gradient. The applied computational code (SAILOR) is based on a high-order compact finite difference scheme on half-staggered meshes and is combined with an Immersed Boundary (IB) method. The IB method is implemented using the direct forcing approach in a simplified stepwise variant. The results are verified using the experimental data and the general agreement is good. The observed discrepancies are small close to the inlet section and increase along the channel length where strong flow separation occurs.

1. Introduction

Finite volume methods are the most popular in industry applications. Their flexibility and possibility of the use of various type unstructured grids allow carrying out simulations in very complex geometries. Unfortunately, a generation process of high quality unstructured meshes is often very difficult task. The level of complexity increases when one want to generate an unstructured mesh for Large Eddy Simulation, as in this case a required large number of the computational cells makes this process very time consuming. Additionally, considering cases with moving elements with dynamically modified meshes the need to generate a new mesh every time step complicates and extends the overall solution times significantly. A relatively simple solution of these problems consist in the use of an Immersed Boundary (IB) method [1], which allows performing the computations in geometries that do not exactly match the grid nodes, as shown in Fig. 2. In this work we combine the IB method with a high-order LES code and apply the LES-IB approach to study a problem of flow separation in a channel with an adverse pressure gradient.

Understanding of the separation phenomenon is an important and ambitious task. The regions of separation include wide ranges of turbulent high-energy structures influencing on turbulence production near the walls, which makes the flow in the separation region very difficult to analyse experimentally. In this view the LES method not only complements the measurements but also gives deep insight into every detail of the flow. The LES requires the use of non-dissipative high order methods [2] and a compact finite difference method used in the applied code (SAILOR) belongs to this family. The weak point of these methods is that they can be applied in cases with rather simple flow domains. To avoid this limitation in the present work we use LES-IB approach, which, as mentioned above enables performing simulations in almost arbitrary domain shapes.

2. Governing equations and solution algorithm

For incompressible Newtonian fluid flow in the LES framework the momentum and continuity equations are given as:

$$\frac{\partial \bar{u}_j}{\partial x_j} = 0 \tag{1}$$

$$\frac{\partial \bar{u}_i}{\partial t} + \bar{u}_j \frac{\partial \bar{u}_i}{\partial x_j} = \frac{1}{\rho}\frac{\partial \bar{P}}{\partial x_i} + \frac{\partial}{\partial x_j}\left((\nu + \nu_t)\left(\frac{\partial \bar{u}_i}{\partial x_j} + \frac{\partial \bar{u}_j}{\partial x_i}\right)\right) + \bar{f}_i^{IBM} \tag{2}$$

where the bar symbol denotes spatially filtered variables [3]; u - velocity components, P - pressure, ρ,ν,ν_t - density, viscosity and turbulent viscosity. The term $\bar{f}_i^{IBM}$ is a body-force term, which is used in IB method based on a direct forcing approach [4]. The spatial discretization is performed using 6th order compact difference scheme on half-staggered meshes. The time integration is performed with 2nd order Adams-Bashfort/Adams-Moulton scheme with the projection method for pressure-velocity coupling. The subgrid viscosity is modeled by the model proposed be Vreman [5]. The correctness of the applied discretization and temporal integration methods and their combination with the IB method was confirmed in laminar and turbulent flow problems [6, 7].

2.1. Solution algorithm

High-order compact finite differences are used for spatial discretization which are based on half staggered meshes. In that approach all components of the velocity are in the same points in space and pressure nodes are shifted half a cell-size in respect to them. More details can be found in [8].

The solution algorithm is schematically presented in Figure 1. It can be seen that the IB method acts four times during one algorithm loop. Its influence is expressed in the equation 2 by $\bar{f}_i^{IBM}$ terms. In the present calculations the simplest stepwise IB approach is used. The schematic view of that method is presented in figure 2. In this approach the solid body is determined by the locations of computational nodes. In the nodes inside the body (Ω_b) velocity is imposed by equalizing it to the object velocity $\bar{u}_\Omega$, in the nodes outside the body (Ω_f) all velocity components are computed using the governing equations. It is easy to see that the shape of the body has a stepwise appearance (dark gray color in figure 2).

2.2. Sub-grid modelling

An appropriate sub-grid modelling is a very important aspect of LES. In wall bounded flows it is required that the turbulent viscosity vanishes in near wall regions. Correctness of the results obtained using a model proposed in [5] within the LES-IB approach was confirmed in the simulations of a flow turbulent in a converging-diverging channel [6]. According to [5] the turbulent viscosity is calculated using the formula:

$$\nu_t = C\sqrt{\frac{B_\beta}{\alpha_{ij}\alpha_{ij}}} \tag{3}$$

$$\alpha_{ij} = \frac{\partial \bar{u}_j}{\partial x_i} \quad , \quad \beta_{kl} = \Delta^2 \alpha_{mk}\alpha_{ml} \tag{4}$$

$$B_\beta = \beta_{11}\beta_{22} - \beta_{12}^2 + \beta_{11}\beta_{33} - \beta_{13}^2 + \beta_{22}\beta_{33} - \beta_{23}^2 \tag{5}$$

where the model constant C is equal to 0.025.

Predictor step

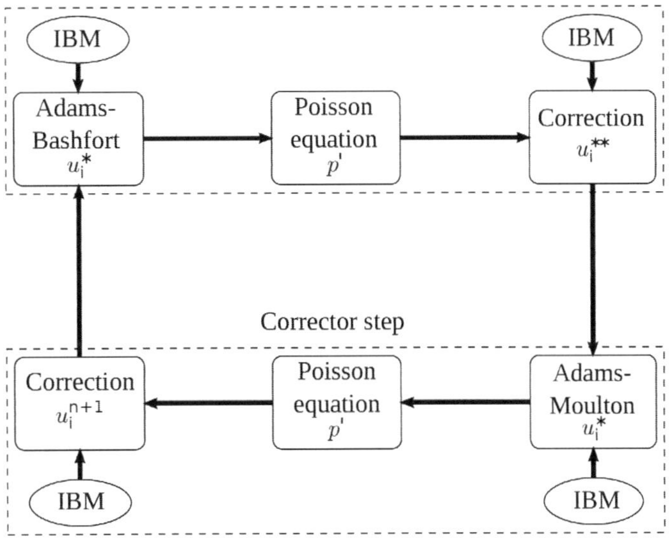

Figure 1. Solution algorithm

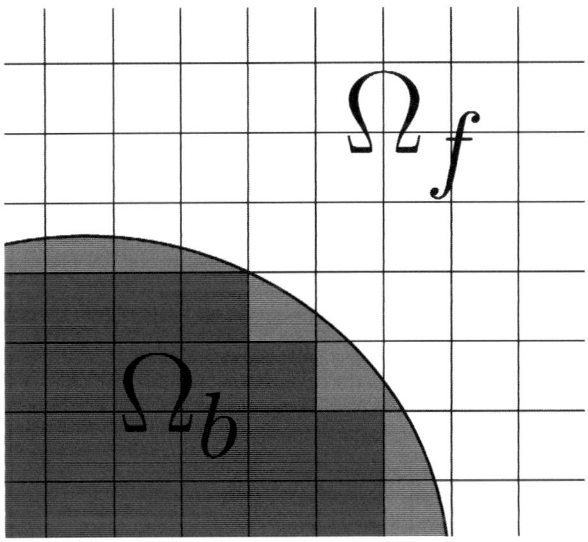

Figure 2. Schematic view of stpewise Immersed Boundary approach.

2.3. Immersed Boundary method

The forcing terms used in the equation 2 are formulated using an assumed velocity of the immersed object, u_i^D, and they take the forms:

$$f^{IBM} = \frac{u_i^D - u_i^n}{\Delta t^n} - \left(1 + \frac{\Delta t^n}{2\Delta t^{n-1}}\right) Res(u^n) + \frac{\Delta t^n}{2\Delta^{n-1}} Res(u^{n-1}) + \frac{1}{\rho}\frac{\partial p^n}{\partial x_i} \tag{6}$$

$$f^{IBM} = u_i^D - u_i^* + \Delta t^n \frac{1}{\rho}\frac{\partial p'}{\partial x_i} \tag{7}$$

in the predictor step, and:

$$f^{IBM} = \frac{u_i^D - u_i^n}{\Delta t^n} - \frac{1}{2}(Res(u^{**}) + Res(u^n)) + \frac{1}{\rho}\frac{\partial p^n}{\partial x_i} \tag{8}$$

$$f^{IBM} = u_i^D - u_i^* + \Delta t^n \frac{1}{\rho}\frac{\partial p'}{\partial x_i} \tag{9}$$

in the corrector step. The forcing terms (6) and (8) are used in the Navier-Stokes equations, whereas (7) and (9) in the velocity correction steps. Using the above formulas, it turns out that in the immersed body region the computed velocity, e.g. u_i^{**}, is equal to the assumed velocity u_i^D. Hence, it is easy to see that there is no need to compute the forcing terms using the equations (6-9), but one can directly impose the velocity u_i^D on $u_i^*, u_i^{**}, u_i^{n+1}$.

3. Results

3.1. Domain and mesh configuration

A computational domain analyzed in this paper is shown in the figure 3. The flow is from the left to right side and the adverse pressure gradient is caused by a curvature of the upper wall and the grid mounted in the outlet section. The complexity of the flow in this seemingly simple configuration is illustrated in the figure 3 by the contours of instantaneous values of the streamwise velocity component and the iso-surfaces of Q-parameter. The Q-parameter is very good indicator of the vortex structures [9] and is defined as $Q = \frac{1}{2}(S_{ij}S_{ij} - \Omega_{ij}\Omega_{ij})$, where S_{ij} and Ω_{ij} are symmetric and antisymmetric parts of the velocity gradient tensor. The walls and the solid region modelled by the IB method are marked by the violet color. The width of domain and the height of inlet section are equal to 0.32 m. In the z direction the periodic boundary conditions are applied. At the inlet, the mean velocity profile of the streamwise velocity component corresponding to experimental data is imposed with 5% noise. In the experiment, in order to increase the total pressure inside the channel a grid with 40% clearance was used at the outlet section. In the performed computations this grid was mimicked using the IB method. The mean Reynolds number based on the friction velocity on the lower wall is equal $Re_\tau = 3000$, whereas the Reynolds number based on the mean inlet velocity and the height of the channel is equal to $Re = 2 \times 10^5$, approximately. The sizes of the domain and the mesh parameters are given in Table 1. In the directions y and z the mesh is uniform, in the direction x the mesh is defined in the following way. In the region near the bottom wall (0 - 0.32 m) the mesh is compacted. The y^+ of the first node is equal 1.5, in the physical coordinates it is 0.0005 m. In the region of the curved wall the mesh is uniform. The smooth transition between these regions was obtained by distributing the mesh nodes using a high-order polynomial functions.

Table 1. Parameters of the computational mesh.

Axis	Length	N	Δ_{uni}	Δ_{min}	Δ_{max}
x	1.09	512	0.003	0.0005	0.003
y	1.8	384	0.005	—	—
z	0.32	64	0.005	—	—

Sample results presented in the figures 4, 5 and 6 show contours of vorticity, mean streamwise velocity and the ratio of the sub-filter viscosity to the kinematic viscosity (ν_t/ν) in the main cross-section plane. Large vortices generated near the curved wall and near-wall the separation

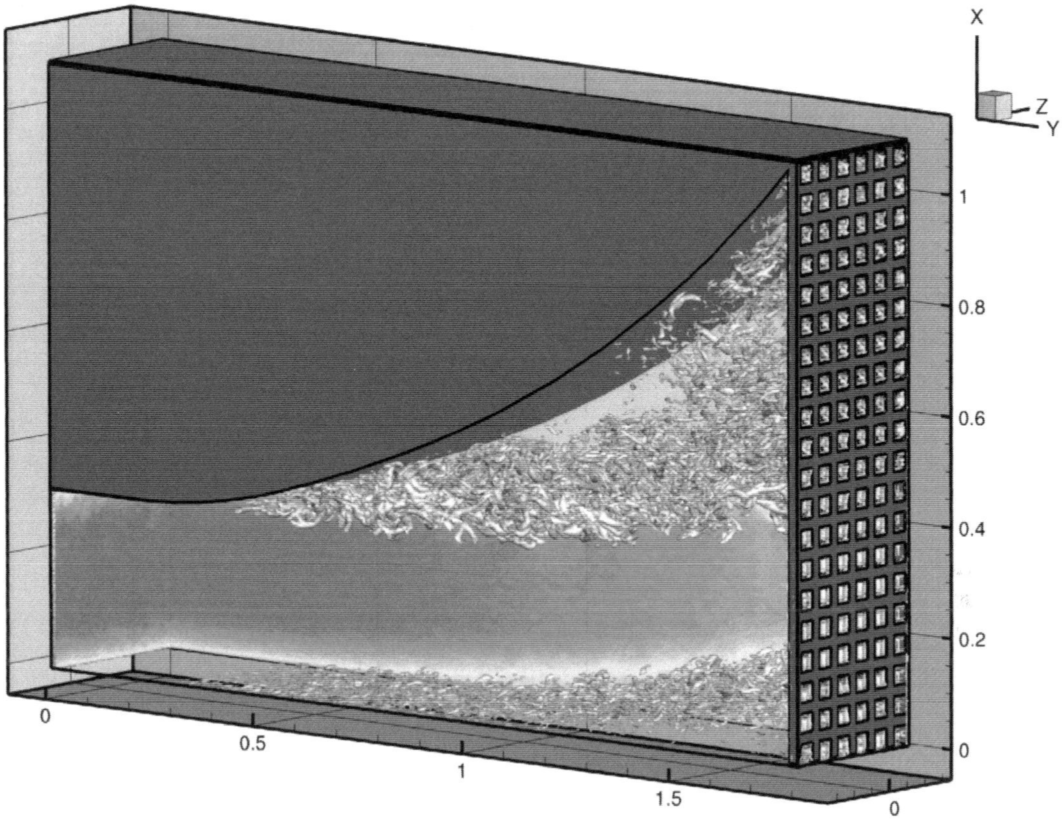

Figure 3. Instantaneous contours of the streamwise velocity and isosurface of the Q-parameter

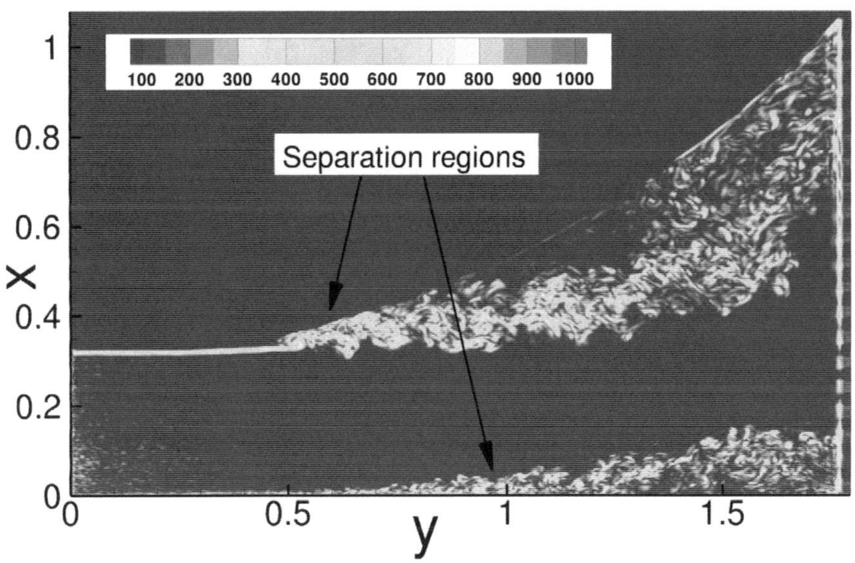

Figure 4. Instantaneous contours of the vorticity. The arrows point the separation regions.

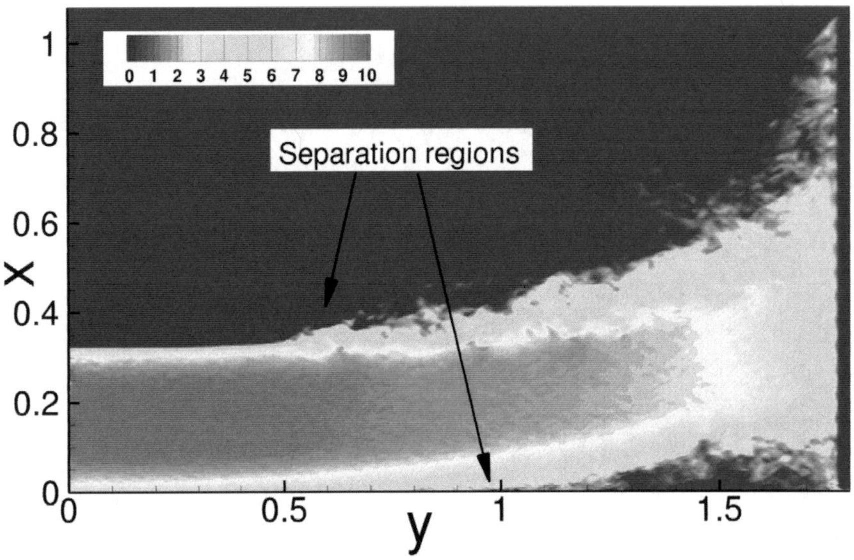

Figure 5. Instantaneous contours of the streamwise velocity. The arrows point the separation regions.

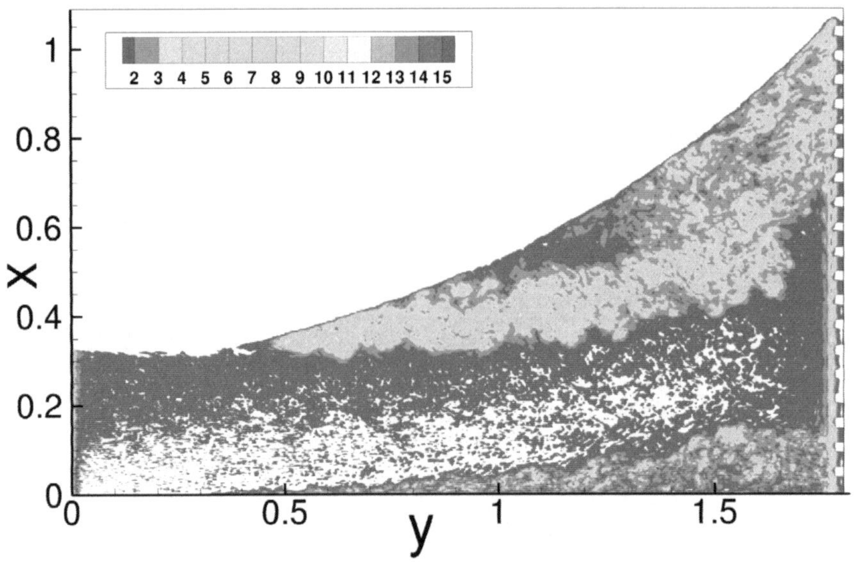

Figure 6. Contours of instantaneous normalized eddy viscosity - $\nu_t/\nu \in (1.2 - 15)$.

regions in the boundary layer close to the bottom and upper walls are apparent. The sub-filter viscosity is calculated in a classical way, i.e. without taking into account the use of the IB method. However, it can be seen that the sub-grid viscosity inside the solid region is virtually equal to zero. This is due the fact that the velocity gradients inside the solid body are very small and as these values are used to calculate the eddy viscosity (see formula 4) it becomes almost zero as well. On the other hand, as one could expect large values of the eddy viscosity occurs in the separation region, where also the vorticity has the biggest value and the flow is regarded as the most turbulent.

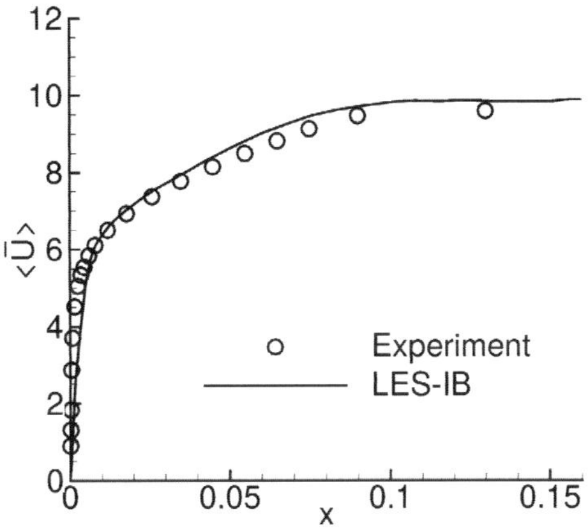

Figure 7. Profiles of the time-averaged mean streamwise velocity along the vertical direction, comparison with experimental data - 0.4 m from inlet.

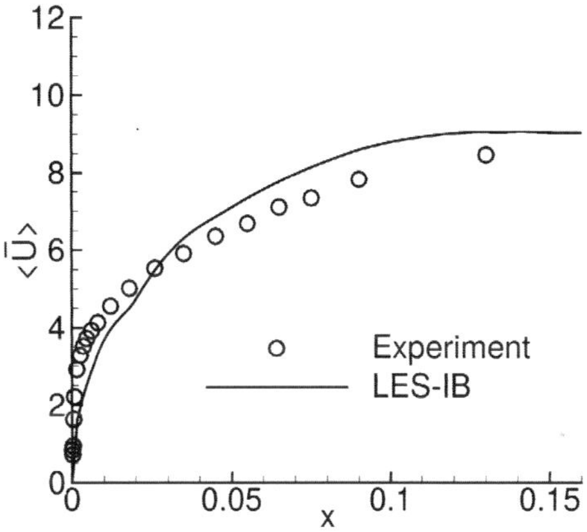

Figure 8. Profiles of the time-averaged mean streamwise velocity along the vertical direction, comparison with experimental data - 0.8 m from inlet.

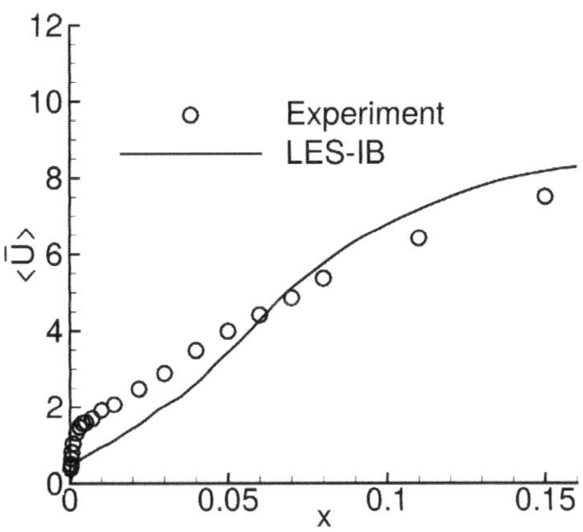

Figure 9. Profiles of the time-averaged mean streamwise velocity along the vertical direction, comparison with experimental data - 1.2 m from inlet.

3.2. Comparison with experimental data
The comparison with experimental data are presented for three locations from the inlet, i.e. at 0.4, 0.8 and 1.2 m. The time averaging procedure started when the flow in the channel was fully developed and lasted for 6 seconds. Figures 7, 8 and 9 show the profiles of streamwise velocity and the symbols correspond to the experimental data. In general the results obtained using the LES-IB method are close to the experimental points. As can be seen the best results are for the first line (0.4 m from inlet) and they get worse along the channel length. This is probably due to the fact that in the location of first line the channel is straight and the influence of the wall curvature on the solution is insignificant. On the other hand, in the regions where the separation occur on the upper and lower wall the discrepancies are clearly visible. Most likely, the number of points used in the present studies is insufficient to correctly capture the physics of the flow in the presence of separation. The grid sensitivity studies are planned for future, however, the present results have already proven that applying LES-IB allows analysing the flow separation phenomena caused by the adverse pressure gradients.

4. Conclusions
This paper presented the results of LES computations of the flow in the channel with adverse pressure gradient. The applied computation code was based on high-order compact methods in combination with the Immersed Boundary method. The simulation conditions and domain shape corresponded to experimental configuration and the obtained results were in acceptable agreement with the experimental data. The discrepancies close to the inlet section were small and increased along the channel length. It is likely, that this could result from to coarse computational mesh or not exact representation of the experimental inlet velocity fluctuations. In the computations the inlet fluctuations were modelled by a white noise. This usually causes damping of the turbulent intensity downstream, which in the present case could result in inaccurate prediction of the separation point. Future research will focus on this subject and on mesh sensitivity studies.

Acknowledgments
The investigation was supported by Polish National Science Center under Grant no. DEC-2012/07/B/ST8/03791. The computations were performed using an infrastructure of PL-Grid (Poland).

References
[1] Peskin C S, 1972, Flow patterns around heart valves: a numerical method, *J.Comp.Phys.* 10:252-271
[2] Wang Z J, Fidkowski K, Abgrall R, Bassi F, Caraeni D, Cary A, Deconinck H, Hartmann R, Hillewaert K, Huynh H T, Kroll N, May G Persson P-O van Leer B, Visbal M 2012. High-order CFD methods: current status and perspective *Int. J. Numer. Meth. Fl.* 00:1-42.
[3] Geurts B J, 2004, Elements of Direct and Large-Eddy Simulation *Edwards Publishing, Philadelphia, USA*
[4] Mohd-Yosuf J, 1997 Combined immersed-boundary/B-spline methods for simulations of flow in complex geometries Ann. Res. B. Cent. Turb. Res.
[5] Vreman A W, 2004. An eddy-viscosity subgrid-scale model for turbulent shear flow: Algebraic theory and applications *Phys. Fluids* 16:36703681
[6] Ksiezyk M, Tyliszczak A 2016. LES of a Converging-Diverging Channel Performed with the Immersed Boundary Method and a High-Order Compact Discretization *Progress in Wall Turbulence 2* Springer, ERCOFTAC Series 191200
[7] Ksiezyk M, Tyliszczak A, 2014 Immersed boundary method combined with a high order compact scheme on half-staggered meshes. *J. Phys. Conf. Ser.* 530, 012066
[8] Tyliszczak A, 2014 A high-order compact difference algorithm for half-staggered grids for laminar and turbulent incompressible flows" *J. Comput. Phys.* 276:438–467.
[9] Hunt J C R, Wray AA, Moin P, 1988 Eddies, streams and convergence zones in turbulent flows. *In: Proceedings of the Summer Program, Cent. Turb. Res.* p. 193

Near-blade flow structure modification

T Kura and E Fornalik-Wajs

AGH University of Science and Technology, Department of Fundamental Research in Energy Engineering, al. Mickiewicza 30, 30-059 Krakow, Poland

E-mail: kura@agh.edu.pl

Abstract. In this paper, the importance of near-blade flow structure influence on the performance of a centrifugal compressor was discussed. The negative effects of eddies and secondary flows appearance were described, together with the proposal of their reduction. Three-dimensional analyses were performed for the rotors. Focus was placed on the blade's 3D curvature impact on the efficiency of compression, and the influence of blade-shroud tip existence. A few design proposals were investigated – their performance maps were the basis of further analysis. Proposed modification of blade shape changed the near-blade flow structure and improved the compressor performance.

1. Introduction

Turbomachinery has been a field of research since the beginning of modern engineering. Experimental data is widely available, but the researchers are still trying to improve designs of particular units and to reduce negative effects of the swirls or secondary flows – in modern world more efficient and, as a result, more ecological machines are desirable. The need for constant improvement is also connected with the fact, that each year, millions of new turbomachinery devices are being produced. Thus the main reason of following studies is a trial to develop appropriate model of centrifugal air compressor.

Progress in new design proposals is possible thanks to CFD methods. Due to them influence of geometry on the flow can be checked with many shape factors which help to obtain flow structure for any examined model. Author has taken the advantage of CFD in [1], where the impact of blades' height, curvature and number on compressor's efficiency was investigated. The results confirmed, that the leaks and backflows in compressor's area can cause significant pressure losses. Those backflows were caused by some design issues, for example the blade-shroud clearance presence. Comprehensive analysis on mentioned flow behavior was done by Brun and Kurz [2]. They have presented the method to predict the direction and strength of secondary flows, as well as the pressure loss caused by them. In [3] Eum et al. have investigated the entropy generation in such areas. However only a few articles from the last couple of years cover the topic of losses caused by mentioned clearance [4, 5, 6]. While it is possible to avoid the issue by designing machines consisted of only one solid part, that process is more difficult and expensive. Apart from flow issues, such intervals between machine's parts are the source of undesirable noises [7]. And, due to the high temperature gradients in that areas, shape deformations may occur, being an additional issue to consider during design process [4, 8].

Many of the equations proposed for evaluation of mentioned losses are empirical, and so they can be used only in particular cases. Therefore, analyses, such as these presented in the following paper, are necessary to verify them or become the basis of the alternative ones. To check, if it is possible to find any relation that would serve as foundation of those, characteristics of compression rate versus mass flow rate depending on a few geometry types as well as overall performance of proposed rotors, together with pressure fields of compressors configurations will be presented in this publication.

2. Mathematical model

The analyses were based on the Reynolds-averaged (RANS) mass, momentum and energy conservation laws, coupled with the ideal gas law. Modeled flow was three-dimensional, stationary and compressible. Results were verified with those coming from theoretical considerations included in [9].

Continuity equation was considered in following form [10]:

$$\frac{\partial}{\partial x_j}(\rho U_j) = 0, \tag{1}$$

where ρ is the fluid density, kg/m^3 and U is its velocity, m/s.

Momentum equation was represented by [10]:

$$\frac{\partial}{\partial x_j}\left(\rho U_i U_j\right) = -\frac{\partial p}{\partial x_i} + \frac{\partial}{\partial x_j}\left(\tau_{ij} - \rho\overline{u_i u_j}\right) + S_{Cor} + S_{cfg}, \tag{2}$$

$$S_{Cor} = -2\rho\omega \times U_r \tag{3}$$

$$S_{cfg} = -\rho\omega\times(\omega\times r) \tag{4}$$

where p is the pressure, Pa, τ is the molecular stress tensor, U_r is the relative frame velocity, u is the fluctuating velocity, S_{Cor} is the source term accounting for the effect of Coriolis force and S_{cfg} is the centrifugal force source. ω is the angular velocity, rad/s and r is the location vector.

Energy equation was applied in the form of total energy [10]:

$$\frac{\partial}{\partial x_j}\left(\rho i_{tot} U_j\right) = \frac{\partial}{\partial x_j}\left(\lambda\frac{\partial T}{\partial x_j} - \rho\overline{u_j i}\right), \tag{5}$$

$$i_{tot} = i + 0.5U_i^2 + k, \tag{6}$$

where T is the temperature, K, λ is the fluid thermal conductivity, W/(m·K) i_{tot} is the specific total enthalpy, J/kg i is the specific static enthalpy, J/kg and k is the turbulence kinetic energy, J/kg.

Analyses were conducted with Ansys 14.5 software, particularly CFX and Turbogrid modules. ANSYS CFX solves all equations with coupled solver method. It is useful due to fast computing speed, yet powerful hardware is needed, with high memory resources. Following simulations were conducted in Academic Computer Centre CYFRONET AGH. High resolution second order upwind advection scheme was utilized, to avoid numerical diffusion and obtain more certain results. To solve turbulence, SST k-ω turbulence model was chosen, since as the combination of typical k-ε and k-ω models, is considered one of the best for simulations [10], because it is stable when high pressure gradients or flow separation areas occur – such as in centrifugal compressors. Most important feature of that model, however, is the ability to generate appropriate results in the boundary layers – which was one of the main goals for Authors.

The air was chosen as the working fluid. It had the density value of 1.225 kg/m^3, the thermal conductivity –value of 0.0242 W/(m·K), the viscosity value of 1.79E-05 kg/(m·s) and the specific heat – 1006.4 J/(kg·K).

A few design proposals were investigated, with rotational speed of the rotor being changed for each case. Its particular values were chosen in the basis of previous experience [1]. The initial temperature in the whole control volume was equal to 288 K. At the inlet, pressure value of 100 kPa was chosen. Due to high rotational speeds in all cases, supersonic conditions were expected throughout the model – therefore adequate outlet boundary condition type was chosen [10]. Inlet turbulence intensity value was set at the low level equal to 5%, as the main reason of the turbulence was the rotation, not the inflow. Table 1 presents boundary conditions implemented in all simulations.

Table 1. Boundary conditions.

Initial temperature [K]	288		
Inlet pressure [kPa]	100		
Rotational speed [rpm]	28 000	30 000	32 000

3. Studied cases and applied solutions

Three cases were investigated for the purposes of the article. Starting from the typical radial blade and finishing on two examples of centrifugal blades. No diffusers were analysed, as the output of the studies was only the rotor performance. Therefore only they were discretized. Software was capable of interpolating results from just one compressor's passage into the whole machine, so that more accurate mesh was generated. In figure 1, first blade, B1, is presented. It was around 10 cm long and its height varied from 3 cm on the end to 2 cm close to the rotor axis. Its shape was developed basing on data from [9] and was used mainly for purposes of mesh verification. In figure 2 second blade proposal, B2 is shown. While its curvature along the length was the same as for B1 case, the curvature along the height was added as the second shape function. It was shorter than the B1 blade, and, at the same time, higher in the inlet zone. In figure 3 third blade design, B3, is shown. Its main dimensions are the same as for the B2 case, apart from the curvature along height. B3 blade is more straight. In figures 1, 2 and 3 the axis of the view is slightly different, to show changes of the shape. Figure 4 shows B2 and B3 blades cut planes to show mentioned difference in that curvature. Considered shapes and the number of blades per rotor, 15, were chosen due to previous experience, supported by research, such as [4,9]. Rotation of all impellers occurred around the axis visible in figure 1. One of the article purposes was to investigate the impact of blade-shroud clearance on the machine's efficiency. Typically, they are no bigger than a few millimetres. For the following publication, two lengths were taken into consideration – 1 mm and 2mm, constant along the blade. Blade-shroud tip position is shown in figure 5.

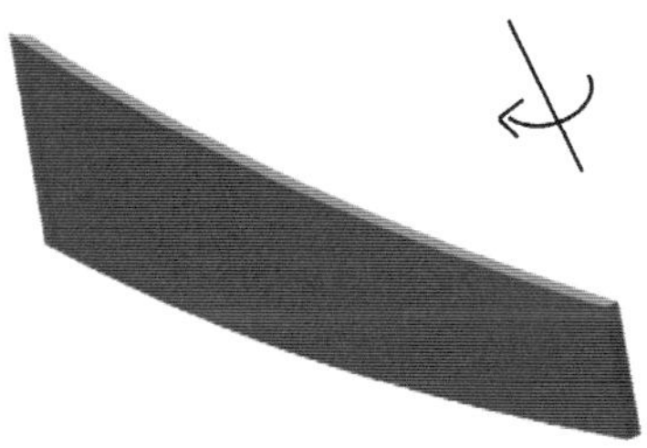

Figure 1. Design of B1 blade.

Figure 2. Design of B2 blade.

Figure 3. Design of B3 blade.

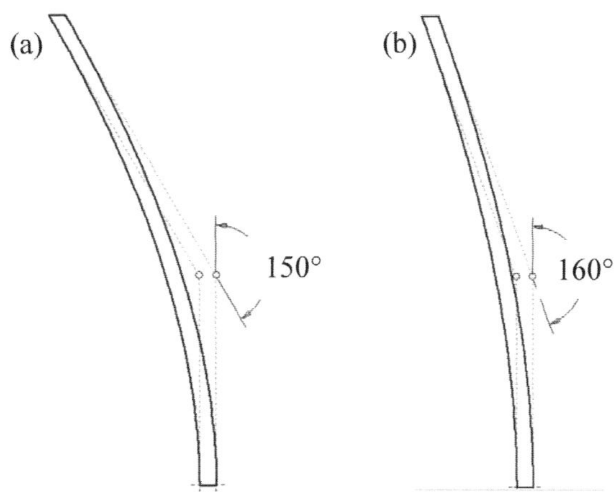

Figure 4. Blades curvature along the height, (a) B2 and (b) B3.

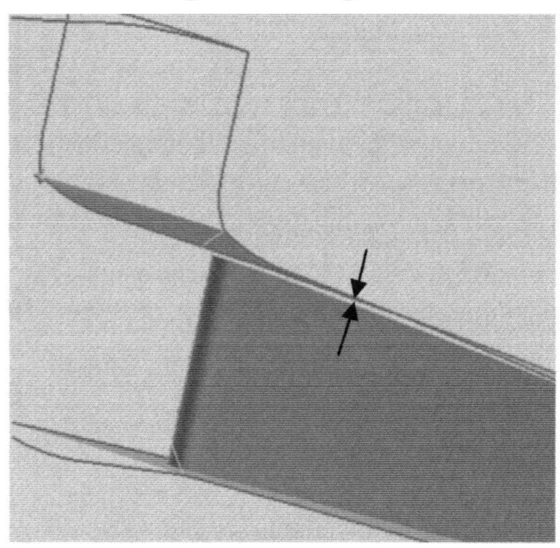

Figure 5. Blade-shroud tip.

Discretization of geometries was performed with Ansys Turbogrid, in the basis of finite volume method. For the quantitative numerics, regular, mostly hexahedral elements were generated. As investigation of near wall flow behaviour was of great importance, refined mesh was necessary in the boundary layer around the blade. Also, due to dimensions of the blade-shroud tip, refining was obligatory in that area. Basing on boundary conditions and therefore approximate Reynolds number, y^+ parameter was calculated, as the main factor that influences boundary layer shape. For the Reynolds number value equal to approximately $0.5x10^7$, y^+ did not exceeded 2 – value considered maximal for proper calculations [10]. Figure 6 presents overall look of the mesh, with visible refining around the blade. Figure 7 represents the close-up of blade-shroud tip mesh. The view is located above the blade, and due to the software characteristics such shape of the mesh is necessary for reliable computing.

To verify, whether generated mesh is correct, its independence test was obligatory to conduct. Various number of elements and nodes was taken into account, from about $3x10^5$ to $1.5x10^6$. However, after reaching $1.3x10^6$, further increase did not visibly changed the obtained results – and that number was chosen as the proper one. Quantity of elements and nodes for each case is presented in table 2. All designs are also divided depending on the blade-shroud tip size. Afterwards, to validate the results, comparison was performed on the basis of empirical formulas taken from [9]. The results are shown in table 3. Difference of numerical and analytical results did not overstep 5% and, as the purpose of the research was to check general tendency and behaviour of flow in described rotors, such discrepancy was considered sufficient and the chosen mesh was decided to be fine enough to continue the studies.

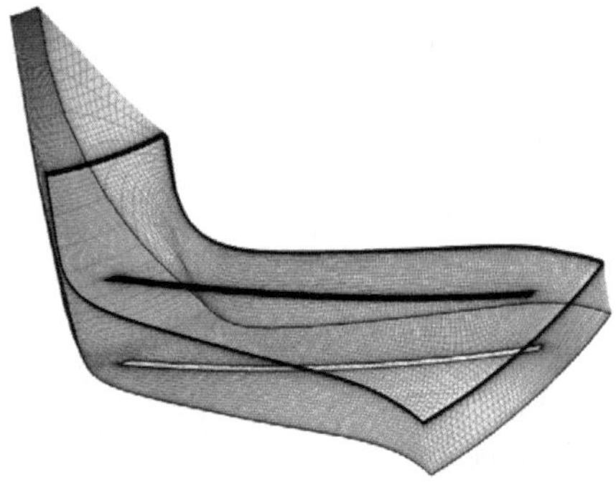

Figure 6. Overall picture of discretized volume.

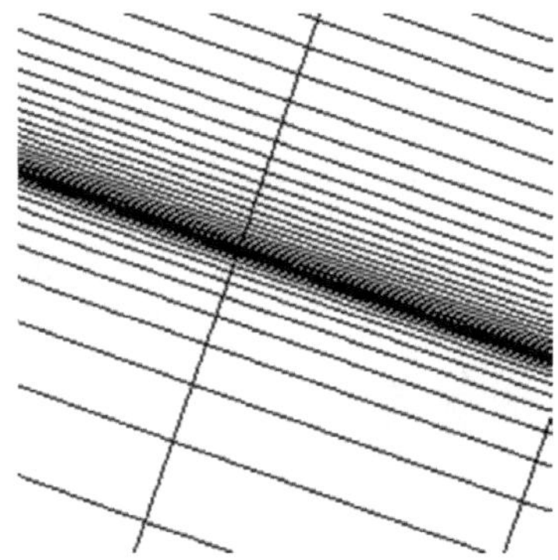

Figure 7. Close-up of blade-shroud tip mesh.

Elementary energy balance of the compressor has the following form:

$$dl_i - dq = di_{tot}, \quad (7)$$

where l_i is the internal work done on gas J/kg, q is the energy loss J/kg. Based on it, the static and total pressures were used for the verification, and were calculated using Bernoulli's equation. Total parameter was the sum of its dynamic and static part. Total temperature was evaluated in the same manner but the equation of state had to be included as well.

Table 2. Elements and nodes number for each case.

Case	B1		B2		B3	
	1mm	2mm	1mm	2mm	1mm	2mm
Elements	1 251 166	1 258 965	1 238 344	1 245 666	1 224 274	1 233 906
Nodes	1 311 860	1 319 916	1 305 557	1 313 399	1 290 864	1 301 248

Table 3. Validation of proposed mesh. Compatibility of analytical and numerical results.

Variable	Static pressure	Total pressure	Static temperature	Total temperature	Density
Accordance, %	96.1	94.6	99.6	99.2	96.7

4. Results and discussion

For the purposes of presented work, all investigated rotors were numerically tested with varying rotational speeds. One of the goals of such behaviour was to examine, if the compression ratio of any machine would change with the change of blade's shape. Compression ratio (CR) is one of the main factors that indicates the performance of the machine. It is being defined as [10]:

$$\mathrm{CR}=\frac{p_{t,out}}{p_{t,in}}, \tag{8}$$

where $p_{t,in}$ is the total pressure at the inlet and $p_{t,out}$ is the total pressure at the outlet – they were calculated by software as the area integral in two mentioned zones. Figures 8 and 9 present values of compression ratio for each case, apart from B1. That design turned out to be inaccurate for proposed boundary conditions and did not give any compression. Its shape was however designed for lower rotational speeds [9] and that situation was expected. Therefore, B1 blade was not included in further analyses. B2 and B3 rotors gave quite similar CR values, up to almost 2.2 for 5 out of 6 situations, however, B3 blades turned out to be slightly better, with higher CR numbers. Moreover, increase of rotational speed led to better compression in all calculations – connected though with higher input energy needs. The noticeable fact is, that it is hard to define any pattern connected with the size of blade-shroud tip, even though it reached 10% of the blade's height in the most narrow part of blade – which is rather significant value. To compare the proposed designs another parameters were chosen – the relative head coefficient, the relative flow coefficient and the relative isentropic efficiency. The head coefficient is a dimensionless factor which relates the adiabatic head capability of a wheel with its peripheral velocity requirement and the flow coefficient is a non-dimensional unit which takes flow into consideration for a given tip speed and impeller diameter [11]. Isentropic efficiency measures the irreversibility of the process. They are defined, as:

$$\Psi=\frac{3600gH}{(N\pi D)^2}, \tag{9}$$

$$\Phi=\frac{240\dot{V}}{\pi^2 D^3 N}, \tag{10}$$

$$\eta=\frac{I_{t2s}-I_{t1}}{I_{t2}-I_{t1}} \tag{11}$$

where Ψ is the head coefficient, Φ is the flow coefficient, η is the isentropic efficiency, g is the gravity constant, 9.91 m/s^2, H is the generated head per impeller, m, N is the rotational speed, rpm, D is the diameter of the rotor, m, $\dot{V}$ is the volume flow rate, m^3/s and I_t is the total enthalpy, J. Indexes mean: 1 – the inlet, 2 – the outlet, s – the value in isentropic process. Relative amounts of those non-dimensional parameters were obtained by dividing their values for each case by the maximal value.

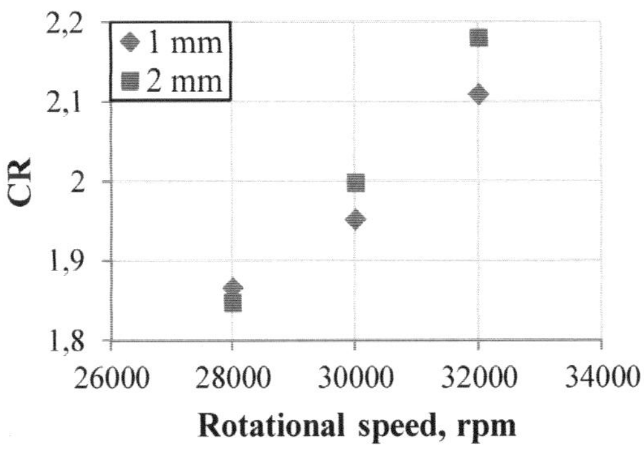

Figure 8. B2 rotor compression ratio.

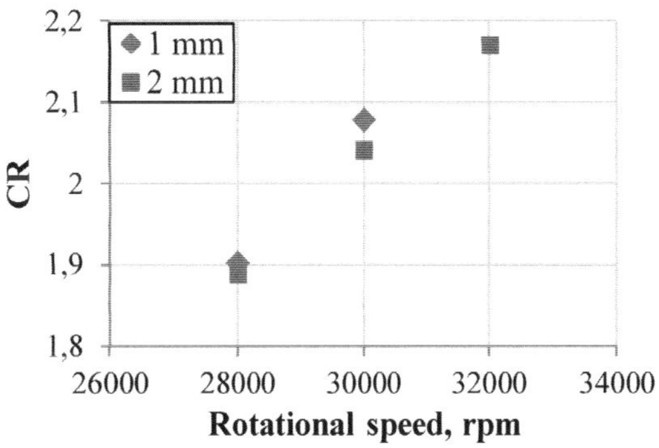

Figure 9. B3 rotor compression ratio.

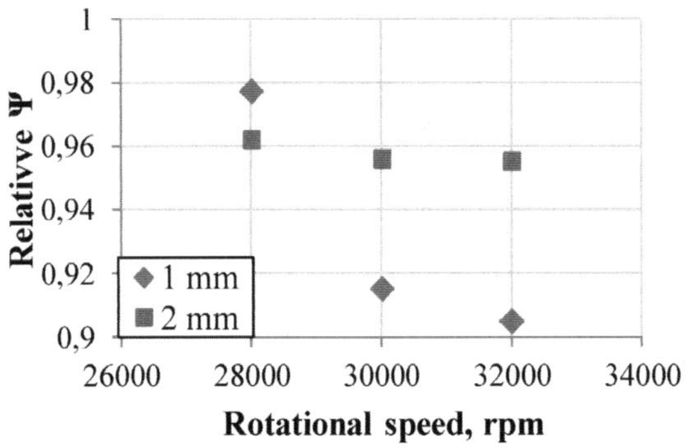

Figure 10. B2 rotor relative head coefficient.

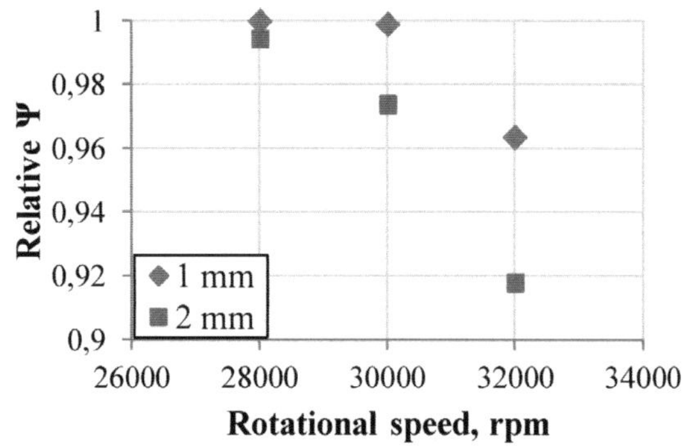

Figure 11. B3 rotor relative head coefficient.

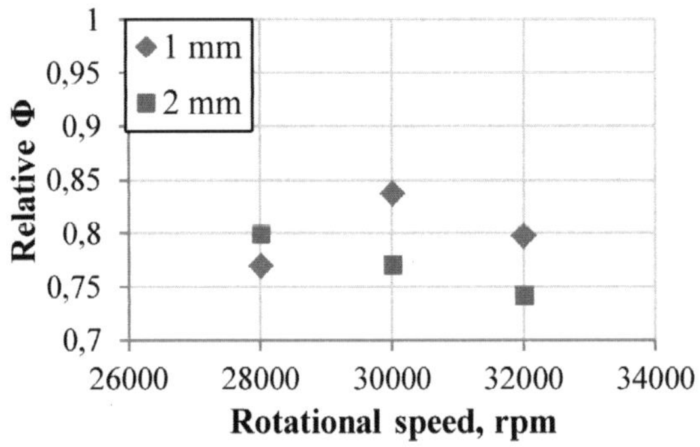

Figure 12. B2 rotor relative flow coefficient.

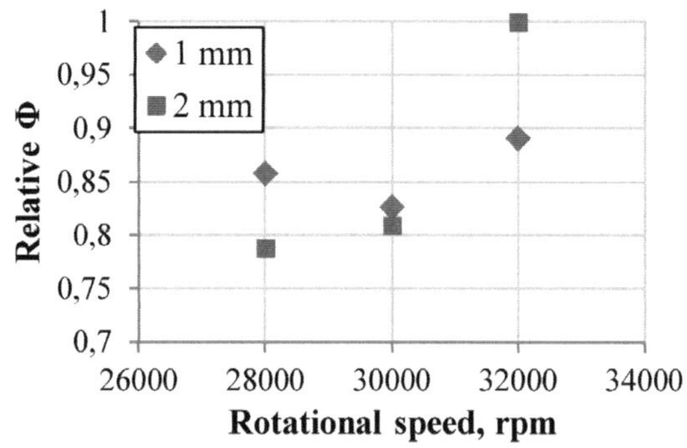

Figure 13. B3 rotor relative flow coefficient.

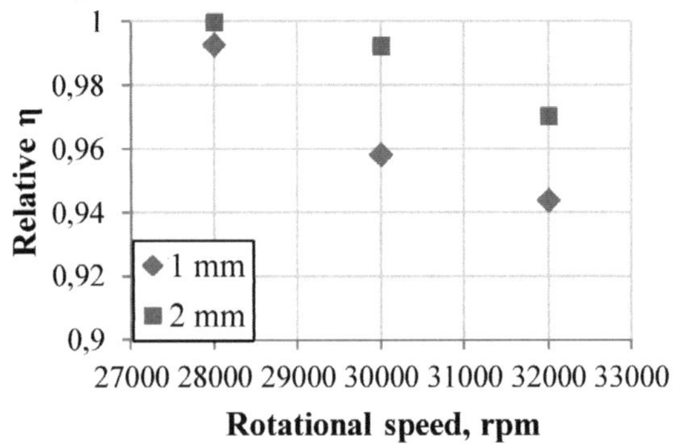

Figure 14. B2 rotor relative isentropic efficiency.

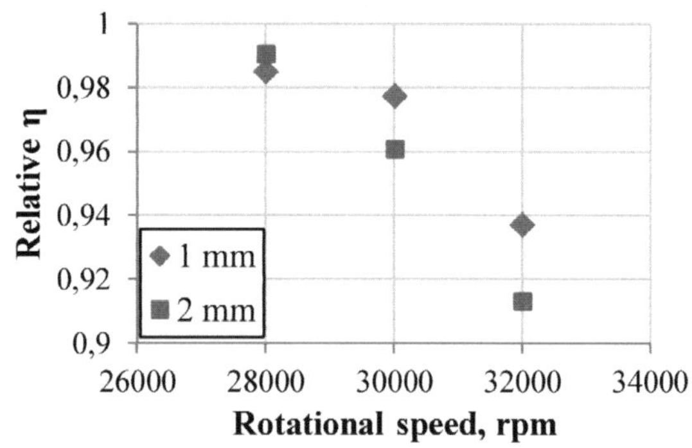

Figure 15. B3 rotor relative isentropic efficiency.

Figures 10 and 11 present relative head coefficient values, for B2 and B3 case, respectively. Figures 12 and 13 present relative flow coefficient values, for B2 and B3 case, respectively. Figures 14 and 15 present relative isentropic efficiency values, for B2 and B3 case, respectively. Analysis of obtained Φ values does not indicate clear pattern. However, there is a clear decrease of Ψ and η parameters with the increase in rotational speed, which means, that further acceleration would move each machine from the design conditions. The next step in comparison process needed analysis of pressure fields. Figures 16 and 17 represent the pressure distribution on cross-section in the middle of device passage. For all images, constant rotational speed, 30000 rpm, was chosen, as representative one. In such planes, slightly higher values of outlet pressure can be noticed for impellers with 2 mm blade-shroud tip, but the overall compression ratio, already mentioned, was higher only for B2 case. For both designs pressure distribution along the blade seemed similar, with visible increase of that parameter along it. Two areas, where pressure increase occurred, can be noticed – closer to inlet and closer to outlet. Black arrows above the figures indicate the flow direction.

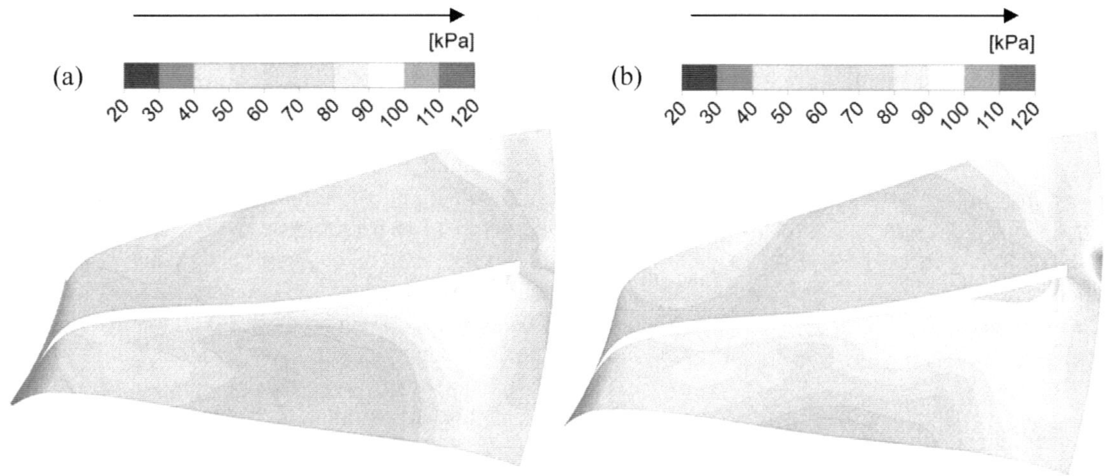

Figure 16. Pressure fields, B2 case, 30000 rpm, blade-shroud tips: (a) 1mm and (b) 2 mm.

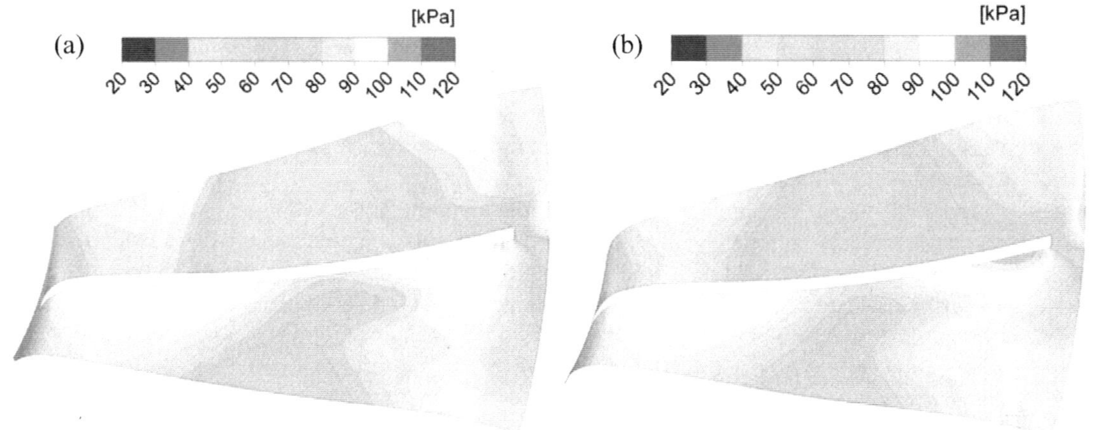

Figure 17. Pressure fields, B3 case, 30000 rpm, blade-shroud tips: (a) 1mm and (b) 2 mm.

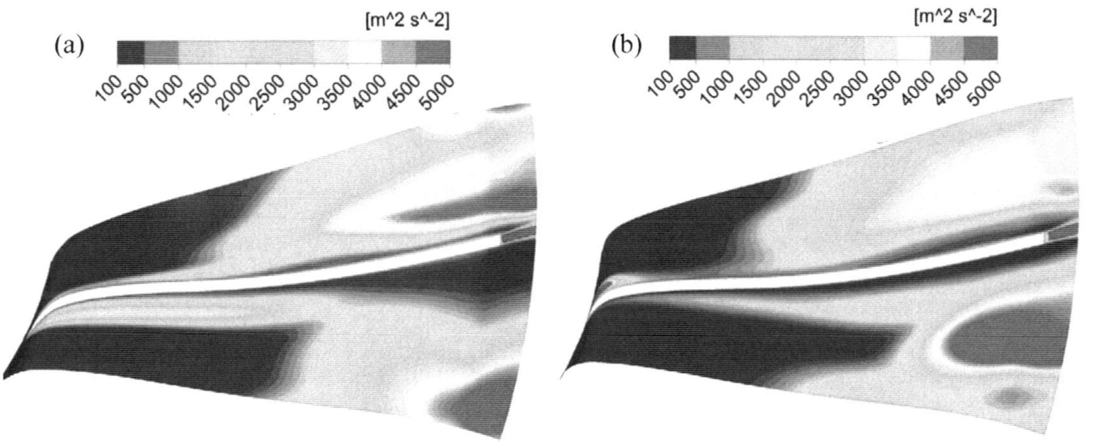

Figure 18. TKE distribution, 30 000 rpm, (a) B2 case and (b) B3 case.

Another possibly important parameter, that could have help receive the whole image of examined impellers was the turbulent kinetic energy, TKE, which is defined as the kinetic energy per unit mass of the turbulent fluctuations and is written as [12]:

$$k = \frac{1}{2}\left(\overline{u_x^{'2}} + \overline{u_y^{'2}} + \overline{u_z^{'2}}\right), \tag{12}$$

where $u_x^{'}$, $u_y^{'}$, $u_z^{'}$ are the velocity fluctuations of the flow in *x, y, z* directions, respectively. Its values for B2 and B3 cases, again for rotational speed of the rotor set to 30000 rpm, are presented in figure 18. While in general both distributions of TKE were similar, differences can be seen near the outlet zone. TKE corresponded with the velocities fields in both cases – and with its values increasing along the flow direction, it indicated increasingly turbulent and possibly irregular flow closer to the outlet.

5. Summary

Various impellers of centrifugal compressor were investigated to examine the impact of blades shape and shroud-tip size on their performance. All boundary conditions remained the same, apart from rotational speeds. Air was being compressed in each case. While B1 design turned out to be absolutely inaccurate for as high rotational speeds as in simulations, B2 and B3 rotors gave very similar results, mainly compression ratios of about 2. Both blades, B2 and B3, depending on speed of rotation, had slightly different parameters: head and flow coefficients and isentropic efficiencies, yet disparities of their values were not very high. It was impossible to deduce any convincing relation between them.

For both cases, B2 and B3, pressure distribution along the passages was not ideal. Due to blades shape, two areas of more significant pressure increase appeared. That fact may have influenced outlet pressure values. Analysis of TKE distribution led to opinion, that the biggest turbulence was noticed in the outlet area, corresponding with the highest velocities.

It was hard to conclude any impact of different blade-shroud tip size on devices performance. It was surprising, considering the fact, that in the most narrow area, it was around 10% of the blade's height. Analyzing the obtained results, it is impossible to find any correlation. Presented results can be the basis of further research. While blade-shroud tip size impact was not confirmed, it is considered a fact in other publications [3]. However, as they reported the various height of it along the blade, more insight investigation should be performed. One of possible explanations of different results can come from the shape of the clearance.

Acknowledgments

The present work was supported by Academic Computer Centre CYFRONET AGH, under award no. MNiSW/IBM_BC_HS21/AGH/021/2015 and by the Polish Ministry of Science (Grant AGH No. 15.11.210.344).

References

[1] Kura T 2015 *Numerical analysis of velocity and temperature fields of compressible fluid flowing through a rotor* (MSc Thesis, in Polish)
[2] Brun K, Kurz R 2005 *Int. J. of Rotating Machinery* **1** 45-52
[3] Hark-Jin E, Young-Seok K, Shin-Hyoung K 2004 *KSME Int. J.* **18** 979-998
[4] Kim C, Lee H, Yang J, Son C, Hwang Y 2016 *Int. J. of Refrigeration* **65** 92-102
[5] Swamy S M, Pandurangadu V 2013 *Int. J. of Research in Engineering and Technology* **2**(9) 445
[6] You D, Wang M, Moin P, Mittal R 2006 *Physics of Fluids* **18**
[7] Galindo J, Tiseira A, Navarro R, López M A 2015 Int. J. of Heat and Fluid Flow **52** 129-139
[8] Jung Y, Choi M, Oh S, Baekl J 2012 *Int. J. of Rotating Machinery* **2012**
[9] Witkowski A 2013 *Sprężarki wirnikowe Teoria, konstrukcja, eksploatacja* (Wyd. Pol. Slaskiej) (in Polish)
[10] *ANSYS® CFX-Solver Theory Guide 2013* (ANSYS Inc., Cecil Township, 2013)
[11] Khan M O 1984 *Int. Compressor Engineering Conference* **509**
[12] Wilcox D 2006 *Turbulence Modeling for CFD* (DCW Industries, Inc.)

Vehicle wheel drag coefficient in relation to travelling velocity – CFD analysis

P Leśniewicz[1], M Kulak and M Karczewski
Institute of Turbomachinery, Lodz University of Technology, Poland

E-mail: lesniewicz.pawel@gmail.com

Abstract. In order to understand the aerodynamic losses associated with a rotating automobile wheel, a detailed characteristics of the drag coefficient in relation to the applied velocity are necessary. Single drag coefficient value is most often reported for the commercially available vehicles, much less is revealed about the influence of particular car components on the energy consumption in various driving cycles. However, detailed flow potential losses determination is desired for performance estimation. To address these needs, the numerical investigation of an isolated wheel is proposed herein.

1. Introduction

In recent years an automotive industry faced the problem of vehicle fuel consumption affected by unstable situation on the petrol market and European regulations concerning greenhouse gas emission. Problem of efficiency concerns not only cars fuelled by gasoline, but also hybrid and electric vehicles where the goal is to extend distance covered on a single battery charge. That is why the investigation of influence of different vehicle components, initially as independent, later as part of a system, on fuel or energy consumption is needed to evaluate the areas in which further improvements will be possible. Results of Euromix test cycle depicted that the leading factor is an aerodynamic drag influencing the vehicles fuel consumption by 40% [1]. This suggests that further attention should be paid to the overall vehicle aerodynamics as the key factor to diminish the energy usage (it is estimated that reduction of drag coefficient by 8% can e.g. decrease the fuel consumption by 0.2-0.3 L/100 km or increase the range of vehicle by 30-40 km).

When repartition of the overall aerodynamic drag is analysed, it can be stated that one of the significant contributions is due to the vehicle wheels, estimated at around 20% or more of the entire car aerodynamic drag [2, 3, 4]. However, a number of investigations dedicated to profound understanding of wheels' contribution to a passenger vehicle aerodynamic behaviour is still insufficient. To change this course a number of publications devoted to the topic of aerodynamics of wheels started to increase in the last decade [5, 6, 7, 8, 9, 10]. An experimental explanation of the flow phenomena created near the isolated rotating wheel in contact with the ground was firstly investigated by [11]. It is a reference point for many CFD and experimental studies conducted nowadays and, therefore, is also included herein for exactly this reason. Another experimental work on isolated wheel in a contact with the ground was done by [7, 12], who, among many tests, presented a comprehensive look on the importance of flow characteristics inside the tyre grooves in comparison to slick tyres. CFD methods were also introduced into engineering problems oriented on explaining the flow

[1] To whom any correspondence should be addressed.

intricacies of rotating wheels. One of the first CFD studies was performed by [13], to be followed by other researchers to mention only [8, 9], where studies of different numerical models were undertaken or [2] who presented an investigation of tyre deformation near the contact patch. Recent works done by [6, 14, 15] confirm that the numerical analysis of the isolated wheel as well as a wheel inside the wheel arch are indispensable part of tyre research. Today, CFD-based tyre models can predict the most important phenomena of the flow with satisfactory accuracy and are a useful tool to complement any experimental study.

Fuel consumption of a vehicle is calculated according to New European Driving Cycle (NEDC) defined by United Nations Economic Commission for Europe. NEDC cycle consists of urban (ECE 15) as well as high-speed (EUDC) parts. For an urban driving cycle test, maximum achieved speed equals 50 km/h (13.9 m/s), while for high-speed driving - 120 km/h (33.3 m/s) [16]. However, due to the increasing number of hybrid and electric vehicles the UN ECE GRPE (Working Party on Pollution and Energy) group is working on the new kind of a test, which, will complement the NEDC cycle. The new test is called Worldwide Harmonized Light Vehicles Test Procedures (WLTP) and includes three driving classes which depends on power to weight ratio. Class 3 cycle being representative of the 3rd class of the power to weight ratio will be used for currently produced vehicles. The speed range for this class is between 56 km/h (15.5 m/s) and 130 km/h (36.1 m/s) depending on the phase of the test [16]. The highway-specific phase of Cycle 3, called extra-high, has an average speed of 94 km/h (26.1 m/s). The new test thresholds, in light of the current passenger car performance, suggest that a tremendous push is required from future vehicles in order to remain competitive in the market. This will require lowering an overall energy consumption of the hybrid and all electric vehicles by smart actuation techniques, power management, but, above all, continuous reduction of energy losses, including aerodynamic drag of the wheels.

The aim of this paper is to analyse aerodynamic differences of the flow around the wheel moving with the speeds pertinent to the NEDC and WLTP test cycles, as well as to investigate the importance of wheel drag in relation to the applied velocities. This is done precisely for the purpose of the first time look at the importance of tyre designs in terms of their aerodynamic performance (energy requirement) for a range of travelling velocities observed in the aforementioned test procedures. Presented work includes numerical analysis of an isolated wheel equipped with two types of tyres - slick and tyres with longitudinal grooves. It aims at exemplifying the tendencies and similarities which could be later mapped on non-isolated cases, in order to decrease the time and computational power necessary for further research. The study includes explanation the influence of tyre tread pattern on part of the energy intake associated especially with overcoming aerodynamic resistance of all-electric vehicles in high-speed traffic. The secondary goal is to instigate discussion about importance of inspecting the aerodynamic resistance of wheels in the function of various travel velocities. Such characteristics could find use in future smart actuation systems responsible for energy distribution in all electric vehicles.

2. Model description and validation

The above research aims at establishing a set of guidelines for flow around wheel simulation, proposing the approach with investigation of isolated wheel only (without necessity of including car body to the virtual model). Observing the presence of phenomena and differences resulting from geometrical features e.g. wheel tread, existing in both cases, one can significantly decrease the cost and time of computation by research of isolated wheel.

Mlinaric [2] has given a broad look at computational methods of simulating flow around a vehicle wheel as based on passenger cars manufactured by Volvo Car Corp. His work describes the importance of utilising a more robust technique of simulating the motion of vehicle wheels. It was decided to use MRF technique in simulations presented in the following paper, mainly to limit the computational time and resources needed to perform a vast number of computations for the analysed travel velocities.

The investigated wheel geometry was based on the tyre analysed by [11]. Diameter of the wheel was equal to 416 mm, while the breadth was 191 mm. The same tyre with three longitudinal grooves evenly distributed with respect to the wheel's centreline was used to study the effect of tread geometry. Dimensions of the grooves are as follows: 6.35 mm wide and 3.175 mm deep. In order to obtain good relation between mesh elements amount and accuracy of the results, comparison of the drag coefficient and mesh size was done for both types of tyres. Independence of drag values was investigated by changing size of the elements and calculating drag values for different mesh sizes. Computationally solid results were observed for the meshes with number of elements around 6 mln for slick tyre, and 13 mln for a tyre with grooves (for more details see [14]). SST turbulence model was applied as the best compromise between accuracy of the results and time needed for simulation. This model was previously validated for tyre studies [6, 14], and benchmarked for the modelling of aerodynamic flow phenomena induced by rotating machine elements [17]. Flow was treated as isothermal and fully turbulent with air as ideal fluid at temperature equal to 25°C. Spatial discretization of governing equations along with turbulence closure was second order accurate. Boundary conditions applied to the model are presented in Fig. 1. Opening boundary condition indicates that the fluid can freely move in and out of the domain. Tyre surfaces were defined to be smooth with no slip boundary conditions applied. At the inlet, uniform flow velocity condition was applied with turbulence intensity of 0.2%. The ground was modelled as no slip wall with wall velocity equal to travel velocity. At the outlet, average static pressure with relative pressure equal to 1 atm and pressure profile blend 0.05 was applied. To enable analysis of tyre aerodynamics at various travel velocities, the free stream flow speed was changed depending on the analysed case. All other simulation parameters were then held constant. Because the analysed speed range expands considerably, the Reynolds number effects on flow behaviour have to be considered. For this reason, the numerical model at lower and upper ranges of analysed velocities was validated against available experimental data. The details regarding the validation at low speed conditions can be accessed in [14]. The validation at the upper range of the considered driving cycles is presented below.

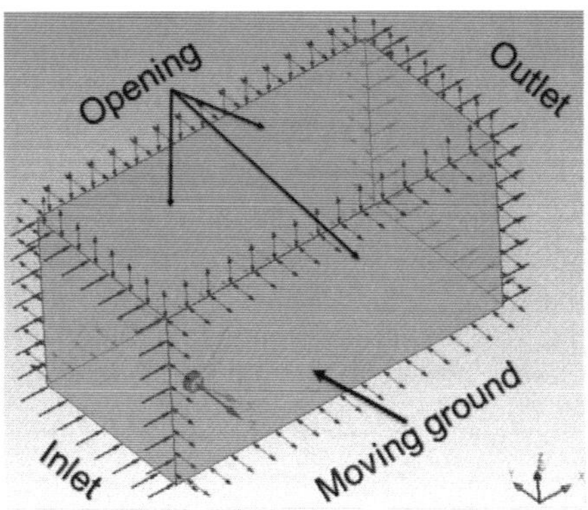

Figure 1. Boundary conditions.

In light of the absence of experimental tests performed at velocities in the upper portion of the NEDC and WLTP tests (33-36 m/s), the results from experimental campaign by [7] were used, where aerodynamic performance of tyre with 4 grooves was assessed at 40 m/s for a generic isolated racing car tyre positioned at 2° camber angle.
Results of the experimental investigation presented in [7] were converted form force values to drag and lift coefficients. Table 1 presents results of Cx - drag and Cz -lift coefficients obtained from experimental work, CFD done by [18] and own numerical simulations.

Table 1. Results of experimental investigation in relation to the CFD simulations in terms of lift and drag coefficients obtained for the upper velocity range of analysed driving cycles.

Parameter	Experiment [7]	CFD [18]	Own CFD
Drag coeff. (Cx)	0.444	0.590	0.482
Lift coeff. (Cz)	0.058	0.235	0.244
% Difference Cx	-	32.8	8.6

Such low value for experimental Cz was caused by the lack of tyre deformation in the contact area due to the materials that were used to make the wheel model (aluminium and composite materials). The wheel was slightly elevated above the belt used to simulate the moving ground condition. That caused the pressures to equalise what was manifested in the unusually small lift force. On the other hand, the drag coefficient is likely to be slightly higher shall the wheel camber be reduced to 0°. At the same time, own simulation results for the tyre with grooves produced Cx = 0.487 and Cz = 0.252 at V = 41.7 m/s. Although a number of grooves is different in own study (herein 3 grooves) from the model in [7] (4 grooves), the obtained result of Cx shows signs of convergence with the work presented by [7]. As the objective of the above article is to identify general tendencies in drag – velocity relations, the level of similarities between both models was considered as acceptable for the purpose of initial comparison. Presented research is aiming at in-house experiment covering both isolated and non-isolated wheel supported with corresponding CFD simulations. The same group of researchers in [18] also performed numerical study applying the identical boundary conditions as in own experiment [7]. Differences between drag values for the experimental and numerical work in [18] were explained by "slightly higher pressures behind the wheel" in case of the experiment which led to the smaller drag force. Comparing results of this investigation with the experimental data presented in [18] it can be observed that, own model is closer to the presented work and can be considered as representative for the upper range of the analysed velocities.

3. Results and discussion

3.1. Drag and lift

In the following section results of the numerical simulations made using ANSYS CFX ver.14 will be included. Speed ranges were selected based on the NEDC and WLTP tests described in the introduction section and vary from 13.1 m/s up to 36.1 m/s. Detailed description of applied velocities, equivalent Reynolds number (Re) together with values of global quantities such as drag (Cx) and lift (Cz) coefficients are presented in Table 2.

Table 2. Results of lift and drag coefficients of isolated slick and grooved wheel model, results derived from own CFD simulations.

V [m/s]	Re [-]	Cz [-] slick	Cz [-] grooved	Cx [-] slick	Cx [-] grooved	ΔCx (with respect to slick)
13.1	3.48E+05	0.287	0.234	0.541	0.528	-0.013
15.8	4.19E+05	0.293	0.239	0.538	0.521	-0.017
18.6	4.93E+05	0.295	0.261	0.535	0.520	-0.015
21.4	5.68E+05	0.294	0.277	0.532	0.520	-0.012
24.2	6.42E+05	0.292	0.279	0.530	0.520	-0.010
26.9	7.14E+05	0.289	0.274	0.526	0.517	-0.009
30.6	8.11E+05	0.288	0.271	0.520	0.508	-0.012
33.3	8.84E+05	0.288	0.264	0.514	0.502	-0.012
36.1	9.58E+05	0.288	0.256	0.509	0.492	-0.017
38.8	1.03E+06	0.290	0.253	0.510	0.486	-0.024
41.7	1.11E+06	0.288	0.252	0.505	0.487	-0.018

What is more, ΔCx which shows differences in drag values between the slick tyre and the tyre with grooves was also estimated. This drag delta is quite consistent also in the experimental and numerical tests of a passenger vehicle model as it was presented in [6]. Analysing relation between the drag versus applied velocities presented in Fig. 2 as well as in the table above, it can be observed that Cx decreases while the speed is increasing. This phenomenon occurs in case of slick and grooved tyres. Results allow to state that the grooves lower the aerodynamic drag independently from velocity applied. The difference in aerodynamics-based energy loss is nearly constant when both characteristics are compared. Many phenomena will be better clarified when planned own experiments for the studied wheel models at a reduced scale will be performed in the wind tunnel facility using PIV (Particle Image Velocimetry) to visualize flow structures. To the authors' best knowledge, such extended analysis, be it numerical or experimental have not been conducted.

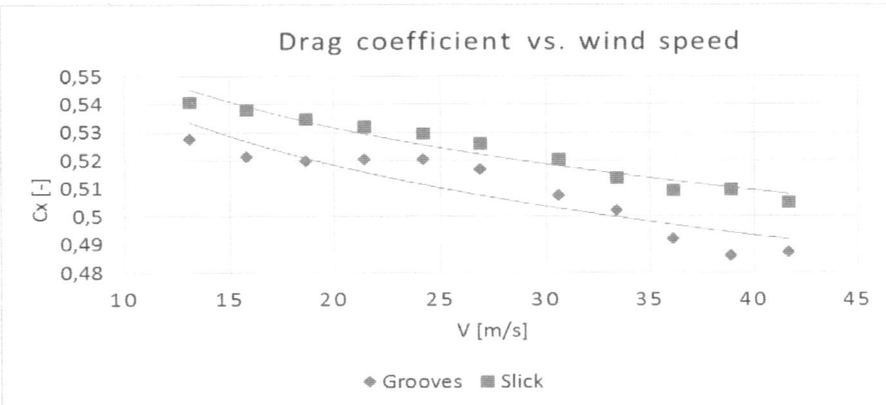

Figure 2. Characteristics of an isolated slick and grooved wheel's drag coefficient in relation to the travel velocities found in European driving test cycles

3.2. Estimation of the energy needed to overcome the aerodynamic drag of the wheels

Using travel velocity data of the analysed driving test cycles it is possible to estimate the energy usage needed to overcome the aerodynamic drag of the wheels. Table 3 includes information about purely aerodynamic-based average energy needed for 4 wheels, with and without the grooves, to cover the distance used in WLTP and EUDC test procedures. All input data were taken from [16]. Calculation procedure was based on power determination achieved by multiplication of the velocity and aerodynamic drag force in the travel direction. What is more, times needed to cover the distances mentioned in the test procedures were included into calculations to obtain relevant required power. Results were presented in the form of watt-hour as a unit of energy.

Table 3. Average energy consumption associated with aerodynamic losses of the analysed vehicle wheel types estimated as based on three European driving cycles.

Tyre type	High part of WLTP (16.9 m/s)	Extra high part of WLTP (26.1 m/s)	EUDC part of NEDC (19.3 m/s)
Slick tyres [Wh]	82	171	72
Tyres with grooves [Wh]	81	169	71
Δ%	1.2	1.2	1.4

It can be observed that more energy is needed to overcome the aerodynamic drag of the slick tyre what is connected with the drag difference observed on the earlier plot. For a drive with the average velocity of 90 km/h (extra high part of WLTP) this difference is twice as big as for the high part of the same cycle. This is especially meaningful at much longer distances and higher velocities (even up to

130 km/h) that are being covered during a motorway traffic. When analysing differences between wheels having slick and grooved tyres, it can be observed that only by changing the shape of the tread (adding three longitudinal grooves) it is possible to reduce the energy usage by 1.2 % up to 1.4 %.

3.3. Pressure distribution

The pressure distribution on a wheel's surface is being presented in the form of 3D visualization. Figures 3 and 4 show pressure coefficient Cp distributions from the bottom (contact region side, air flow from left to right in order to preserve the same direction in the simulations and the experimental work) and the upstream view perspectives. Results of own numerical investigations for tyre with grooves were compared to the experimental study [7]. Slightly different geometry was investigated by Dimitriou et al., where four grooves were used, and the wheel was positioned at the 2° wheel camber. What is more, during the experiment, the model was equipped with outriggers to hold it, what could have potentially caused additional flow differences.

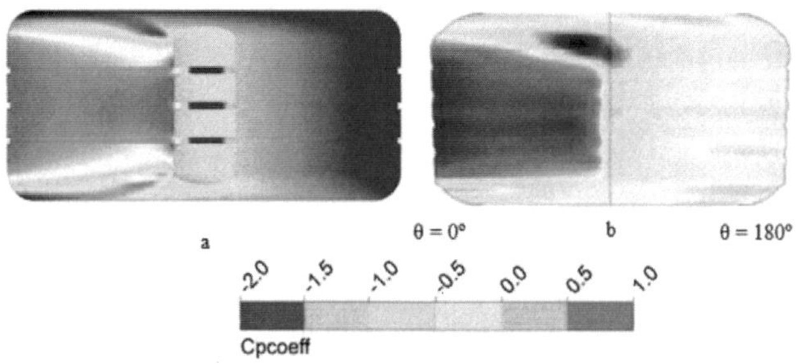

Figure 3. Bottom view of Cp pressure distribution for the tyre with the grooves – CFD investigation with velocity 36.1 m/s (a) and experiment made by [7] at 40 m/s (b)

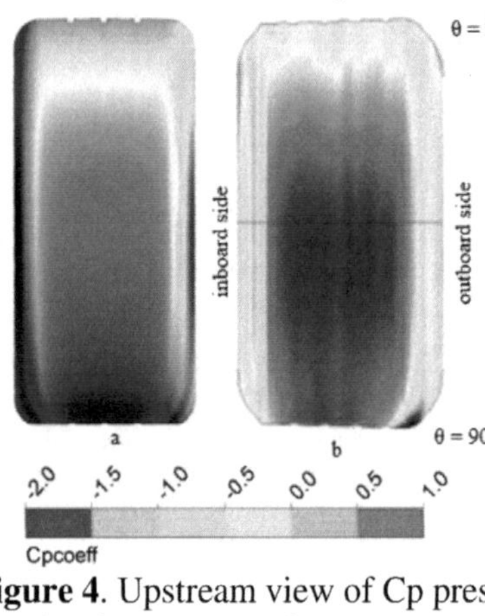

Figure 4. Upstream view of Cp pressure distribution for the tyre with the grooves – CFD investigation with velocity 36.1 m/s (a) and experiment made by [7] at 40 m/s (b)

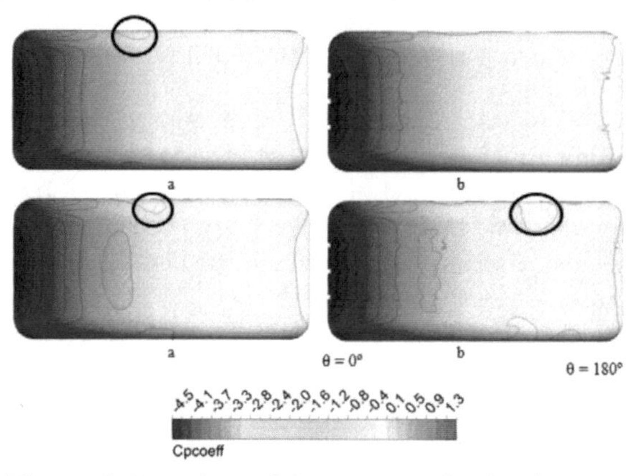

Figure 5. Top view of Cp pressure distributions for the slick tyre (a) and tyre with grooves (b) at velocities of 13.1 m/s (upper row) and of 36.1 m/s (bottom row)

Nevertheless, the upstream view of the pressure distribution shows similarities between numerical and experimental investigations around the stagnation region (θ angle equal to 0°), where the highest values of pressure can be observed. Pressure increase can be noticed in the middle part of the tyre surface, while close to the sides the decrease of pressure can be observed. The distribution differs between numerical and experimental results especially in the upper part of the visible wheel section. Boundary layer is formed from a stagnation point (θ = 0°) and separates behind the top of the wheel (around θ = 270°). The decrease of static pressure in the separation region leads to the acceleration of the stream. The small wheel camber used for the model tested experimentally causes asymmetric pressure distribution. At the outboard side the pressure drop associated with the jetting is also present in the CFD simulation. On the inboard side this is no longer true as applying camber diminishes the contact patch size and tends to equalize the pressures (Cp value is likely to approach 0 while wheel camber increases), what is visible in case of the wheel on the imprint 4b. With camber equal to 0° in the simulations, distribution remains symmetric there. In Fig. 3b it can be further observed that the small camber angle causes the pressure distribution around the wheel to be no longer symmetric. Additional discrepancies may be connected with experimental stand configuration in which the contact between wheel and the ground was not fully developed. Hence the pressure drop visible in the three grooves (Cp around -2.0) finds no similarities in the experimental results. In case of the experimental work, the decrease of pressure was also observed, however its intensity was much smaller (Cp around -1.0) due to the under-developed contact patch - wheel model made form aluminium and composite materials was unloaded and had to be kept at a certain minimal distance away from the moving ground belt. However, for the major part, pressure distributions in the centre part of the wheel from the stagnation point towards the contact region ($\theta = 0^{\circ}$ to 90°) look very similar (CFD vs. experiment). In both cases increased value of pressure is observed. Apart from the exact value of pressure coefficient in the contact region, both models show pressure decrease in that area as well as Cp increase behind the contact patch.

In order to analyse the pressure differences on the wheel surface between slick tyre and tyre with grooves in relation to the minimum and maximum velocity applied, it was decided to further compare the distribution of pressure at the top surface of the wheel. Figure 5 presents top views of pressure distribution. In the upper part of the figure results of simulations with velocity 13.1 m/s are presented, while in the bottom part those with V = 36.1 m/s. The pressure drop after passing the $\theta = 270^{\circ}$ angle is more severe for the higher values of velocity. In case of the slick tyre pressure drop near the separation region shows more regular shape than for the tyre with grooves what is connected to the fact that some portion of the air travels through the grooves. The second difference is related with the sides of the tyre (inboard and outboard). In case of the slick tyre, the pressure increase can be observed on the side of the wheel, the only difference was connected with the size which in case of higher velocity was larger (region marked in black). When comparing tyre with grooves it can be observed that for velocity equal to 13.1 m/s such a pressure increase was not observed. Results of simulation with higher velocity show increase of pressure near to the side of the wheel. Its structure and location differs in relation to the slick tyre. In this case higher pressure region is moved to the lower value of theta angle and its influence on the tread part of the tyre is more significant.

4. Conclusions

Numerical investigation was carried out to characterise the relation between Reynolds number and the value of drag for isolated rotating wheel equipped with slick tyre and tyre with three longitudinal grooves. Presented model was verified based on the experiments performed by [7,11,12]. CFD analysis showed flow characteristic for different velocities applied to the model. Relation between drag, lift coefficients and pressure force was also presented. Differences in pressure distribution between analysed cases were identified and explained in relation to the experimental studies. Due to the obtained characteristics, attention was also paid to the velocity distribution at the top of the wheel ($\theta = 270^{\circ}$). What is more, estimation of the energy usage needed to overcome the aerodynamic drag of the wheels was presented to shed the light on the energy management in terms of electric vehicle. Presented calculation may be a useful step for the introduction of the new category in the tyre labelling regulations related to the energy needed to overcome the tyre's aerodynamic resistance. Above work is only the first step in discussion about new tyre labelling categories in which significant contribution will be connected with the WLTP driving cycle and the new vehicle class division based on power to weight ratio. Tyre manufacturers may be forced to add new categories to the labelling system, not only in relation to aerodynamic resistance, but also to aquaplaning or driving stability. Nowadays customers' needs are the key factors in automotive industry and authors expect that in the near future car buyers would like to know how individual car parts (such as wheels and tyres) over which they hold the purchasing power help them save energy and extend the distance covered on a single battery charge. That is why, it is high time to open the discussion about methods and procedures of additional tests useful in establishing new labelling categories for the tyres.

The second main point is related to the creation of first aerodynamic characteristic of the wheel (equipped with slick tyre and tyre with grooves) in the wide range of the velocities. Such a characteristics performed on the experimental stand surely constitute a very important contribution to the development of the energy distribution control algorithms. It is planned to support the numerical investigations with in-house experiment, focusing on isolated as well as non-isolated wheel. Developed methodology, while verified by wind tunnel tests, could be applicable both to scientific researches and to industrial purposes.

References

[1] Piechna J 2000 Podstawy aerodynamiki pojazdów WKŁ Warszawa
[2] Mlinaric P 2007 Investigation of the influence of tyre deformation and tyre contact patch on CFD predictions of aerodynamic forces on a passenger car (Chalmers University of Technology)
[3] Regert T and Lajos T 2007 Description of flow field in the wheelhouse of cars *Int. J. of Heat and Fluid Flow* **28** 616-629
[4] Croner M, Bezard H, Sicot C and Mothay G 2013 Aerodynamic characterization of the wake of an isolated rolling wheel *Int. J. of Heat and Fluid Flow* **43** 233-242
[5] Diasinos S 2009 The Aerodynamic interaction of a rotating wheel and a downforce producing wing in ground effect (University of New South Wales)
[6] Kulak M, Karczewski M and Spolaore G 2014 Flow around rotating wheels and its interaction with vehicle aerodynamics – CFD vs. wind tunnel tests *FISITA world congress* (Maastricht) F2014-LWS-068
[7] Dimitriou I and Klussmann S 2006 Aerodynamic forces of exposed and enclosed rotating wheels as an example of the synergy in the development of racing and passenger cars *SAE Int.* **2006-01-0805**
[8] McManus J and Zhang X 2006 A Computational study of the flow around an isolated wheel in contact with the ground *J. of Fluids Eng.* **128** 520-530
[9] Ramachandran D and Doig G C 2012 Unsteady flow around an exposed rotating wheel (University of New South Wales)
[10] Landstrom C, Josefsson L, Walker T and Lofdahl L 2012 Aerodynamic effects of different tire models on a sedan type passenger car *SAE Int. J. Passeng. Cars - Mech. Syst.* **5** 136-151
[11] Fackrell J E 1974 The Aerodynamics of an isolated wheel rotating in contact with the ground (Imperial College London University of London)
[12] Mears A P, Dominy R G and Sims-Williams D B 2002 The Air flow about an exposed racing wheel *SAE Int.* **2002-01-3290**
[13] Axon L, Garry K and Howell J 1999 The Influence of ground condition on the flow around a wheel located within a wheelhouse cavity *SAE Int.* **1999-01-0806**
[14] Lesniewicz P, Kulak M and Karczewski M 2014 Aerodynamic analysis of an isolated vehicle wheel J. of Physics: Conf. Series **530** 012064
[15] Hobeika T, Sebben S and Landstrom C 2013 Investigation of the influence of tyre geometry on the aerodynamics of passenger cars *SAE Int.* **2013-01-0955**
[16] https://www.dieselnet.com/standards/cycles
[17] Lipian M, Karczewski M and Olasek K 2015 Sensitivity study of diffuser angle and brim height parameters for the design of 3 kW diffuser augmented wind turbine *Open Engineering* **5.1**
[18] Hashmi A A and Dimitriou I 2006 Needs & possibilities for the correction of drag and lift wheel forces which have been derived by integrating its static pressure distribution *SAE Int.* **2006-01-3623**

Assessment of turbulence-chemistry interaction models in the computation of turbulent non-premixed flames

M T Lewandowski[1,2] and J Pozorski[1]

[1] Institute of Fluid-Flow Machinery, Polish Academy of Sciences, Gdańsk, Poland
[2] Conjoint Doctoral School of Gdańsk University of Technology and IFFM, Gdańsk, Poland

E-mail: michal.lewandowski@imp.gda.pl, jp@imp.gda.pl

Abstract. The present work reports on the assessment of different turbulence-chemistry interaction closures for the modelling of turbulent non-premixed combustion. Two-dimensional axisymmetric simulations have been carried out based on three different laboratory flames. The methane fueled, piloted jet flame Sandia D, the simple jet syngas flame and the so-called Delft Jet-in-Hot Coflow flame are studied. All the flames can be characterised as non-premixed but differ by some features which are taken into account through appropriate modelling approach.

1. Introduction

Turbulent combustion is an interdisciplinary and broad topic. It combines turbulent flow, described by the fluid dynamics equations, and complex chemical kinetics. Turbulence enhances mixing of the reactants; on the other hand, the chemical reactions, involving the temperature rise, change density and affect the flow itself. Therefore, aside of mass and heat transfer (including radiation), the coupling between turbulence and chemistry plays a crucial role in turbulent reactive flows. Due to the multi-scale character of the phenomena in space and time [1], combustion appears as a challenging task for numerical predictions. Nowadays, computational tools are of great importance for the design of fluid-flow machinery [2]. The accurate control of turbulent flames is required, e.g., to meet the environmental restrictions on the air pollution emissions, for optimal combustion chamber design, and for the safety issues. Development of various alternative technologies can be observed in the fields of combustion systems in the energy industry. Examples are fluidized bed combustors, oxy-fuel combustion, Moderate and Intense Low Oxygen Dilution (MILD) combustion or technologies based on production and utilization of low calorific gas from biomass or different kind of waste.

The Reynolds-Averaged (RANS) methods have been widely used in industrial applications for more than 20 years. Now, due to the increase in computational capabilities, the Large Eddy Simulations (LES) are getting more popular as well. The unsteady effects are meaningful in the reactive flows, especially in complex geometries of engines or turbine combustion chambers. Therefore, the LES method seems to be much more suitable than the RANS. Although the turbulent combustion including complex chemical mechanisms is still computationally very expensive, the LES method is being applied and developed in real life applications by world leading research institutions supported by the top automotive and aerospace industries.

However, in many technical applications of local industry the cost of LES approach for combustion is prohibitive, therefore we still see the need of development for RANS methods, especially in case of specific conditions resulting from the use of new low calorific fuels or technologies such as MILD combustion. The need for comprehensive understanding of combustion leads to the study of chemical reactions in flames using experimental and numerical tools. Experimental investigations that aim for numerical model validation are often conducted on the flames issuing from the laboratory burners with the use of laser diagnostic techniques. In this work we present a numerical study of three non-premixed flames and an assessment of four turbulence-chemistry interaction (TCI) models with the use of AnsysFluent (AF) and OpenFOAM (OF) CFD codes.

2. Theory and modelling

2.1. Turbulent flow and turbulence-chemistry interaction

In order to deal with the problem of turbulent reactive flow, one has to solve the system of closed equations of motion, species transport and energy conservation [3]. In the present work the Navier-Stokes equations are subject to decomposition and Favre averaging. The closure for the turbulent fluxes can be obtained by statistical modelling based on turbulent viscosity which can be calculated as $\mu_t = \bar{\rho}\nu_t = C_\mu \bar{\rho} k^2/\epsilon$. The turbulent scalar fluxes are modelled with a gradient closure. The turbulence kinetic energy k and its dissipation rate ϵ are obtained with a two equation $k - \epsilon$ turbulence model [4]. The key difficulty in mathematical modelling of turbulent combustion is the source therm in the species transport equation. The reaction rate is based on the Arrhenius law which is highly non-linear, and it is not easy to express it as a function of mean values. Expanding the mean reaction rate as a Taylor series of the temperature fluctuation leads to various difficulties and is not commonly used [3]. Therefore existing models are based on physical analysis, comparing chemical and turbulent time scales, and most of them can be classified as one-point statistics, geometrical analysis or turbulent mixing approach [2].

2.1.1. Assumed Probability Density Function. It is one of the simplest approaches for TCI. The idea lies behind defining a mixture fraction ξ in the way that it is linked with instantaneous mass fractions Y_k in the relation $Y_k(\xi)$ and takes values 0 in the oxidizer stream inlet and 1 at the fuel inlet. This approach is valid for infinitely fast, irreversible chemistry and the Lewis number equal to one. Then, the mean species mass fractions $k = 1, ..., N$ are obtained from [3]:

$$\bar{\rho}\tilde{Y}_k = \int_0^1 Y_k(\xi)\rho(\xi)p(\xi)d\xi, \tag{1}$$

where $p(\xi)$ is the probability density function which is often assumed to be parameterised by the mean mixture fraction and its second moment (the mixture fraction variance). Then instead of N species mass fraction transport equations only those for $\tilde{\xi}$ and $\widetilde{\xi''^2}$ are solved.

2.1.2. Steady Laminar Flamelet. This model assumes that the structure of turbulent flame can be reconstructed by one-dimensional laminar flames embedded in a turbulent flow [5]. In order to do that, a so-called flamelet equation needs to be solved for a laminar flame:

$$\rho\frac{\partial Y_k}{\partial t} = \frac{1}{2}\rho\chi\frac{\partial^2 Y_k}{\partial \xi^2} + R_k \tag{2}$$

where $\chi = 2D(\nabla\xi)^2$ is the scalar dissipation rate, and its stoichiometric value may be used as a parameter for correct representation of the turbulent flow. The local derivative is neglected in a steady state approach. Then, functions $Y_k(\xi, \chi_{st})$ are stored in the flamelet libraries [3] and used to evaluate mean values of mass fractions similarly to equation (1).

Journal of Physics: Conference Series **760** (2016) 012015 doi:10.1088/1742-6596/760/1/012015

2.1.3. Eddy Dissipation Concept. The EDC is based on turbulent mixing approach. The idea is that the reactions take place there where the reactants are mixed at the molecular level and the turbulence energy dissipation takes place in so-called fine structures whose size is of the order of the Kolmogorov scale [6, 7]. Fine scales are determined through the cascade model of energy dissipation [8] from larger to smaller scales with the use of turbulence model quantities, e.g. k and ϵ. Then the relations for the mass transfer rate $\dot{m}^*$ between fine structures and their surroundings (divided by the fine-structure mass) and the ratio of the fine structure mass to the total mass γ^* are known. The final expression for the mean reaction rate is given as [7]:

$$\bar{R}_k = \frac{\bar{\rho}\gamma^*\dot{m}^*}{1-\gamma^*}(Y_k{}^* - \widetilde{Y}_k), \tag{3}$$

where Y_k^* is a mass fraction of a specie k in the fine structure which should be obtained from the solution of ordinary differential equation describing the perfectly stirred reactor (PSR).

2.1.4. Partially Stirred Reactor Model. The PaSR model recalls a detailed chemistry approach of EDC [9]. Here each computational cell is also split into two zones: reacting and non-reacting. A reacting zone has a homogeneous composition which can be treated as PSR. The mean reaction rate for the species k is given by the following expression:

$$\bar{R}_k = \kappa_k R_k \tag{4}$$

where R_k is the reaction rate of the species according to the used kinetic mechanism. The reactive fractions κ_k depend on the chemical time scale τ_{ch} and the micro-mixing time scale τ_m.

3. Experimental flames

In the current study we have simulated three flames. A commonly studied piloted flame Sandia D and a simple jet syngas flame CHN case B are from the TNF Workshop data archive [10]. The third experiment has been conducted at the Delft University of Technology by Oldenhof et al. [11]. The scalar data of Sandia D are from the work of Barlow and Frank [12] at Sandia National Laboratories, whereas velocity measurements were done by Schneider et al. [13] at Darmstadt University of Technology. Corresponding measurements for the flame CHN were conducted by Barlow et al. [14] and Flury [15] at ETH Zurich, respectively. The characteristic features of the three flames are presented in table 1. For the flame Sandia D there are available axial and six radial measurements, for CHN B axial and five radial and for DJHC there are axial and six radial. Additionally in the case of CHN and DJHC velocity measurements were done at the jet exit and 3 mm above it, respectively. For the DJHC burner also the radial temperature profile was available. These data were used for inlet boundary conditions in the simulations. For the flame Sandia D the uniform values of velocity and temperature were set as a boundary condition on the pre-inlet pipe in order to achieve a proper profile. Grid size independence studies were performed in the all three cases. We have used RMS values to plot the error bars of the experimental data presented in the next sections.

Table 1. The characteristic parameters of the flames [12, 14, 11]: fuel composition, burner type, jet Reynolds number and diameter of the fuel nozzle.

	Sandia D	CHN B	DJHC
Fuel	25% CH_4, 75% air	40% CO, 30% H_2, 30% N_2	81% CH_4, 4% C_2H_6, 15% N_2
Type	Piloted jet flame	Simple jet flame	Jet-in-Hot-Coflow
Jet *Re*	22400	16700	4100
Nozzle d	7.2 mm	7.72 mm	4.5 mm

Journal of Physics: Conference Series **760** (2016) 012015 doi:10.1088/1742-6596/760/1/012015

4. Results

We have run a large number of simulations for the three non-premixed flames from which we present only the chosen ones. We have focused on the axial and radial distributions of scalar values as well as on locating the position of the highest deterioration of the temperature value, which may be an important information of practical meaning in the real life applications. In table 2, simulation series are summarized together with the maximum error in temperature prediction with its relative error and position. In the following sections, quantitative results are shown for the axial and radial distributions of temperature, velocity and OH radical mass fraction.

Table 2. The locations of the highest deviations of numerical results of temperature from the experimental values together with the relative error obtained with different approaches on the three flames. AF and OF denote AnsysFluent and OpenFOAM simulation, respectively.

Flame	Solver	Turbulence model	TCI model	$[z/d; r/d]$	ΔT [K]	rel. err. [%]
SandiaD	AF	standard k-ϵ	β PDF	[15; 1.9]	818	122
	AF	standard k-ϵ	SL Flamelet	[15; 1.9]	835	125
	AF	realizable k-ϵ	EDC	[15; 1.9]	872	130
	AF	s. k-ϵ, $C_{\epsilon 1} = 1.6$	β PDF	[15; 1.7]	589	62
CHN B	AF	s. k-ϵ, $C_{\epsilon 1} = 1.587$	β PDF	[20; 2.3]	523	60
	AF	s. k-ϵ, $C_{\epsilon 1} = 1.587$	SL Flamelet	[20; 2.5]	499	66
	OF	standard k-ϵ	EDC	[20; 2.3]	715	83
	OF	standard k-ϵ	PaSR	[20; 2.3]	855	100
	AF	standard k-ϵ	β PDF	[20; 2.7]	771	124
DJHC	AF	realizable k-ϵ	EDC	[20; 2.9]	385	27
	AF	realizable k-ϵ	EDC modified	[20; 7.6]	147	21
	OF	standard k-ϵ	EDC	[13.3; 2.2]	329	25
	OF	standard k-ϵ	EDC modified	[13.3; 8.7]	173	34

4.1. Sandia D flame

Three different TCI models were used in the simulation of Sandia D, namely β pdf, steady laminar flamelet and the EDC. All were used within AF code on a 2352-cell 2-dimensional anxisymmetric grid with a steady state simulation [17]. SIMPLE algorithm was used for pressure-velocity coupling and a second order discretisation method were applied. In the case of β PDF a 17 species PDF table was created, the flamelet library was obtained with GRI 3.0 detailed chemical mechanism and skeletal mechanism for CH_4 with 41 reactions and 16 species was applied in case of EDC. For turbulence closure the two equation $k - \epsilon$ model was used for β PDF and flamelet and realizable $k - \epsilon$ in case of EDC. Additionally we have performed a simulation for β PDF with standard $k - \epsilon$ model with $C_{\epsilon 1}$ constant changed from 1.44 to 1.6

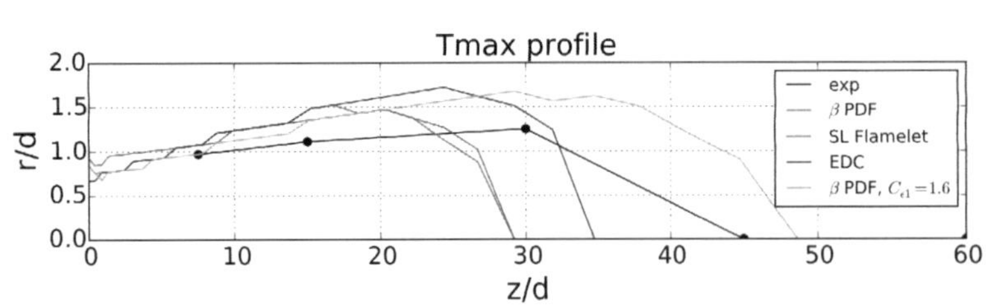

Figure 1. Positions of the peak temperature in the radial distributions presented in the non-dimensional axial and radial coordinates for the four simulations of the flame SandiaD.

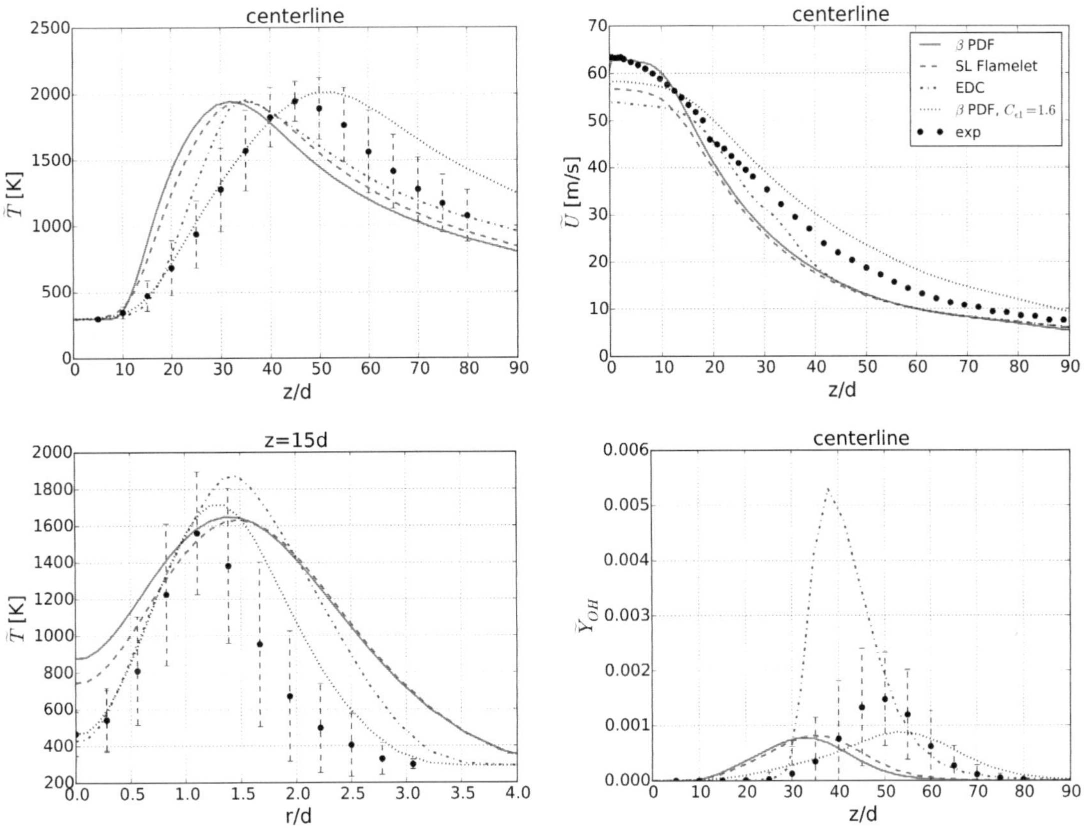

Figure 2. The results of simulations with AF for the flame Sandia D with three different turbulence-chemistry interaction models. The temperature distribution along the axis (top left) and the radial profile at the position $z = 15d$ (bottom left), the axial velocity values along the centerline (top right) and the mean mass fraction of OH radical (bottom right).

as it is common practice in the case of round jets [6, 16]. The effect of this correction is visible by a prediction of higher velocity values along the centerline in comparison to the standard model, what in fact shows better agreement with the experiment. On the other hand, it has a very strong impact on the axial temperature distribution by shifting the maximum value more downstream the flame. The upstream part predictions are very good but the downstream part temperature is slightly overestimated. Besides, predictions of the temperature distribution with the β PDF and flamelet models are quite similar whereas the EDC model shows a narrower profile of axial temperature what better corresponds to the experiment. The position of the maximum temperature might be adjusted with the use of $C_{\epsilon 1}$ constant as long as the flow results are in agreement with the experimental data. Results in fig. 1 show positions of the maximum temperature value in the radial distribution in the function of axial distance from the nozzle, what roughly indicates the flame zone and near stoichiometric region. It can be seen that the numerically predicted flame is always slightly wider and shorter without correction to $k - \epsilon$ model for the round jet. The mass fractions of OH radical distributions are predicted poorly with all the three models what is visible in fig. 2. For the β PDF and flamelet the maximum value in the axial distribution is underestimated whereas for the EDC it is highly overestimated. In the case of the latter model this might be improved by the use of a more detailed chemical mechanism.

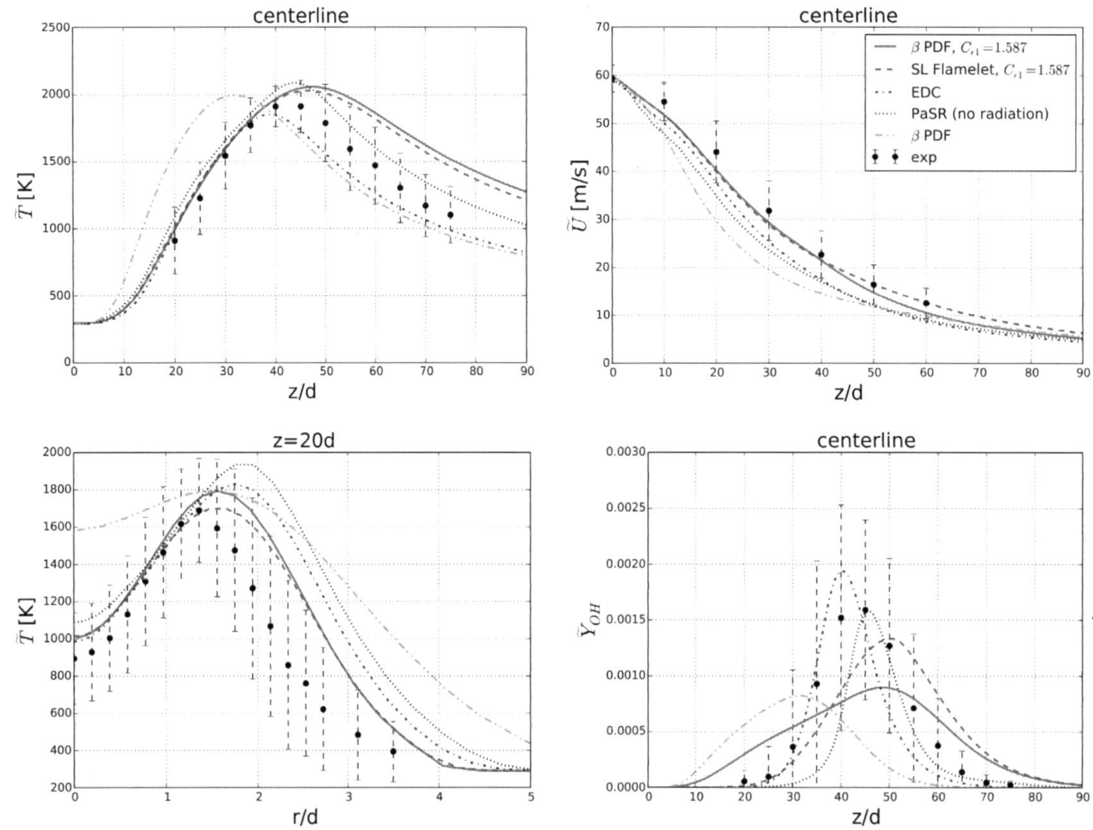

Figure 3. The results of AF and OF simulations for the syngas flame CHN B with four turbulence-chemistry interaction models. The temperature distribution along the axis (top left) and the radial profile at the position $z = 20d$ (bottom left), the axial velocity values along the centerline (top right) and the mean mass fraction of OH radical (bottom right).

4.2. Syngas CHN B flame

This flame was modelled with four TCI models: β pdf and steady laminar flamelet used within AF code and EDC and PaSR model used within OF code (edcPisoFoam [6] and reactingFoam solver, respectively). Simulations were carried out on axisymmetric 9949-cell grid as a steady state in AF and transient in OF [18]. Therefore the pressure-velocity coupling in the first case was done with the SIMPLE algorithm and with PISO in OF. In OF simulations a first-order implicit Euler method for the time integration was used and in all the simulations second order methods were used for space discretisation. The PDF table was constructed with 17 species and flamelet library was obtained with GRI 3.0 detailed chemical mechanism.

For the two mixing models, EDC and PaSR, several reduced chemical mechanisms were tested from which the optimal choice appeared to be the one proposed in [19] which consists of 13 species and 34 reactions. We have set the value of $C_{\epsilon 1}$ constant to 1.587 according to the formula given in [20]. This modification was applied to the AF simulations but not in case of OF where the corrected value of $C_{\epsilon 1}$ led to overestimation of the velocity centerline profile and drastically shifted the axial temperature distribution more downstream the jet leading to serious discrepancies from the experiment. The results are shown in fig. 3, where again the axial temperature distributions from β PDF method and SL flamelet are comparable. Similar situation was observed in the case of EDC and PaSR when no radiation model was applied.

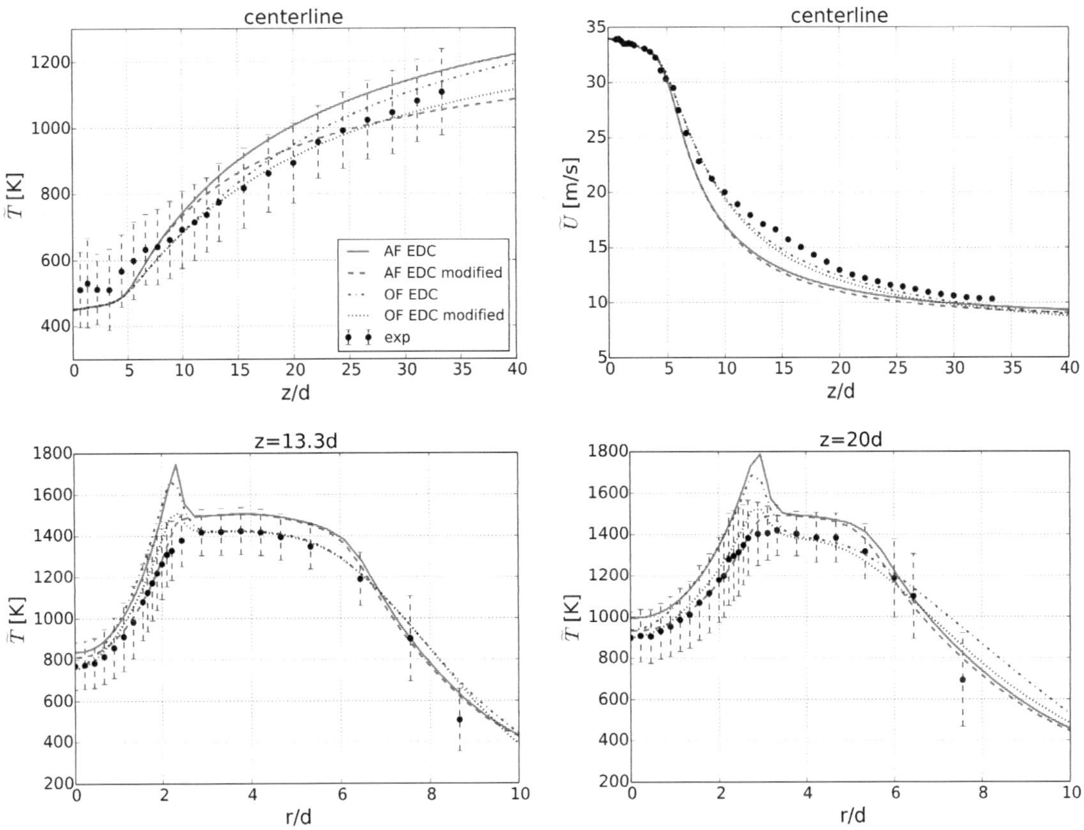

Figure 4. The EDC simulations results with AF nad OF for the flame DJHC-I $Re = 4100$ with and without modifications. The temperature distribution along the axis (top left) and two radial profiles at the position $z = 13.3d$ (bottom left) and $z = 20d$ (bottom right). Top right plot presents the axial velocity values along the centerline.

4.3. Delft-Jet-in-Hot-Coflow flame

Similar analysis as in the previous cases cannot be easily performed for the DJHC-I flame. Three different stream inlets are present here what would require the two mixture fraction approach. In MILD combustion regime the reaction rates are slower and the Damköhler number approaches unity. This requires including the finite rate chemistry. Besides, the reaction zone resembles well stirred reactor rather than laminar flamelets embedded in a turbulent flow. Therefore we have decided to use the EDC model with DRM19 chemical mechanism which consists of 19 species and 84 reactions as it proved to perform relatively well in MILD conditions in other investigations [16]. As it can be seen in fig. 4, the standard formulation of the EDC leads to a high overestimation of the peak temperature in the radial distributions. The specific features of MILD combustion are claimed [16] to be responsible for this effect. In order to overcome this problem, several modifications were proposed [16, 21, 22, 23] and most of them concern the change in the EDC model constants. We present the results of four simulations obtained with the two codes, AF and OF, with and without modifications. In the case of AF the modification concerns the change in EDC constants. Both values of C_γ and C_τ were changed to 1 and 3, respectively, as proposed by De et al. [16] and Evans et al. [21]. This approach, however, rises concerns about its generality and consistency with the turbulence models. Therefore in case of OF simulation we have applied clipping of γ^* greater than 0.5, what reduces the reaction rate for low Reynolds numbers ($Re_T < 84$) but does not impact the model at fully turbulent conditions. It must be noted that due to the implementation differences this approach is not equivalent to the default clipping mentioned by De et al. [16] and discussed by Shiehnejadhesar et al. [24]. The adequacy of the proposed approaches is the subject of our other ongoing investigations. It should be also noted that the effect of radiation model in the case of OF simulation is stronger than in AF as evidenced by a lower temperature in the coflow region in radial distributions presented in fig 4. It is also found that both modifications reduce the peak value of the temperature so that the highest error is shifted off the jet to the outer edge of the coflow.

5. Conclusions

In general, the predictions of temperature for the Sandia D and CHN flames are in relatively good agreement with the experimental data for used TCI models. The highest deviations from the experiment occurred downstream the flame. Though, the largest temperature error was observed at the axial location 15-20 diameters downstream of the nozzle and 2.5-3 diameters off the jet axis for all the models. This is due to the shift of reaction zone more outside the axis upstream the jet and the predicted flames were wider. This discrepancy concerns the location of the flame zone and not the temperature value itself. In the case of DJHC burner this kind of deviation was not observed due to the presence of hot coflow. Yet, the maximum temperature was highly overestimated, what on the other hand was not the problem in the two previous flames. Two different modifications were applied to solve this problem: change in the EDC constants (as in the literature) and newly proposed clipping of γ^*. Additionally, we have observed some differences between the AF and OF results of velocity distributions that might be caused by the round jet configuration. As presented, the chosen TCI models are able to properly simulate different non-premixed flames that differ in fuel composition and specific features. The future work will focus on the modelling of combustion of low calorific fuels in MILD regime.

Acknowledgment

We are indebted to Prof. Dariusz Kardaś for his courtesy to access PC21 cluster for AnsysFluent runs. Calculations with OpenFOAM were carried out at the Academic Computer Centre in Gdańsk.

References

[1] Peters N 2009, *P. Combust. Inst.* **32** 1-25
[2] Veynante D and Vervisch L 2002 *Prog. Energ. Combust.* **28** 193-266
[3] Poinsot T and Veynante D 2012 *Theoretical and Numerical Combustion*, Third Edition, Bordeaux
[4] Launder B E and Spalding D B 1974 *Comput. Method. Appl. M.* **3** 269-289
[5] Peters N 2000 *Turbulent Combustion*, Cambridge University Press
[6] Lysenko D A, Ertesvåg I S and Rian K 2014 *Flow Turbul. Combust.* **93** 577-605
[7] Magnussen B 2005 *ECCOMAS Thematic Conference on Computational Combustion*, Lisbon
[8] Ertesvåg I S and Magnussen B F 2000 *Combust. Sci. Technol.* **159** 213-235
[9] Golovitchec V I and Chomiak J 2001 *The 4th International Symposium on HiTAC and Gasification*, Rome
[10] http://www.sandia.gov/TNF/abstract.html (10.06.2016)
[11] Oldenhof E, Tummers M J, van Veen E H and Roekaerts D J E M 2011 *Combust. Flame* **158** 1553-1563
[12] Barlow R S and Frank J H 1998 *P. Combust. Inst.* **27** 1087-1095
[13] Schneider Ch, Dreizler A, Janicka J and Hassel E P 2003 *Combust. Flame* **135** 185-190
[14] Barlow R S, Fiechtner G J, Cartner C D and Chen J Y 2000 *Combust. Flame* **120** 549-569
[15] Flury M 1998 *Exp. Analyse der Mischungstruktur in turbulent nicht vorgemischten Flammen* ETH Zurich
[16] De A, Oldenhof E, Sathiah P and Roekaerts D 2011 *Flow Turbul. Combust.* **87** 537-567
[17] Lewandowski M T and Pozorski J 2014 in Szewczyk M (Eds), *Applications of Thermodynamic Analysis in Description of Physical Phenomena In Energy Devices* (in Polish), OWPRz, Rzeszów
[18] Lewandowski M T and Pozorski J 2015 *P. Eur. Combust. Meet.* art. **P4-58**, Budapest
[19] Frassoldati A, Faravelli T and Ranzi E 2007 *Int. J. Hydrogen Energ.* **32**:3471-3485
[20] McGuirk J J and Rodi W 1979 *1st Symp. on Turbulent Shear Flows* 71-83, Pennsylvania State University
[21] Evans M J, Medwell P R and Tian Z F 2015 *Combust. Sci. Technol.* **187** 1093-1109
[22] Parente A, Malik M R, Cantino F, Cuoci A and Dally B B 2016 *Fuel* **163** 98-111
[23] Aminian J, Galletti C and Tognotti L 2016 *Fuel* **165** 123-133
[24] Shiehnejadhesar A, Scharler R, Goldin G M and Obernberger I 2014 *Fuel* **126** 177-187

A study of geometrical structure of packed beds using flow simulation with the immersed boundary method

M Marek
Faculty of Mechanical Engineering and Computer Science, Częstochowa University of Technology, al. Armii Krajowej 21, 42-200, Częstochowa, Poland

E-mail: marekm@imc.pcz.czest.pl

Abstract. In this work, a new method is proposed for calculation of one of the essentials geometrical parameters characterising a packed bed – tortuosity. The geometry of random packings of cylinders and rings (with various packing densities) is numerically generated with Discrete Element Method approach. For flow calculations in a highly complex geometry of a random packed bed, immersed boundary method is employed (with direct forcing variant). The obtained flow velocity field allows for calculation of tortuosity using two methods: by passive advection of massless particles and by comparison of the mean flow velocity magnitude with the mean streamwise velocity component. It is demonstrated that for the packings of low density these two methods are in a good agreement but may diverge for dense packings. Frequency distribution of tortuosity values for individual flow paths is also examined. The results show more frequent occurrence of low-tortuosity paths for coarse packings.

1. Introduction

Random packed beds – fixed packings of randomly oriented particles – are widely employed in many areas, especially in chemical and process engineering as a porous medium of a large specific surface enhancing the mass transfer or reaction rate between reacting substances (e.g. in trickle bed reactors or absorbers). The flow characteristics in such beds are thus of a significant practical importance but also are interesting from purely scientific point of view. According to the well-known Darcy formula average flux $\mathbf{q}$ of a fluid in the low-Reynolds number regime is proportional to the imposed pressure gradient:

$$\mathbf{q} = -\frac{k}{\mu}\nabla p \tag{1}$$

where μ - dynamic fluid viscosity and k - *permeability* of the porous medium directly related to its geometrical structure. In one of the first formulas for k, introduced by Kozeny, permeability depended only on the *porosity* ε of the medium (or *void fraction* in the context of packed beds – ratio of volume of empty space between the particles to the total volume of the bed) and its specific surface S (pore surface area per unit volume of the bed) but later was modified by Carman to include another geometrical parameter – *tortuosity* τ [1,2]:

$$k = \frac{\varepsilon^3}{cS^2\tau^2} \tag{2}$$

where c - structural parameter depending on the cross-section of the pore channels.

Tortuosity can be defined in many different ways generally expressing the degree of deformation of flow paths within the medium. For an individual path we define its tortuosity as [2]:

$$\tau = \frac{L_{path}}{L} \tag{3}$$

where L_{path} is the actual length of the path and L - the thickness (linear size) of the porous medium. When there are no obstacles in the medium, the path length coincides with L giving $\tau = 1$ but, as in packed beds only the void space can be penetrated by the flow, the paths follow a very complex pattern and $\tau > 1$ (for some structures even $\tau \gg 1$). Tortuosity as a parameter of a medium, not just of an individual path, takes into account an average path length representative to the whole volume of the medium.

In the literature, there are many attempts to find a relation between porosity (void fraction) ε and tortuosity (see a short review in [1]). Such a relation cannot be, of course, universal as it is possible to provide examples of porous media with the same porosity and arbitrary large tortuosities. However, for structures of particular type – e.g. media consisting of randomly distributed grains of a given regular shape – tortuosity can be shown to be a function of porosity through numerical simulations [1,2] or theoretical considerations [3].

One of the approaches to calculation of tortuosity is based on purely geometrical considerations as determination of shortest path between different points of the medium [4,5]; a little different approach is used in [6]. These methods have relatively low computational cost but their main disadvantage is that they do not take into account the physical flow structure in the medium. In principle, it is possible that pore channels comprising the shortest paths are characterised by large flow resistance and the major part of the fluid is transported through totally different paths. In such a case, geometrically-based estimations of tortuosity may be significantly distorted. It seems that physically well-grounded calculations of tortuosity for general packings have to include the information about the real flow.

However, simulation of a fluid flow in such a complex geometrical structure as a random packed bed is definitely not an easy task. Although flow analysis in porous media can be efficiently performed at the level of macroscopic modelling and averaged flow equations [7,8], calculation of tortuosity requires modelling at the microscopic level (of individual bed particles). Application of classical finite-volume solvers for unstructured grids fitted to the particles is not straightforward, still has been tried by some researchers [9,10,11] and for dense packings, particularly of spheres, modifications of the original geometry (artificial gaps or bridges between particles) are necessary for generation of computational grids with sufficient quality [12,13]. Other popular approach is a simulation of flow with Lattice Boltzmann Method (LBM) [14,15]; see particularly [1,16] where LBM was applied to supply flow information for calculation of tortuosity of simple porous media.

In the present work, another method is proposed in which flow through a random packing is simulated with the classical second-order finite-volume solver for structured grids. The complex domain geometry is handled with the immersed boundary method originated by Peskin [17]. The paper is organised as follows. Section 2 contains the description of the employed methods – for numerical generation of the geometry of random packings (cylinders and rings), for flow calculations and estimation of the tortuosity parameter given the flow velocity field. In Section 3 sample results are presented and discussed. The paper closes with Summary in Section 4.

2. Description of the methods

2.1. Numerical generation of a random packed bed geometry

The geometry of random packed beds can be obtained experimentally from real beds with the use of computer tomography techniques [18] but here a different approach is employed – Discrete Element Method (DEM) simulation using method presented in [19]. The main idea of the method is the simulation of the process in which the particles are sequentially dumped into a cylindrical container (column) and move due to gravity and reaction forces from other particles or the column wall. When mechanical equilibrium is attained, a particle is incorporated into the bed structure and remains at rest to the end of the simulation. Sample random packings of cylinders and rings are shown in Figure 1.

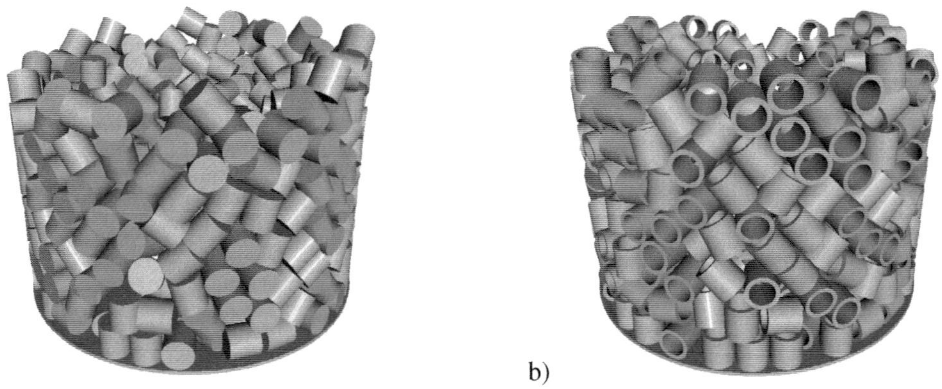

a) b)

Figure 1. Sample numerically generated random packed beds of cylinders (a) and rings (b).

The density of the packing can be increased by inclusion of additional force acting on the falling particles in the direction from the column's axis to its walls. Without this force the void fraction of the generated beds is relatively large, in fact, the largest possible to obtain in experiments. As demonstrated in [19,20] the algorithm is able to reproduce the structure of real beds, in particular void fraction distribution [18,21], with a sufficient accuracy.

For further calculations, at first, beds consisting of 2000 particles of diameter D_p and height H_p (in case of rings the inner diameter was D_i) were generated in a cylindrical container of diameter D_c. As the base of the container introduces a distinct ordering of the particles [20], for the flow simulation only a part of the bed of height L was selected positioned over $5D_p$ above the container's bottom to obtain more representative sample of the packing.

Four random packed beds were analysed denoted as C0, C1, R0 and R1, where the letter corresponds to type of the particle – cylinder (C), ring (R) – and number denotes packing density – coarse packings (0), dense packings (1).

2.2. *Flow solver and immersed boundary method*

The fluid flow is solved with the in-house finite volume code developed at the Institute of Thermal Machinery based on regular MAC staggered grid and projection method [22,23,24]. Denoting flow velocity of the incompressible flow as $\mathbf{u}=[u_x,u_y,u_z]$, the flow equations take the form:

$$\rho\frac{\partial\mathbf{u}}{\partial t}+\rho\mathbf{u}\cdot\nabla\mathbf{u}=-\nabla p+\nabla\cdot\mu\left(\nabla\mathbf{u}+\nabla\mathbf{u}^T\right),\quad \nabla\cdot\mathbf{u}=0 \tag{4}$$

where ρ - density of the fluid.

Time discretisation of equation (4) is performed with second order Runge-Kutta scheme. Spatial derivatives are also of second order – upwind scheme for the advection term and central scheme for the viscous term. The pressure equation in the projection method is solved with preconditioned conjugate gradient method (IPCG). More detailed description of the solver can be found in [25].

As calculations on regular structured grids are usually limited to simple domain geometries, the simulations in such complex domains as random packings require special treatment. In this case so-called immersed boundary method (IBM) is used [17,26,27,28]. The idea of the method is to take as the computational domain a regular area (e.g. a box) encompassing the actual flow domain and to modify equation (4) by inclusion additional source terms in order to enforce correct flow velocity on the „immersed" boundaries (here, on the particles' surface and the column wall). When the immersed boundaries are stationary with no-slip boundary condition, the flow velocity should be damped on them and it can be done after discretisation of the flow equation without explicit additional terms (direct forcing approach [26,28]). In the current solver, the damping of flow velocity is attained with the use of smooth damping function constructed from the actual geometry of the packing. As demonstrated in [25] this model leads to calculated values of pressure drop in packed beds of rings in a good agreement with empirical formulas.

The flow configuration is shown in Figure 2. A gas characterised by density $\rho = 1.15$ [kg/m^{3}] and viscosity $\eta = 1.85 \times 10^{-5}$ [Pa·s] enters the domain at the top with uniform velocity profile (v_g=0.005 [m/s]). In order to separate the inlet and outlet planes from the bed, empty buffer zones were introduced above and below the packing. The calculations were performed on cartesian grid with 160x192x160 nodes in respective directions until the steady state was reached

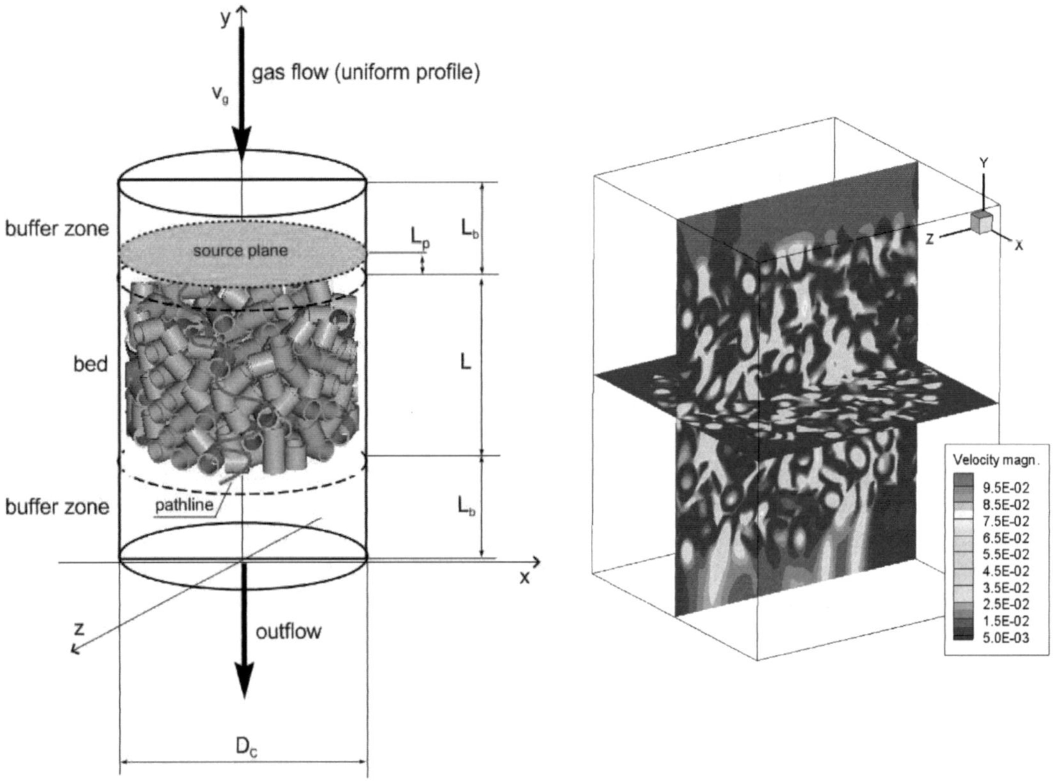

Figure 2. Configuration of the system and sample velocity field obtained with the flow solver.

2.3. Passive advection of massless particles

When the steady state flow velocity **u** is known, the tortuosity of the streamlines can be found by passive advection of massless particles and calculation of the length of their trajectory (path). In this method the following equation is integrated:

$$\frac{d\mathbf{x}_p}{dt} = \mathbf{u}(\mathbf{x}_p) \tag{5}$$

where $\mathbf{x}_p$ is the position vector of a given particle. Initially the particles are distributed on a regular dense grid in the source plane located at distance L_p over the top of the bed (see Figure 2). The equation (5) is integrated with 4th order Runge-Kutta method and time step $\Delta t = 10^{-3}$s. As the flow velocity is available only at the faces of the computational cells (staggered arrangement), trilinear interpolation was used to estimate the velocity in arbitrary points of the domain. It should be noted that the trajectories of different particles can significantly vary in length. Some of the paths end prematurely (see Figure 3a) because of the errors of integration or the fact the velocity field is extremely complex and not exactly divergence-free (Figure 3b). For the analysis of tortuosity only the trajectories covering at least 2/3 of the bed height were taken into account (about 80% of the total number of paths). The tortuosity calculated with this method will be further denoted as τ_{adv}.

2.4. Method of mean velocities

Explicit calculation of trajectories is not the only way of estimation of bed tortuosity. In the literature another method was proposed in which tortuosity can be obtained directly from the flow velocity field. According to [16]:

$$\tau_{mvel} = \frac{\langle u \rangle}{\langle u_y \rangle} = \frac{\int_V u(\mathbf{r})dV}{\int_V u_y(\mathbf{r})dV} \tag{6}$$

where the integration is performed over the void space in the packing (as the velocity equals 0 inside immersed boundaries integration over the whole computational domain gives the same result). This method will be further referred to as the *method of mean velocities* and the value estimated from formula (6) will be denoted as τ_{mvel}.

3. Results and discussion

The methodology presented in Section 2 was used for calculation of tortuosities of four random packings: C0, C1, R0, R1 (see 2.1). The dimensions of the particles, the container and other characteristic lengths are shown in Table 1.

Table 1. Dimensions of the particles and the container.

	Size	Unit
D_p	0.025	[m]
D_i	0.019	[m]
H_p	0.025	[m]
D_c	0.30	[m]
L	0.30	[m]
L_b	0.06	[m]
L_p	0.02	[m]

In Table 2 the results were gathered for the tortuosities calculated with both methods: advection of particles (averaged value for sufficiently long paths) and mean velocities. We can see that for packings of lower density (C0 and R0) the agreement between the methods is very good. However, in the case of dense packings (C1 and R1) the discrepancies cannot be neglected. It must be mentioned that increased packing density is related not only to the reduced void fraction but also introduces additional ordering of the particles [18,20] which could influence the tortuosities of individual flow paths and the result of the averaging process, especially for the method based directly on the tortuosity definition (3).

In order to test the influence of grid density on the results, τ_{adv} for the most demanding case R1 was calculated on four different grids: 80x112x80, 120x168x120, 160x192x160 and 200x280x160 giving 1.257, 1.291, 1.306 and 1.309 respectively. Good grid convergence justifies the use of the selected grid (160x192x160) also for other considered cases.

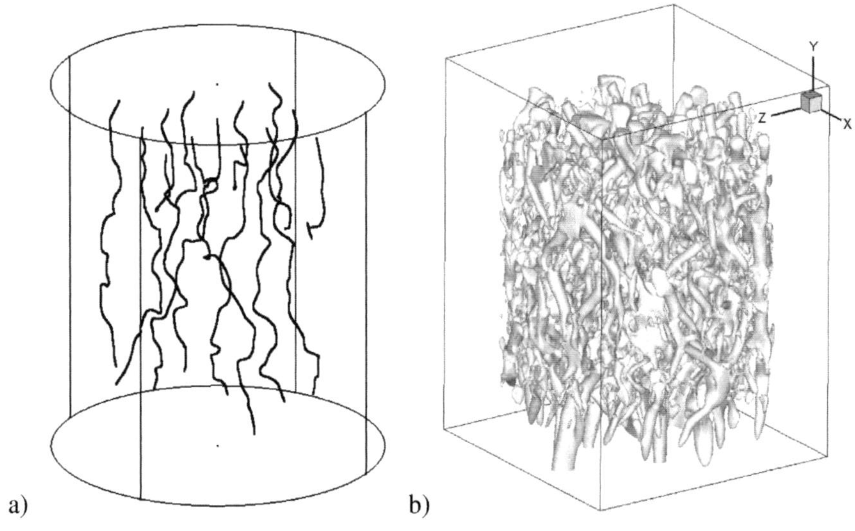

Figure 3. Sample particles' trajectories (a) and iso-surface of the flow velocity for R1 case (b)

Table 2. Calculated tortuosities for the considered random packings.

Case	ε	τ_{adv}	τ_{mvel}
C0	0.469	1.293	1.308
C1	0.390	1.433	1.337
R0	0.765	1.291	1.299
R1	0.728	1.306	1.348

In the context of works connecting void fraction with tortuosity, it is worth noting that although void fraction of random packing of rings is prominently larger than that of packing of cylinders, the tortuosities are in fact comparable (see C0 and R0 cases). However, rings (and other hollow particles, in general) are definitely not typical particles considered in the search for $\tau = \tau(\varepsilon)$ formula. As the obtained results show, the void space inside the rings contributes to the total void fraction of the bed but the complexity of flow paths is similar as in the case of full cylinders.

The method based on the advection of particles give information about tortuosity of each individual path, so beside the calculation of the averaged value representative to the whole bed more detailed characteristics, like the frequency distribution of particular tortuosity values, can also be determined (Figure 4). In all analysed cases, the calculated values of τ_{adv} are in major part in the range 1-1.6 (however, for C1 case values even as high as 6 can be found). Values in the range 1.2-1.25 are generally the most frequent. As could be expected, for the packings of lower density (C0, R0) the contribution from low values of τ_{adv} (below the maximum) is more pronounced. This could result from wider pore channels extending over larger parts of the bed, in comparison with dense random packings.

4. Summary

The method proposed in this work can be employed for physically-base calculation of tortuosity of random packings using detailed information about flow velocity field in the low Reynolds number regime. The IBM method allows for application of a classical MAC solver to solution of the flow in complex geometries by simple modification of the projection method and generation of unstructured

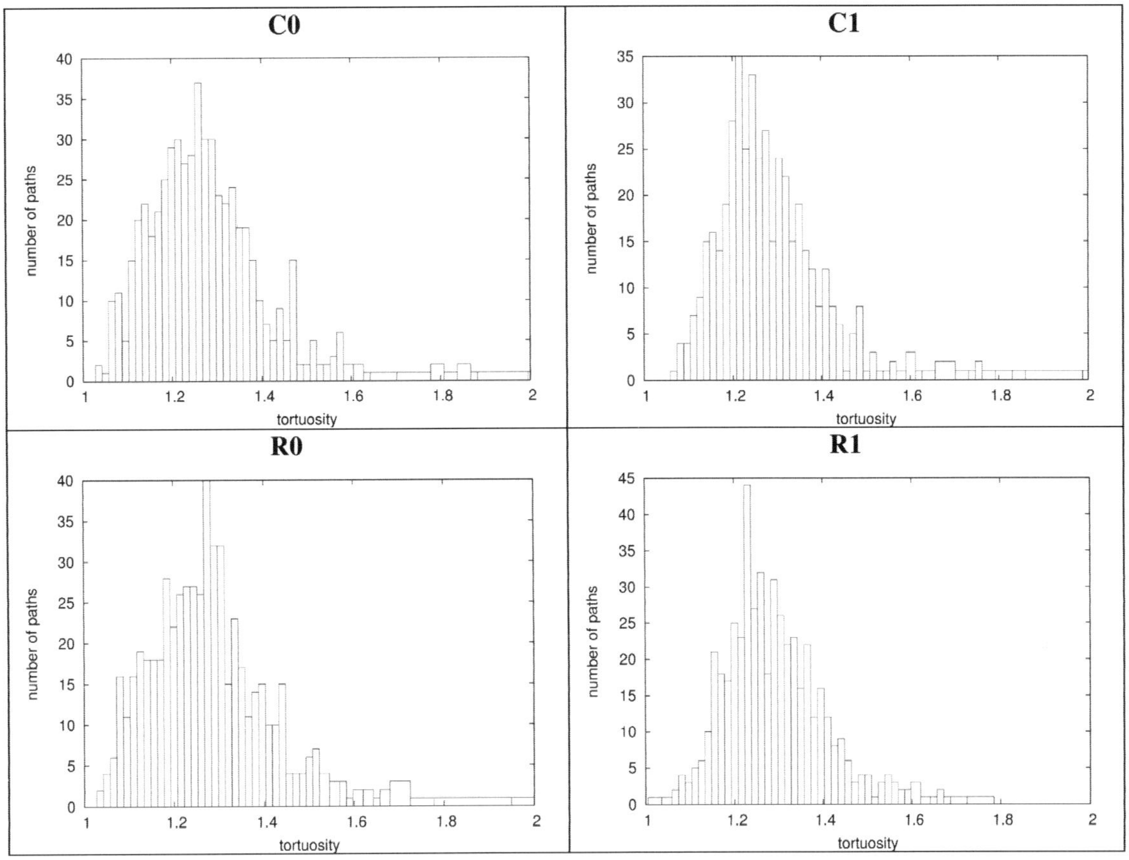

Figure 4. Frequency distribution of tortuosities calculated for individual particle trajectories in various random packings.

grid fitted to the boundaries is not necessary. Given steady state flow velocity field the tortuosity can be estimated directly from this field and for coarse packings of rings and cylinders the obtained values coincide with the ones determined explicitly from the definition of tortuosity. For dense packings, however, the results are markedly different.

References

[1] Matyka M, Khalili A and Koza Z 2008 Tortuosity-porosity relation in porous media flow, *Physical Review E* **78** 026306

[2] Koponen A, Kataja M and Timonen J 1996 Tortuous flow in porous media *Physical Review E* **54**(1) 406-410

[3] Lanfrey P Y, Kuzeljevic Z V and Dudukovic M P 2010 Tortuosity model for fixed beds randomly packed with identical particles *Chemical Engineering Science* **65** 1891–1896

[4] Wua Y S, van Vliet L J, Frijlink H W and van der Voort Maarschalk K 2006 The determination of relative path length as a measure for tortuosity in compacts using image analysis *European Journal of Pharmaceutical Sciences* **28** 433–440

[5] Gommes C J, Bons A J, Blacher S, Dunsmuir J H and Tsou A H 2009 Practical Methods for Measuring the Tortuosity of Porous Materials from Binary or Gray-Tone Tomographic Reconstructions *AIChE Journal* **55**(8) 2000-2012

[6] Sobieski W, Zhang Q and Liu C 2012 Predicting Tortuosity for Airflow Through Porous Beds Consisting of Randomly Packed Spherical Particles *Transp Porous Med* **93** 431–451

[7] Bazmi M, Hashemabadi S H and Bayat M 2012 CFD simulation and experimental study of liquid flow mal-distribution through the randomly trickle bed reactors *Int. Commun. Heat Mass* **39** 736-743
[8] Asendrych D, Niegodajew P and Drobniak S 2013 CFD modelling of CO_2 capture *Chem. Proc. Eng.* **34**(2) 269-282
[9] Augier F, Idoux F and Delenne J Y 2010 Numerical simulations of transfer and transport properties inside packed beds of spherical particles *Chemical Engineering Science* **65** 1055-1064
[10] Nijemeisland M, Dixon A G and Stitt E H 2004 Catalyst design by CFD for heat transfer and reaction in steam reforming *Chemical Engineering Science* **59**, 5185-5191
[11] Robbins D J, El-Bachir M S, Gladden L F, Cant R S and von Harbou E 2012 CFD modeling of single-phase flow in a packed bed with MRI validation *AIChE Journal* **58**(12) 3904-3915
[12] Bai H, Theuerkauf J and Gillis P A 2009 A coupled DEM and CFD simulation of flow field and pressure drop in fixed bed reactor with randomly packed catalyst particles *Ind. Eng. Chem. Res.* **48**(8), 4060-4074
[13] Finn J, Apte S 2013 Relative performance of body fitted and fictitious domain simulations of flow through fixed paced beds of spheres *International Journal of Multiphase Flow* **56** 54-71
[14] Manjhi N, Verma N, Salem K and Mewes D 2006 Simulation of 3D velocity and concentration profiles in a packed bed adsorber by lattice Boltzmann methods *Chemical Engineering Science* **61**, 7754-7765
[15] Rong, L, Dong K and Yu A 2013 Lattice-Boltzmann simulation of fluid flow through packed beds of uniform spheres: Effect of porosity *Chemical Engineering Science* **99**, 44-58
[16] Duda A, Koza Z and Matyka M 2011 Hydraulic tortuosity in arbitrary porous media flow *Physical Review E* **84** 036319
[17] Peskin C S 1972 Flow patterns around heart valves: a numerical method *Journal of Computational Physics* **10** 252-271
[18] Zhang W, Thompson K, Reed A H and Beenken L 2006 Relationship between packing structure and porosity in fixed beds of equilateral cylindrical elements *Chem. Eng. Sci.* **61** 8060
[19] Marek M 2013 Numerical generation of a fixed bed structure *Chem. Proc. Eng.* **34** (3) 347-359
[20] Niegodajew P and Marek M 2016 Analysis of orientation distribution in numerically generated random packings of Raschig rings in a cylindrical container *Powder Technology* **297** 193-201
[21] Giese M, Rottschafer K and Vortmeyer D 1998 Measured and modeled superficial flow profiles in packed beds with liquid flow *AIChE J.* **44** 484-490
[22] Chorin A J 1968 Numerical solution of the Navier-Stokes equations *Mathematics of Computation* **22** (104) 745-762
[23] Tryggvason G, Scardovelli R and Zaleski S 2011 *Direct numerical simulations of gas-liquid multiphase flows* (Cambridge: Cambridge University Press)
[24] Ferziger J H and Peric M 2002 *Computational methods for fluid dynamics* (Berlin: Springer-Verlag)
[25] Marek M 2014 CFD modelling of gas flow through a fixed bed of Raschig rings *Journal of Physics: Conference Series* **530** 012016
[26] Iaccarino G and Verzicco R 2003 Immersed boundary technique for turbulent flow simulations *Appl. Mech. Rev.* **56**(3) 331-347
[27] Mittal R and Iaccarino G 2005 Immersed boundary methods *Annu. Rev. Fluid. Mech.* **37** 239-261
[28] Mohd-Yosuf J 1997 Combined immersed-boundary/B-spline methods for simulations of flow in complex geometries *Annual Research Briefs, Center for Turbulence Research* 317-328

Acknowledgments
This study was performed within the framework of the contract: DEC-2014/15/B/ST8/04762 funded by National Science Centre.

Particle-fluid interaction inside a beater mill

M J Marijnissen[1] and J Rojek[1]

[1]Institute of Fundamental Technological Research, Polish Academy of Sciences, Pawińskiego 5b, 02-106 Warsaw, Poland

E-mail: mmarijn@ippt.pan.pl

Abstract In this work a trajectory study of particles through a beater mill (sometimes referred to as a fan mill) was performed. The attention was focused on particle behaviour. We used the CFD-DEM (Discrete Element Method) commercial code, ANSYS Fluent. Particles of different sizes were analysed. Results highlight particle behaviour, fluid flow conditions and mark places requiring geometrical improvements. Most of the improvements needed are in the mill's classifier, where perpendicular walls and large vortices hinder the filtration process, by suspending the particles inside the classifier.

1. Introduction

We can create a general division in the processes of comminution based on particle velocities. On one side of the scale we can put standard ball mills, with slow velocities. On the other side of the spectrum we can put beater mills. Standard ball mills have been known to take up to 50% of the ore production process's energy requirement [1]. Beater mills on the other hand seem to compete in energy efficiency whilst achieving high internal fluid velocities due to a spinning flywheel with its axis of revolutions normal to the flow. After hitting the flywheel the particles are shed upward into a classifier, where particles small enough are passed further on towards the next process. Particles considered too big are recirculated back onto the flywheel. Particle-fluid interaction plays a grand role in high speed comminution. Any improvements made into this process have an immense impact on cost reduction. The analysis of particle trajectories allows for insight in the current design of beater mills and provides ground for further improvements. With high internal flow velocities a particle-fluid coupling is required to simulate the appropriate behaviour of the ore. This was achieved with a CFD-DEM coupling which has been used in simulations and validations of particulate flow in a channel [2], fluidized beds [3] and slurry flow [4]. Additionally it has successfully been tested against analytic and experimental data [5].

2. Model description

We considered a multiphase fluid model where the fluid was treated as a continuous phase and the solid phase as discrete particles. The fluid was considered compressible and the flow turbulent. The k-ω turbulence model was used to take account for turbulent flow. Particle volume fraction were taken into consideration during calculations. Mass and heat exchange between phases were omitted. Discrete particles are considered undeformable and their drag coefficient is unaffected by neighbouring particles or compressibility effects. A two-way coupling is used between the continuous and the discrete phase.

2.1. Fluid flow

The Finite Volume Method is used to calculate the fluid taking into consideration of the particle volume fraction in given cells. A multiphase fluid is modelled with the use of conservation equations for a particular phase as given by

$$\frac{\partial \alpha_q \rho_q}{\partial t} + \nabla \cdot (\alpha_q \rho_q \vec{u}_q) = 0 \tag{1}$$

$$\frac{\partial \alpha_q \rho_q \vec{u}_q}{\partial t} + \nabla \cdot (\alpha_q \rho_q \vec{u}_q \vec{u}_q) = \alpha_q \rho_q \vec{g} - \alpha_q \nabla p_r + \nabla \cdot \bar{\tau}_q + \vec{F}_{add,q} \tag{2}$$

$$\bar{\tau}_q = \alpha_q \mu_q (\nabla \vec{u}_q + \nabla \vec{u}_q^{\,T}) + \alpha_q (\lambda_q - \frac{2}{3}\mu_q) \nabla \cdot \vec{u}_q I \tag{3}$$

$$\frac{\partial h_q \rho_q}{\partial t} + \nabla \cdot (\alpha_q \rho_q \vec{u}_q h_q) = \alpha_q \frac{\partial p_{rq}}{\partial t} + \bar{\tau}_q \cdot \nabla \vec{u}_q - \nabla \cdot \vec{q}_q . \tag{4}$$

In these equations α is the phasic volume fractions, $\vec{u}$ is the velocity of the fluid, g gravity, ρ density, p_r pressure and subscript q denotes the phases respectively. Additionally λ and μ denote the shear and bulk viscosity. Moreover $\bar{\tau}$ is the stress tensor and $\vec{F}_{add}$ represents additional forces (e.g. contact, coupling, pressure gradient) and $\vec{g}$ is the gravitational acceleration. Lastly h is the specific enthalpy and $\vec{q}$ is the heat flux, while t is the time variable. The model presented here holds for a n-phase fluid, while we only used one fluid phase with the volume fractions taken from the discrete phase. Details of the k-ω turbulence model can be found in [6].

2.2. Discrete Element Method

The DEM was first introduced by [7]. Since then the processing power of computers has increased allowing for increasing use of this method. In DEM we track the position of discrete particles in a Lagrangian reference frame as given by

$$\rho_p V_p \frac{d\vec{u}_p}{dt} = \vec{F}_{drag} + \vec{g}(\rho - \rho_p) V_p + \vec{F}_{add.} \tag{5}$$

$$\vec{F}_{drag} = \frac{3\mu C_d \,\mathrm{Re}}{4 d_p^2} \tag{6}$$

$$\mathrm{Re} = \frac{\rho d_p \,|\vec{u} - \vec{u}_p|}{\mu} \tag{7}$$

$$C_d = \begin{Bmatrix} 0.424 & : \mathrm{Re} > 1000 \\ \frac{24}{\mathrm{Re}}\left(1 + \frac{\mathrm{Re}^{2/3}}{6}\right) & : \mathrm{Re} \le 1000 \end{Bmatrix}. \tag{8}$$

Here equation (5) represents the forces acting upon the particles, where $\vec{u}$ is the fluid velocity, ρ indicates the fluid density, μ the molecular viscosity. V is the particle volume. All variables containing the subscript p consider the particle. Moreover $\vec{F}_{add.}$ represents additional forces acting upon the particle (ex. contact, coupling, pressure gradient) and $\vec{F}_{drag}$ can be described as shown in equations (6) and (7). What remains is the definition of the particle drag coefficient. We did not take into account for other particles shapes than spherical, so we used the formula presented by equation (8) without any shape coefficients.

2.3. Particle contact

Contact between particles is considered. The particles are allowed to overlap for a small distance. This overlap is the deciding factor while calculating the contact force. For the overlap to be small it is crucial to use a small time step. When an overlap is detected contact forces are calculated and applied to the particles. The governing equations are given by

$$\vec{e}_{12} = \frac{x_2 - x_1}{\|x_2 - x_1\|} \tag{9}$$

$$\delta = \|x_2 - x_1\| - (r_2 + r_1) \tag{10}$$

$$\vec{F}_1 = -\vec{F}_2 = (k\delta + \gamma\vec{u}_{12} \cdot \vec{e}_{12})\vec{e}_{12} \tag{11}$$

$$\vec{F}_{\tan gential} = \beta(u_r)\vec{F}_{normal}. \tag{12}$$

Where δ is the overlap, x_1 is the coordinate position of the centre of the particle. r_1 and $\vec{F}_1$ its radius and force acting upon the particle. Index 2 refers to the second particle whereas k and γ are the spring and damping coefficients. Finally $\beta(u_r)$ is the tangential friction coefficient dependent on radial velocity.

2.4. Coupling

After a velocity field is obtained the forces are calculated on the DEM particles which change their position accordingly. New particle coordinates as well as forces, are passed back on to the fluid flow part of the simulation. This process creates a 2-way coupling, were the fluid acts upon the particles and likewise, the particles on the fluid. The coupling is achieved by the α and $\vec{F}_{add.}$ variables presented in equation (1)-(5). For stability purposes, it is recommended for the particle time step to be small.

3. Geometry and mesh

The geometry was obtained from the producers of the mill. The gross dimensions of the whole mill are 5.5 m of height, 3 m of depth and 2.5 m of width. The mill's flywheel has a diameter just below 2 m and spins at 1000 RPM. The mesh consists of approximately 600.000 elements and is presented next to the geometry in figure 1. Additionally the inlet surface is marked with a green colour, whilst the outlet surfaces are marked in red. During simulations the angle of the flaps inside the classifier as well as before the flywheel were held constant.

4. Boundary and initial conditions

For the inflow a mass flow inlet boundary condition was used. The mass flow was set to match the parameters given by the producers. On the outlet a pressure-outlet condition was assigned with a -50 Pa pressure recreating the conditions behind the mill. The flywheel and its nearest fluid was set in a rotary frame. A time step of $5 \cdot 10^{-5}$ s was used to insure stability. After the simulation stabilized 360 particles were injected evenly on the inlet. The mass flow inlet of particles was much smaller than intended but was limited to save computational time and to increase the tracking ease of single particles. There were 3 groups of particles injected separately upon simulation conditions stabilized with a diameter of 10^{-6}, 10^{-5} and 10^{-4} m. The density for each particle was set the same at 1550 kg/m^3. Particles were injected with the same velocity as the fluid on the inlet surface, after which the simulation was run for another 0.72 seconds. This procedure was done to ensure no interaction between bigger and smaller particles and to study the behaviour of each size fraction during transition. Particles upon reaching the outlet were deleted from the computational domain. The amount of particles was compared after the same simulation run time. The fluid flow conditions in which the particles were injected are shown in figure 2 with 2 cross-sections and a close up of the classifier. This close up shows a surprising amount of big vortices and a complicated flow pattern. These conditions pose a problem for particle size filtration.

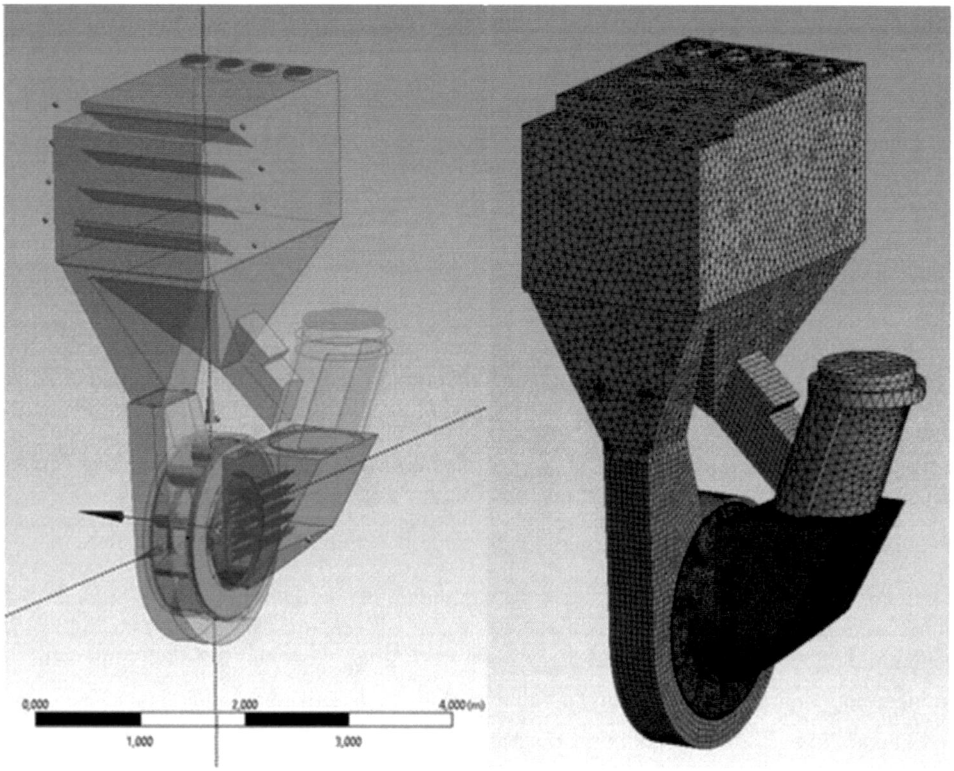

Figure 1. Geometry (left) and mesh (right) shown. Mesh consisted of approx. 600.000 elements.

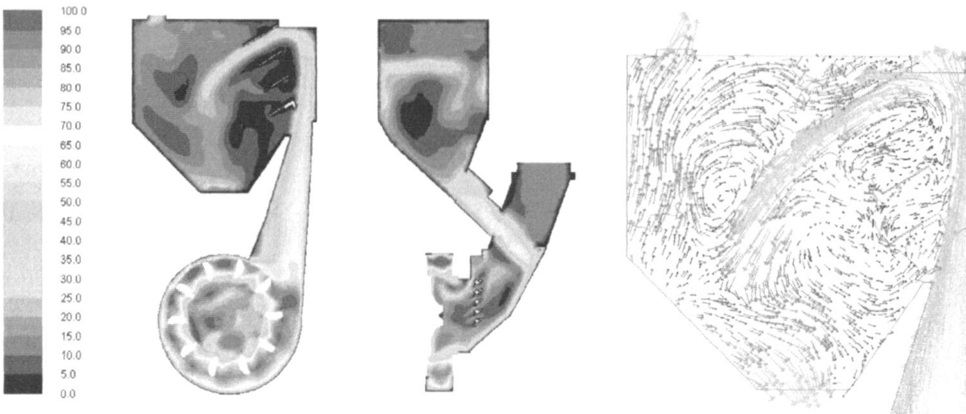

Figure 2. Velocity contours in 2 cross-sections (left) and velocity vectors in the classifier. Coloured by fluid velocity magnitude [m/s].

5. Obtained results

Figures 3-6 present the results obtained from 3 different perspectives; the side (figure 3), the back (figure 4 and 6) and from the top (figure 5). Each figure show the evolution of the particle movement in the mill with a 0.1 s interval. The first figure presents the particles 0.22 seconds after injection. The left side of each figure shows the smallest particles whilst the right side, the biggest. After 0.72 seconds the particles escaped from the domain were counted, these results are presented in figure 7 in the form of a line chart. The results show a dominance of small particles to stick towards the front of the mill (towards the inlet), while big particles stay towards the side. This can be observed in figure 3. After being shed into the classifier the smaller particles tend to change their direction swiftly with the fluid. The bigger particles hit the ceiling of the mill's classifier, after bouncing from it, they are shed upwind with the fluid or are bounced from the lower flaps into the classifier. This depletes those particles of momentum, which they slowly regain due to the fluid's velocity. This can be observed in figure 4.

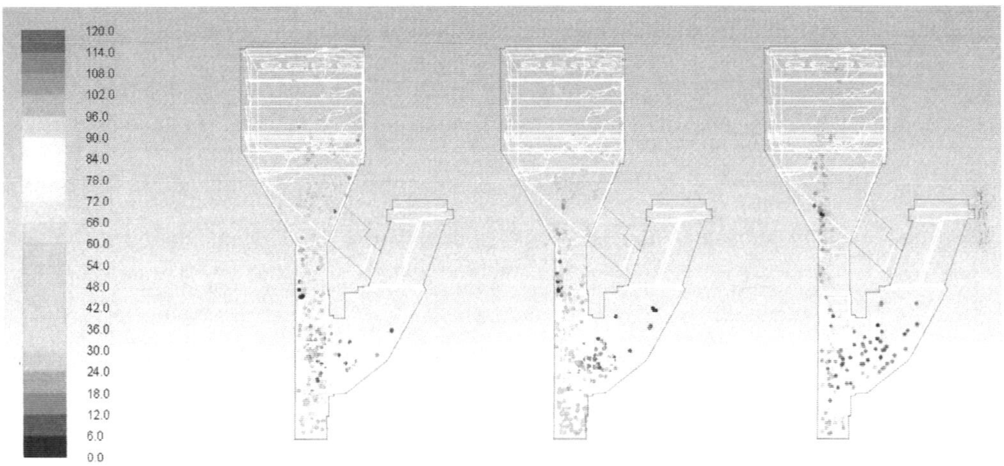

Figure 3. Particles coloured by particle velocity [m/s]. Particle diameter from left to right is appropriately; 10^{-6}, 10^{-5} and 10^{-4} m. Time step shown: 0.22 seconds after particle injection.

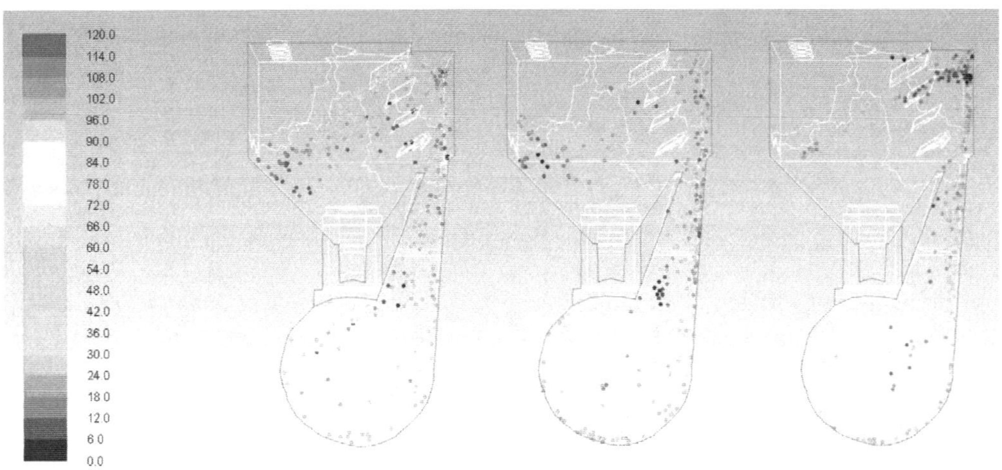

Figure 4. Particles coloured by particle velocity [m/s]. Particle diameter from left to right is appropriately; 10^{-6}, 10^{-5} and 10^{-4} m. Time step shown: 0.32 seconds after particle injection.

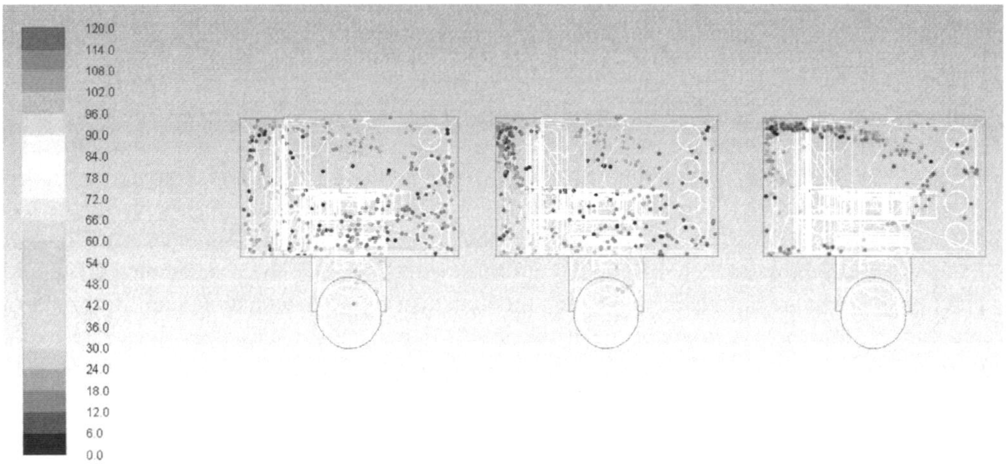

Figure 5. Particles coloured by particle velocity [m/s]. Particle diameter from left to right is appropriately; 10^{-6}, 10^{-5} and 10^{-4} m. Time step shown: 0.42 seconds after particle injection.

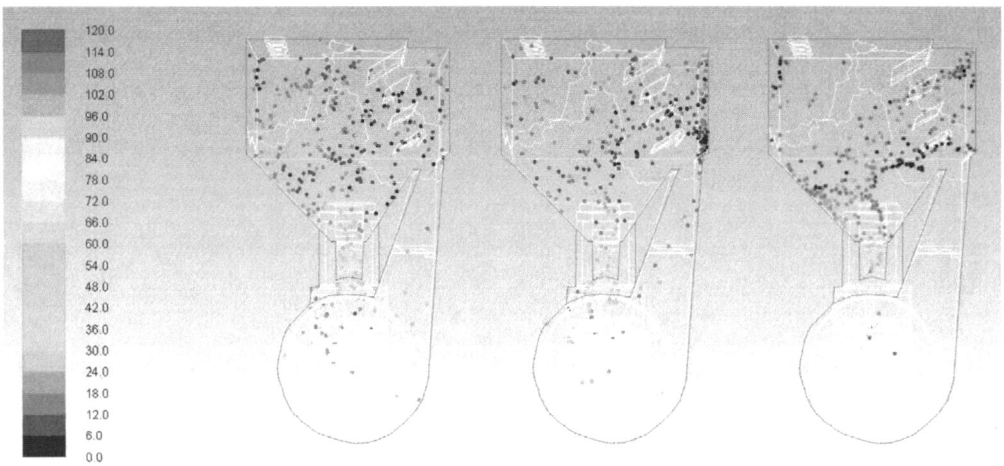

Figure 6. Particles coloured by particle velocity [m/s]. Particle diameter from left to right is appropriately; 10^{-6}, 10^{-5} and 10^{-4} m. Time step shown: 0.52 seconds after particle injection.

Figure 5 present a curvilinear motion of all the particles: after they are shed up they turn towards the injection side wall and start their descent towards the recirculation channel. This movement is found dominant for the biggest particles studied. In figure 6 we can observe an dominance for the biggest particles in the lower regions of the classifier (near the recirculation channel). This happens due to – after regaining momentum due to the fluid – the particles hit the oncoming wall of the classifier and lose their momentum. This happens for all the particles only the bigger particles are more affected due to a longer time needed to regain momentum in the surrounding fluid velocity field. In the following step gravity takes its toll and pulls them into the recirculation channel. Those particles were observed to recirculate the most, while the smallest particles remained suspended in the vortices of the classifier. This suspension would not occur for particles that remained near the wall. As can be seen in figure 2 the velocity vectors of the fluid near that region would drag the particles towards the outlet if they would stay near the wall. However they can get very easily caught in the stream leading them towards the suspension vortices. A change of outlet size and position could affect this behaviour and allow for better particle filtration. Moreover, clogging of particles of 10^{-5} m size has been observed near the entry of the classifier. The widening of the stream seams to affect only particles of this size. The amount of escaped particles shown in figure 7 is surprisingly small but one has to keep in mind that this is a time period of 0.72 seconds. The shape of the plot is considered correct due to a higher percentile of smaller particles filtered.

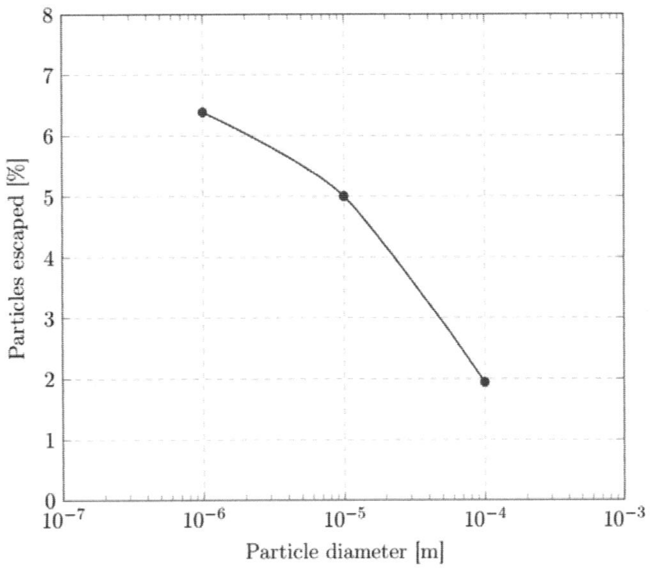

Figure 7. Particles escaped from domain in function of particle diameter after 0.72 seconds after injection.

6. Conclusions

We conducted a study of particle-fluid interaction inside a beater mil with the use of a CFD-DEM code. We chose 3 representative group sizes, which showed different behaviour. Upon analyses of the obtained data we concluded that the effectiveness of the mill can be increased by applying the following geometrical improvements:

- By creating a channel of fluid moving from the flaps towards the outlet at a high velocity. The presents of such a channel should provide a sufficient curvature to shed out the bigger particles, while retaining the smaller ones, thus improving the effectiveness of the classifier.
- The absence of large low-velocity vortices that would not allow the suspension of smaller particles. This can be achieved by compacting the classifier and changing the geometry to create a channel mentioned above.
- The 2 lower flaps of the classifier seem to have none effect on the fluid flow and can be with ease deleted or moved towards a place, where they can contradict the large low-velocity vortices. They do however bounce some of the larger particles inside the classifier's domain.
- The ceiling part of the classifier between the upper flap and the right side (while looking at figure 6) should be angled to allow the particles to bounce off from the wall towards the inside of the classifier, instead of clogging up this region.
- Additional clogging has been observed near the classifier entry surface. It seems to mostly affect particles of the 10^{-5} m size group.
- A change of position and/or size of the outlets should improve the filtration process.

Acknowledgment

The authors would like to acknowledge the financial support from the Polish NCBiR and KGHM granted upon the contract number CuBR/I/3/NCBR/2014.

References

[1] Cleary P W 1998 *Predicting charge motion, power draw, segregation and wear in ball mills using discrete element methods* Minerals Engineering 11 (11)

[2] Azimian M Lichti M and Bart H J 2014 *Investigation of particulate flow in a channel by application of CFD, DEM and LDAPDA* The Open Chemical Engineering Journal 8 1–11

[3] Yang C and Duan Y 2013 *CFD-DEM model for simulating solid exchange in a dual-leg fluidized bed* Chemical Engineering and Technology. 36 (11), 1907–1914

[4] Kubicki D and LO S 2012 *Slurry transport in a pipeline Comparison of CFD and DEM models* Ninth International Conference on CFD in the Minerals and Process Industries 10-12 December

[5] Kloss C Goniva C Hager A Amberger S and Pirker S 2012 *Models, algorithms and validation for opensource DEM and CFD-DEM* Progress in Computational Fluid Dynamics 12 (2/3) 140–152

[6] D C Wilcox 1998 *Turbulence Modelling for CFD* DCW Industries Inc. La Canada, California

[7] Cundall P A Stark O D 1979 *A discrete numerical model for granular assemblies* Géotechnique, 29, 47-65

Streamwise vortex generator for separation reduction on wind turbine profile

J Martinez Suarez[1], P Flaszyński[1] and P Doerffer[1]
[1] Aerodynamics Department, The Szewalski Institute of Fluid-Flow Machinery (IMP PAN), 14 Fiszera Street, Gdansk 80-231, Poland

E-mail: javier.martinez@imp.gda.pl

Abstract. High angles of attack of the wind turbine blades induce severe flow conditions which lead to flow separation and, as the consequence, aerodynamic performance reduction. Implementation of a new type of passive streamwise vortex generator (Rod Vortex Generator - RVG), on a wind turbine profile in order to reduce the flow separation is presented. Numerical model validation is carried out for the S809 aerofoil and a wide range of angles of attack (AoA) employed as reference for flow control cases. Investigation of proposed passive control method involves attached as well as incipient and massive flow separation. A study of chordwise location of RVGs for different inflow conditions is performed. The numerical and experimental results are in good agreement. Obtained numerical results based on the RANS approach reveal a large potential of selected passive devices in reduction of flow separation and increase of aerodynamic performance.

1. Introduction

The interest in application of flow control devices on wind turbine blades has increased in last decades. One of the design objectives and turbine operation requirements is the reduction of blade vibration amplitude and improvement of turbine blade performance. Development is also driven by continuous increase of blade length. Compared with constant rotational speed and fixed/collective pitch-regulated rotors, current use of variable speed rotors combined with individual blade pitch control have proven capability to increase global energy extraction, but yet failed to handle the non-uniform inflow conditions along blade span, leading to presence of flow separation on the rotor blade. Separated boundary layer leads to increased aerodynamic losses, noise generation and fatigue loads. Different flow control devices have been developed and implemented on the blades during the last few decades in order to mitigate flow separation and/or blade loading. Different flow control devices and strategies have been proposed; e.g.: trailing/leading edge flaps [1], turbercled leading edge [2], plasma actuators [3] or vortex generators [4].

Vortex generators (VGs) are designed to create streamwise vorticity, which by convection mixes the flow, entraining high momentum fluid from the undisturbed air stream and enforcing transport toward the surface. It finally affects the higher shear stress and decreases separation size. The first experiments on VGs were reported by Taylor in the 1940s (vane vortex generators, VVGs) [5]. These first vortex generators consisted of thin solid plates mounted normal to the surface, of a height (h) of the order of the boundary layer, set at an angle with respect the local flow. Already in the 1958 Wallis demonstrated that a very similar wall shear stress could be obtained by the action of an inclined vortex jet blown from a designed orifice (air jet vortex generators, AJVGs) with a much lower drag penalty [6]. Dimensions and geometry of the hole, as well as, optimum orientation angles and air supply strategy have been widely investigated [7, 8, 9]. Nevertheless, vane vortex generators are the most popular and lot of results of experimental and numerical investigations are published.

More recent studies on wall vortex generators were focused on the parasitic drag induced and its minimisation. In the early 1970s, Kuethe [10] studied non-conventional VGs with height to boundary layer thickness ratios inferior to 0.5. Research carried out by Rao in the 1980s [11] stated that low profile vortex generators ($h/\delta \leq 0.65$) have the potential of exceeding the performance of conventional VGs ($h/\delta \sim 1$) due to much smaller parasitic drag. Nevertheless, low-profile VGs are more sensitive to positioning than classical VGs as consequence of less intense streamwise vortices generated. Since then many researchers have conducted investigations regarding shapes and dimensions of VGs. A comprehensive review on low profile VGs was presented by Lin [12].

The presented investigation is focused on the application of a new type of passive VGs, Rod Vortex Generator (RVG) [13], on a wind turbine profile, in order to study an aerodynamic enhancement possibility. The main advantage over air jet vortex generator is no need to supply air. Compared with classical vanes parasitic drag is much smaller. Besides, RVGs simple geometry allows the combination with MEMS technology and, as consequence, an option to deploy when needed. RVGs are defined by five parameters: skew angle (θ), pitch angle (α), diameter (D), height (h), and spacing (W). The θ and α orientation angles are set equal to their optimum values (maximum streamwise vorticity), $\theta = 45°$ and $\alpha = 30°$, whereas rod diameter, height and spanwise spacing are proportional to the boundary layer thickness, δ (see Figure 1). Optimisation of the rod angles have been done as the preliminary investigations for the simplified configuration with the similar flow conditions and boundary layer at RVG location. Streamwise vorticity generated by a RVG with optimum skew and pitch angles is observed in Figure 1 (right). Displayed streamlines and contour map of streamwise vorticity reveal the creation of a strong primary vortex. Above the main vortex, a counter rotating secondary weak one is also created.

A validation of employed numerical model for wind turbine profile investigations is done as the first step. The NREL S809 aerofoil is chosen for research purposes, not only due to be representative of wind turbine profiles, but preferably because it constitutes a basic cross-section of the well experimentally documented NREL Phase VI wind turbine rotor [14]. The S809 aerofoil was extensively tested at the Ohio State University (OSU) [15], Colorado State University (CSU) [16], Delft University of Technology (DUT) [17] and more recently at the Glasgow University (GU) facilities [18]. From the amount of available data only the fixed pitch, free transition configurations with Reynolds number of 1 million, from the Somers' [17] and Ramsay's [15] experiments are harnessed for validation purposes.

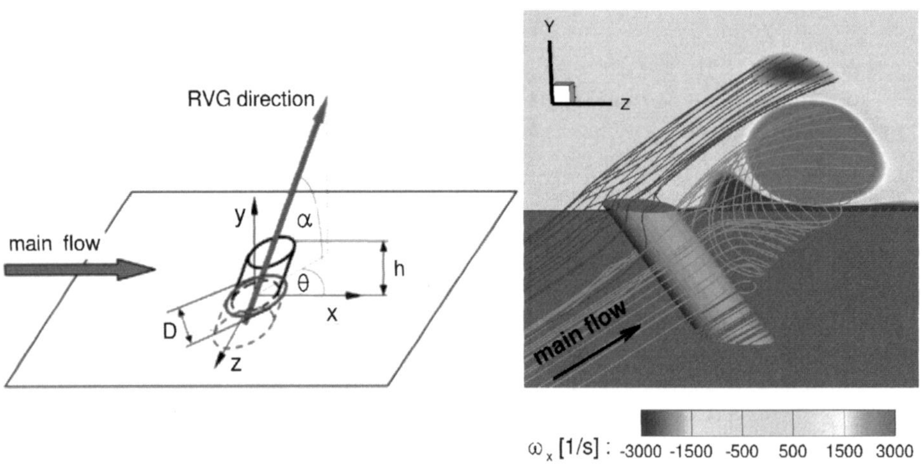

Figure 1. Rod vortex generator configuration with retraction possibility (left) and induced streamwise vorticity with streamlines (right)

2. Numerical model description

Numerical RANS (Reynolds Averaged Navier-Stokes) simulations are carried out by means of the 3D block structured Euranus flow solver from the FINE^{TM}/Turbo package (Numeca Int.). The compressible, mass-weighted equations are solved adopting a Hakimi preconditioning scheme to increase the convergence rate and accuracy of the solution. The two equation nonlinear eddy viscosity turbulence model of the low-Reynolds Explicit Algebraic Reynolds Stress Model (EARSM) is applied. The solver uses a finite volume approach for a spatial discretization of the RANS equations (central, 2nd order with scalar artificial dissipation) and Runge-Kutta type integration in time. A 2 step multigrid approach with 2 grid levels is applied in order to accelerate the convergence. The CFL number is set to 2. A perfect gas model is employed for closure of the system of differential equations. The dynamic viscosity is calculated using the Sutherland's law.

The aerofoil surface and rod vortex generators are modelled as adiabatic walls. The profile chord length is taken as the unit, c = 1 m. Farfield boundary conditions are located at 40 chords from the aerofoil to avoid influence of the boundary conditions on the flow around the profile. At the farfield, inflow velocity V_∞ and static pressure P_∞ and temperature T_∞ conditions are imposed. Two sets of C type grids are employed and normal distance to the first grid points layer above the surface is set to $3\cdot10^{-5}$ m ($y^+ < 1$). The mesh is also refined close to the leading and trailing edges, as well as, in the proximities of implemented rod vortex generators for the case with RVGs and in the suction side mid-chord (initial rod location) for clean configuration. The S809 aerofoil surface is defined by 737 x 145 nodes in the wrap around and normal directions, respectively.

For the clean case a 2D grid is generated with 0.3 million computational cells distributed into 13 blocks. In order to limit needed computational time and required resources, the 3D computational domain for the RVG case is reduced to a slice of the original experimental model, inside which a single rod is placed and periodic boundary conditions applied at lateral surfaces. Further reduction of number of cells is possible by the application of non-matching boundary conditions (NMB) at a distance of 0.13·c normal to the wall, so the final mesh size is $3.3\cdot10^6$, distributed among 29 blocks. The mesh for clean and RVGs cases is the same in the streamwise and normal directions, apart from the blocks affected by RVG presence. Proximities of the RVG are resolved by means of a butterfly mesh topology with 33 nodes in the spanwise direction (Figure 2).

For the reference case the dimensionless parameters Re and M_∞ are kept constant compared to the Ramsay's experimental model [15] and the ambient temperature T_∞ is set to 288 K. Thus the free stream density ρ_∞ and pressure P_∞ are rescaled ($\rho_\infty = Re\ \mu\ /\ V_\infty$, $P_\infty = \rho_\infty\ R_{gas}\ T_\infty$). The free stream speed V_∞ and dynamic viscosity μ do not need to be rescaled, $V_\infty = M_\infty\ [(\gamma\text{-}1)\ C_p\ T_\infty]^{1/2}$ and $\mu = \mu\ (T)$ (Sutherland's law). Turbulence intensity at the farfield applied is such that level of turbulence is lower than 0.1% upstream of the aerofoil.

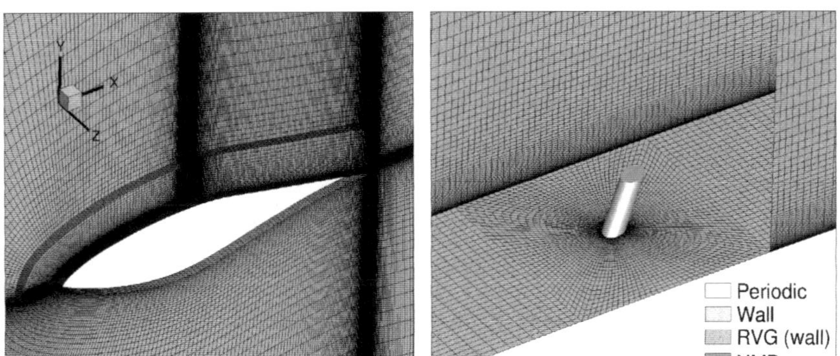

Figure 2. Grid and boundary conditions for RVG and S809 aerofoil

3. Reference configuration

Lift C_l and drag C_d coefficients as a function of angle of attack AoA are depicted in Figure 3. The EARSM model provides an accurate prediction of lift for moderate angles of attack, but overestimates its value for larger ones. A more detailed insight into the flow behaviour may be obtained by comparing the predicted pressure coefficient C_p distributions with the experimental data. In Figure 4, C_p distributions for two selected angles of attack, AoA = 4.1° and 13.3°, are compared with available experimental data and a reasonably good agreement is obtained.

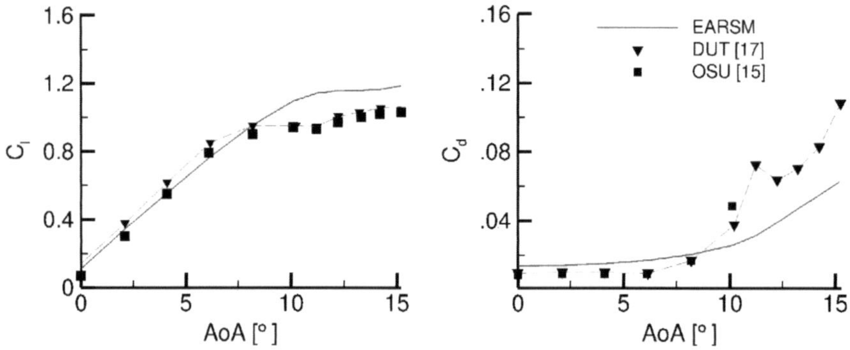

Figure 3. Force coefficients versus angle of attack for S809 aerofoil at Re = $1 \cdot 10^6$

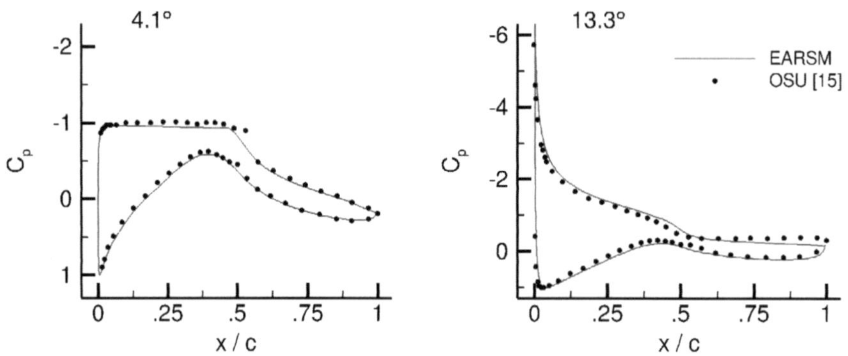

Figure 4. Pressure coefficient C_p distributions for S809 aerofoil at Re = $1 \cdot 10^6$

4. Flow control case – Rod Vortex Generator application

Angles of attack ranging from 4° to 14° are considered for flow control study and the highest inflow angle, with separation onset located at $x/c \approx 0.53$, taken as reference. Rods are initially located at mid-chord and dimensioned according to parameters established in Table 1, where δ_{base} is set equal to mid-chord boundary layer thickness at reference inflow angle ($\delta_{base} \approx 0.022 \cdot c$).

Figure 5 compares lift C_l, drag C_d and pitching moment C_m coefficients for clean and RVGs cases with respect to the angle of attack. At low inflow angles the pressure redistribution induced by presence of rod vortex generators causes a slight lift increase, while aerodynamic performance, given by C_l/C_d, is decreased as a consequence of induced parasitic drag. As the angle of attack increases to 10° separation emerges. With the application of RVGs separation present at angles of attack above 10° is reduced and lift is increased. Thus, despite the parasitic drag introduced by RVGs, implementation of these flow control devices lead to increase of lift to drag ratio as consequence of separation reduction. Due to the pressure redistribution induced by presence of RVGs, a modification in C_m with respect to the clean configuration is observed for the studied range of inflow angles.

Table 1. RVGs parameters value

Parameter		Value
Pitch angle,	θ [°]	45
Skew angle,	α [°]	30
Height,	h [mm]	7.78 ($0.36\cdot\delta_{base}$)
Diameter,	D [mm]	4.32 ($0.20\cdot\delta_{base}$)
Spacing,	W[mm]	43.2 ($2.0\cdot\delta_{base}$)

Aerodynamic enhancement is obtained by means of the implementation of RVGs for inflow angles ranging from 10° to 14° and the maximum improvement is observed for an angle of attack of 12.2°. At such inflow angle, the lift is increased by 5.2% with respect to the clean case and drag reduced by 1.2%, leading to an increase of lift to drag ratio above 11%. Additionally, as shown in Figure 6, the volume of reversed flow is reduced to less than a 3% of the original. Numerical simulations are carried out for one rod only, but for visualization purposes a set of three RVGs are depicted in Figure 6 (total span of $6\cdot\delta_{base}$). In the background of isosurfaces indicating detached flow zone (dark green), contour maps of chordwise velocity V_x above a surface $z = \delta_{base}$ (mid-span of first slice) for clean and flow control cases are presented.

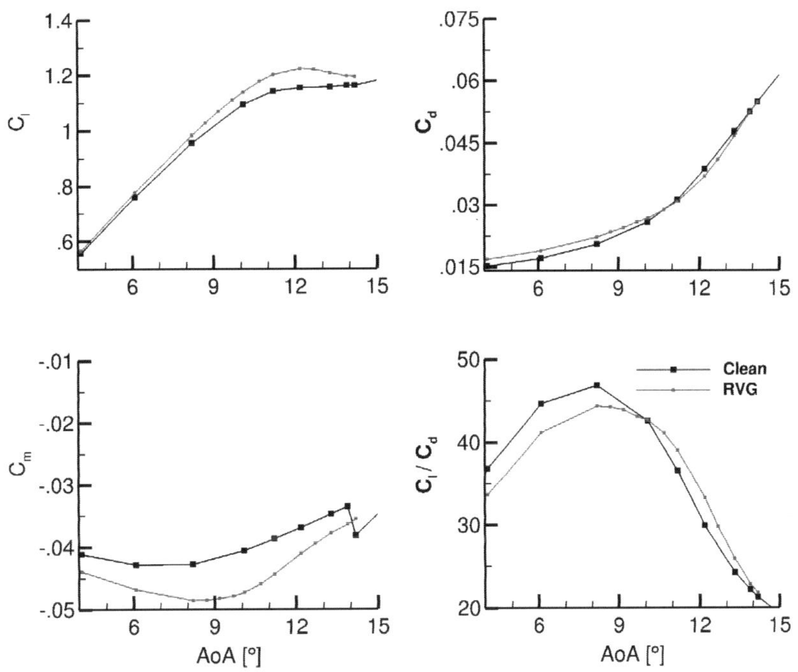

Figure 5. Force coefficient versus angle of attack for clean and flow control cases

An increase of inflow angle above 10° enforces more upstream separation location. Consequently, it can be expected that flow control devices location moved upstream from the initial mid-chord position will lead to a higher improvement of aerofoil performance for higher inflow angles. If RVGs moved more downstream, it may happen that they are located in the separation zone and they are not effective. Thus, notwithstanding the aerodynamic performance improvement obtained by implementing the RVGs at certain location, it is needed to investigate the chordwise location effect on the device effectiveness. Flow control device should be neither positioned too close to the separation (not enough space to develop vortical structures), nor too far away (vorticity diffusion).

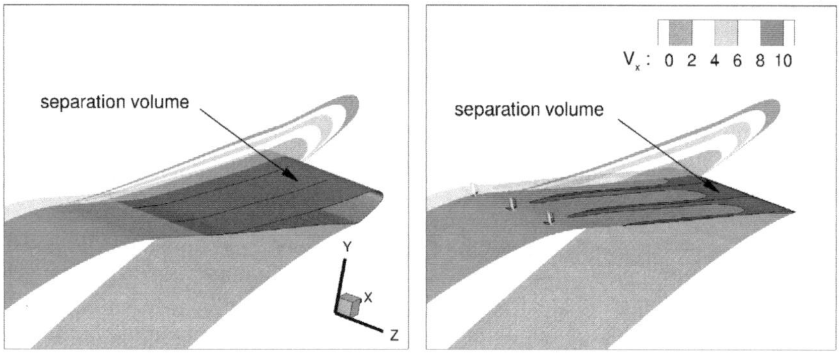

Figure 6. Isosurface of detached flow (dark green) and contour maps of chordwise velocity for clean and flow control cases with $x_{RVG}/c = 0.50$ at AoA = 12.2°

Four additional chordwise locations, $x_{RVG}/c = 0.35$, 0.40, 0.45 and 0.55, and four different flow cases are considered for RVGs investigation. All cases are selected for high inflow angles: 11.2°, 12.2°, 13.3° and 13.9°, with separation onset located at x/c = 0.88, 0.65, 0.55 and 0.53, respectively. Boundary layer thickness at reference inflow angle (AoA = 13.9°) grows rapidly along the suction side of the aerofoil, $\delta/c \approx 0.013$ at $x_{RVG}/c = 0.35$ and $\delta/c \approx 0.035$ at $x_{RVG}/c = 0.55$. Consequently, for the RVG chordwise location investigation, two scenarios are considered: RVGs with dimensions proportional to local reference boundary layer (case 1) and RVGs with height (h) and diameter (D) kept constant (case 2). In the former case local boundary layer thickness δ^* at AoA = 13.9° is taken as reference. RVGs parameters values are summarized in Table 2.

Table 2. RVGs parameters value for cases 1 and 2

Parameter		Case 1	Case 2
Height,	h [mm]	$0.36 \cdot \delta^*$	7.78
Diameter,	D [mm]	$0.20 \cdot \delta^*$	4.32
Spacing,	W [mm]	43.2 (W/D ≈ 6.3-16.5)	43.2 (W/D = 10)
Location,	x_{RVG}/c [-]	0.35 - 0.55	0.35 - 0.55
Inflow angle,	AoA [°]	11.2 - 13.9	11.2 - 13.9

In the Figure 7 contour maps of skin friction coefficient on the S809 aerofoil suction side at the angle of attack 12.2° for case 1 are shown. At the top, the uncontrolled case is shown, below, the cases for increasing chordwise location of RVGs are displayed. The skin friction provides information of the shear stresses distribution and indicates the location of flow separation. Additionally, streamlines plotted allow the assessment of the flow patterns development on the aerofoil and visualisation of the reattachment line.

It is shown in the Figure 7 that skin friction along the chord increases downstream of the RVGs location and it is higher than the one obtained for the clean case. Zones of lower friction are present between rods. This structure is originated by the streamwise vortices and its effect on locally increased and decreased shear stresses in spanwise direction. Additionally, a spanwise velocity component has been generated by the RVGs, rooted in the spanwise migration of the created vortices. As the RVGs location is moved downstream, the penetration of streamwise vorticity generated is increased and the separation further decreased.

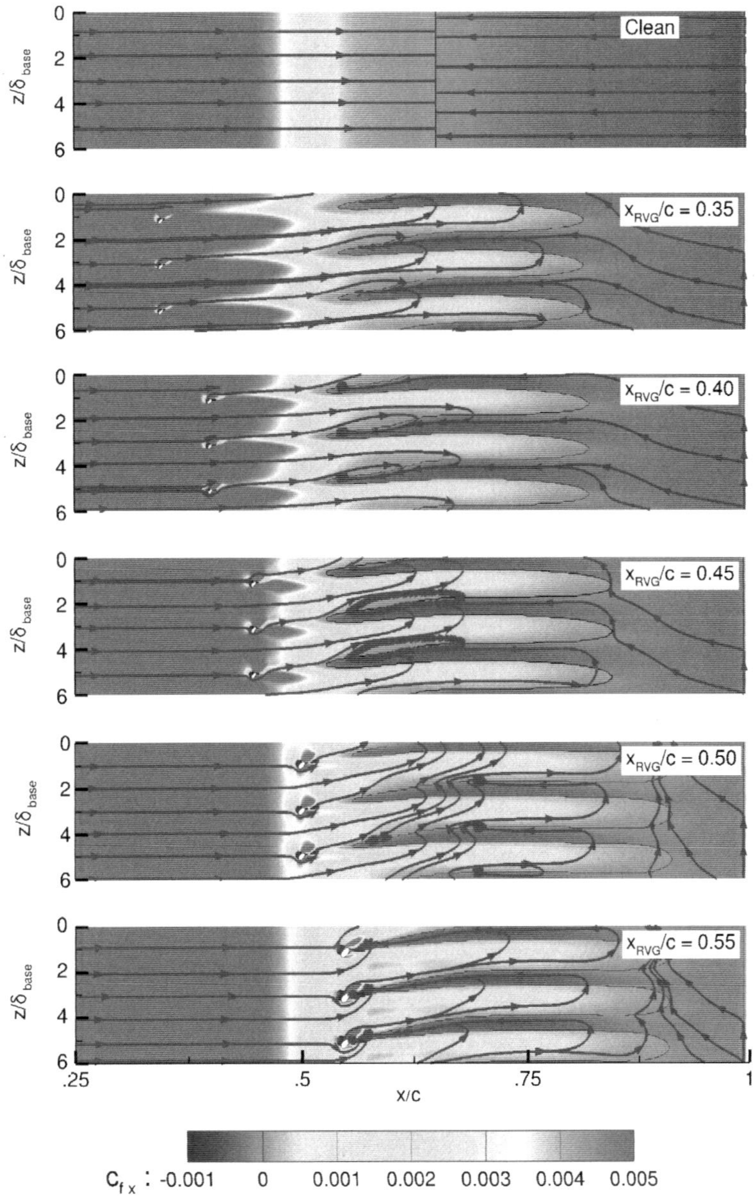

Figure 7. Skin friction coefficient C_{fx} and surface streamlines for different RVGs chordwise locations at AoA = 12.2° for case 1

For case 1 lift C_l and drag C_d with respect to the case without flow control are shown in Figure 8. The variation for the considered RVG locations and inflow angles indicate better or worse cases, which means the positive or negative effect obtained if RVG is applied. It is observed that the lift increases for all locations and inflow angles, but for the angle of attack 13.9° and $x_{RVG}/c = 0.55$, where RVG is located inside the separation bubble. The reduction of drag is obtained for certain rods locations and angles of attack. However, the configuration of maximum lift increase corresponds with the area of drag reduction, pointing the configurations as the aerodynamically improved. In case of angle of attack 11.2° the maximum lift is obtained for RVGs located at $x_{RVG}/c = 0.55$. The lift is decreasing as they are placed closer to the leading edge. Simultaneously, when the RVG location is moved upstream the drag is increased. It means that even if a higher vorticity is created by a RVG located more upstream, the streamwise vortices dissipate along the distance to the separation. Similar effect to the analysed above is obtained for the inflow angle 12.2°. The maximum improvement in aerodynamic performance by the implementation of flow control devices is obtained for x_{RVG}/c = 0.50. As the angle of attack increases to 13.3° the separation line moves to x/c = 0.55 and the optimum RVG location is at the mid-chord. More upstream locations provide smaller increases in lift (vortices are more dissipated) and higher drag. For x_{RVG}/c = 0.55 the lack of distance to the separation onset impedes the development of the vortex structure and a high drag penalty appears. Further increasing of the angle of attack up to 13.9° a more modest profit is obtained with the RVG inclusion. The RVGs located at x_{RVG}/c = 0.55 show a negative effect on the aerodynamic performance as expected since the devices are located inside the region of detached flow (separation onset at x/c = 0.53). At this inflow angle only RVGs located at mid-chord provide a drag reduction. Based on the contour map analysis one can conclude that for the present case the optimum RVG location is at mid-chord, x_{RVG}/c = 0.50.

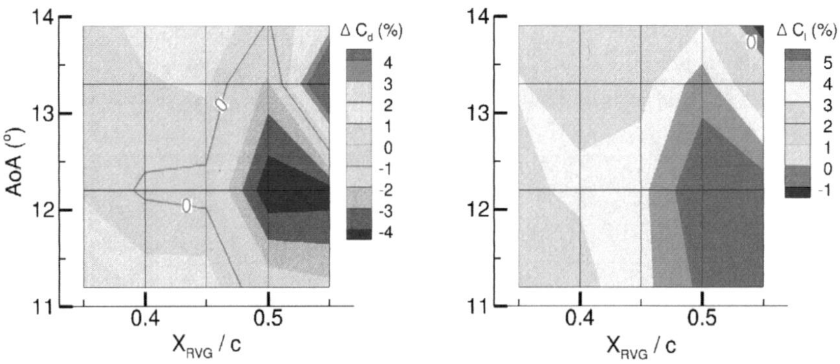

Figure 8. Effect of RVGs chordwise location on C_l and C_d for case 1

In Figure 9, contour maps for variations of lift C_l and drag C_d coefficient for the considered RVG locations and inflow angles, with respect to the case without flow control, are depicted for case 2. Similarly to the case 1, the lift increase is obtained for all locations and inflow angles, but for the angle of attack 13.9° and x_{RVG}/c = 0.55, where RVG is located inside the separation bubble. Compared with the case 1, reduction of drag and further increase of lift are shown for RVGs locations upstream of the mid-chord. If RVGs are moved downstream to x_{RVG}/c = 0.55 the opposite effect is obtained. Variation of lift and drag values is based on the alteration of RVGs height to local boundary layer thickness ratio h/δ, keeping rod height and diameter unchanged.

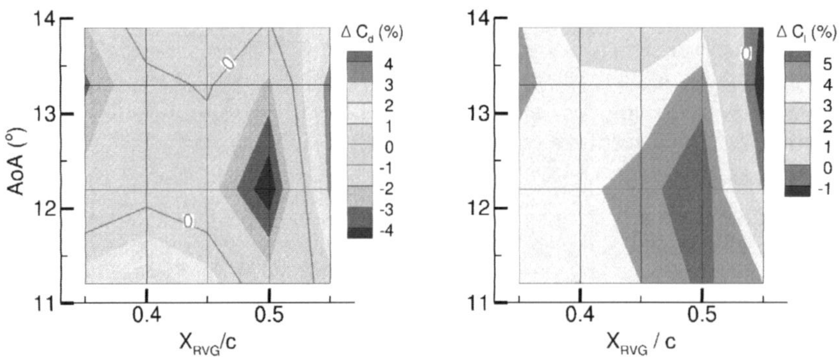

Figure 9. Effect of RVGs chordwise location on lift and drag coefficients for case 2

If RVG designed for initial location at mid-chord are moved upstream, then the rods are located at thinner boundary layer what increases the h/δ ratio, leading to stronger streamwise vortices in comparison with the same rod location for case 1. Notwithstanding this fact, when initial location $x_{RVG}/c = 0.5$ is taken as the reference, penetration of the streamwise vorticity is shortened and interaction with detached flow is weaker for upstream locations. It leads to smaller values of lift and higher drag. This effect can be observed in Figure 10 where streamwise values of C_{fx} together with streamlines are depicted on the suction side of the S809 aerofoil for an angle of attack AoA = 12.2° and the range of RVGs locations studied, $x_{RVG}/c = 0.35$-0.55. When compared with Figure 7 for case 1 vorticity penetration is extended for rods locations upstream of initial configuration and shortened at $x_{RVG}/c = 0.55$, as mentioned above.

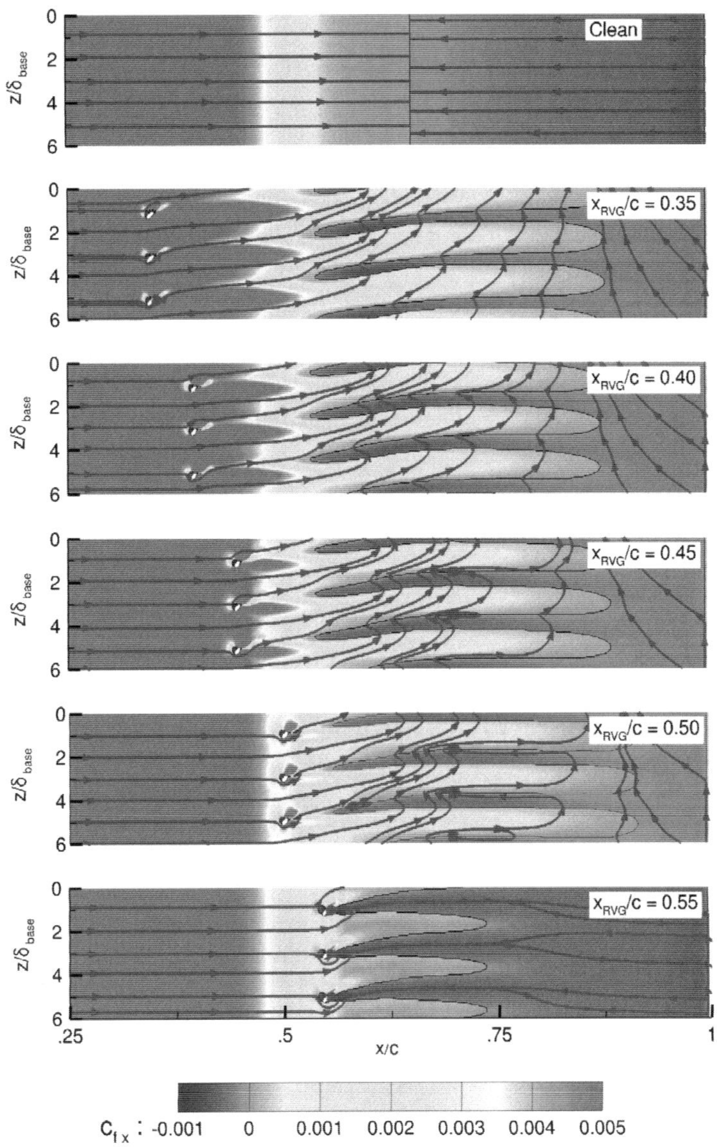

Figure 10. Skin friction coefficient C_{fx} and surface streamlines for different RVGs chordwise locations at AoA = 12.2° for case 2

5. Conclusions

Numerical model was satisfactorily validated and employed for the investigation of aerodynamic improvement by means of rod vortex generators. Numerical results obtained for a clean S809 aerofoil were considered as reference for the flow control analysis. RVGs were initially located at mid-chord and dimensioned in relation to the local boundary layer thickness. A location study was performed and effect on aerodynamic forces analysed. Introduction of RVGs have proven to lead to improvement of aerodynamic performance and decrease of flow separation. Optimum configuration was provided by a mid-chord location. At mid-chord location the maximum improvement is observed for an angle of attack of 12.2°, with an increase of lift to drag ratio above 11% and separation is nearly totally eliminated. The successful implementation of rod vortex generators to the wind turbine profile points RVGs as possible flow control devices for the wind energy industry.

Acknowledgment

The authors acknowledge the European Commission for their research grant under the project FP7-PEOPLE-2012-ITN 309395 "MAREWINT". Research was supported by the PL-Grid Infrastructure.

References

[1] Li CF, Xu Y, Zhao XL and Xu JZ 2013 Influence of trailing edge flap on wind turbine blade using three-dimensional computational fluid dynamics method. *IOP Conference Series: Materials Science and Engineering* **52** 5
[2] Zhang RK and Wu VDJZ 2012 Aerodynamic characteristics of wind turbine blades with a sinusoidal leading edge *Wind Energy* **15** 3 pp 407-424
[3] Potočar E, Širok B, Hočevar M and Eberlinc M 2012 Control of separation flow over a wind turbine blade with plasma actuators. *Strojniški vestnik-J. Mech. Engineering* **58** 1 pp 37-45
[4] Shun S and Ahmed NA 2012 Wind turbine performance improvements using active flow control techniques *Procedia Engineering* **49** pp 83-91.
[5] Taylor HD 1947 The elimination of diffuser separation by vortex generators *Research department rep. No. R-4012-3. United Aircraft Corporation* (East Hartford, Connecticut)
[6] Wallis RA and Stuart CM 1958 On the control of shock induced boundary layer separation with discrete jets. *Aeronautical Research Council Current Paper* 494
[7] Zhang X 2000 An inclined rectangular jet in a turbulent boundary layer-vortex flow. *Experiments in Fluids* **28** 4 pp 344-354.
[8] Freestone M Preliminary tests at low speeds on the vorticity produced by air-jet vortex generators *Research Memo. Aero* 85/01
[9] Flaszynski P and Szwaba R 2008 Optimization of streamwise vortex generator *Developments in Mech. Engineering* 2
[10] Kuethe AM 1972 Effect of streamwise vortices on wake properties associated with sound generation *J. Aircraft* **9** 10 pp 715-719.
[11] Rao DM and Kariya TT 1988 Boundary-layer submerged vortex generators for separation control-an exploratory study *AIAA paper* 3546
[12] Lin JC 2002 Review of research on low-profile vortex generators to control boundary-layer separation *Progress in Aerospace Sciences* **38** 4 pp 389-420
[13] Doerffer P, Flaszynski P and Szwaba R 2009 *Polish Patent P.38968*
[14] Martinez J, Doerffer P, Szulc O and Tejero F 2015 Aerodynamic analysis of wind turbine rotor blades *TASK Quarterly* **19** 2 pp 129-140
[15] Ramsay R, Hoffmann M and Gregorek G 1995 Effect of grid roughness and pitch oscillations on the S809 airfoil: airfoil performance report *Tech. Rep. NREL/TP-442-781*
[16] Butterfield C, Musial W and Simms D 1992 Combined experiment phase 1 final report *Tech. Rep. NREL/TP-257-4655*
[17] Somers D 1997 Design and experimental results for the S809 airfoil *Tech. Rep. NREL/SR-440-6918*
[18] Sheng W, Galbraith RAM and Coton FN 2009 On the S809 airfoil's unsteady aerodynamics characteristics *Wind Energy* **12** 8 pp 752-767

Comparison of different approaches in the Trefftz method for analysis of fluid flow between regular bundles of cylindrical fibres

M Mierzwiczak[1], J K Grabski[1] and J A Kołodziej[1]

[1] Poznan University of Technology, Institute of Applied Mechanics, Jana Pawla II 24, 60-965 Poznan, Poland

Abstract. In the paper three different approaches for the Trefftz method are compared in analysis of the fluid flow between regular bundles of cylindrical fibres. The approximate solution is a linear combination of such trial functions which fulfil exactly the governing equations. The trial functions can be defined in the Cartesian coordinate system (the first approach), in the cylindrical coordinate system (and can fulfil also some boundary conditions – the second approach) or be defined as a fundamental solutions (the third approach – the method of fundamental solutions).The average velocity and the product of the friction factor and the Reynolds number $f \cdot \mathrm{Re}$ are compared for selected parameters of a considered region.

1. Introduction

The Trefftz method (TM) was proposed in 1926 by Erich Trefftz [1]. In this method an approximate solution is a linear combination of trial functions which fulfil exactly the governing equation of the considered problem. The trial functions in the method are called the Trefftz functions or the T-functions. However real development of this method came only with development of computer techniques. One of the version of the Trefftz method is the boundary collocation technique. In the method the governing equation is fulfilled exactly by the trial functions while the boundary conditions are satisfied in an approximate way. Different applications of the method can be found in [2].

Sometimes the T-functions fulfil also some of the boundary conditions. In such case the method is often called the Trefftz method with the special purpose Trefftz functions (SPTF) [3]. Then the boundary collocation technique is applied only for unfulfilled boundary conditions. This approach was successfully applied for different problems in applied mechanics, e.g. conductive heat flow [4], fluid flow in conduits with polygonal cross-section [5] or elastic torsion of bars [6].

Another version of the Trefftz method is the method of fundamental solutions (MFS). The MFS was proposed in 1964 by Kupradze and Aleksidze [7]. The numerical implementation of the MFS was given by Mathon and Johnston [8]. In the MFS the approximate solution is a linear combination of fundamental solutions which are functions of distance between the points and the source points. The fundamental solutions fulfil exactly the governing equation and the unknown coefficients of the approximate solution are calculated using the boundary collocation technique [2]. The source points are located outside the considered region which has similar shape to the boundary of the domain. Some interesting review of applications of the MFS can be found in [9-11].

The purpose of the present paper is to compare these three approaches in analysis of the longitudinal fluid flow between bundles of cylindrical fibres. The filtration velocity and the product of the friction factor, f and the Reynolds number, Re are compared for a different arrangement of parallel cylindrical fibers.

2. Formulation of the problem

Consider the steady, fully developed, laminar, isothermal flow of an incompressible viscous fluid driven by a constant pressure in a system of regular parallel fibers. The flow is longitudinal with respect to fibers which are arranged in a regular hexagonal (Fig. 1a), square (Fig. 1b) and triangular (Fig. 1c) array. The radius of the fibers is equal to *a*, and the distance between the fibers is equal to 2*b*.

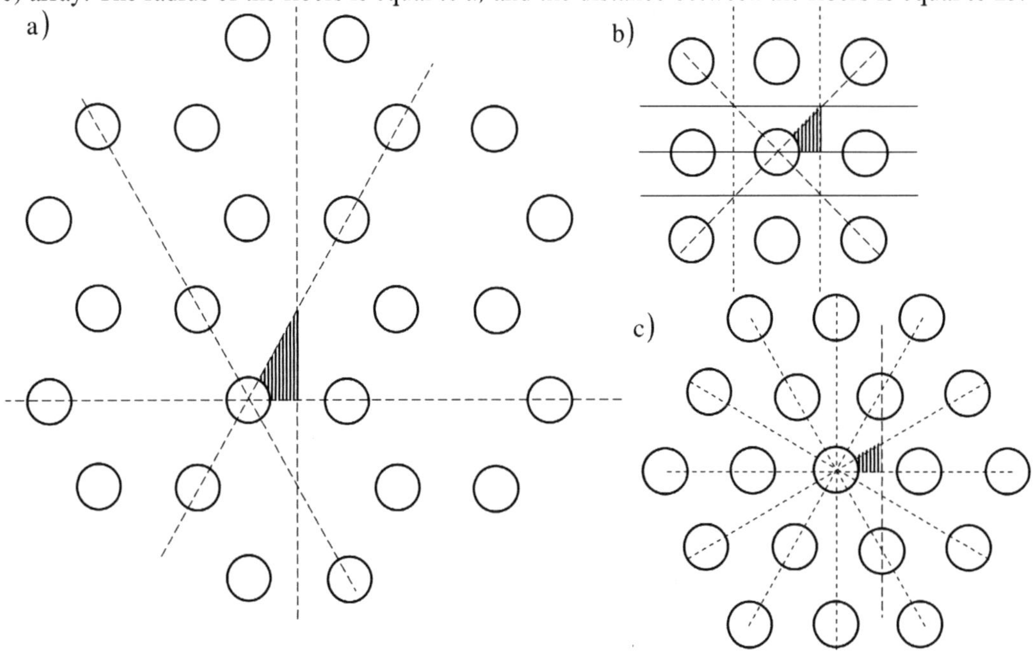

Figure 1. Three different kinds of a regular array of cylindrical rods (*a*) $L = 6$ hexagonal array, (*b*) $L = 4$ square array, and (*c*) $L = 3$ triangular array.

The equation of motion is reduced to a single partial differential equation in the form

$$\frac{\partial^2 w}{\partial x^2} + \frac{\partial^2 w}{\partial y^2} = \frac{1}{\mu}\frac{\mathrm{d}p}{\mathrm{d}z} \quad \text{in } \Omega_f \tag{1}$$

where w is the velocity component in the z direction, $\mathrm{d}p/\mathrm{d}z$ is the constant pressure gradient, μ is the dynamic viscosity, and Ω_f is the repeated element of a fluid array (Fig 2a).
Equation (3) is solved with the following boundary conditions in a repeated element array (Fig. 2a)

$$w = 0 \text{ for } \sqrt{x^2 + y^2} = a, \tag{2}$$

$$\frac{\partial w}{\partial y} = 0 \text{ for } y = 0,\ a \le x \le b, \tag{3}$$

$$\frac{\partial w}{\partial x} = 0 \text{ for } x = b,\ 0 \le y \le b \cdot \cot\left(\frac{\pi}{L}\right), \tag{4}$$

$$\frac{\partial w}{\partial n} = 0 \text{ for } y = x \cdot \cot\left(\frac{\pi}{L}\right),\ a \cdot \sin\left(\frac{\pi}{L}\right) \le x \le b. \tag{5}$$

It is convenient to introduce the following dimensionless variables

$$X=\frac{x}{b},\ Y=\frac{y}{b},\ E=\frac{a}{b},\ W=\frac{\mu\cdot w}{-b^2\cdot\frac{dp}{dz}}, \tag{6}$$

Now, the governing equation (3) has the dimensionless form

$$\frac{\partial^2 W}{\partial X^2}+\frac{\partial^2 W}{\partial Y^2}=-1 \text{ in } \Omega_f, \tag{7}$$

Eq. (9) is solved with boundary conditions in a dimensionless form

$$W=0 \text{ for } \sqrt{X^2+Y^2}=E, \tag{8}$$

$$\frac{\partial W}{\partial Y}=0 \text{ for } Y=0,\ E\le X\le 1, \tag{9}$$

$$\frac{\partial W}{\partial X}=0 \text{ for } X=1,\ 0\le Y\le \cot\left(\frac{\pi}{L}\right), \tag{10}$$

$$\frac{\partial W}{\partial n}=0 \text{ for } Y=X\cdot\cot\left(\frac{\pi}{L}\right),\ E\cdot\sin\left(\frac{\pi}{L}\right)\le X\le 1. \tag{11}$$

The non-linear governing Eq. (9) with boundary conditions (10-13) yields a micro structural boundary value problem (MBVP) in a repeated element of an array of fibers.

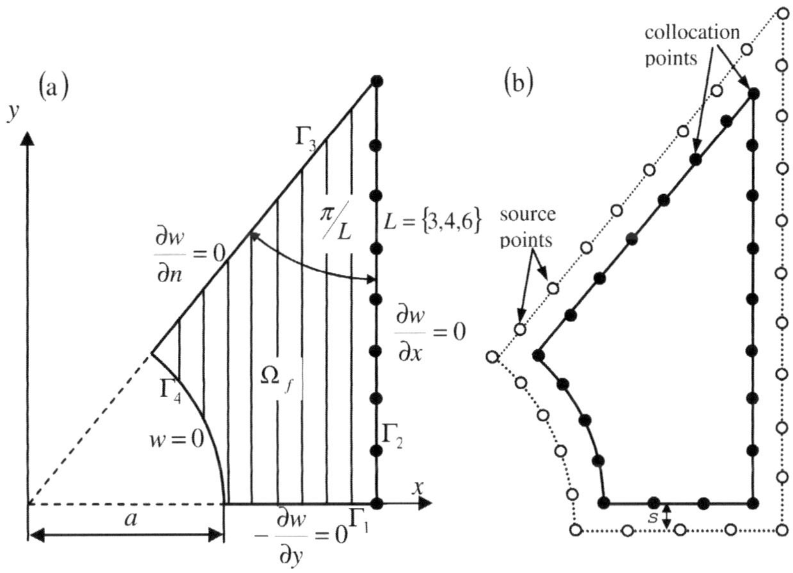

Figure 2. The repeated element a) the formulation of the boundary value problem, b) the distribution of source, collocation and interpolation points

3. Method of solution of micro structural boundary value problem

The MBVP problem can be solved using the meshless method. The approximate solutions can be expressed as a sum of general and particular solution

$$W(X,Y)=W_g(X,Y)+W_p(X,Y) \tag{12}$$

The particular solution for Eq. (9) has a form

$$W_p(X,Y) = -\frac{1}{4}\left(X^2 + Y^2\right) \tag{13}$$

The general solution $W_g(X,Y)$ fulfills the Laplace equation and can be solved using one of the three Trefftz method. The detailed differences are presented in this section.

3.1. The Trefftz method - TM

The approximate general solution using the Trefftz method can be written in the following form

$$W_g(X,Y) = \sum_{n=0}^{M} c_n F_n(X,Y) + \sum_{n=1}^{M} d_n G_n(X,Y) \tag{13}$$

where $F_n(X,Y)$ and $G_n(X,Y)$ are the trial functions defined in the Cartesian coordinate system

$$F_n(X,Y) = \sum_{k=0}^{\left[\frac{n}{2}\right]} (-1)^k \frac{X^{n-2k} Y^{2k}}{(n-2k)!(2k)!}; \quad G_n(X,Y) = \sum_{k=0}^{\left[\frac{n-1}{2}\right]} (-1)^k \frac{X^{n-2k-1} Y^{2k+1}}{(n-2k-1)!(2k+1)!}. \tag{14}$$

The above trial functions fulfill the Laplace equation in 2D. The unknown coefficients $\{c_n\}_{n=0}^{M}$ and $\{d_n\}_{n=1}^{M}$ are calculated by fulfilling the boundary conditions (8-11) in the collocation sense [2]. In this way we obtain a system of linear equations. The number of collocation points must be greater or equal to the number of unknowns coefficients ($2 \cdot M+1$).

3.2. The Trefftz method with special purpose Trefftz functions - SPTF

As the general solution of the Laplace equation in 2D the following expression in the polar coordinate system can be used

$$W_g(R,\theta) = A_0 + A_1\theta + A_2\theta \ln R + A_3 \ln R + \sum_{n=1}^{M}\left(B_n R^{\lambda_n} + C_n R^{-\lambda_n}\right)\cos(\lambda_n\theta) + \sum_{n=1}^{M}\left(D_n R^{\lambda_n} + E_n R^{-\lambda_n}\right)\sin(\lambda_n\theta) \tag{15}$$

Applying it for the specific region, some of the boundary conditions can be fulfilled exactly by calculating some constants in the above solution. After that the approximate solution fulfils the governing equation and some of the boundary conditions (8, 9, 11).

$$W_g(R,\theta) = \frac{1}{4}E^2 + A_1 \ln\frac{R}{E} + \sum_{n=2}^{M} A_n\left(R^{L(n-1)_n} - \frac{E^{2L(n-1)}}{R^{L(n-1)}}\right)\cos(L(n-1)\theta). \tag{16}$$

The constants $\{A_n\}_{n=1}^{M}$ are calculated from Eq. (10) using the boundary collocation technique [2]. The number of collocation points located on Γ_2 must be greater or equal M.

3.3. The method of fundamental solutions - MFS

In the MFS the approximate solution is assumed as a linear combination of fundamental solutions. The fundamental solution fulfils exactly the governing equation and it is a function of distance between the point inside the considered region and the source point. Using the MFS the approximate solution for the Laplace equation takes the form

$$W_g(X,Y) = \sum_{j=1}^{M} c_j \ln(r_j), \tag{17}$$

where $\{c_j\}_{j=1}^{M}$ are unknown coefficients, r_j is the distance between the point (X, Y) and the j-th source point (Xs_j, Ys_j). To avoid the singularities sources points are located outside the considered domain on a fictitious boundary (see Fig. 2b). The distance between the source boundary and the real boundary is equal to s. Fulfilling the boundary conditions (8-11) in collocation points (Xc_i, Yc_i) we obtain a system of linear equations

$$\begin{aligned} &\sum_{j=1}^{M} c_j \ln\left(r_{ij}\right) = W_b + \frac{1}{4}\left(Xc_i^2 + Yc_i^2\right), \quad i = 1,\dots,N_1 \\ &\sum_{j=1}^{M} c_j \left(\frac{\partial \ln\left(r_{ij}\right)}{\partial X} nx + \frac{\partial \ln\left(r_{ij}\right)}{\partial Y} ny\right) = \frac{\partial W_b}{\partial n} + \frac{1}{2}\left(nx \cdot Xc_i + ny \cdot Yc_i\right), \quad i = N_1 + 1,\dots,N \end{aligned} \tag{18}$$

where $r_{ij}^2 = \left(Xc_i - Xs_j\right)^2 + \left(Yc_i - Ys_j\right)^2$, $\mathbf{n} = [nx, ny]$ is a unit normal vector to the boundary. The number of collocation points N must be greater or equal to the number of unknowns coefficients M.

After determined the velocity field, the filtration velocity can be calculated as

$$W_b(X,Y) = \frac{1}{\Omega}\int_{\Omega_f} W \, d\Omega_f \,, \tag{19}$$

where $\Omega = \frac{1}{2}\mathrm{ctg}\left(\frac{\pi}{\mathrm{L}}\right)$ is a total dimensionless area of the element, $\Omega_f = \frac{1}{2}\left(\mathrm{ctg}\left(\frac{\pi}{\mathrm{L}}\right) - \pi\frac{L-2}{2L}E^2\right)$ is a flow area.

Used the Darcy-Weisbach equation

$$\frac{\Delta p}{L} = f\frac{1}{4}\frac{\rho}{a}w_b \approx -\frac{dp}{dz}, \tag{20}$$

the Reynolds number definition

$$\mathrm{Re} = \frac{2a\rho w_b}{\mu}, \tag{21}$$

and the definition of the non-dimensional variables (6), the product of friction factor, f and Reynolds number, Re can be express by a dimensionless form

$$f \cdot \mathrm{Re} = \frac{8E^2}{W_b}. \tag{22}$$

For the fibrous porous medium the fiber volume fraction is defined as

$$\varphi = \pi E^2\left(\frac{L-2}{2L}\right)\tan\left(\frac{\pi}{L}\right). \tag{23}$$

where E is the non-dimensional radius of the fiber and $L = \{3, 4, 6\}$.

4. Numerical experiments

In numerical experiments, the non-dimensional field of the velocity for longitudinal flow are calculated. The cylindrical fibers are arranged regularly in a triangular $L = 3$, square $L = 4$ and hexagonal $L = 6$ arrays.

At first we test the influence of the number of trial function N for a value of the error of fulfilling the boundary conditions. The calculation was performed for a square $L = 4$ arrays of fibers, $E = \{0.2, 0.4, 0.6, 0.8\}$ and results are presented on Figure 3. The smallest values of the error were obtain for the SPTF. Figure 4 shows the value of the condition number of the collocation matrix, *CN*. The condition number was calculated as a relation of the maximal to the minimal singular value of the matrix. The lowest value of *CN* was obtained for the TM and the greatest for the MFS.

Journal of Physics: Conference Series **760** (2016) 012019 doi:10.1088/1742-6596/760/1/012019

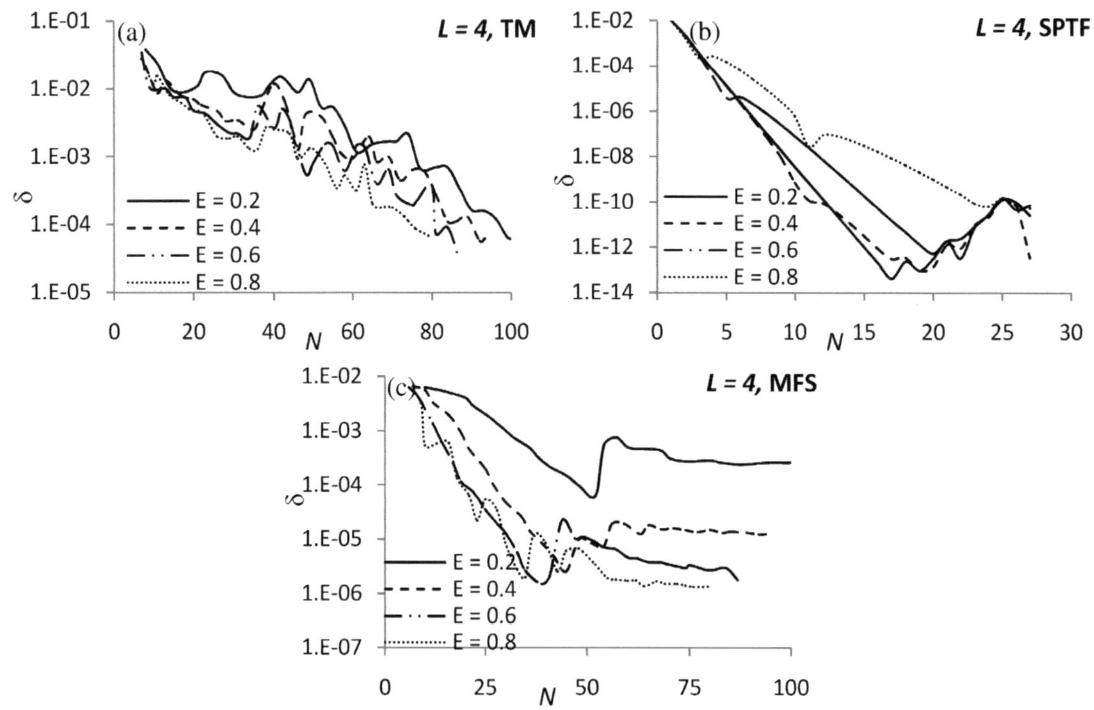

Figure 3. The error of fulfilling the boundary conditions for the square array of fibrous $L = 4$ and for three methods: TM, SPTF and MFS as a function of N

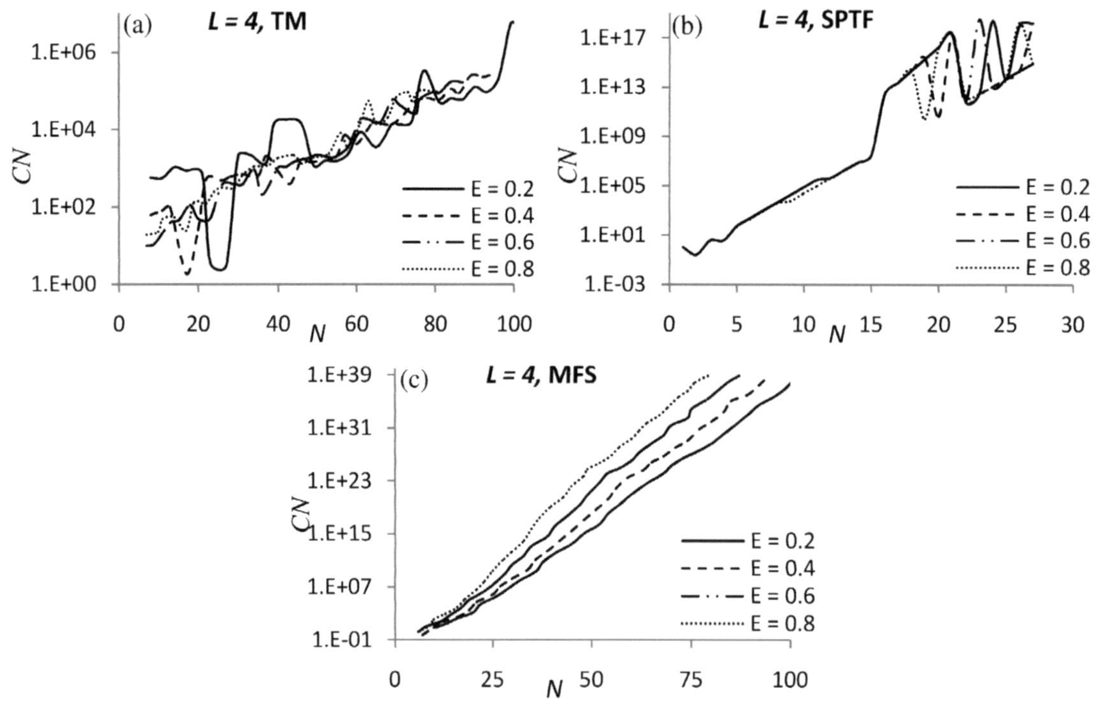

Figure 4. The condition number of collocation matrix for the square array of fibrous $L = 4$ and for three methods: TM, SPTF and MFS as a function of N

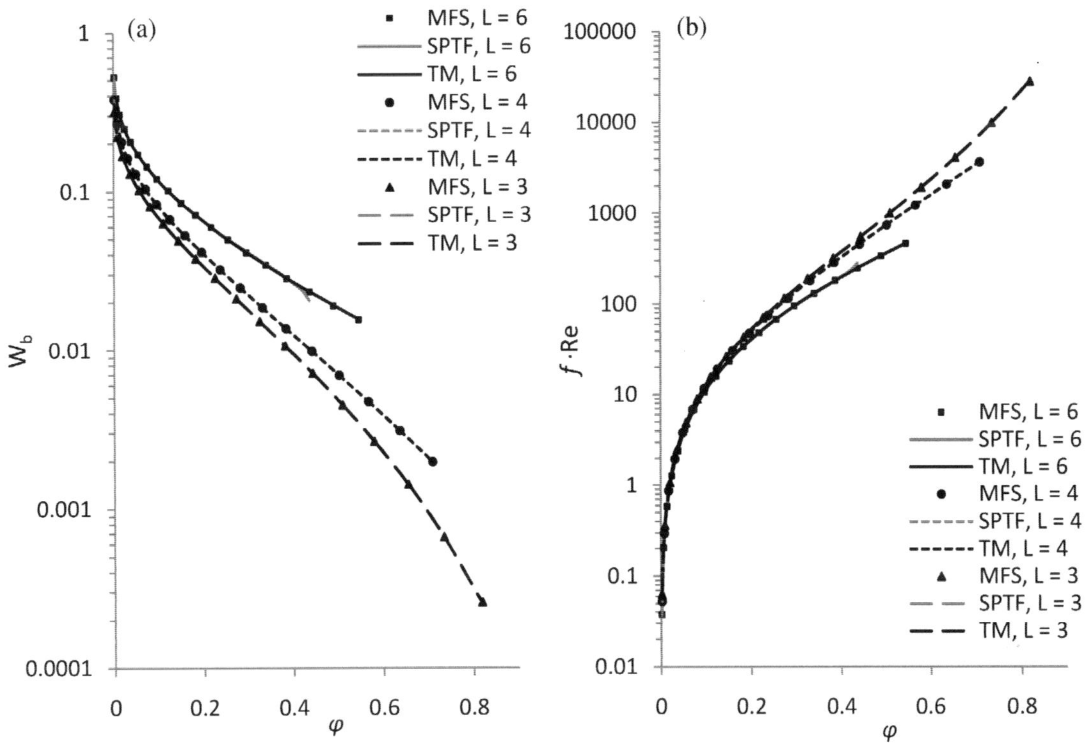

Figure 5. The filtration velocity W_b and the product $f \cdot$ Re for three different kind of fibrous array $L = \{3, 4, 6\}$ and three methods: TM, SPTF and MFS as a function of fibrous volume fraction φ.

Figure 5 shows the value of the filtration velocity, W_b and the product, $f \cdot$ Re as a function of the fiber volume fraction, φ. The result obtained by the TM, the SPTF and the MFS were shown for hexagonal ($L = 6$), square ($L = 4$), and triangular array ($L = 3$) of fibers. For all methods the results are the same. For the Trefftz method we used 100 collocation points and $M = 16$. For the SPTF we used the $M = 15$, and the number of collocation points was also equal to 15. For the MFS the number of collocation point was equal to 86 while the number sources points was equal to 58. The distance between source contour and the boundary was equal $s = 0.15$.

5. Conclusions

Three different approaches for the Trefftz method were compared in analysis of the longitudinal Newtonian fluid flow through the regular bundles of cylindrical fibres. The cylindrical fibres were arranged regularly in a hexagonal, square and triangular array. The numerical algorithm was based on the boundary collocation method. For a repeated element the Trefftz method, the special purpose Trefftz functions and the method of fundamental solutions were used to determined the velocity field. Presented methods are easy to implement, accurate and meshless. The filtration velocity and the product of the friction factor and the Reynolds number were compared for those meshless methods. The results of $f \cdot$ Re and the filtration velocity obtained for all methods were the same. The SPTF is more accurate and less time consuming, because some of the boundary conditions are fulfilled exactly by the trial functions. This method need less number of collocation points, hence the collocation matrix has smaller size and is better conditioned. But we should notice that not for all the boundary value problem this method is appropriated. We can use this method easy for a domain with symmetry line and with simple boundary at which the boundary condition we can fulfil exactly.

Acknowledgments
The paper is financially supported by the Grant 02/21/DSMK/3480.

References

[1] Trefftz E 1926 Ein Gegenstück zum Ritzschen Verfahren, Proceedings of the 2nd International Congress of Applied Mechanics (Zurich) Orell Fussli Verlag pp 131-137

[2] Kołodziej J A and Zieliński A P 2009 Boundary collocation techniques and their application in engineering WIT Press

[3] Kołodziej J A and Uściłowska A 1997 Trefftz-type procedure for Laplace equation on domain with circular holes, circular inclusions, corners, slits and symmetry *CAMES: Computer Assisted Methods in Engineering and Science* vol 4 pp 507-519

[4] Kołodziej J A and Stręk T 2001 Analytical approximations of the shape factors for conductive heat flow in circular and regular polygonal cross-sections *International Journal of Heat and Mass Transfer* vol 44 pp 999-1012

[5] Kołodziej J A, Uściłowska A and Ciałkowski M 2002 Semi-analytical approximations of the laminar friction coefficients for flow in conduits with polygonal cross-section *Acta Mechanica* vol 158 pp 127-144

[6] Kołodziej J A and Fraska A 2005 Elastic torsion of bars possessing regular polygon in cross-section using BCM, *Computers and Structures* vol 84 pp 78-91

[7] Kupradze V D and Aleksidze M Q 1964 The method of functional equations for the approximate solution for certain boundary value problems *USRR Computational Mathematics and Mathematical Physics* vol 4 pp 82-126

[8] Mathon R and Johnston R L 1977 Thc approximatc solution of clliptic boundary value problems by fundamental solutions *SIAM Journal of Numerical Analysis* vol 14 pp 638-650

[9] Fairweather G and Karageorghis A 1998 The method of fundamental solutions for elliptic boundary value problems *Advances in Computational Mathematics* vol 9 pp 69-95

[10] Fairweather G, Karageorghis A and Martin P A 2003 The method of fundamental solutions for scattering and radiation problems *Enginering Analysis with Boundary Elements* vol 27 pp 759-769

[11] Karageorghis A, Lesnic D and Marin L 2011 A survey of applications of the MFS to inverse problems *Inverse Problems in Science and Engineering* vol 19 pp 309-336

Journal of Physics: Conference Series **760** (2016) 012020 doi:10.1088/1742-6596/760/1/012020

Reconstruction of three-dimensional velocity vector maps from two-dimensional PIV data

D Pavlík[1], P Procházka[2] and V Kopecký [1]

[1]Institute of New Technologies, Technical University of Liberec, Studentská 2,
[2]Liberec, Czech Republic Institute of Termomechanics, Academy of Sciences of the Czech Republic

Abstract. In this paper a three-dimensional reconstruction of flow field in a tunnel with the Hump profile is presented. The reconstruction is always performed for pair of 2D vector maps obtained by 3D PIV with two cameras which record measurement area from different locations. Three-dimensional reconstruction can be obtained in various ways. This paper summarizes two: the reconstruction based on the known correspondence between images, and the reconstruction based on the knowledge of intrinsic and extrinsic parameters of cameras. The both methods can be used in the cases when it is impossible to use a calibration pattern or when reconstruction by commercial software fails.

1. Introduction

The 3D PIV is a contactless measurement method which is advisable to use for the comprehensive description actions inside a tunnel. The method is based on the records of displacements of seeding particles carried by the fluid stream. Satiating particles are very small and their ideal size or other properties are determined by many experiments. During a measurement pulsed lasers are used in order to illuminate particles in predetermined measuring area. For recording of the particle images two cameras are mostly used. The optimal angle between axies of the cameras lens is 90° [1]. Two sets of images are used for computation of two-dimensional vector maps of flow field corresponding to each cameras. The first set is captured at the time t while second set of images is recorded several microseconds later $(t+\Delta t)$. The calculation is based on correlation methods. Subsequently, there are reconstructed 3D vector maps from the two-dimensional vector maps, which characterize fluid flow in the measurement area. The reconstruction can be done in several ways.

A calibration procedure of stereo-view has to precede each measurement. The calibration can be done based on the knowledge of geometry layout and parameters of cameras or by recording of a calibration pattern. The calibration pattern may be a plate with black/white dots on white/black background and with defined dimensions, see Fig. 3. This type of pattern is firstly recorded in the place of the laser cut and furthermore in several positions in front of the cut and behind the cut, see in paragraph 2.1. Another possibility is the calibration pattern in the form of a chessboard, which is positioned and recorded in place of the laser cut, as in the previous case, and is turned into random positions around the laser cut. The calibration pattern called multilevel target is another frequently used form of pattern. On the surface of the multilevel target, as the name suggests, dots are placed at least at two levels (eg. 0 mm and 1 mm). The multilevel target does not require multiple recording during the reconstruction procedure. The recording at laser cut position is sufficient for the complete calibration [2].

The vast majority of laboratories, which deals with measurement method 3D PIV, use a commercial software, like Dynamic studio, for the evaluation of measured data. Despite their many advantages, in

some cases the three-dimensional reconstruction algorithms are not sufficiently robust in order to correctly evaluate the measurements. This is mainly in cases where the calibration target is recorded through components which significantly distort the resulting image of the pattern or particles, see paragraph 2.2. A typical and practical example can be measurement of flow within a cylindrical pipe.
Another problem arises when it isn't possible to place the calibration pattern into the measurement area (eg. in a long pipe, tunnel) or its movement isn't possible. The reconstruction by standard commercial programs, fails in these cases, see paragraph 3.2.
Two approaches for extraction of 3D information from 2D PIV data have been developed and they are introduced in this paper. The approaches have been used and tested in experimental research of flow in the tunnel with Hump profile in cooperation with Institute of Termomechanics, Academy of sciences of the Czech Republic. The measurement was done in distortion free environment in order to have relevant data for comparison with results obtained by commercial available Dynamic Studio software. The applicability of approaches was verified in the other measurement when the distortion in the observation direction was presented (the experimental research of flow inside cylindrical pipe).

1.1. Hump profile (Glauert-Goldschmied body)
The Hump profile is used among scientists for experimental research or numerical simulations in the case of an active flow control. Main objective of scientists is the research of separation point and re-attachement point of boundary layer separation. The flow separation creates an area of reverse flow, which increase pressure resistance (separation bubble, separation zone). Therefore it is desired to reduce area of the reverse flow and thereby reduce the resistance of the profile by active methods (especially synthetic jet, plasma actuator). In ideal case the boundary layer separation does not occur at all. The Hump profile is shown in Fig. 1. Such profile was used to test active methods for flow control at positions between 55-75% of the profile length [3].

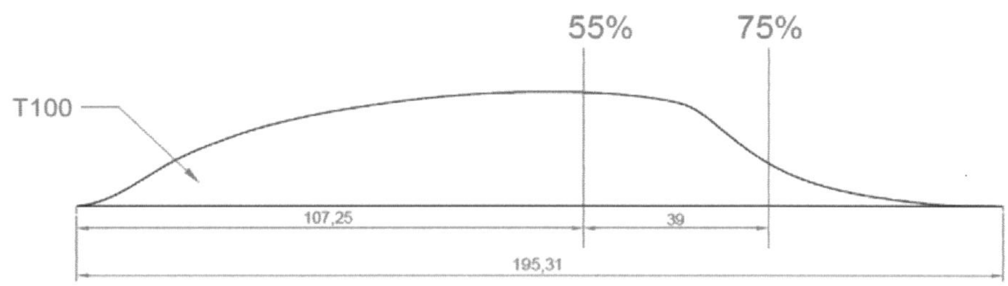

Figure 1. Hump profile.

2. Reconstruction based on known correspondences between images
For the three-dimensional reconstruction of vector maps a method based on the knowledge of the point positions in object coordinate system (X,Y,Z) and the positions of their projections in the image planes (u,v) was used [4]. The object coordinate system is also called a scene. For that were used calibration patterns and their recordings. This approach can be used in cases where records of measured area are subject to great optical distortion. A preparation of the experiment is described in following paragraph.

2.1. Experimental setup
The setup is schematically shown in Fig.2. The cameras used in measuring are Nano HISENSE 3 (Dantec) with parameters: CCD, maximal frame rate 600 Hz at resolution 1280x1024 px, objectives Micro Nikor 60mm. Cameras were mounted on a traverse system, which allowed precise shift in three axes. Angle between axes of their objectives were 60 °.

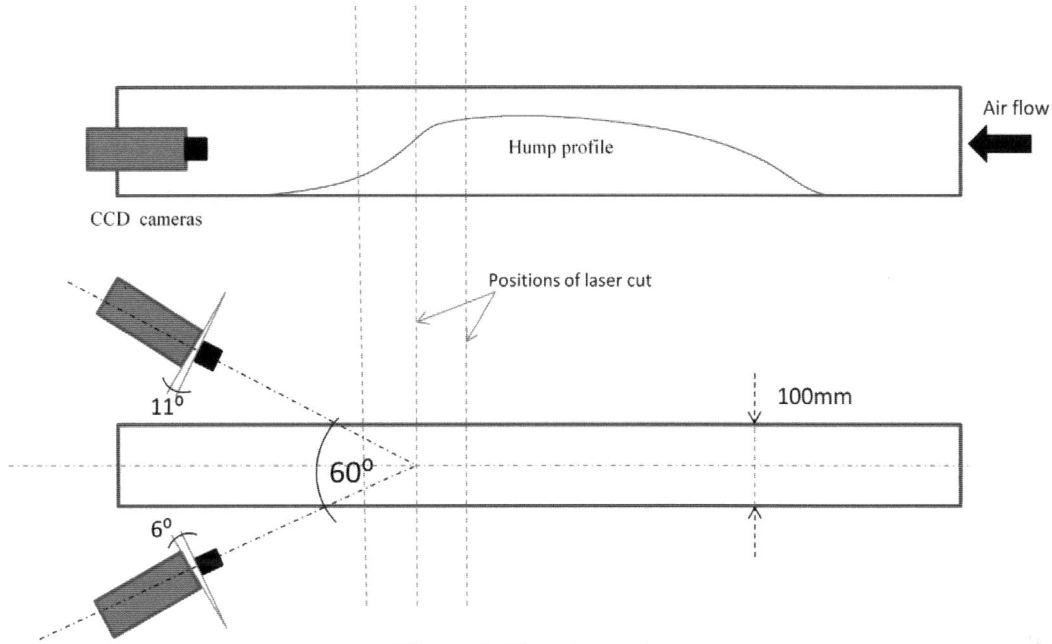

Figure 2. Experimental setup.

The cameras have a very small depth of field. If we want to focus the whole field of view corresponding to the measuring plane (object) on the image plane (CCD detector), the Scheimpflug condition must be met, see Fig. 3. The condition is met if the image plane, object plane and camera lens plane cross each other at a straight line in space [5]. A special camera mount, which allows rotation of camera body towards camera lens, was used for this purpose.

A laser beam of double cavity pulse laser New wave Pegasus PIV was shaped into a so-called laser sheet by cylindrical lens. The laser sheet was a vertical plane with thickness of 1 mm and width of 100 mm. Parameters of the laser: 10mJ at 1KHz, maximal 10KHz, wavelength 527nm. The laser sheet defines a measuring plane (area) in the flow field. The measurement was carried out in several cross sections along the tunnel. Dantec Dynamic components were used to synchronize cameras with the laser.

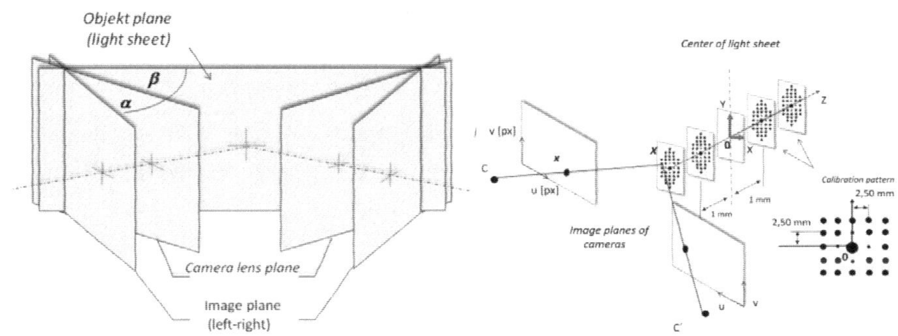

Figure 3. Scheimpflug condition (left), recording of calibration pattern and coordinate systems(right).

The first step of measurement was to place calibration pattern, with small traverse system into the tunnel. Black dots form a pattern on a white background. Pattern was illuminated by a very powerful source of light. This was done because we need uniform distribution of light intensity along a pattern.

The calibration pattern was recorded in five positions: in front of and behind the laser sheet. The pattern was moving along axis perpendicular to the plane of the measurement area, see Fig. 3.
The calibration pattern was removed after recording and the tunnel was saturated by particles (Safex). The measurement was performed at air flow speed: $5ms^{-1}$ .The raw images were processed in Dantec software, where vector maps of flow velocity were calculated and validated.

2.2. Reconstruction

For the reconstruction the relation between positions of points in the scene *(X,Y,Z)* and positions of their projections in the image planes *(u,v)* was used, as mentioned above. The calibration patterns and their recordings were used for this purpose. Positions of calibration points in the scene were described in the object coordinate system. The origin of the coordinate system was identical to the intersection of the central axis of laser sheet and line from zero points (calibration images centers placed in defined positions inside tunnel (0,0,Z)). The orientation of the object coordinate system and the image coordinate system is shown in Fig.3. Further, the dimensions of the calibration pattern, as well as an example of the projection of one point into image planes are shown in the same figure.
The relation between positions of points in scene *(X,Y,Z)* and positions of their projections in the image plane *(u,v)* is expressed by polynomial model n-th order [4], which is able to describe nonlinearities caused by the image distortion:

$$F(u,v) = \sum_{i=0,j=0,k=0}^{i=n,j=n,k=n} A_{i,j,k} x^i y^j z^k \tag{1}$$

The polynomial model of the third order (in X,Y, in Z second order), which is sufficient for most experiments, was used:

$$\begin{aligned}\begin{bmatrix} u \\ v \end{bmatrix} = A_{000} &+ A_{100}X + A_{010}Y + A_{001}Z \\ &+ A_{110}XY + A_{101}XZ + A_{011}YZ \\ &+ A_{200}X^2 + A_{020}Y^2 + A_{002}Z^2 \\ &+ A_{300}X^3 + A_{210}X^2Y + A_{120}X^2Z \\ &+ A_{030}Y^3 + A_{120}XY^2 + A_{021}Y^2Z \\ &+ A_{120}XZ^2 + A_{012}YZ^2 + A_{111}XYZ\end{aligned} \tag{2}$$

In (2) coefficients A are two-dimensional vectors. Nineteen unknown vectors for each camera were necessary to be calculated.
First step of reconstruction was to find positions of the calibration points in the targets images *(u,v)* and assign them to corresponding points in the object coordinate system *(X,Y,Z)*. The coefficients were calculated by last squares method for each camera.
Next step was the reconstruction of a three-dimensional flow model from the vector maps. The reconstruction was resolved by triangulation. For this it was necessary to identify a common field of view of the cameras. A vector from two-dimensional velocity vector map, computed from the particles images recorded by left camera, was assigned to the corresponding vector from the two-dimensional velocity vector map captured by the right camera. It resulted in a set of four nonlinear equations with three unknowns, see (2). Solution of the equations was a three-dimensional velocity vector in a certain interrogation area. Same procedure was used to all correspondences and finally we calculated the three-dimensional velocity vector map.

3. Reconstruction based on knowledge of cameras intrinsic and extrinsic parameters

Reconstruction based on knowledge of cameras intrinsic and extrinsic parameters enables to reconstruct the three-dimensional flow velocity vector maps in cases where the calibration target cannot be put into the measuring area or where it isn't possible to move with the calibration target inside tunnel or pipe. However the approach cannot be used in the presence of the optical distortions. This reconstruction is based on a linear transformation unlike to the polynomial model.
The experimental setup was the same as in previous case (Fig.2), except for a placement of the calibration target into tunnel. Before starting the measurement it was necessary to get parameters which are listed below.

3.1. Theory of pinhole camera model

A pinhole camera model (also perspective camera) approximates very well properties of real camera and in this approximation is sufficient in this case. The camera transforms points *X(X,Y,Z)* from the object coordinate system to points *x(u,v)* in image plane, such that:

$$\boldsymbol{x} = P\boldsymbol{X}, \tag{3}$$

In (3) is P a camera projection matrix [6], with dimensions 3*x*4 and rank 3.
The projection matrix of cameras can also be expressed by equations:

$$P = \begin{bmatrix} f\alpha & 0 & p_x & 0 \\ 0 & f\beta & p_y & 0 \\ 0 & 0 & 1 & 0 \end{bmatrix} = \begin{bmatrix} f\alpha & 0 & p_x \\ 0 & f\beta & p_y \\ 0 & 0 & 1 \end{bmatrix} [\,\mathrm{I}\,|\,0\,] = K\,[\,\mathrm{I}\,|\,0\,] \tag{4}$$

In (4) p_x and p_y are coordinates of principal point in the image plane, α is the width and height ratio of pixels, β is degree bevel of pixels, *f* is focal length. These parameters are collectively called Intrinsic camera parameters and are expressed by calibration matrix *K* [6]. Situation, when during projection the optical center is not located at the origin of the object coordinate system and main axis of camera is not identical with the axis Z, is described by the extrinsic parameters (5). The extrinsic parameters accordingly define orientation and position of camera coordinate system in the scene. The orientation represents matrix *R*, with dimensions 3*x*3 (rotation of the camera around all three axes of object coordinate system) and translation vector *t* which specifies position of camera optical center (focus of lens).

$$P = KR\,[\,\mathrm{I}\,|\,-t\,] \tag{5}$$

3.2. Reconstruction

The aim of this approach was to find the projection matrices of both cameras. It was necessary to determine the matrices *K,R* and vector *t*. The matrices K_1, K_2 (for each camera) were calculated from several records of chessboard with defined grid, see Fig.4. Camera calibration toolbox for Matlab was used for calculation [7]. The recording was done before the measurement, when cameras were located outside traverse system. Next step was to determine the extrinsic parameters. The orientation of cameras coordinate systems represents matrices R_1, R_2. The translation vectors t_1, t_2 specify the position of the optical centers of cameras (focus of lens), as noted above. The center of the coordinate system was identical with center of measurement area, see Fig. 3.

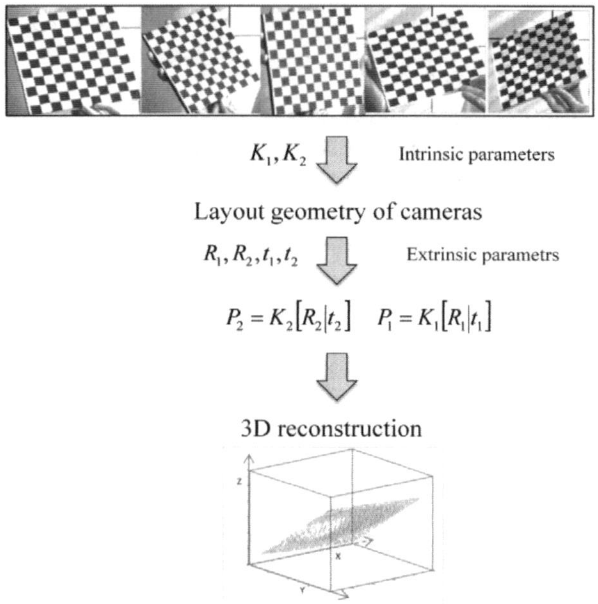

Figure 4. The reconstruction process.

The vector of the two-dimensional velocity vector map, computed from the particles images recorded by left camera, was assigned to corresponding vector from the right camera two-dimensional velocity vector map, same as in paragraph 2.2. The three-dimensional velocity flow vector in certain interrogation area was calculated according to equation (3) and subsequent triangulation [6]. Same approach was used for all correspondences and it was possible to reconstruct the three-dimensional vector map. The reconstruction is not invariant to rotation, translation and evinces certain scale to reality. The scale is possible to determine from other measures, eg. from 2D PIV method or Pitot tube point measurement.

4. Results of reconstructions

The results of the reconstructions are presented by graphs of individual components of the velocity vectors. The measurement was performed at air flow speed $5ms^{-1}$, and at 75% of the profile length.

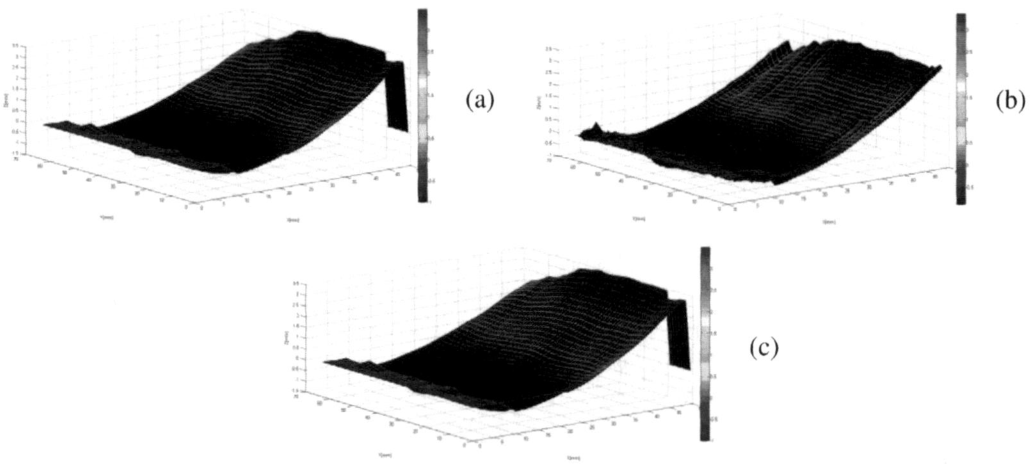

Figure 5. Flow velocity vectors: Z component (a) the reconstruction in paragraph 2., (b) the reconstruction in paragraph 3., (c) Dynamic studio

Journal of Physics: Conference Series **760** (2016) 012020 doi:10.1088/1742-6596/760/1/012020

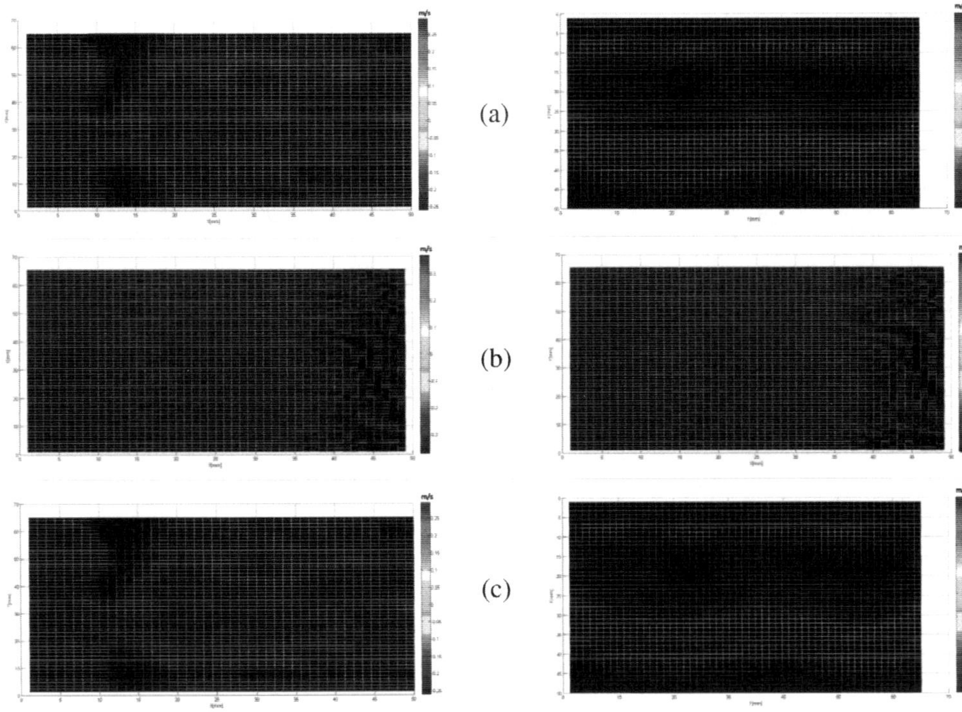

Figure 6. Flow velocity vectors: X (left), Y (right) component (a) the reconstruction in paragraph 2., (b) the reconstruction in paragraph 3., (c) Dynamic studio

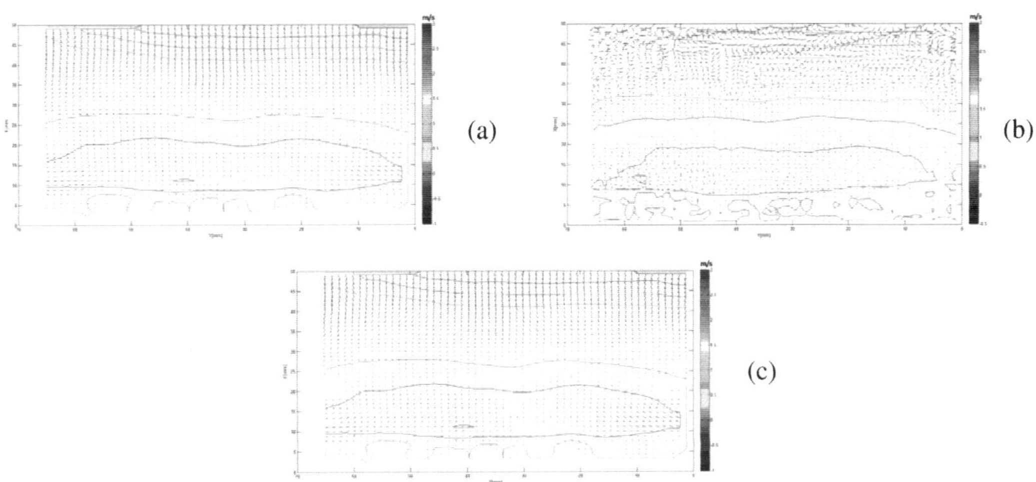

Figure 7. 2D vector map of flow velocity with scalar map (Z component): (a) the reconstruction in paragraph 2., (b) the reconstruction in paragraph 3., (c) Dynamic studio

The accuracy of the reconstruction method based on the known correspondence between images was approximately +/- 1% (for all components) compared with the results from Dynamic Studio.
The accuracy of the reconstruction method based on the knowledge of intrinsic and extrinsic parameters of cameras was approximately +/- 2% in Z component, and +/- 30% in X, Y component compared with the results from Dynamic Studio. The error in the results is mainly caused by high sensitivity to inaccuracy of the measured parameters *(R,t,K)*.

5. Conclusion

The paper describes two approaches for evaluation of data obtained by 3D PIV method that can be used in situations where commercial available softwares can fail. The reconstruction based on known correspondence between images (paragraph 2.) is applicable in cases where the measurement area recorded through objects (eg. cylindrical, convex, concave glass) which significantly distort the images of calibration pattern and satiating particles. The reconstruction by commercial software (eg. Dynamic Studio) often fails in these cases. It was shown that the third order of the polynomial fitting model (paragraph 2.2) is sufficient in most cases. Otherwise it is possible to choose a higher order.
The reconstruction based on intrinsic and extrinsic parameters of cameras (paragraph 3.) is applicable in case where the calibration target can´t be put into the measuring area or where is not possible to move with the calibration target inside the area (eg. long pipe, tunnel). The reconstruction cannot be performed by conventional commercial softwares in such cases. The error in the results is mainly caused by high sensitivity to inaccuracy of the measured parameters *(R,t,K).*

Acknowledgment

This work was supported by the Ministry of Education of the Czech Republic within the SGS project no. 21066/115 on the Technical University of Liberec and by the Grant Agency of the Czech Republic, project number GP14-25354P.

References

[1] Tropea C., Yarin A. L., Foss J. F., *Handbook of experimental fluid mechanics*, Volume 1, Springer, 1556 pgs, 2007.
[2] Suastegui L. M., *Overview on stereoscopic Particle Image Velocimetry,* Instituto Politécnico Nacional Santa Catarina, Mexico, 16 pgs, 2012.
[3] Prochazka P., Uruba V., *On the flow around Glauert-Goldschmied body in the narrow channel and separation control strategy*, Proceedings in applied mathematics and mechanics, 2 pgs, 2016.
[4] Bsibsi M., *Advances in stereoscopic Particle Image Velocimetry and application to supersonic base flow exhaust plume interaction*, TUDelft, 129 pgs, 2004.
[5] Kopecky V., *Laser anemometry in fluid mechanics*, Ed. TRIBUN EU s.r.o., 204 pgs, Brno 2008.
[6] Turonova B., 3D reconstruction of scene from stereo images, 48 pgs, Charles Univesity in Prague, 2008.
[7] Bouguet J. Y., *Computational vision* [online] Camera Calibration Toolbox for Matlab, 2005, URL: http://www.vision.caltech.edu/bouguetj/calib_doc/ .
[8] Soloff S. M., Adrian R. J., Liu Z.C., *Distoetion compensation for generalied stereoscopic Particle Image Velocimetry*, Meas. Sci. Technol 8, 1454 pgs, 1997.

Journal of Physics: Conference Series **760** (2016) 012021 doi:10.1088/1742-6596/760/1/012021

Spectral analysis of self-sustained oscillations for the free jet under variable outlet conditions

A Pawłowska[1], S Drobniak[1] and P Domagała[1]

[1]Częstochowa University of Technology, The Faculty of Mechanical Engineering and Computer Science, Institute of Thermal Machinery, Al. Armii Krajowej 21, 42-200 Częstochowa, Poland

e-mail: pawlowska@imc.pcz.czest.pl

Abstract. The paper presents results of research devoted to the experimental verification of the new regime of self-sustained oscillations in the free axisymmetric jet, which was found numerically at the Institute of Thermal Machinery. Self-sustained oscillations may only appear when two conditions are fulfilled: the exit shear layer should be sufficiently thin and the level of turbulence intensity at the nozzle exit should be sufficiently low. The hot-wire measurements were performed for wide range of exit shear layer thicknesses and level of initial turbulence intensity, which allowed to formulate the preliminary conclusions regarding the conditions required for the appearance of the new phenomenon and its scaling as well as to verify the results of numerical modelling.

1. Background and motivation

Since the fundamental works of Crow and Champagne [1] and Brown and Roshko [2] it has been well known, that high-Reynolds number free shear flows are dominated by large-scale coherent structures. One of the most important shear flow types is a round free jet used in many technical applications. Dynamics of large scale structures in free jets determines the mixing quality important for example in chemical reactors and non-premixed combustion chambers. Extensive experimental work on dynamics of coherent structures performed in the 1980s summarized in reviews by Hussain [3] and Fiedler [4], allowed for understanding the instability mechanism leading to coherent vortices development, vortex pairing and break-up. It was found in particular, that coherent structures are generated by convective instability of Kelvin – Helmholtz type occurring in shear layer with inflection point in velocity profile. The early stage of this process may be described with good accuracy by linear stability theory, while later stages are dominated by nonlinear interactions leading to large scale vortex structures formation. The dynamics of round jet instability was also investigated at the Institute of Thermal Machinery (ITM) Częstochowa during the 1990s by Drobniak et al. [5] and [6]. Despite this knowledge there were still some open questions concerning even very simple flow types like round jet but new light on the shear flow instabilities was shed by numerical works. One of the first DNS studies of the round jet was performed by Verzicco and Orlandi [7] who used finite differences to simulate the temporally evolving round jet at a low Reynolds number $Re = 500 \div 1\,000$. Their simulations of Kelvin-Helmholtz roll-up showed that the growth of the jet largely depends on the initial shear layer thickness.

The round jet received a renewed interest since the work of Huerre and Monkewitz [8] who showed a possibility to find absolute instability in shear flows. In the case of absolute instability a small perturbation grows in time by contrast to convectively unstable flows where the growing perturbation is swept away from the place of generation. The growing in time amplitude of absolute unstable mode is limited only by non-linear interactions and may attain the highest possible amplitude without external stimulation. Monkewitz and Sohn [9] performed linear stability analysis, showing that low-density jets undergo absolute instability when the jet density to ambient fluid ratio is lower than the critical one $S_{cr} \approx 0.71$. The two principal experimental works on air-helium round jet by Kyle & Sreenivasan [10] and on hot-air round jet by Monkewitz et al. [11] delivered the most important data to characterize the instability of variable density jets. Despite the long time since the earliest papers on absolute instability in flow with variable density, this phenomenon is still far from being fully understood, in particular the results of experiments performed in hot air and in helium jets reveal substantial discrepancies. Experimental investigations of absolute instability were also performed at ITM by Bogusławski [12], which confirmed the existence of one of the modes discovered by Monkewitz [11] but also revealed the existence of strong oscillations in isothermal flows.

The discrepancies in experimental results were a motivation for numerical simulations initiated at ITM. The DNS studies are still extremely demanding as far as the computational resources are concerned and so the Large Eddy Simulation (LES) seems to be the only method possible. That is why the LES simulations were performed by Bogusławski et al. [13] which revealed that in isothermal jets a new phenomenon was found. According to this new idea, for sufficiently thin shear layer (i.e. $r/\theta > 25$ where $r -$ jet exit radius and $\theta -$ momentum thickness of the shear layer) the initial perturbation grows sufficiently fast to generate large scale vortex in the jet near field. As a result, the self-sustained oscillations were obtained, which scaled on momentum thickness θ as a convective instability, but were growing locally in time up to high amplitudes limited by non-linear interactions, being in this feature similar to the self-sustained absolute unstable mode present in hot and plasma jets. The physical mechanism and governing parameters of this new type of instability were investigated both numerically and experimentally at ITM and numerical simulations confirmed the existence of self-sustained oscillations [14]. The experimental verification of this idea, which is presented in the paper required to build a test rig, that would allow to generate a jet with sufficiently thin exit shear layer and very low turbulence level.

2. Experimental test rig and flow conditions.

As it was already stated, the necessary conditions for the appearance of self-sustained oscillations are the sufficiently thin exit shear layer and possibility to achieve sufficiently low value of turbulence intensity Tu. All these requirements were fulfilled by the design of the test rig, which is shown at Figure 1.

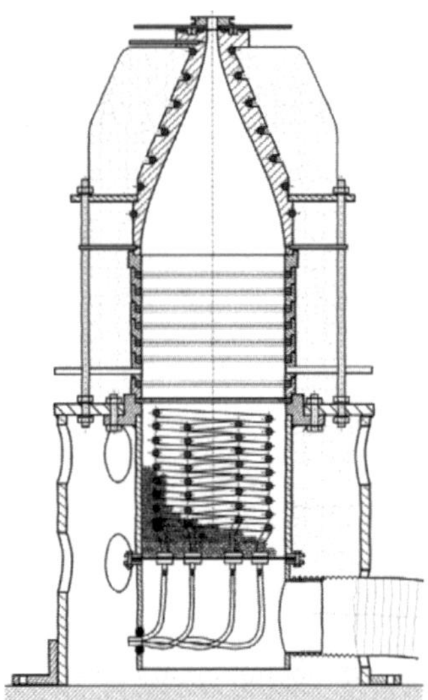

Figure 1. Sketch of the test rigs.

The air is supplied at the bottom, the fan, filter, ducts and acoustic silencer are located outside the measuring area and are not shown at the sketch. The air first passes through the heater, where electric heaters are filled with iron pellets which are designed to increase the heat capacity and to make the flow more uniform. The heaters allow to change the density ratio S down to value S = 0.5, however in the present paper only the isothermal results (S = 1) are presented and discussed. After passing the heaters the air enters the settling chamber where six wire gauzes are located, which makes the flow more uniform and reduces the turbulence intensity at the outlet. The rig allows to perform measurements at the range of Reynolds number from 5 000 to 20 000 for test rigs denoted as Exp. 2 and Exp.3 and from 5 000 to 40 000 for first experiment (Exp.1). The Reynolds number is based on the outlet nozzle diameter and the mass flux at the jet exit. Table 1 summarizes the main flow parameters at the jet exit plane applied both in computations and experiment.

Table 1. Parameters characterizing the nozzle and flow at the jet exit plane.

	$(D/d)^2$	H	r/θ
LES^*	-	1.38	16 - 32
LES^{**}	-	2.59	16 - 32
$Exp.1$	128	2.50 – 2.66	28 , 47
$Exp.2$	144	2.49 – 2.70	12 - 81
$Exp.3$	225	2.55 – 2.66	14 - 68

Three different designs which are addressed in the paper differ mainly by the surface contraction ratio $(D/d)^2$, where $D, d-$ inlet and exit diameters of the nozzle, all test rigs utilized the cubic nozzle profile well proven in previous ITM research. The first experiment (Exp.1) was used for primary research reported in [13], as may be noticed only two values of r/θ parameter could be obtained in this case, which did not allow to demonstrate the change in jet behaviour when passing through the $r/\theta = 25$ critical value found in numerical computations. That is why the two new test rigs (denoted as Exp. 2 and 3) were developed, which enabled to cover the much wider range of r/θ parameter and what is the most essential the critical value $r/\theta = 25$ was contained within this range. One should notice the large contraction ratio of nozzle in Exp. 3, which is much larger than in most experiments described in the literature. Such a high value of contraction was deliberately applied in Exp. 3 to obtain the lowest possible value of turbulence intensity Tu, that in turn allowed to compare the results from Exp. 2 and 3 and to investigate the influence of Tu, which was identified as one of key parameters. The shape parameter H obtained in all three experiments is identical with only one series of computational series. In the first series of computations which were presented at [13] and are denoted as (LES*) the hyperbolic tangent was used as the initial velocity profile while in the second series reported at [14] and denoted as (LES**) the Blasius profile was applied. The values of shape parameter H in second series of computations agree with all experiments and confirm, that the shear layer considered is laminar. As it was mentioned before the exit shear layer thickness is one of the most important parameter in self-sustained regime so it was essential to control this parameter. The cylindrical nozzle extensions were used to vary the exit shear layer thickness and consequently r/θ parameter. In such a way that the longer extension tube (L/d denotes hereinafter the normalized length of extension tube) the thicker is the shear layer that allows to reach and exceed the critical value $r/\theta = 25$. As it was already mentioned the self-sustained oscillations were found extremely sensitive to external disturbances (e.g. the presence of measuring sensor), so CTA and LDA were used as measuring techniques. However, the comparative test of CTA and LDA (not reported here) performed in isothermal flow revealed, that the statistics and spectra obtained with these two methods were exactly the same, it allowed to conclude, that the presence of CTA probe does not affect the jet behaviour. That is why it was decided that CTA will be used as a measuring technique for constant density (isothermal) flow and LDA will be a choice for future measurements in variable density (non-isothermal) flow (not discussed here). The accuracy of CTA probe positioning was better than 0.01 mm. The flow statistics were calculated using the author's Fortran programs while spectral analysis was performed with MatLAb software.

3. Experimental results and discussion.

The first problem that had to be solved during the preliminary investigations was related to comparison of numerical and experimental initial flow conditions. In LES calculations which allowed to establish the idea of self-sustained regime the moderate degree of co-flow was used, because periodic boundary conditions were applied [13]. There were some numerical tests suggesting that the co-flow up to the 5% of mean velocity does not influence the results of computations but it had to be verified experimentally, whether this numerical trick may exert some influence upon the behaviour of the free jet. As it was shown by Pawłowska et al. [15] the jets behaviour was essentially the same regardless on "free - slip" (no co-flow) or "no-slip" (with co-flow) initial conditions applied. The next task concerned the possibility to investigate the influence of initial turbulence intensity by comparison of flow characteristics obtained in two test rigs (Exp. 2 and 3). Figure 2 presents the radial distributions of turbulence intensity Tu, where normalized distance r/d = 0 corresponds to the jet axis, while r/d = 0.5 corresponds to the nozzle edge. The Tu profile is flat within the potential core and shows a slightly higher values of Tu near the wall , which is a typical feature of a laminar exit shear layer.

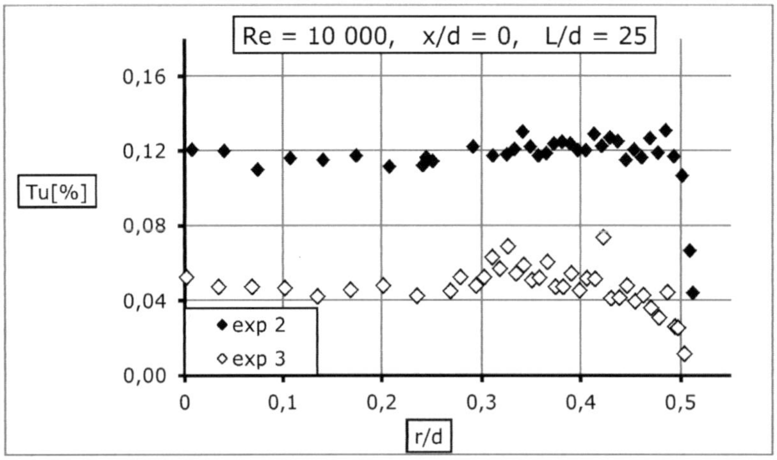

Figure 2. Comparison of turbulence intensity Tu profiles at the jet exit for two designs of test rig.

The comparison of turbulence intensity Tu radial distributions measured at the jet exit plane and shown at figure 2 reveals, that the disturbance level in potential core, which was already low in the initial design ($Tu \approx 0.1$ % for Exp. 2) became even lower ($Tu \approx 0.05$ % for Exp. 3). The summary of turbulence intensity measurements are presented at table 2, which contains the averages values of turbulence intensity at the potential core (Tu_{MEAN}) and the maximum values encountered in the laminar shear layer (Tu_{MAX}) for all values of Re and r/θ investigated within the paper.

The results presented at Figure 2 and Table 2 allow to expect, that proper conditions for self-sustained oscillations were achieved, bearing in mind, that the low level of disturbances at the jet exit plane is among the primary requirements for the appearance of self-sustained oscillations. To clarify this problem in more detail, the analysis of jet evolution was performed based on the analysis of mean velocity and turbulence intensity distributions along the jet axis and the results are shown at Figure 3 for Re = 5 000 and the same relative shear layer thickness r/θ provided by the same normalized length of extension tube L/d = 7. The mean velocity U is normalized by the mass flux velocity U_0 at the jet exit, while the downstream distance x along the jet axis is measured from the jet exit and normalized by the jet exit diameter d.

Table 2. Values of turbulence intensity for both test rigs.

	exp. 2				*exp. 3*	
r/θ	Tu_{MEAN} *[%]*	Tu_{MAX} *[%]*	$Re * 10^3$	*r/θ*	Tu_{MEAN} *[%]*	Tu_{MAX} *[%]*
36	0.26	0.30	**5**	35	0.19	0.90
22	0.21	0.25	**5**	21	0.13	0.17
15	0.21	0.25	**5**	15	0.13	0.17
12	0.16	0.19	**5**	14	0.13	0.18
54	0.12	0.19	**10**	49	0.12	0.27
31	0.15	0.17	**10**	27	0.09	0.30
20	0.14	0.17	**10**	19	0.06	0.25
15	0.12	0.13	**10**	15	0.05	0.14
81	0.05	0.79	**20**	68	0.05	0.30
39	0.08	0.18	**20**	41	0.04	0.25
31	0.09	0.32	**20**	31	0.04	0.30
23	0.08	0.12	**20**	23	0.03	0.25

As can be seen at Figure 3a the mean axial velocity is constant within the potential core, which is the longer the lower is the turbulence level. As the radially growing shear layer reaches the jet axis the increase of Tu is observed at Figure 3b, which develops into the characteristic "plateau" at the distance $x/d \approx 6$. As it was shown by Crow & Champagne [1] the location of the "plateau" marks the region where the dominant mode reaches the maximum amplitude and the pairing of coherent vortices starts.

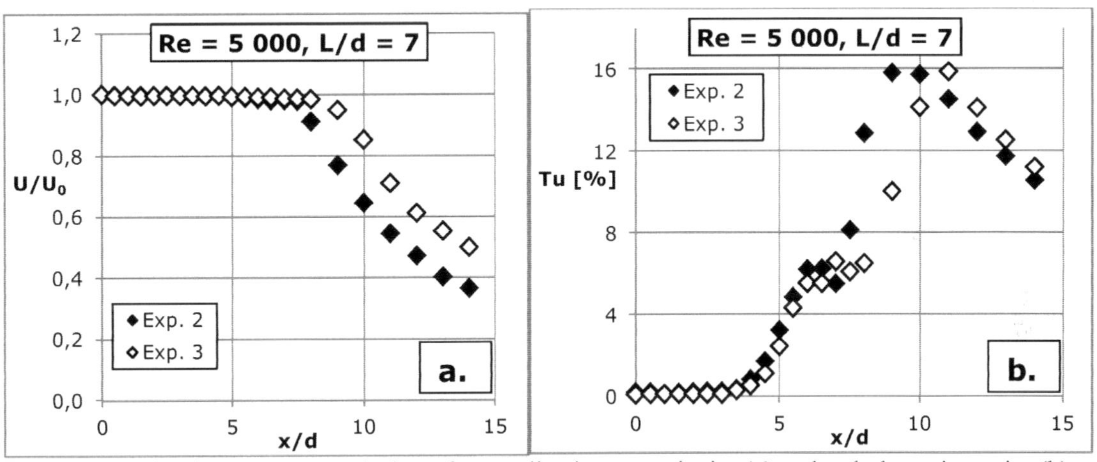

Figure 3. The downstream evolution of normalized mean velocity (a) and turbulence intensity (b) along the jet axis for Re = 5 000 and normalized length of extension tube L/d = 7.

The lower value of Tu obtained in Exp. 3 increases the axial extent of the "plateau" and makes it more pronounced. Furthermore, for Exp. 3 the location of maximum value of turbulence intensity Tu is shifted downstream in comparison with Exp.2, it should be noted however, that the maximum value of Tu is identical in two experiments. The results obtained confirm, that the initial level of turbulence is an important factor governing the behaviour of coherent structures, that agrees with findings from numerical experiments [13]. The important question that must now be answered is to what extent the change of initial disturbance level affects the development of coherent structures. In particular, one should find an answer whether this change is qualitative (i.e. the radical change of coherent structure pattern) or quantitative (i.e. the change of growth rate of the same pattern).To answer this question the spectral analysis of velocity fluctuations along the jet axis was performed for the flow case presented at Figure 3. The amplitude spectra are presented at Figure 4 for three sample points located along the jet axis within (x/d = 5.5 and 6.5) and immediately behind the "plateau" (x/d = 9).

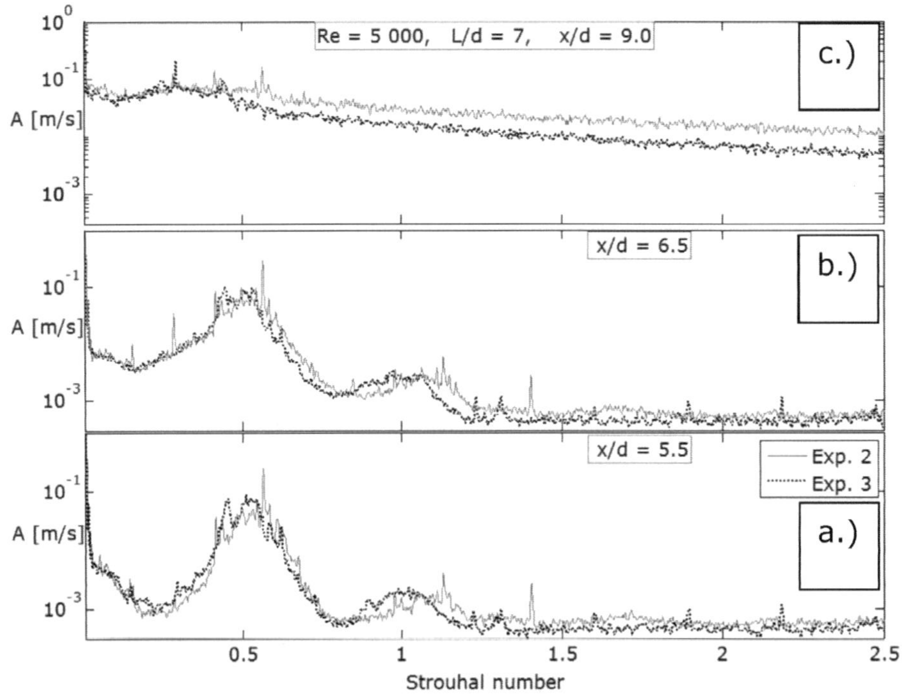

Figure 4. Amplitude spectra of velocity fluctuations along the jet axis at x/d = 5.5 a.); x/d = 6.5 b.) and x/d = 9.0 c.) for Re = 5 000 and normalized length of extension tube L/d = 7.

The amplitude spectra at Figure 4 are presented versus the non-dimensional frequency – Strouhal number based on the outlet nozzle diameter ($St_d = \frac{f*d}{U_0}$, where $f -$ frequency, $d -$outlet nozzle diameter, $U_0 -$ mass flux velocity at the jet exit). One may easily notice that spectra for both Exp. 2 and 3 are essentially identical, that in turn suggests that in both test rigs the sufficiently low level of disturbances was achieved, that creates favourable conditions for existence of the same formation pattern of coherent structures. The analysis of initial region of Tu growth (up to x/d = 6.0) shown at Figure 3b reveals that the same growth rate is achieved for both experiments and only the consecutive stages of development are affected by the initial disturbances level. It means therefore, that results of both Exp. 2 and 3 are meaningful with only quantitative difference resulting from the different initial disturbances levels encountered in both experiments.

As it was stated above, the self-sustained oscillations obtained as a new free jet regime, which should scale on momentum thickness θ as the convective instability, but they should grow locally in time up to high amplitudes limited by non-linear interactions, being in this respect similar to absolutely unstable modes present in hot and plasma jets, which scale on the jet diameter. The scaling of dominant modes for all the flow cases was therefore performed, the frequencies of dominant modes were either expressed as Strouhal number St_d based on the outlet nozzle diameter (Figure 5) or as Strouhal numbers St_θ based on the exit shear layer momentum thickness ($St_\theta = \frac{f*\theta}{U_0}$, where $f -$ frequency, $\theta -$ exit shear layer momentum thickness, $U_0 -$ mass flux velocity at the jet exit) see Figure 6. At both figures the results of LES computations and the results of preliminary experiment (Exp.1) were also shown, to enable the trend analysis. Both scaling attempts are presented versus r/θ, which was already identified as a key parameter governing the appearance of self-sustained oscillations. As one may notice at Figure 5 the results of experiments do not reveal any clear tendency and do not show even a qualitative similarity with LES computations. The thick broken line was added for $r/\theta = 25$, which was found as a critical value in numerical computations. Unfortunately, even the addition of critical r/θ value does not allow to find any pattern that would suggest the possible scaling. One should notice however, that the results of preliminary Exp. 1 (black triangles) seem to agree with numerical results, but it is nothing more than an accidental coincidence resulting from the limited experimental evidence. The results shows at Figure 5 may be regarded as the further development of results presented in [16], where only the results of Exp. 1 and a limited number of results from Exp. 2 were available.

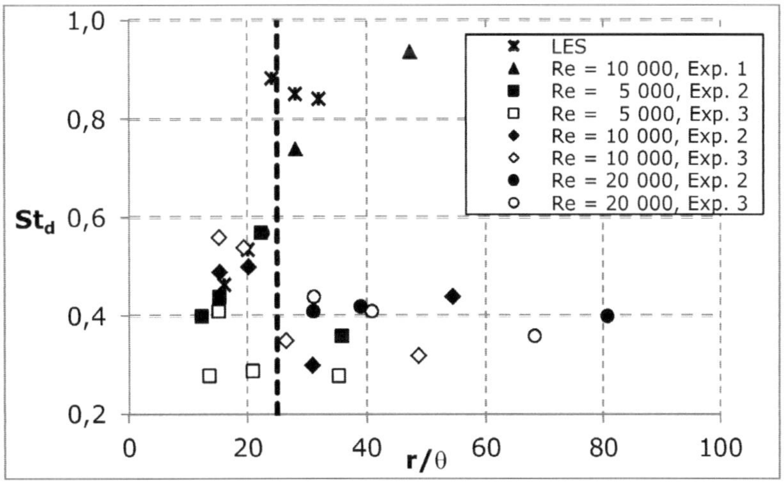

Figure 5. Scaling of dominant modes on jet diameter and comparison with preliminary experiment and numerical computations.

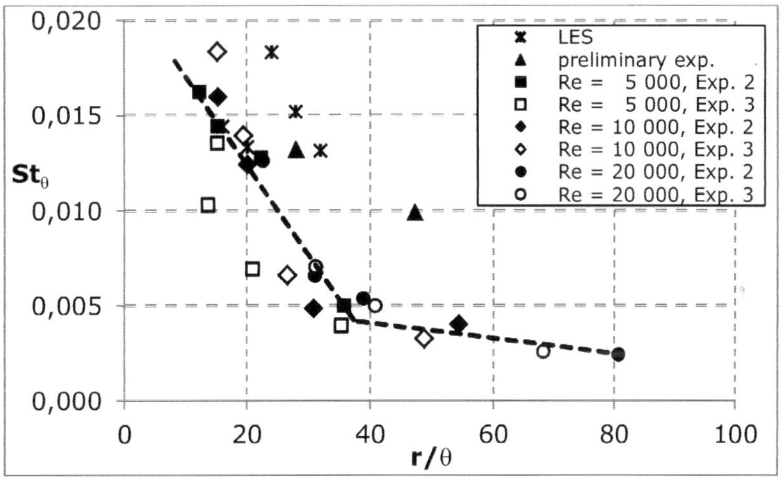

Figure 6. Scaling of dominant modes based on exit shear layer momentum thickness and comparison with preliminary experiment and numerical computation.

Quite a different picture was obtained with scaling of dominant modes based on exit shear layer momentum thickness, presented at Figure 6, where a clear trend may be observed. For comparison, the results of numerical computations were also added, the two broken lines present a least square fit (LSF) to experimental results calculated separately for $r/\theta < 25$ and $r/\theta > 25$. One may easily notice that they intersect at $r/\theta \approx 30$, which is very close to numerical prediction of critical value equal $r/\theta = 25$. All experimental results reveal a clear pattern, furthermore the inclination of LSF line for $r/\theta < 25$ is almost identical as the trend outlined by the results of numerical computations. One may conclude therefore , that since the dominant modes scale on exit shear layer momentum thickness so most probably they are of convective nature, as predicted by Bogusławski [13]. In order to complete the experimental verification the next step is needed, i.e. the proof that self-sustained oscillations grow locally in time up to high amplitudes in a manner that does not allow to increase their growth rate by the external forcing, which is currently being performed at ITM.

4. Conclusions

The paper deals with the experimental verification of the new mechanism of laminar – turbulent transition in round free jets, which has been found numerically at ITM. Instability processes in shear flows are of primary importance in fluid dynamics since transitional free flows are present in many technical applications, but first of all it is a scientific challenge, since the explanation of the transition mechanisms may result in better understanding of turbulence and vortex dynamics. The set of specially designed test rigs allowed to obtain proper conditions for the existence of self–sustained oscillations, in particular the low level of disturbances at the jet exit plane and sufficiently thin exit shear layer. The results obtained so far confirm, that the initial level of turbulence is an important factor governing the behaviour of coherent structures, that agrees with findings from numerical experiments. The scaling tests revealed, that the dominant modes of instability scale on exit shear layer momentum thickness, that in turn suggests that they most probably are of convective nature, as it was predicted numerically at ITM. In order to complete the experimental verification the next step is needed, i.e. the proof that self-sustained oscillations grow locally in time up to high amplitudes in a manner that does not allow to increase their growth rate by external forcing, that will constitute the next stage of research.

Acknowledgements
The investigations presented in this paper have been performed with funding from Polish National Science Centre within the grant DEC 2011/03/B/ST8/06401.

References

[1] Crow SC Champagne FH 1971 Orderly structure in jet turbulence *J. Fluid Mech.* **48** pp 547-592
[2] Brown GL Roshko A 1974 On density effects and large structure in turbulent mixing layers *J. Fluid Mech.* **64** part 4 pp. 775-816
[3] Hussain AKMF 1986 Coherent structure and turbulence *J. Fluid Mech.* **44** pp. 1349-1357
[4] Fiedler HE 1988 Coherent structures in turbulent flows *Prog. Aerosp. Sci.* **173** pp. 303-356
[5] Drobniak S Elsner JW Abou-El-Kassem E 1998 The relationship between coherent structures and heat transfer processes in the initial region of a round jet *Exp. Fluids* **24** pp. 225-237
[6] Drobniak S Elsner JW Abou-El-Kassem E 1998 Experimental analysis of internal energy transport in the presence of coherent motion *Int. J. Heat Mass Tran.* **41** No. 3 635-643
[7] Verzicco R Orlandi P 1994 Direct simulations of the transitional regime of a circular jet *Phys. Fluids* **6** (2)
[8] Huerre P Monkewitz PA 1985 Absolute and convective instabilities in free shear layers *J. Fluid Mech.* **159** pp. 151-168
[9] Monkewitz PA Sohn KD 1988 Absolute instability in hot jets *AIAA J.* **26** No. 8 pp. 911-916
[10] Kyle DM Sreenivasan KR 1988 The instability and breakdown of a round variable density jet *J. Fluid Mech.* **249** pp. 619-664
[11] Monkewitz PA Bechert DW Barsikow B Lehman B 1990 Self-sustained oscillations and mixing an a heated round jet *J. Fluid Mech.* **213**
[12] Bogusławski A 2002 Convective and absolute instability of free round jet with uniform density *seria monografie* **85**
[13] Bogusławski A Tyliszczak A Drobniak S Asendrych D 2013 Self-sustained oscillations in a homogenous density round jet *J. Turbul.* **14:4** pp. 25-52
[14] Wawrzak K Bogusławski A Tyliszczak A 2015 LES predictions of self-sustained oscillations in homogenous density round free jet *Flow. Turbul. Combust.* **95** pp. 437-459
[15] Pawłowska A Domagała P Wysocki M 2014 Hot-wire study on the axisymmetric free jet dependence on the nozzle shape *J. Phys.: Conf. Ser.* **530**
[16] Pawłowska A Drobniak S Domagała P 2016 Self-sustained oscillations of the axisymmetric free jet at low and moderate Reynolds numbers *Adv. Mech: Theor. Comput. and Interdi. Issues Proc. of 3rd PCM and 21st CMM*

Journal of Physics: Conference Series **760** (2016) 012022 doi:10.1088/1742-6596/760/1/012022

Highly deformable nanofilaments in flow

S Pawłowska[1]

[1] Institute of Fundamental Technological Research PAS, Department of Mechanics and Physics of Fluids, Pawińskiego 5B, 02-106 Warszawa, Poland

Abstract. Experimental analysis of hydrogel nanofilaments conveyed by flow is conducted to help in understanding physical phenomena responsible for transport properties and shape deformations of long bio-objects, like DNA or proteins. Investigated hydrogel nanofilaments exhibit typical macromolecules-like behavior, as spontaneous conformational changes and cross-flow migration. Results of the experiments indicate critical role of thermal fluctuations behavior of single filaments.

1. Introduction

Predicting behavior of deformable objects carried by the flowing fluid is necessary for understanding underlying physics of fibrous suspensions and transport properties of biological macromolecules. Evaluating conformational flexibility of long biomolecules and polymers suspended in fluid permits to determine their structure and mechanical properties. However, the intriguing and important interplay between polymer conformation and fluid dynamics remains still unrevealed. Its knowledge is crucial for understanding their dynamics, transport properties, and complex behavior like folding - unfolding sequences. Despite decades of theoretical studies, the fundamental dynamics of such systems remains a mystery. In addition there is still lack of experimental investigations to validate assumptions of the theoretical and numerical models. Experimental studies on molecules are difficult and rather inaccurate, mainly due to the necessary compromise between spatial and temporary resolution of optical methods at nanoscales. To systematically investigate the influence of flexible objects on their interaction with given flows and the resulting macroscopic properties, a synthetic experimental model of such nano-objects could be useful. It can help in future designing of objects transported by body fluids in order to perform targeted drugs release or as model object for biomolecules, which observation and analysis is difficult due to their small size. Introduction of microscale experimental model allows for precise optical measurements and the use of simple hydrodynamic models that give possibility for the description of dynamics of elastic bio-objects such as DNA or protein.

Theoretical assumptions, especially in the field of polymer physics, use coarse-grained models to study the folding process of nucleic acids and proteins. Most existing models describing the dynamics and properties of flexible objects use complex systems of interconnected identical spherical particles (worm-like chain) that illustrates the two particles configuration of two ends of the polymer which form the filament connected by a spring [1-5]. The possibility to describe this phenomenon on molecular scale is still limited to very small length and time scales. In these scales it is very difficult (or impossible) to find answers to basic questions about the potential effects of interactions or complex hydrodynamic behavior of long biological molecules. Additionally, so far available experimental data are not sufficiently precise to verify numerical models. Most of them are focused on experimental studies of the dynamics of DNA molecules [6] or other bio-objects like actin filaments [7], not accurate enough for detailed analysis.

Journal of Physics: Conference Series **760** (2016) 012022 doi:10.1088/1742-6596/760/1/012022

Migration of fibers or other long objects in the Poiseuille flow is one of the fundamental problems of modern rheology with important impact on a variety of biological, medical and industrial contexts, such as Brownian dynamics of proteins, DNA or biological polymers, cell movement or drugs delivery. Simulations conducted for this type of long objects confirmed that stiff fibers tend to be accumulated near the wall and are slightly bended. Flexible fibers undergo large deformations and accumulate far from the wall [3-5]. Analysis of cross flow migration of micro sized objects conveyed by flow of viscous fluid in a tube have attracted considerable efforts, for example drag reduction observed for flow of blood in capillaries [8]. Experimental studies performed for solid spherical particles [9] and droplets [10,11] confirmed presence of their radial migration, however its theoretical interpretation still remains far from being completed. Despite the common wisdom on reversibility of creeping flow, groups of suspended particles or single deformable objects conveyed by Poiseuille flow may experience lift force even for negligible inertial effects. The effect depends on several factors, beside wall and velocity profile interactions. Also object deformations influence the direction of migration and the equilibrium position that is finally reached in a tube [12-14]. In case of long deformable objects like molecules and filaments, the cross-flow migration problem becomes quite complex, additionally conformational changes, tumbling and their elongation could influence it.

There are very little experimental data documenting details of the migration for long deformable nanoobjects. To the best of our knowledge, the present study offers unique possibility to evaluate effects of object deformability on its behavior in a channel flow. The knowledge gained may help to validate existing modeling attempts [3-5], perform cell sorting, improve DNA sequencing tools, and interpret actin or fibrin flow induced conformational changes.

For the purpose of our studies we have developed a new method for fabrication of highly deformable hydrogel filaments with the typical diameter of 100 nm and contour length ranging from a single micrometer to millimeters [15]. These objects are characterized by similar mechanical properties (ratio of persistence length to contour length) as DNA molecules, which supports the possibility of using them as model objects of biomolecules. An important issue is to understand hydrodynamic confinement effects. By analyzing performance of long filaments interacting with microchannel walls it will be possible to resolve effects of hydrodynamic screening, dramatically changing Stokesian description of such flow [16]. Understanding the relationship between the filaments behavior and flow properties opens the possibility of designing nano-objects capable to travel in crowded body fluids in order to selectively delivery drugs into the target area or to reach specific local tissues with the aim of helping in their regeneration.

2. Materials and Methods

2.1. Materials used

Poly(L-lactide-co-caprolactone) (PLCL), chloroform ($CHCl_3$), N,N-dimethylformamide (DMF), Bovine Serum Albumin conjugated with fluoresceine (BSA-FITC), N,N-isopropylacrylamide (NIPAAm), N,N'-methylene bisacrylamide (BIS-AAm), ammonium persulfate (APS), N,N,N',N'-tetramethylethylenediamine (TEMED).

Permanent epoxy negativ photoresist Su-8 2075, Developer mr-Dev, Silicon wafer, Sylgard 184 Silicone elastomer KIT polydimethylsiloxane (PDMS).

2.2. Preparation of flexible hydrogel nanofilaments

In first step, core-shell nanofibers were produced using coaxial electrospinning technique [17]. The shell is formed by PLCL that is produced using a polymer solution in DMF and $CHCl_3$ in proportion 1:1:9 (w/w). To prepare core solution we used 10 wt.% NIPAAm/BIS-AAm dissolved in deionized water (mass ratio of NIPAAm to BIS-AAm was 37.5:1). The core solution includes also components, which initiate the polymerization reaction (10 μl APS and 1 μl TEMED for 1ml polymer solution), and fluorescent markers enabling material imaging (BSA-FITC). In order to release the nanofilament constituting the core of electrospun nanofiber, a suitable solvent, e.g. DMF, dissolved its shell. This

Journal of Physics: Conference Series **760** (2016) 012022 doi:10.1088/1742-6596/760/1/012022

process should not destroy the structure of the hydrogel nanofilaments. More details about experimental procedure of preparing hydrogel nanofilaments are given in our previous paper [15].

2.3. Nanofilaments in pulsating flow

The fluorescent hydrogel nanofilaments were imaged through a fluorescent microscope (Leica AM TIRF MC) using mercury lamp as a light source. Single nano-objects were observed using a 20x 0.40 DRY objective lens. The movement of nanofilaments was recorded using a high-gain EM-CCD camera (C9100-2, Hamamatsu) with typical framing time of 10Hz.

The experiment is based on the observations of hydrogel nanofilaments with length varying from a few of micrometers to tens micrometers, conveyed in the channel under the influence of pulsating fluid motion generated by two syringe pumps (Nemesis, Celoni GmbH). The length of the microchannel, which was used to analyze filaments flow, is 500 μm, channel width is 200 μm, and its depth is 60 μm. Figure 1 presents schematic of the described measuring system. Observations were made perpendicularly to the plane in which the objects were analyzed. The controlling software allowed the synchronized operations of the pumps in both directions. Flow rate was set to be below 100 μl/s. The presence of the objects in the analyzed area was checked by varying position of the observation, limiting area of interest to the central cross section of the channel. Only clearly distinguishable objects, remaining in focal plane and exhibiting deformations within 2-D plane of observation were taken into account (Fig. 1). For this purpose, constancy of apparent contour length of visualized filaments is used as a control parameter. Changes of observed projected filament contour length are interpreted as out of plane deformations and such sequence is not analyzed.

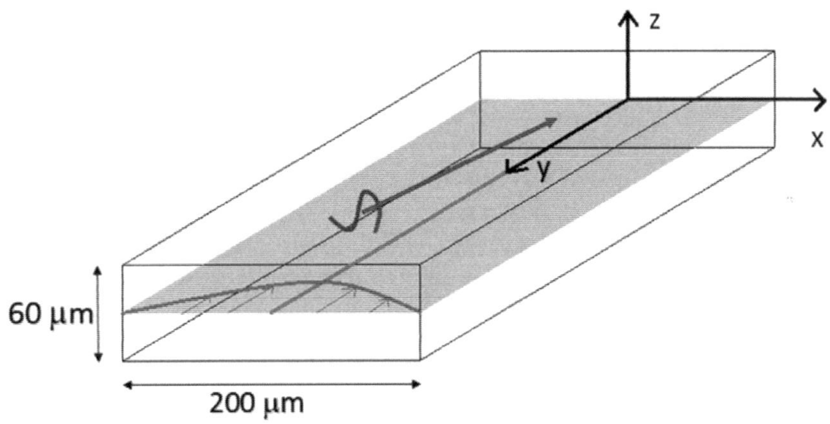

Figure 1. Schematic of the measuring system for nanofilaments conveyed by pulsatile flow. Gray marked is the analyzed area. Red line represents microchannels center.

3. Results

3.1. Deformation dynamic of flexible nanofilaments

Images of nanofilaments are recorded at constant framing rate, set depending on the flow velocity. Usually one experimental sequence consisted of nearly thousand images with resolution of 1M pixels. Selected sequences of images were used to localize filament, analyze its shape, and to describe their deformation using specially developed Matlab software [15]. Mechanical properties of the analyzed filament (Young modulus) were evaluated with help of experimentally determined persistence length. Table 1 collects typical data for filaments observed in the experiments. We may find that their mechanical properties resemble that of long biomolecules.

Table 1. Typical experimental parameters: mean flow velocity (V_f) and flow Reynolds number (Re); filament contour length (L), radius (R), persistence length (Lp), and Young modulus (E). Reynolds number is based on the channel width (200 μm).

No	V_f [μm/s]	L [μm]	R [nm]	Re	Lp [μm]	E [kPa]
1	59.02	71.91	81	$1.51 \cdot 10^{-2}$	17.62	2.17
2	77.78	72.87	81	$1.98 \cdot 10^{-2}$	12.38	1.53
3	39.65	34.49	45	$1.01 \cdot 10^{-2}$	3.44	4.45
4	68.87	54.31	45	$1.76 \cdot 10^{-2}$	7.51	9.72
5	250.43	75.58	45	$6.39 \cdot 10^{-2}$	9.56	12.37
6	199.64	82.90	81	$5.09 \cdot 10^{-2}$	14.34	1.77
7	124.24	32.12	81	$3.17 \cdot 10^{-2}$	25.26	3.12
8	134.70	29.62	81	$3.44 \cdot 10^{-2}$	2.89	0.35
9	119.20	26.84	81	$3.04 \cdot 10^{-2}$	3.75	0.46
10	59.02	71.91	81	$1.51 \cdot 10^{-2}$	17.62	2.17
11	77.78	72.87	81	$1.98 \cdot 10^{-2}$	12.38	1.53

Our experiments indicated that mobility of hydrogel filaments strongly depends on their length. It is also well visible that long filaments exhibit high bending flexibility (Fig. 2). It is interesting to note that despite negligibility of ionic interactions, bending followed by characteristic coiling and knots formation is observed. For long filaments their conformational changes are dominated over translational or rotational diffusion. Classical description of diffusion of long molecules using 'so called' hydrodynamic diameter appears in such case to be very far from the physical phenomena. For short filaments (10–20 μm) translational and rotational diffusion coefficients could be evaluated from their Brownian motion, and such filaments tend to rotate, apparently similar to elongated spheroids (Table 2). Thermal fluctuations of the filament shape observed in quiescent fluid are indicators of their mechanical properties [15]. These data can be used to evaluate interactions of deformable objects with pulsatile flow in the channel, as it is done in the next step.

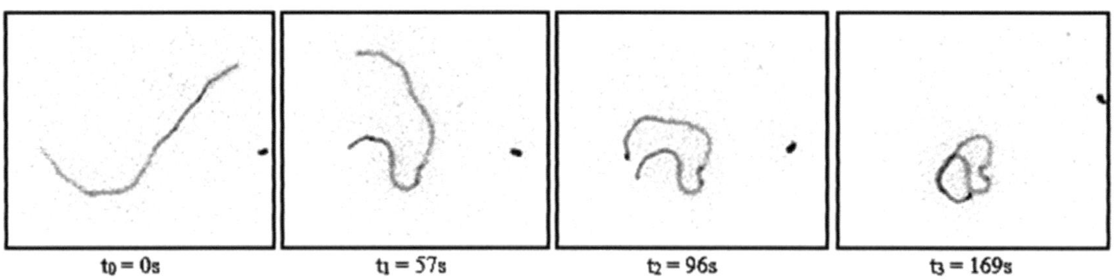

Figure 2. Sequence of images (negatives) for a single and long flexible nanofilament suspended in water, bending under influence of Brownian motion (R= 90 nm, L= 60 μm); image height 35 μm.

Table 2. Comparison between experimental and theoretical diffusion coefficients for selected nanofilaments: L – contour length; 2R – diameter of filament; D_a, D_b – translational diffusion coefficients; D_r – rotational diffusion coefficient; Theoretical diffusion values obtained modeling filaments geometry as that of elongated droplet [15]. Vm – migration velocity calculated for filament approximated by elongated droplet.

	L/2R	D_a (m^2)	D_b (m^2)	D_r (rad^2)	L (μm)	Vm [m/s]	Motion	Shape
exper.	29.8	0.059	0.140	0.0029	14.9	$2.1 \cdot 10^{-8}$	Rotation	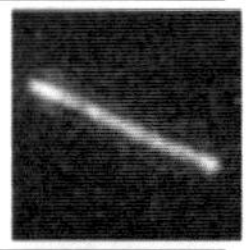
theor.		0.189	0.121	0.0051				
exper.	31.2	0.093	0.067	0.0025	15.6	$4.3 \cdot 10^{-8}$	Rotation	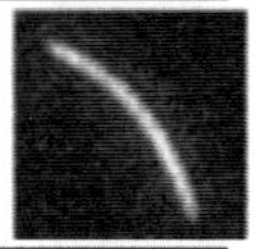
theor.		0.182	0.116	0.0045				
exper.	43.0	0.052	0.035	0.0006	21.5	$1.3 \cdot 10^{-7}$	Bending	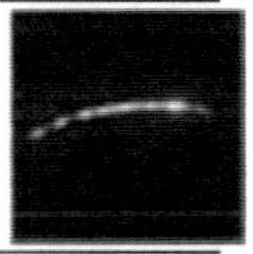
theor.		0.144	0.090	0.0019				
exper.	48.4	0.112	0.105	0.0026	24.2	$6.7 \cdot 10^{-9}$	Bending	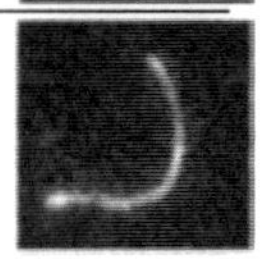
theor.		0.132	0.082	0.0013				

Furthermore we can observe rotation and bending for filaments in pulsating flow. Short filaments move along the channel and rotate while long filaments have the tendency to coil and form knots. This effect remains prevalent for objects conveyed by Poiseuille flow (Figure 3).

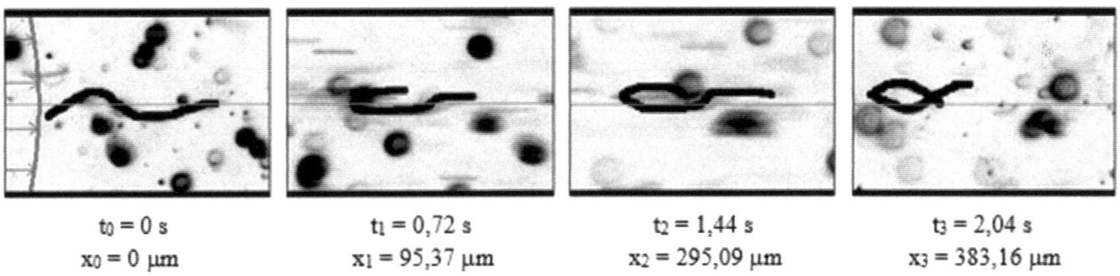

Figure 3. Sequence of images for a single flexible nanofilament conveyed by Poiseuille flow (no. 6 in Table 1), green line represents the center of the channel, t – time stamp, x – distance traveled from the channel entry (picture frame moves with the object), image height 77 μm; negative of the images shown.

Comparison of the measured fluid velocity (by nanoparticles tracing) and that of the filament center of mass observed during pulsatile flow shows slight differences in their waveforms (Figure 4). The differences result from changes in conformation of the analyzed filament as it is moved along the channel conveyed by pulsatile flow, as well as from the relative *"slip"* velocity characteristic for droplets conveyed by Poiseuille flow [13].

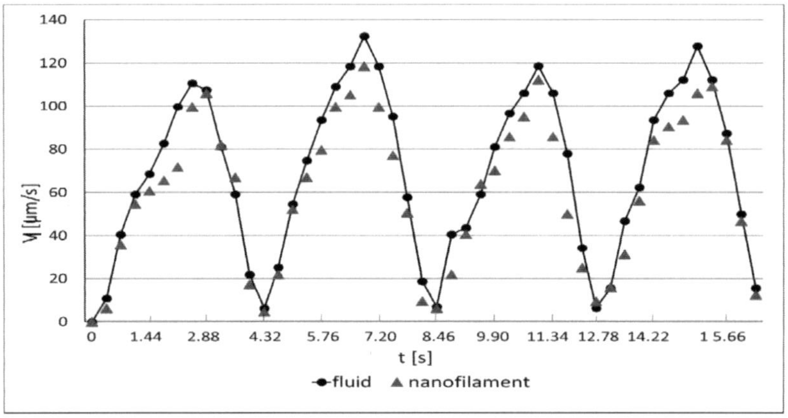

Figure 4. Absolute value of the fluid velocity and velocity of the filament (no 11 in Table 1) conveyed in the channel by the pulsatile flow; filament velocity calculated for center of mass.

3.2. Cross-flow migration

The analysis of the nanofilaments behavior in a pulsating fluid flow allowed observing migration of the objects into the center of microchannel, present in majority of cases. The observations of cross flow migration were carried out for the longest possible time in which the tested filament remained in the field of view of the camera, usually for 5 to 10 minutes. We have determined the variation of the filament distance from the centerline of the microchannel, evaluated for its center of mass. In some cases, we also have determined the distance of both ends of the object from the center of a microchannel (Figure 5). The typical duration of the imposed velocity pulse was two seconds.

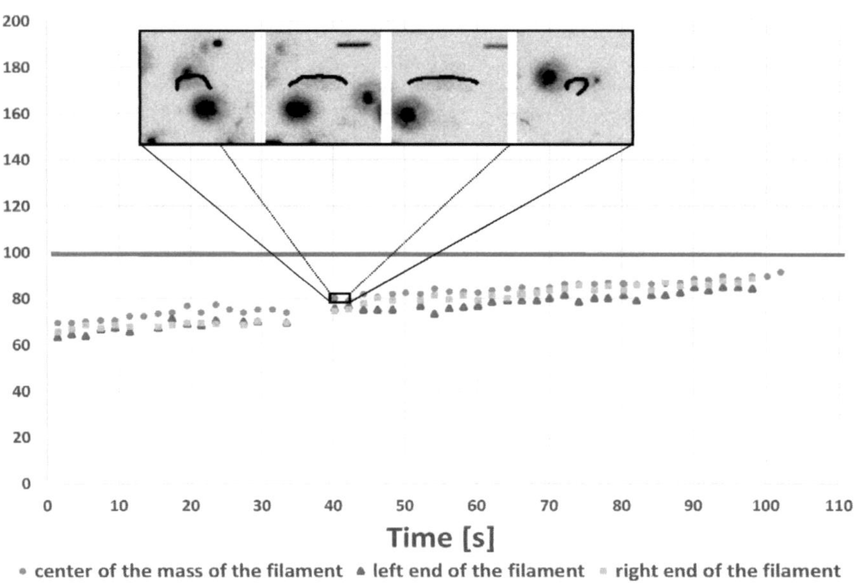

Figure 5. Cross-flow migration of nanofilament into the center of channel. Red line represents the center of the microchannel. Distance from the side wall in μm. Inserts show typical images of filaments during single velocity pulse.

Approximate value of the migration velocity for filament with geometry like elongated droplet (Vm) was calculated from the following equation [11] applying parameters of a typical hydrogel filament.

$$Vm = -2 \cdot \left(\frac{a}{d}\right)^3 \cdot \frac{\upsilon_m \cdot \mu}{\gamma} \frac{\upsilon_m \cdot \bar{x}}{(1+\lambda)^2 \cdot (2+3\lambda)} \cdot \left[\frac{16+19\lambda}{42 \cdot (2+3\lambda) \cdot (4+\lambda)} \cdot (13 - 36\lambda - 73\lambda^2 + 24\lambda^3) + \frac{10+11\lambda}{105} \cdot (8 - \lambda + 3\lambda^2)\right]$$

In our analysis, for the calculation of the objects hydrodynamic radius, the experimental and theoretical diffusion coefficients (Table 2) were used to obtain value of physical radius to be set in the formula. Hence data set in in the formula are: a = 0.25 μm; d = 100 μm – half of width of channel; υ_m = 0.0001 m/s - flow velocity at the channel axis; $\bar{x}$ = x/d = 0.1 – distance from the channel center line; λ = 21 - viscosity ratio of dispersed phase to viscosity of continuous phase (μ = 7.98 · 10^{-4} kg/ms); interfacial tension γ = 0.07 kg/s^2.
Migration velocity strongly depends on the viscosity ratio between material of the suspended object (droplet) and medium. In this estimation we took λ equal 21. It is only an approximate value for such type of materials. For rheological liquid, like hydrogel, its viscosity varies with shear rate and it has to be measured. It will be performed in the next stage of work.
Average value of the migration velocity of hydrogel nanofilaments evaluated in our experiments during pulsatile flow is around 10^{-7} m/s. We may observe that migration velocities of elongated droplets evaluated for the same flow parameters (Table 2) have similar value. It can be assumed that migration velocity of short filaments with geometry approximated by an elongated droplet is relatively well described by existing models for droplets [13].
Analyzing the migration curves of the center of mass and the two ends of the filament we may find that for most of these waveforms there is an overlap (Figure 5). However, during a single pulse the nanofilaments stretches and bends. Small deviations of the migration curves are probably associated with bending of the elongated objects during pulsatile flow.
We also observed cases where the test filament moved with carried liquid parallel to the walls of the channel all the time during the observation - without cross-flow migration (Figure 6 a). Therefore, we conducted additional test to evaluate impact of the flow velocity on the migration of filaments. Preliminary results show that the occurrence of nanofilaments migration in the analyzed channel geometry does not depend on the applied liquid flow velocity, at least in the small range of the investigated variation (doubled velocity). Further experiments will allow for verification of the occurring dependence. However, it seems that the conformation of the filament can affect the migration rate of the object (Figure 6 b). It is in line with theoretical predictions [5].

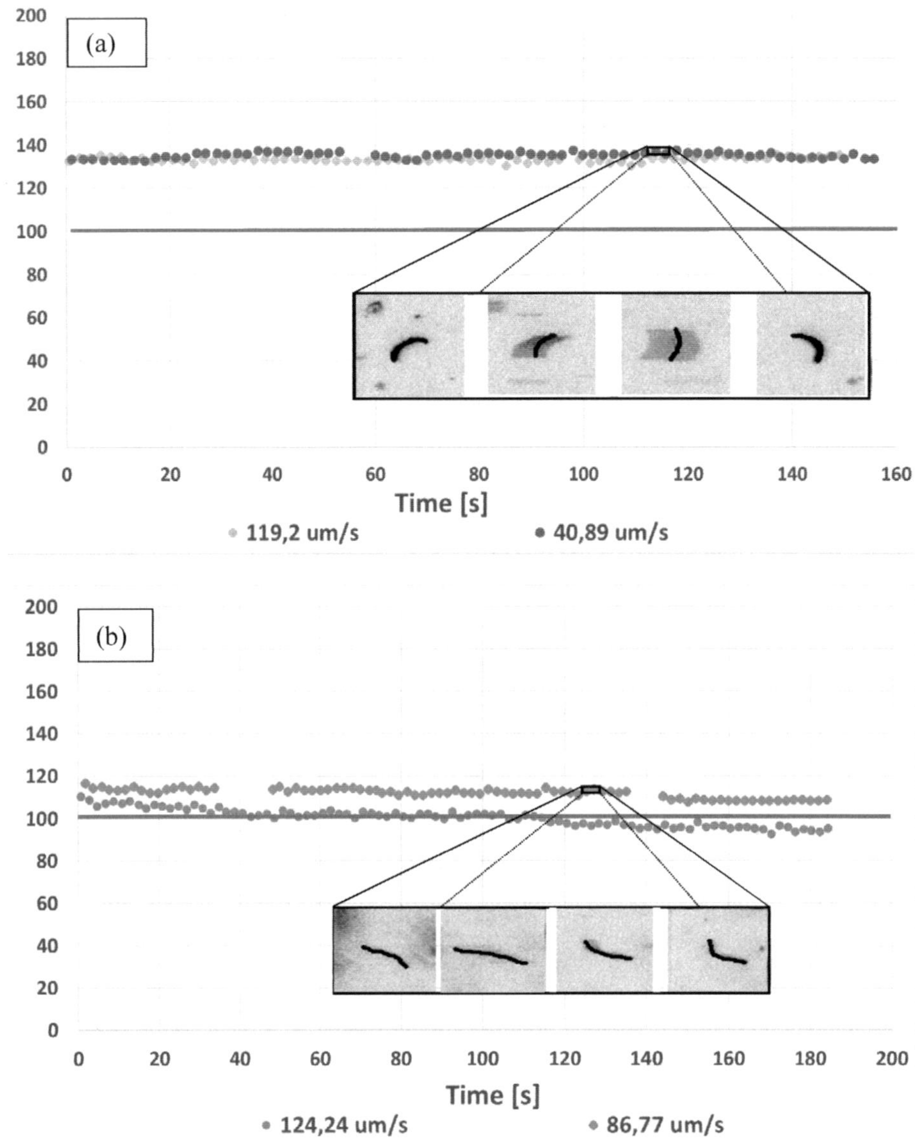

Figure 6. Effect of the flow velocity on the behavior of nanofilament conveyed in microchannel: (a) absence of migration; (b) presence of migration for less deformed filaments. Red line represents the center of the microchannel. Distance from the side wall is given in μm.

There are number of parameters that affect the rate and direction of migration for flexible filaments. Just to list the most important factors, they are: the average velocity of fluid flow (which translates directly to the speed of the filament during pulsatile flow), contour length and persistence length, length to diameter of filament ratio, viscosity of medium, Reynolds number, and position inside the microchannel (distance from the center of the channel). Knowing the contribution of individual parameters on the dynamics of filaments we should be able to control the behavior of such objects in a particular environment. Probably statistical analysis performed for large number of experiments is the only way to obtain plausible average mobility values convenient to predict longitudinal and cross-flow transport properties of such objects.

4. Discussion

Presented work demonstrates possibility of using hydrogel nanofilaments to characterize dynamics of deformable objects conveyed by pulsatile flow. Depending on the length of the filaments we observed their rotation and / or bending during short sequences of time. Especially long objects have a tendency to change configuration during flow. Comparing local velocity of fluid and that of the conveyed filament we can see that bending dynamics slightly changes mean velocity of the suspended objects.

Pulsatile flow well simulates intercellular movements of fluids inside living organisms. Present study confirms the hypothesis that the hydrogel nanofilaments may be used to model transport properties of biological molecules as well as vesicles. However, it is necessary to complete the characterization of these objects behavior, to gain full support about their potential applications, particularly in biomedical fields [18].

The phenomenon of cross-flow migration during pulsatile flow inside the microchannel is strongly associated with the bending dynamics of a nanofilament [19-21]. The issue of migration is important to describe the transport of deformable macromolecules in the capillaries. In the majority of the observed cases a tendency to migration of tested objects toward the center of the channel was found. As the rate of migration depends on several factors a deeper study is necessary to determine their identification and impact necessary to have a possibility to control the nanofilaments behavior. Several studies have shown that the occurrence of rotation, bending, and knots formation for filaments may influence the presence and rate of migration across the channel. The present study is trying to confirm such relationships by comparing the results obtained with the data evaluated for filaments with geometry approximated by elongated droplet. However, this comparison does not confirm unequivocally initial hypothesis formulated on the basis of experimental data, that the tendency of the filaments to bend favors the occurrence of migration, whereas objects showing only the rotation do not migrate across the channel or such migration is imperceptibly small. These preliminary results confirm the need for a larger number of experiments, showing various properties and behavior of the filaments.

5. Conclusions

Highly deformable hydrogel nanofilaments have demonstrated the possibility to be used to study the hydrodynamic interactions of long objects. Study of the flow characteristics for suspensions of highly deformable nano-objects has fundamental importance for the understanding and prediction of the rheological properties of typical biological fluid (cytoplasm, plasma). These data can be fundamental to create biocompatible nano-objects that can become tools for the regeneration of body tissues, e.g. neural tissue. There is still open issue of hydrodynamic effects for confined polymer solutions. In our case the confinement (channel depth) is of the same order of magnitude as contour length of the filament. The equilibrium conformation of the object can be expected to change substantially, influencing both transport as well as migration velocity. How far these effects influence observed filaments dynamics is not clear at the moment.

Acknowledgements

The support of NCN grant no. 2015/17/N/ST8/02012 is acknowledged. Author gratefully acknowledge her supervisors Tomasz A. Kowalewski and Filippo Pierini and colleagues Paweł Nakielski, Krzysztof Zembrzycki and other for their help.

References

[1] Jendrejack RM, Schwartz DC, de Pablo JJ and Graham MD 2004 Shear-induced migration in flowing polymer solutions: simulation of long-chain DNA in microchannels *J. of Chemical Physics* **120**, pp. 2513–2529

[2] Szymczak P, Cieplak M 2007 Influence of hydrodynamic interactions on mechanical unfolding of proteins *J. Phys.: Condens. Matter* **19**, 285224

[3] Sadlej K, Wajnryb E, Ekiel-Jezewska ML, Lamparska D and Kowalewski TA 2010 Dynamics of nanofibres conveyed by low Reynolds number flow in a microchannel *Int. J. Heat Fluid Flow* **31**, pp. 996–1004

[4] Słowicka AM, Ekiel-Jeżewska ML, Sadlej K and Wajnryb E 2012 Dynamics of fibers in a wide microchannel *J. of Chemical Physics* **136**, 044904

[5] Słowicka AM, Wajnryb E and Ekiel-Jeżewska ML 2013 Lateral migration of flexible fibers in Poiseuille flow between two parallel planar solid walls *Eur. Phys. J. E* **36**, 31

[6] Stein D, van der Heyden FHJ, Koopmans WJA and Dekker C 2006 Pressure-driven transport of confined DNA polymers in fluidic channels *PNAS* **103**, pp. 15853-15858

[7] Steinhauser DR 2008 Actin filaments and bundles in flow, PhD thesis, Gottingen

[8] Herczynski R 2016 Fåhræus effect revisited *Arch. Mech.* **68**, pp. 81–93

[9] Segre G and Silberberg A 1961 Radial particle displacements in Poiseuille flow of suspensions *Nature* **189**, 209

[10] Kowalewski TA 1984 Concentration and velocity measurements in the flow of droplet suspensions *Exp. in Fluids* **2**, pp. 213-219

[11] Hiller W and Kowalewski TA 1987 An experimental study of the lateral migration of a droplet in a creeping flow *Exp. in Fluids* **5**, pp. 43-48

[12] Gauthier FJ, Goldsmith HL and Mason SG 1972 Flow of suspensions through tubes. X. Liquid drops as models of erythrocytes *Biorheology* **9**, pp. 205-224

[13] Chan PCH and Leal LG 1979 Motion of a deformable drop in a second order fluid *J. Fluid Mech.* **92**, 131-170

[14] Stone HA 1994 Dynamics of drop deformation and breakup in viscous fluids *Annu. Rev. Fluid Mech.* **26**, pp 65-102

[15] Nakielski P, Pawłowska S, Pierini F, Liwinska V, Hejduk P, Zembrzycki K, Zabost E and Kowalewski TA 2015 Hydrogel nanofilaments via core-shell electrospinning *PLOS ONE* **10**(6):e0129816

[16] Chen YL, Graham MD, de Pablo JJ, Randall GC, Gupta M and Doyle PS 2004 Conformation and dynamics of single DNA molecules in parallel-plate slit microchannels *Phys. Rev. E* **70**, 060901

[17] Bazilevsky AV, Yarin AL and Megaridis CM 2007 Co-electrospinning of core–shell fibers using a single-nozzle technique *Langmuir* **23**, pp.2311–2314

[18] Chilukuri S, Collins CH and Underhill TP 2015 Dispersion of flagellated swimming microorganisms in planar Poiseuille flow *Physics of Fluids* **27**, 031902

[19] Ma H, Michael D and Graham MD 2005 Theory of shear-induced migration in dilute polymer solutions near solid boundaries *Physics of Fluids* **17**, 083103

[20] Kumar A and Graham MD 2012 Mechanism of margination in confined flows of blood and other multicomponent suspensions *Phys. Rev. Lett.* **109**, 108102

[21] Kuei S, Słowicka AM, Ekiel-Jeżewska M, Wajnryb E and Stone HA 2015 Dynamics and topology of a flexible chain: knots in steady shear flow *New J. of Phys.* **17**, 0530095

Journal of Physics: Conference Series **760** (2016) 012023 doi:10.1088/1742-6596/760/1/012023

Numerical investigations of shock wave interaction with laminar boundary layer on compressor profile

M Piotrowicz[1] and P Flaszyński[1]

[1] Institute of Fluid-Flow Machinery, Polish Academy of Sciences,
Fiszera 14, 80-231 Gdańsk, Poland

E-mail: Michal.Piotrowicz@imp.gda.pl

Abstract. The investigation of shockwave boundary layer interaction on suction side of transonic compressor blade is one of main objectives of TFAST project (Transition Location Effect on Shock Wave Boundary Layer Interaction). In order to look more closely into the flow structure on suction side of a profile, a design of generic test section in linear transonic wind tunnel was proposed. The experimental and numerical results of flow structure on a suction side of the compressor profile investigations are presented. The numerical simulations are carried out for EARSM (Explicit Algebraic Reynolds Stress Model) turbulence model with transition model. The result are compared with oil flow visualisation, schlieren pictures, Pressure Sensitive Paint (PSP) and static pressure.

1. Introduction

Shock wave boundary layer interaction is a phenomena existing in all transonic flows. Such interaction can affect the boundary layer separation downstream of the interaction. In the case of a civil turbofan engine operating at particularly high altitudes the Reynolds number can drop by a factor of 4, when compared to the sea level values. The laminar boundary layer on the transonic compressor rotor blades interact with shock waves and as a result a strong boundary layer separation is formed. This can seriously affect the aero-engine performance and operation. One way to avoid strong separation is to ensure that the boundary layer upstream of the shock wave is turbulent. Shock wave boundary layer interaction was widely investigated and lot of results are published [1-2, 11].

Forcing transition within the boundary layer can be achieved by the application of a surface roughness or a turbulator patch. Although such passive control methods are already in use, the mechanism of the shock wave-laminar boundary layer interaction, and in particular the source of the strong shock unsteadiness are still not understood in detail. Furthermore, the benefits of boundary layer control obtained for low Reynolds numbers can turn into loss increase at the higher levels of Reynolds numbers. Another possibility of transition control is to use Vortex Generators driven by Air Jets (AJVGs). In the compressor application the jets may be driven by the pressure difference between the suction and pressure sides of the blade.

In order to analyze the flow structure as the effect of shock wave boundary layer interaction on transonic compressor profile, test section was designed in IMP PAN within a framework of EU FP7 TFAST project [3-4].

The paper presents the comparison of numerical results and experimental data for laminar boundary layer interaction with the shock wave. One has to emphasize that modelling of such phenomenon on transonic compressor profile is very challenging, due to dependence of the shock wave and flow structure on the secondary flows [4]. As the first step, the flow structure details are compared and validation of numerical model is presented.

2. Geometry

Computational geometry (figure 1) consists of several elements, such as convergent-divergent nozzle, transonic profiles, support for holding profiles and slots. The nozzle was designed during EU FP6 UFAST project (Unsteady Effects of Shock Wave Induced Separation) [1, 2]. The main feature of the nozzle is the uniform distribution of the Mach number (M=1.22) at the location of installed compressor profiles. The nozzle was selected for the investigations in order to fulfil design requirements defined by Rolls-Royce Deutschland, one TFAST project partners. Mach number 1.22 is representative for relative Mach number at aircraft engine inlet, upstream of fan blades. Cascade configuration was considered as the reference for test section design, so the relative location of the profiles in the wind tunnel is the same as in the cascade. It is essential condition if one needs to reproduce flow structure as the one obtained in cascade configuration (figure 2). Reference conditions for the single passage are shown in table 1. Additionally in test section, the slots are implemented (marked 1, 2 and 3) and used to control the secondary flows, especially corner flows which have strong impact on the axial velocity density ratio [10].

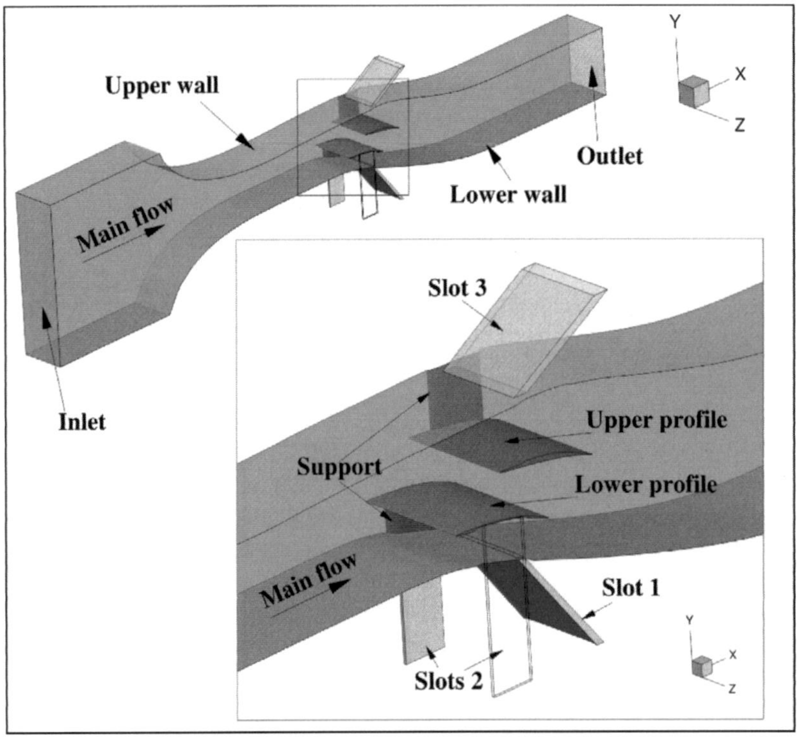

Figure 1. Test section – general view

Journal of Physics: Conference Series **760** (2016) 012023 doi:10.1088/1742-6596/760/1/012023

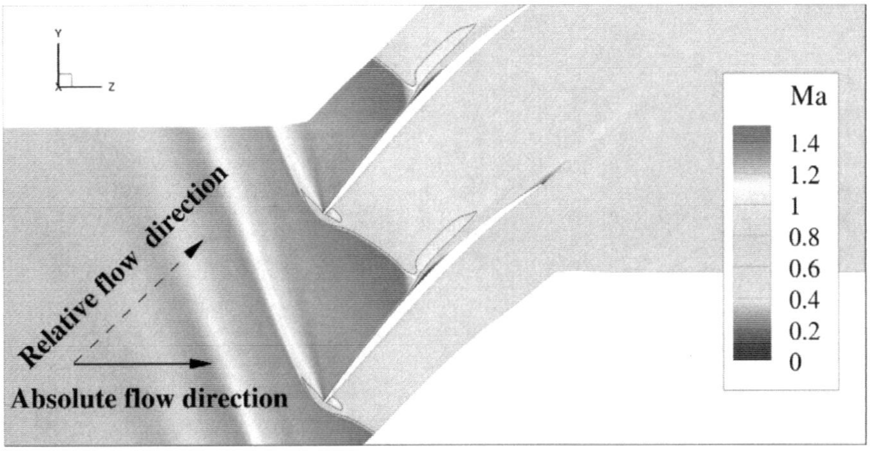

Figure 2. Mach number at mid span on linear cascade system

Table 1. TFAST Compressor Profile.

Name	Value
Inlet Mach number	1.22 [-]
Chord length	100 [mm]
Pitch	60 [mm]
Profile thickness	3 [mm]
Blade inlet angle	50.9 [deg]
Blade exit angle	33.2 [deg]

3. Numerical model description

The numerical simulations were carried out by means of FINE/Turbo Numeca code. The two-equation nonlinear eddy viscosity turbulence model of the EARSM (Explicit Algebraic Reynolds Stress Model) [6] with transition model was applied. The transition model (Menter and Langtry [15]) is based on two transport equations for intermittency and transition momentum thickness Reynolds number. The set of equations is closed by a perfect gas equation and Sutherland`s law for viscosity. It is applied 2nd order central scheme with scalar artificial dissipation for spatial discretization formulated by Jameson, Schmidt and Turkel [7].

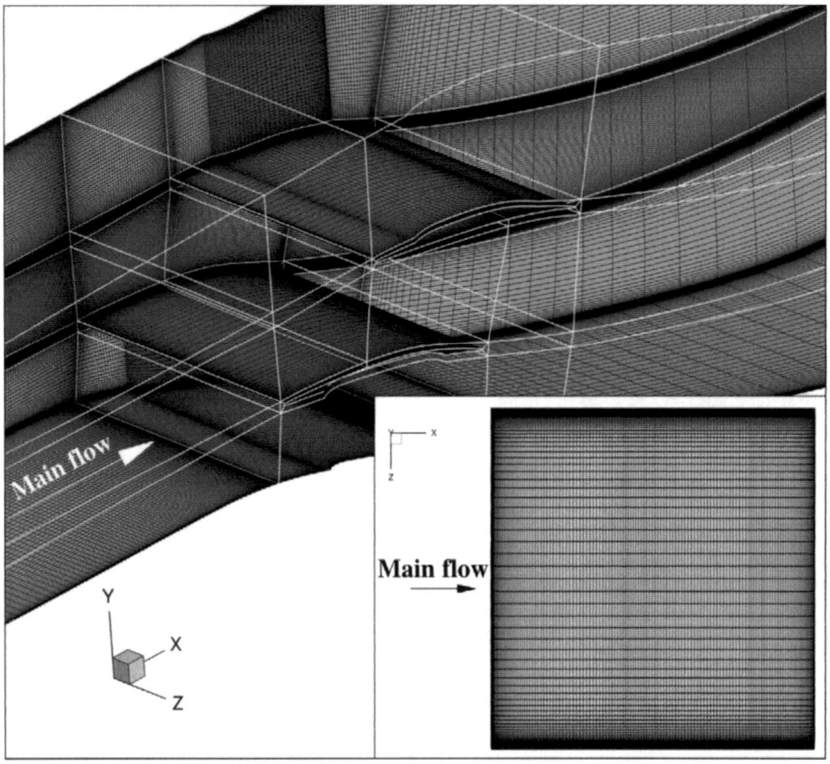

Figure 3. View of 3D mesh for test section and suction side of lower profile

The structured mesh was generated by means of IGG Numeca (Interactive Geometry Modeler and Multi-Block Structured Grid Generator). The multi-block topology shown in figure 3 consists of 35 blocks and the total number of hexahedral cells is $14{\cdot}10^6$. The mesh resolution near the solid wall boundaries was assumed in such a way that y^+ was below 1.

Full non matching boundary connection (FNMB) between blocks with different node distribution where applied. The full non matching connection allows keeping the mass flow conservation, the momentum and energy through the interfaces between the blocks. An example of a mesh connection is shown in figure 4. The red mesh represents the bottom wall and the blue one shows the connection to the additional slot.

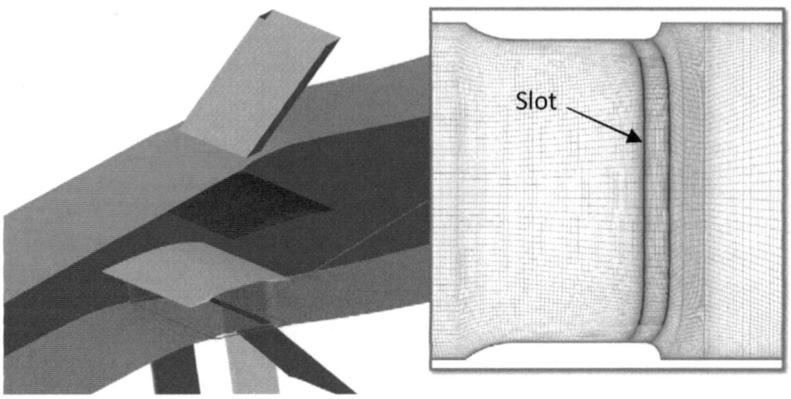

Figure 4. Full non matching boundary connection between bottom wall of test section and slot 1

Total pressure 101 000 Pa and total temperature 293 K are set at the inlet. Turbulence kinetic energy and dissipation rate were set assuming that the viscosity ratio was 10 and the turbulence intensity was equal to 0.5% according to the measurements. The static pressure at the outlet boundary condition was set as shown in Table 2. It is adjusted properly in order to get the required shock wave location.

Table 2. Outlet boundary conditions.

Name	Value
Outlet	76 [kPa]
Slot 1	50 [kPa]
Slots 2	38 [kPa]
Slot 3	10 [kPa]

4. Numerical results

The boundary layer upstream of the shock wave at four location is shown in figure 5. Traverses 1, 2, 3 and 4 are located at relative position x/C=0.089, 0.163, 0.242, 0.355. The integral parameters of the boundary layer profiles are shown in table 3 where δ^* is displacement thickness, δ^{**} momentum thickness and H means shape factor.

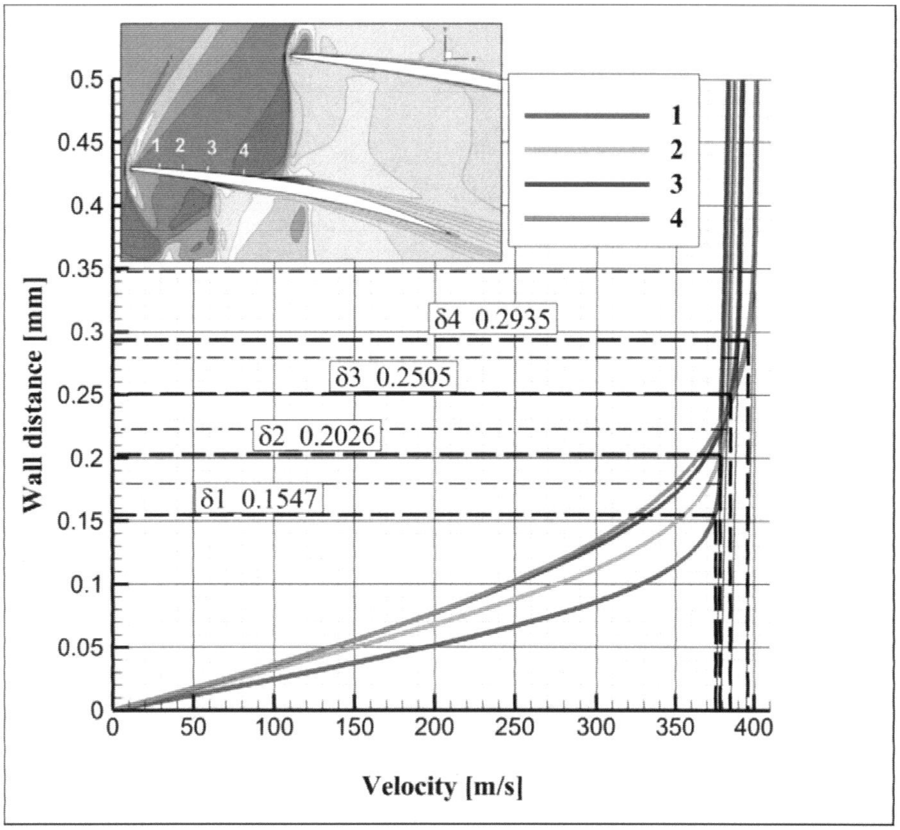

Figure 5. Boundary layer thickness

Table 3. Integral parameters for velocity traverse.

Traverse	δ^*	δ^{**}	H
1	0.063	0.017	3.717
2	0.085	0.022	3.772
3	0.101	0.027	3.726
4	0.112	0.03	3.691

Velocity profiles upstream of the shock wave are compared with laminar boundary layer profile (Blasius) and turbulent boundary layer Prandtl profile [9] on a flat plate (figure 6). In the equation for turbulent profile used coefficient n=7 which is valid for Re numbers in the range of $\mathbf{5 \cdot 10^5 - 10^7}$ (in the investigated case Re=$\mathbf{1,27 \cdot 10^6}$). Boundary layer profile at traverses 1-4 are similar to the laminar one therefore, one can conclude that the boundary layer downstream of the leading edge is laminar and shock wave interacts with laminar boundary layer. In the case of traverse 4, the profile with higher shear stress close to the wall can be noticed, what arises from local expansion zone. Nevertheless, the boundary layer starting at the leading edge up to the shock wave is laminar which is also confirmed by the intermittency distribution shown in figure 7 (0-laminar flow, 1-turbulent flow).

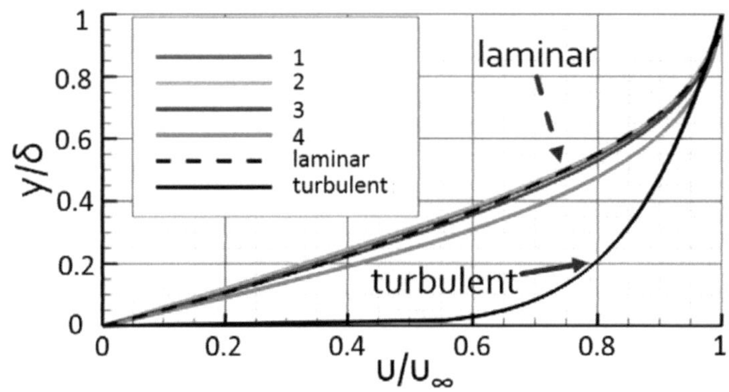

Figure 6. Normalized Velocity distribution

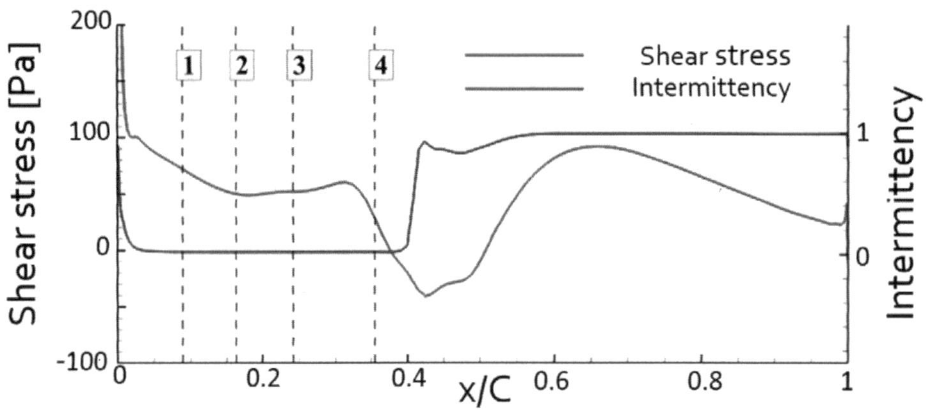

Figure 7. Shear stress distribution and intermittency at mid span on suction side on lower profile

The schlieren picture from experimental data (Figure 8) showing the flow pattern in the investigated passage can be compared with similar picture predicted by numerical simulations (Figure 9). As shown in both figures, the complex flow structure above suction side of lower profile can be reproduced and the position of shock waves predicted numerically is consistent with that shown in measurements. In both cases the shock wave generated at the leading edge of the lower profile is reflected from upper wall of test section and propagates downstream in the passage between the upper profile and the upper wall. The reflected shock wave penetration into the blade passage would be highly undesirable because such interaction does not exist in in cascade passages. In spite of differences of the shock wave interaction with the upper wall, different λ foot height and reflection angle, the shock wave structure between profiles is very similar to the measured one. In Figure 9, thicker boundary layer downstream of the shock than its influence on the wake is shown. The λ foot on the lower profile is not well visible in schlieren picture as in case of numerical results, but its height is similar.

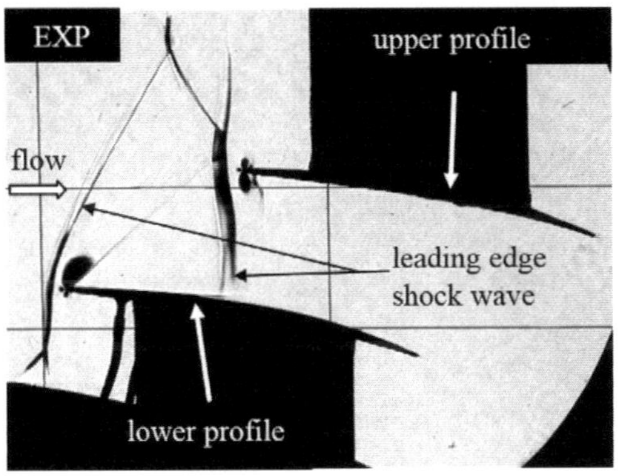

Figure 8. Schlieren picture

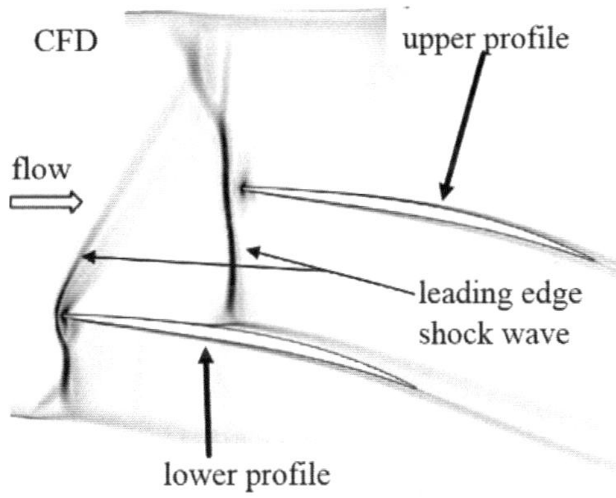

Figure 9. Density gradient magnitude

Isentropic Mach number on suction side of lower profile at middle span is shown in Figure 10 Numerical simulations results (black line) are compared with isentropic Mach number obtained from the static pressure measurements (red dots) and data (blue line) delivered by Rolls-Royce as the reference for the test section design. Downstream of the shock wave, the flow accelerates slightly and Mach number predicted numerically is very close to the measured one. Similar agreement is obtained downstream of the shock wave. In the location shortly upstream of the shock wave, it is difficult to assess the numerical prediction due to the lack of measurement point in range x/C 0.3 – 0.4. However, general picture of the pressure distribution can be compared by means of Pressure Sensitive Paint measurements.

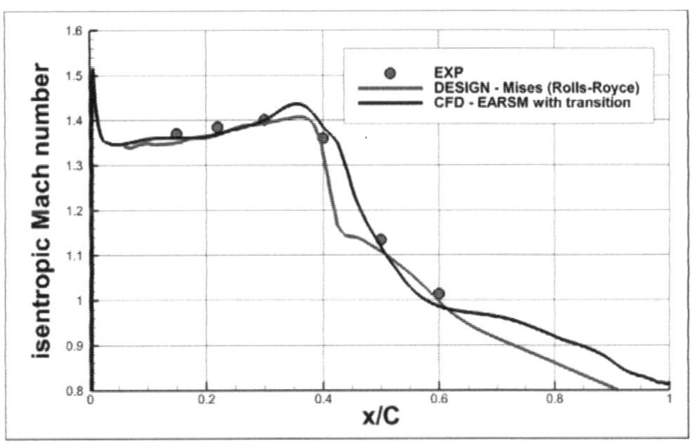

Figure 10. Isentropic Mach number distribution at mid span on suction side on lower profile

Static pressure distribution on the whole suction side of the investigated profile is shown in Figure 11, Pressure Sensitive Paint measurement on the left and numerical results on the right. As already shown in Figure 10, the pressure distribution in the middle section is predicted well. Discrepancy increases closer to the sidewalls, where the effect of suction slots can be noticed. These disparities arises from different size of corner vortices and mass flow at suction slots, which is adjusted in order to get the correct shock wave location and pressure distribution at middle section.

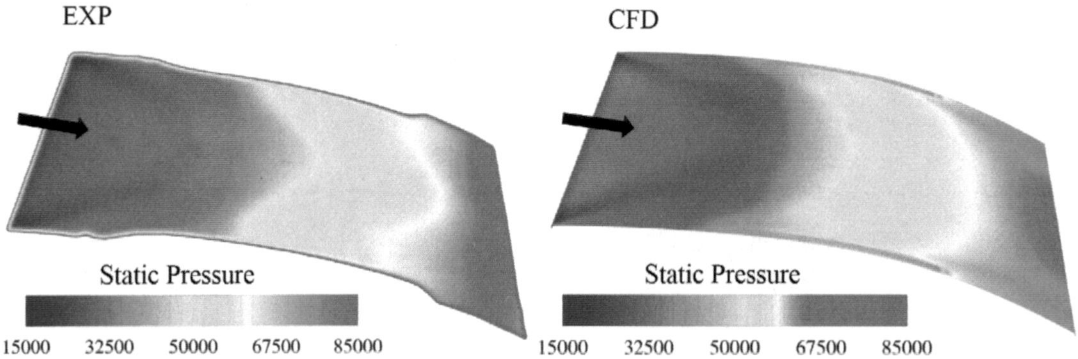

Figure 11. Static pressure on suction side

Different pressure distribution close to the sidewalls influences on the separation downstream of the shock wave, as shown in oil flow visualization (Figure 12 and 13). The separation length (streamwise direction) is the same in both cases, experimental and numerical. In the measurements, one can see a lower spanwise size of the separation bubble which is the effect of more intensive corner flows. At the leading edge, close to the sidewalls, the corner flows are generated and they develop further downstream. As shown by oil flow visualisation, their intensity is higher than predicted numerically what is visualised by streamlines near the wall. Finally, they influence on the separation size in spanwise direction, but its effect on the flow downstream of the shock wave is much weaker.

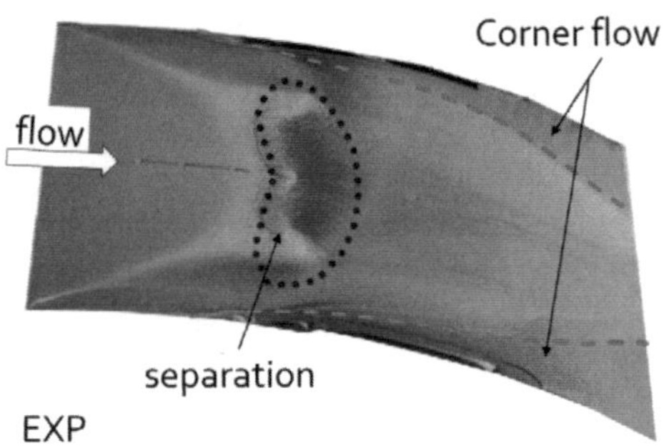

Figure 12. Oil flow visualisation

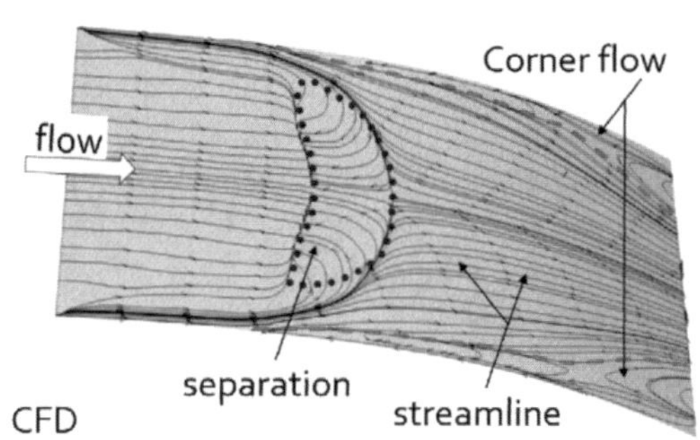

Figure 13. Streamlines near the suction side

5. Conclusions

In the paper, there are presented numerical simulation results carried out for the transonic compressor profile investigated experimentally in IMP PAN wind tunnel. Investigations of flow structure on suction side of compressor profile can be successfully done in single passage test section.

The main objective of the investigations was the shock wave interaction with laminar boundary layer. Numerical prediction of such interaction enforces application of turbulence model including transition modelling effect. As indicated by velocity profiles in the boundary layer upstream of the shock wave,

shear stress distribution on the investigated profile and intermittency, the boundary layer is laminar from the leading edge up to the shock wave.
Separation bubble length downstream of the shock wave is predicted properly.. The difference arises from the corner flow existence and the application of suction at sidewall slots what affects less agreement than obtained in the middle section of the investigated profile.
Anyway, the very good agreement in the middle of the profile allows for further investigations of flow control devices application to transition location control and its influence on shock wave boundary layer interaction.

Acknowledgements
This research was supported by the 7 EU framework project and was carried out within the research project with the acronym **TFAST** (Transition Location Effect on Shock Wave Boundary Layer Interaction).
This research was supported by CI TASK and in part by PL-Grid Infrastructure.

References
[1] Doerffer P 2009 UFAST Experiments Data Bank: Unsteady Effects of Shock Wave Induced Separation *IMP PAN Publishing*
[2] Doerffer P, Hirsch M, Dussauge J, Babinsky H and Barakos G 2010 Unsteady Effects of Shock Wave Induced Separation *Springer*
[3] Piotrowicz M, Flaszyński P, Doerffer P 2014 Investigation of Shock Wave Boundary Layer Interaction on Suction Side of Compressor Profile *Journal of Physics: Conference Series Vol. 530*
[4] Piotrowicz M, Flaszyński P, Szwaba 2015 Influence of Limiting Walls on Shock Wave Structure in Single Passage Test Section with Compressor Profile *Task Quarterly: Scientific Bulletin of Academic Computer Centre in Gdańsk Vol.* ***19****(2) pp.141-152*
[5] Flaszyński P, Doerffer P, Szwaba R, Kaczyński P, Piotrowicz M 2015 Shock Wave Boundary Layer Interaction on Suction Side of Compressor Profile in Single Passage Test Section *Journal of Thermal Science Vol.* ***24****(6) pp. 510-515*
[6] Numeca FINE^{TM}/Turbo v9.1 2014 *Theoretical Manual Documentation v9.1a*
[7] Jameson A, Schmidt W, Turkel 1981 Numerical Solutions of the Euler Equations by Finite Volume Methods Using Runge-Kutta Time-Stepping Schemes *AIAA Paper 81-1259, Palo Alto, CA*
[8] Becker B, Rayer M, Swoboda M, 2007 Steady and unsteady numerical investigation of transitional shock-boundary-layer-interaction on fan blade *Aerospace Science and Technology, Vol. 11, No 7-8, pp. 507-517*
[9] Puzyrewski R, Sawicki J, 1987 Podstawy Mechaniki Płynów i Hydrauliki *Państwowe Wydawnictwo Naukowe, Warszawa*
[10] Bo Song, Wing F. Ng Influence of Axial Velocity Density Ratio in Cascade Testing of Supercritical Compressor Blades 40th *AIAA/ASME/SAE/ASEE Joint Propulsion Conference and Exhibit 11-14 July 2004, Fort Lauderdale, Florida*
[11] Babinsky H, Harvey J 2011 Shock Wave-Boundary-Layer Interaction, *Cambridge University Press*
[12] Langtry R, Menter F, 2005 Transition Modelling for General CFD Applications in Aeronautics *AIAA 2005-522*

Journal of Physics: Conference Series **760** (2016) 012024 doi:10.1088/1742-6596/760/1/012024

Three-dimensional structures behind Glauert-Goldschmied profile under control of plasma actuation

P Prochazka[1] and V Uruba[1]

[1] Institute of Thermomechanics, Czech Academy of Sciences, Dolejškova 5, Praha 182 00, Czech Republic

E-mail: `prochap@it.cas.cz`

Abstract. The effect of plasma dielectric barrier discharge (DBD) to developed boundary layer was investigated previously. It has been shown that plasma actuation can stabilize the flow near to the surface. In the case of streamlined profile, the plasma actuation is able to shift the point of separation and hence to suppress the extent of separation area. This should lead to decrease of the form drag and increasing the total efficiency. An experimental approach is suggested for investigation and time-resolved Particle Image Velocimetry (PIV) is considered as a main measurement technique.

1. Introduction

The plasma actuation belongs to the group of active flow control strategies. These devices are relatively new in this field since their fully integration among research group is not older than one or two decades. These devices are supplied by electric current, no matter whether DC or AC. Nevertheless, the actuators, which are powered by AC current, enable to operate in unstable regime given by modulation of power signal. The laminar-to-turbulent transition (e.g. [1]) can be affected by DBD actuators as well as the shift of the separation point (e.g. [2]) and other flow field features.

The presented DBD (dielectric barrier discharge) actuator has slightly different geometry to be able to produce sufficiently strong plasma-induced flow (ionic wind) on the surface of the profile. This ionic wind affects close to the actuator surface (up to 1 mm) and is able to stabilize or destabilize the boundary layer as it was described in [3]. The actuator can be placed in spanwise orientation to produce parallel ionic with respect to the channel flow as well as in streamwise orientation to produce ionic wind perpendicularly. Normally, the actuator works in steady regime and continuous wall-jet-like flow is generated with maximal mean velocities up to 2 $m \cdot s^{-1}$. After that the rectangular modulation of the high-voltage high-frequency waveform is applied, one can observe the production of pseudo-periodical vortices inside the ionic wind - this is called as unsteady regime. This regime is very convenient due to lower power consumption demand and it allows to reach more sophisticated flow control. The vortical structures exhibit different properties given by modulation parameters and are to be used to affect the changing flow of distinct regimes around the streamlined body.

Since the study on the properties of generated vortical structures by DBD plasma actuator in steady air was performed previously [4], now this knowledge can be utilized to affect the channel flow behind an obstacle. In this study, mainly the vortex diameter, vortex trajectory, distance between two consecutive vortices, the circulation of the vortex and convective velocity of the vortex were parametrically described according to changing modulation parameters as modulation frequency and duty cycle. There is an example in equation:

$$u_{conv} = a_1 + a_2 \cdot DC + a_3 \cdot f + a_4 \cdot DC^2 + a_5 \cdot DC \cdot f + a_6 \cdot f^2 \qquad (1)$$

which expresses the vortex convective velocity in dependency on modulation frequency f and duty cycle value (DC). The coefficients a_i were obtained by detailed experimental investigation.

The Glauert-Goldschmied (GG) type profile was chosen as a test profile because is commonly used for testing of the active flow control capability in numerical simulation as well as in experimental research. Some results presented in the literature will be helpful for comparison with other authors. The approach of this study is only experimental due to complexity of studied phenomenon. The results will be treated from statistical point of view when mean flow field and statistical indicator will be investigated, although the dynamics will be treated also. Since the phenomenon of vortex production is pseudo-periodical, the phase-locked measurement should be applied to highlight the dominant component of the flow.

2. Experimental layout

2.1 Actuator geometry

Every DBD actuator should consist at least of two electrodes and a dielectric layer. The wire type of DBD actuator, used during the study of boundary layer affecting along flat plate [5], has proved the ability to supplied ionic wind with sufficient velocities to influence the boundary layer and moreover its discharge and generated thrust were homogenous.

This actuator had to be rebuilt to allow the assembly on the curved profile. The dielectric layer consists of several layers of kapton foils which can be curved along the profile body. The upper electrode is made from thin tungsten wire (30μm) and is powered by HV. A series of several anchor points is utilized to lead the wire electrode across the actuator body in the specific location. Spiral torsion spring is used to make the wire uptight for all time. The lower electrode is thin layer of copper foil and it is grounded. The horizontal gap between electrodes is set to two millimeters. The actuator creates the surface of the hump profile and does not affect the flow at all.

2.2 Power source

The power source was designed and fabricated in the Institute of Thermomechanics. It is designed as a constant voltage power supply and is conceived as a half-bridge circuit excitating output high voltage ferrite transformer. The voltage value can be set linearly and the maximum peak value is 12 kV. The frequency can be also adjustable (from 10 to 30 kHz) but the constant value of 16 kHz is used as an optimum. The source has so called shutdown function which allows to use the rectangular modulation of high-voltage high-frequency waveform. The modulation brings another two parameters – modulation frequency and duty cycle – which can be used for control of the ionic wind. The velocity magnitude of the ionic wind is linearly dependent on applied voltage.

The actuator working without modulation in steady regime generates continuous wall-jet-like flow. After that the modulation is applied, the unsteady regime is set and actuator can generated a series of vortical structures called as a vortex street. The used modulation frequency determines the diameter of vortices and distance between two consecutive vortical cores. On the other hand, duty cycle determines the rate between working cycle (plasma on) of the actuator and relaxation time (plasma off) during one period. The duty cycle of 30% means that the plasma discharge is switched on for 70% of one period. Duty cycle determines the time during which the vortex is supplied by energy.

2.3 Experimental layout

The experiment was carried out in the test section connected to the blow down wind tunnel. The test section is a narrow rectangular channel with cross section dimensions of 100 x 250 mm and the length of 3000 mm. The walls of the channel had smooth surface with rectangular corners. The channel was constructed from the plexiglass plates to ensure the optical access for the laser and the cameras. The turbulence intensity level is approximately 0.1% behind the contraction. The channel velocities were varied from 5 $m \cdot s^{-1}$ to 20 $m \cdot s^{-1}$ resulting in fully developed turbulent boundary layer in desiderative location of the profile installation. The Reynolds number referred to conventional boundary layer thickness of that location can be seen from the table 1.

The Hump profile was flush-mounted to the bottom wall (the profile width was 100 mm) so that the distance from the leading edge to the contraction was 1800 mm. Two identical profiles differing in only dimensions were tested. The high of the first profile was 50 mm (20% of cross section high) and second one was 25 mm high and 194 mm long (lower level of congestion). Further, only smaller profile will be described.

The naturally developed separation point and the size of the recirculation zone (separation bubble) were evaluated previously. The table 2. shows these results experimental acquired for the smaller hump profile. Many authors have found out that the actuator should be place as close as possible to the point of separation to have the most effective flow control. There is the question what is the exact right point of activity. Four DBD actuator location was tested for spanwise orientation when the ionic wind was generated parallel and in the same direction as the channel flow. The furthest location upstream of the separation was x/L = 0.63. The determining point of the actuator location is the wire powered electrode. This corresponds to the wire electrode location 122 mm downstream from the leading edge in horizontal sense. Next three location are x/L = 0.66, x/L = 0.69 and x/l = 0.72. The first variant has wire electrode upstream the separation, the second one is suggested to have electrode in the point of separation and the last two variants are considered downstream the separation.

The base case is referred as the flow over profile without plasma actuation. To control the size and structure of the separation bubble, the steady and mainly unsteady regimes are investigated. The plasma actuation is conducted under maximal but constant voltage value (8kV for steady and 11 kV for unsteady). The modulation frequency varies from 10 to 50 Hz during modulation. The duty cycle is set to 30 or 70%.

2.4 Measurement technique and data acquisition

To capture wake structure behind the profile, standard time-resolved PIV measurement in longitudinal plane perpendicular to the bottom side was considered. The CCD camera Phantom V711 was used to acquire the images of the flow around the profile in three distinct positions seen from the side. This camera has a resolution of 1280 x 800 pixels and corresponding maximal acquisition frequency is 6 kHz. The camera memory is 8 GB. To illuminate the tracing particles, laser New Wave Pegasus Nd:YLF is utilized. It disposes with double-head and cylindrical optics and emitted coherent light has wavelength of 527 nm. Maximal frequency is 10 kHz and pulse energy is 10 mJ for 1 kHz (10 W per head). The tracing particles were generated from safex device (droplets oil particles with diameter of 1 μm), medium image density was used.

Stereo PIV measurement in cross-section plane was suggested as a complementary measurement. Two CMOS cameras NanoSense MKIII, maximal resolution 1280 x 1024 and corresponding maximal frequency 512 double-images per second, were used. However, the acquisition time is limited due to camera memory of 4 GB. Camera chip, objectives and measurement plane did fulfill the Scheinpflug criterion. The traverse system was used to move the laser with cameras at a constant distance to measure four planes. The planes were situated at x/L = 0.69, 0.87, 1.00 and 1.50 from the leading edge.

The data acquisition was handled from three distinct points of view. The base one is to study statistical characteristic of the flow (mean flow field velocity distribution, standard deviation, etc.). That is why the lower frequency and longer measurement time was suggested (approx. 100 Hz and at least 10 seconds). To capture all fast dynamical phenomenon of boundary layer - ionic wind merging, it is necessary to use higher frequency (2 kHz) but these results will be presented only marginally. Since the production of coherent structures is pseudo-periodical phenomenon, phase-locked measurement is suggested. Then the periodic effect can be seen clearly because all random effects are averaged. Twenty phases per one period, which is given by the modulation frequency, is suggested and averaging over at least 80 periods is considered.

Table 1. The Reynolds number based on horizontal distance inlet-leading edge and on conventional BL thickness.

	Re_x	Re_δ
$V_{ext} = 5\ \mathrm{m \cdot s^{-1}}$	$6 \cdot 10^5$	$10 \cdot 10^3$
$V_{ext} = 10\ \mathrm{m \cdot s^{-1}}$	$12 \cdot 10^5$	$18 \cdot 10^3$
$V_{ext} = 20\ \mathrm{m \cdot s^{-1}}$	$24 \cdot 10^5$	$32 \cdot 10^3$

Table 2. The separation point location and the reattachement point location for different channel velocities, examined on the Hump profile with the actuator.

$V_{ext} = 5\ \mathrm{m \cdot s^{-1}}$	x/L = 0.65	x/L = 1.31
$V_{ext} = 10\ \mathrm{m \cdot s^{-1}}$	x/L = 0.65	x/L = 1.12
$V_{ext} = 20\ \mathrm{m \cdot s^{-1}}$	x/L = 0.65	x/L = 0.97

3. Results

Two different Hump profile sizes were tested in the channel due to congestion. The velocity increment far above the boundary layer over the hump region can be seen at the figure 1. Each color represents a different camera position. This diagram shows that the increase of the u-component is up to 14% and gradually decreases to normal value in x/L = 1.5. This test was also performed for bigger profile (25% of channel height) where the level of congestion was higher and inconvenient for DBD actuator testing. Next paragraphs will be devoted only to the smaller profile.

The figure 2 shows the velocity vector map of the ionic wind produced by the actuator in steady regime without the influence of the channel flow. The wire electrode is situated at x/L = 0.63. The maximal mean velocity is around 1 $\mathrm{m \cdot s^{-1}}$ but peak value is at least two times larger. The topology of the ionic wind at little bit differs from the ionic wind over flat plate. The flow is fully attached to the curved surface due to Coanda effect. The position of maximal velocity area is very close to the surface (1-2 mm) and approximately 5 mm downstream the wire. This implies that the wire electrode position is not so important in relation to the separation point. That is why two positions upstream the separation were chosen to test.

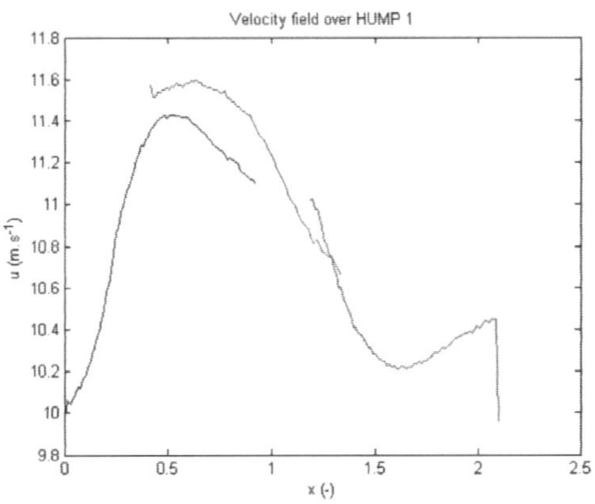

Figure 1. Velocity is increasing above the profile. The height is 25% of channel height.

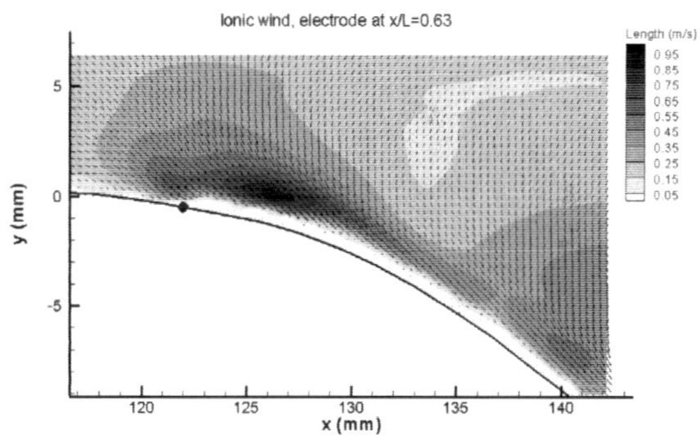

Figure 2. Ionic wind generated in steady air. Wire electrode at x/L = 0.63.

The wake behind the profile is dependent on Reynolds number. The size of the separation bubble is decreasing with increasing channel velocity. The figures 3a,b,c show the flow field made by streamlines visible for the case of no plasma actuation (base case). The flow separates at approx. the distance x/L = 0.64 for all velocities. The recirculation zone is almost negligible for the velocity 20 $\mathrm{m \cdot s^{-1}}$ and the identification from PIV acquired data is not easy task. On the other hand, the separation bubble for 5 $\mathrm{m \cdot s^{-1}}$ is great and easy to influence. The length of recirculation zone is 130 mm, 92 mm and only 63 mm for the channel velocity 5 $\mathrm{m \cdot s^{-1}}$, 10 $\mathrm{m \cdot s^{-1}}$ and 20 $\mathrm{m \cdot s^{-1}}$, respectively. The figure 4 shows the mean velocity vector maps of the wake created by 5 $\mathrm{m \cdot s^{-1}}$.

There can be seen the comparison of the effect of different wire position under 5 $\mathrm{m \cdot s^{-1}}$ at the figures 5a, b, c, d. The corresponding positions x/L are 0.63, 0.66, 0.69 and 0.72. From the mean velocity distribution, it can be determined that there is almost no recirculation for the first position for the steady actuation (5a). The similar situation have been observed during unsteady actuation under 10 Hz and 30%. The actuator placed at the second position can almost fully prevent the creation of the recirculation (there is only small separation bubble up to x/L = 0.77). The situation is completely different when the actuator is further downstream. Since the wire at x/L = 0.69 is behind the natural separation (and the ionic wind origin is even farther), the generated thrust is not able to attached the flow. The energy of the ionic wind is only able to deform the zone of recirculation – separation up to x/L = 1.08 (figure 5c). When the wire is situated at the last position (figure 5d), the effect on the separation bubble size is negligible. However, the topology of the wake structures is changed (e.g. figure 8).

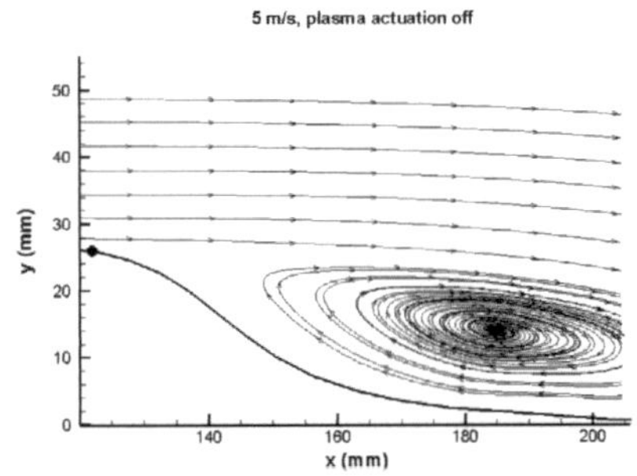

Figure 3a. Wake behind Hump, no actuation, channel velocity 5 m·s^{-1}.

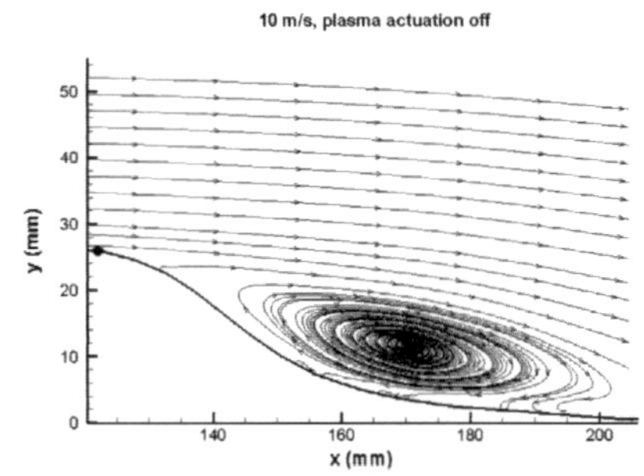

Figure 3b. Wake behind Hump, no actuation, channel velocity 10 m·s^{-1}.

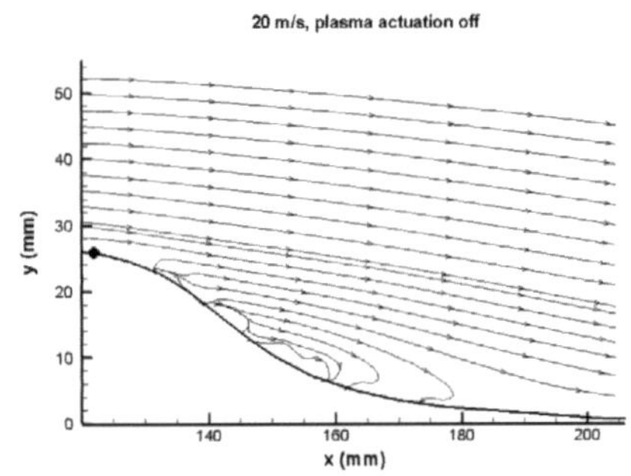

Figure 3c. Wake behind Hump, no actuation, channel velocity 20 m·s^{-1}.

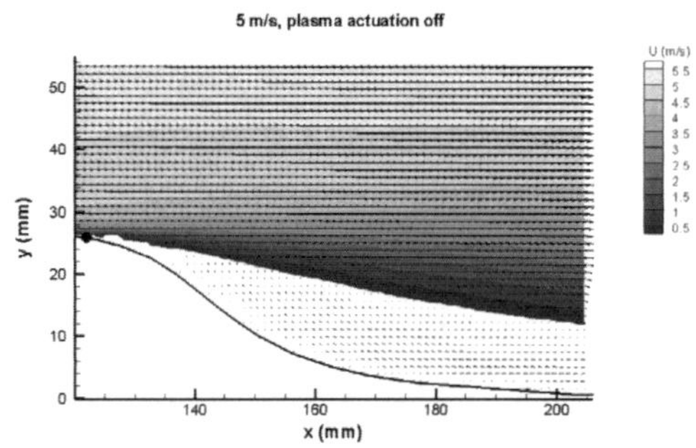

Figure 4. Velocity field, wake behind Hump, no actuation, channel velocity 5 m·s^{-1}.

The influence of the ionic wind becomes less important for the case of higher channel velocities. Only first two position were able to change significantly the wake dimension in the case of 10 m·s^{-1} (reattachement point at x/L = 0.95 or 1.03 instead of 1.12 in base case). The case of the highest velocity seems to be out of the actuator ability. This demonstrated the flow field acquired for the wire in the first two positions. The improvement is present but almost negligible. However, there is significant reduction for the third position. The recirculation zone is already very tiny for base case and it seems that ionic wind occurring in the middle of the bubble can influence the recirculation zone thickness if this zone is not very extensive in normal direction.

The effect on separation point was not observed except the second position. Probably due to suction effect of the actuator, the separation point was shifted about approx. 3 mm downstream.

The effect of the modulation frequency and the duty cycle will be shown in the fourth position of the wire (figures 6a, b and 8). The effect of modulation frequency on separation bubble dimension is tiny. Both the length and the height of the zone are not significantly changed. The effect of duty cycle can be evaluated in dependency of existence of a secondary steady vortex. This steady secondary vortex can be easily evaluated from phase-locked acquired data. This vortex is present in the wake in front of the center of the recirculation and it is cause due to the ionic wind (figure 7). The duty cycle determines the time of plasma actuation and hence the duration time of this secondary vortex during one period. This will appear in mean flow field (figure 6a and 6b).

Journal of Physics: Conference Series **760** (2016) 012024 doi:10.1088/1742-6596/760/1/012024

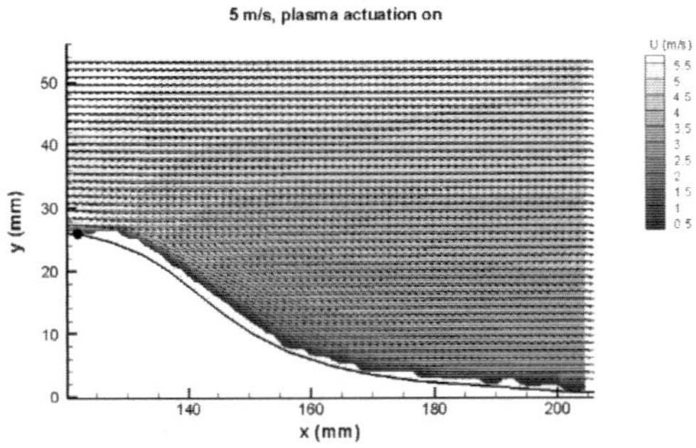

Figure 5a. Effect of ionic wind behind the Hump, channel velocity 5 $m \cdot s^{-1}$, wire electrode at x/L = 0.63.

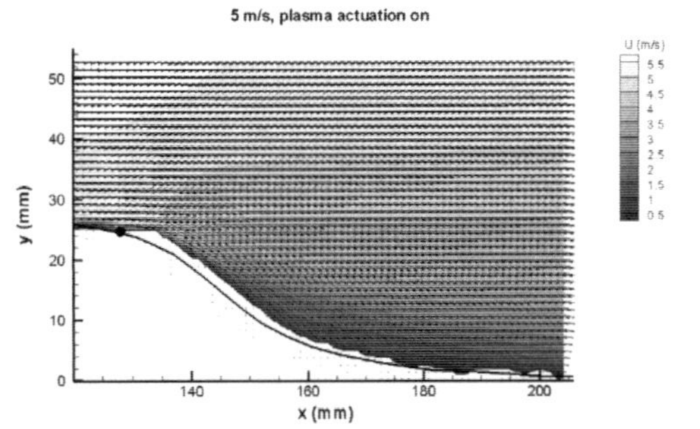

Figure 5b. Effect of ionic wind behind the Hump, channel velocity 5 $m \cdot s^{-1}$, wire electrode at x/L = 0.66.

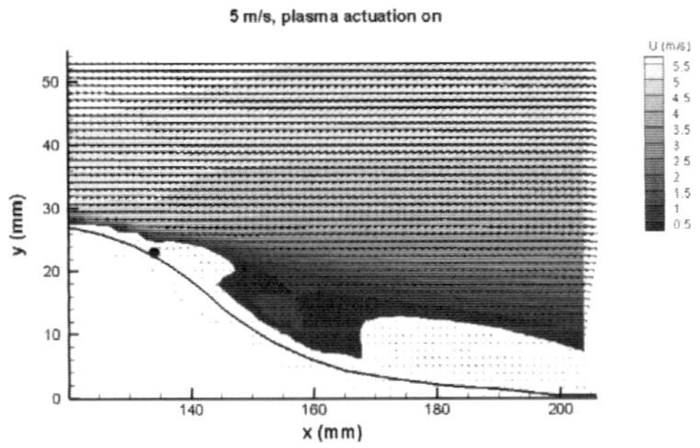

Figure 5c. Effect of ionic wind behind the Hump, channel velocity 5 $m \cdot s^{-1}$, wire electrode at x/L = 0.69.

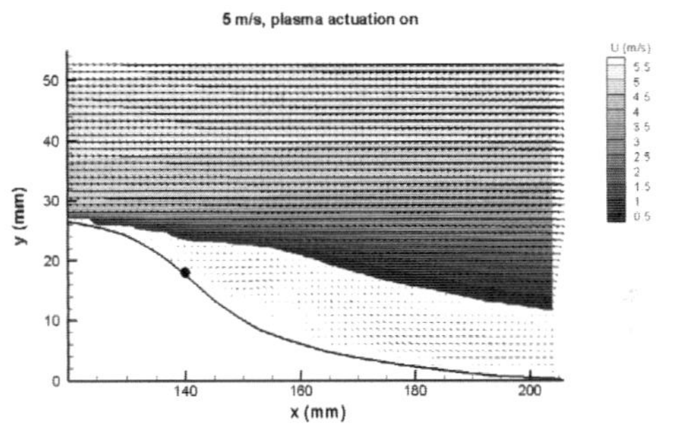

Figure 5d. Effect of ionic wind behind the Hump, channel velocity 5 $m \cdot s^{-1}$, wire electrode at x/L = 0.72.

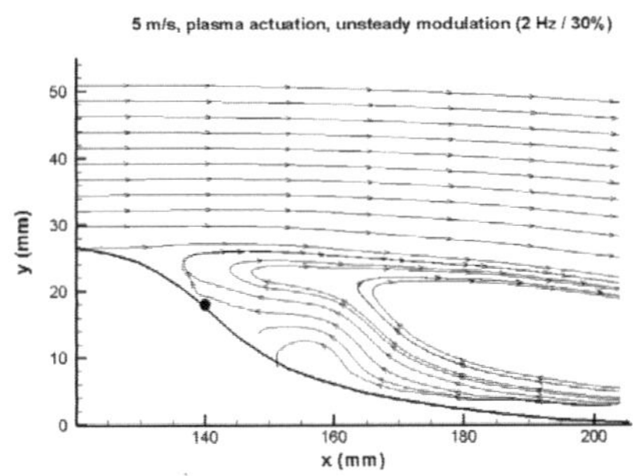

Figure 6a. The effect of duty cycle. Longer time duration of plasma. The secondary vortex is more obvious close to the surface.

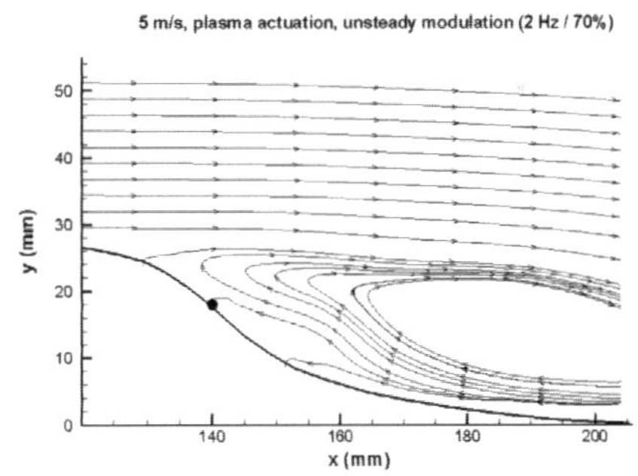

Figure 6b. The effect of duty cycle. Shorter time duration of plasma. The secondary vortex is almost not present close to the surface.

The figure 7 shows the averaging image of the flow field in the seventh phase from twenty. Notice that the secondary structure has opposite sense of rotation. More, this structure is not steady in time, but it is shifting during one period. The convective velocity of the vortex is constant and for the case of 10 Hz and 30% is 0.07 $m \cdot s^{-1}$. Of course, the shape of primary recirculation structure is also changing due to the influence of periodical occurrence of the secondary vortex. This vortex always appears at the beginning of the period.

Journal of Physics: Conference Series **760** (2016) 012024 doi:10.1088/1742-6596/760/1/012024

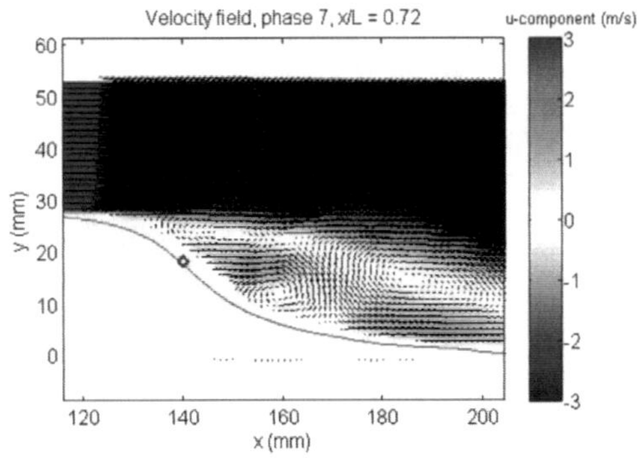

Figure 7. 7th phase of the periodical process – unsteady regime (10 Hz, 30%). The secondary vortex influences the recirculation zone.

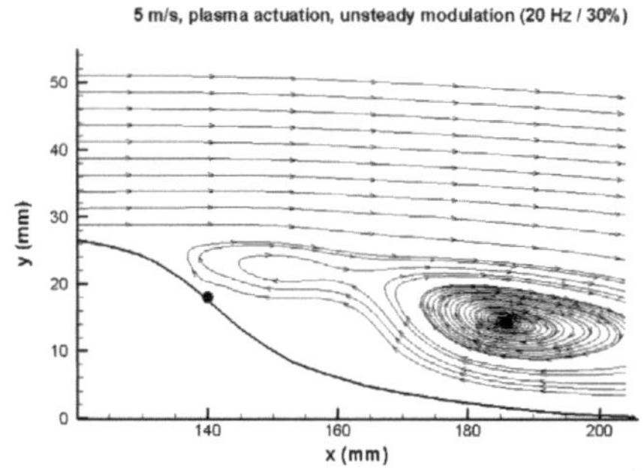

Figure 8. Mean flow field, Unsteady regime (20 Hz, 30%). The secondary vortex influences the recirculation zone.

Stereo PIV measurement in cross-section planes was conducted in order to have the information in the z-coordinates. Since the ionic wind affects the channel flow in parallel sense, there are no important velocity components present in cross-section. This auxiliary measurement has also proved the dimension of the wake under different regimes. The figure 9 reveals the wake under no actuation in x/L = 0.87. POD analysis of this data is considered in the future research.

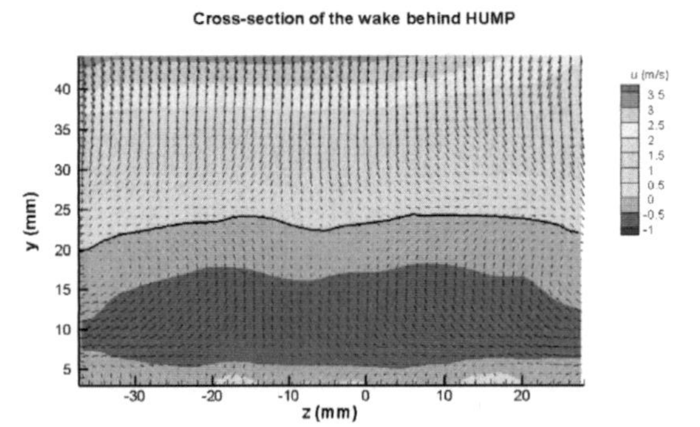

Figure 9. Cross-section plane by stereo PIV measurement at x/L = 0.87. Black solid line denotes the zone of recirculation. 5 $m \cdot s^{-1}$, no actuation.

4. Conclusion

The wake behind the Hump profile (GG body) was experimentally influenced by plasma-induced ionic wind. The plasma actuator was operated in steady as well as in unsteady regime. The unsteady regime was driven by rectangular modulation of voltage waveform. The dimension of the recirculation zone is dependent on Reynolds number.

The wake can be significantly reduced by plasma actuation especially under lower channel velocities. There is also an effect for higher velocities. The separation point is not affected but the reattachement point is shifted upstream significantly. Also the topology of the wake can be influence especially when the actuator is placed farther downstream from the natural separation point. The plasma actuation in x/L = 0.66 and especially in x/L = 0.63 has very positive influence and can reduced the wake and the form drag. The effect of modulation frequency seems to be not crucial. However, this should be investigated also for the first two positions more in detail because the effect will differ there more. Duty cycle value will affect the origin and time of duration of the secondary pseudo-periodical vortical structures. Detailed phase-locked measurements as well as mode analysis have to be done to better investigate this phenomenon.

Especially analysis of dynamics, the measurement with hot wire probes (to determine e.g. frequency spectrum of higher frequencies present in the flow) and energetic balance calculation will be performed in the future. The comparison of the actuator efficiency with other authors will be assigned.

Acknowledgement
The authors gratefully acknowledge financial support of the Grant Agency of the Czech Republic, project No. GP14-25354P.

References

[1] Kurz A et al 2013 *Journal AerospaceLab.* Issue 6 Al06-02
[2] He Ch, Corke T C and Patel M P 2007 *AIAA 45th Aerospace Sciences Meeting*, Paper 2007-0935
[3] Prochazka P and Uruba V 2015 *EPJ Web of Conferences* vol 92 Article number 02074
[4] Prochazka P 2013 Dissertation thesis. Praha ČVUT
[5] Prochazka P and Uruba V 2014 *Journal of Physics: Conf. Ser.*, 530 Article number 012008

Journal of Physics: Conference Series **760** (2016) 012025 doi:10.1088/1742-6596/760/1/012025

Influence of a strong magnetic field on a paramagnetic fluid's convection in systems with different aspect ratios

L Pyrda[1], A Kraszewska[1] and J Donizak[1]

[1]AGH University of Science and Technology, Faculty of Energy and Fuels, Al. Mickiewicza 30, 30-059 Kraków, Poland

E-mail: kraszka@agh.edu.pl

Abstract. Influence of a strong magnetic field gradients on a natural convection of a paramagnetic fluid in enclosures with aspect ratios (AR=height/width) 0.5 and 2.0 was investigated. Performed experiments allowed to determine the impact of strong magnetic field on heat transfer and the nature of the flow, showing that strong magnetic field enhances Nusselt number and rapidly changes the character of the flow. Conducted Fast Fourier analysis also allowed to determine the aspect ratio influence on investigated systems.

1. Introduction

Heat transfer due to convection phenomenon was investigated in three dimensional enclosure of various shapes and aspect ratios by numerous researchers over decades, starting from XX century. With introduction of superconducting materials, it was possible to construct superconducting magnets, which allowed obtaining strong and constant magnetic fields. By application of strong magnetic field gradients to non-electrically conductive fluids with paramagnetic or diamagnetic properties, a thermo-magnetic phenomenon could be observed. Thermo-magnetic convection could be utilized in diverse industrial applications: heat exchangers, magnetic mixers and micro gravity conditions.

First reported magnetic convection of paramagnetic fluid was reported by Braithwaite et al., where magnetic field gradient was used both to enhance and to suppress buoyancy-driven convection. Huang and Edwards, provided the theory of magnetically controlled convection in a horizontal paramagnetic fluid layers heated from either above or below [1]. Bednarz et al. performed numerical and experimental analysis of laminar thermo-magnetic convection of paramagnetic fluid inside cubical enclosure [2]. Kenjeres et al. investigated transients and turbulence pockets [3], oscillatory states [4] in thermo-magnetic convection of paramagnetic fluids. Kenjeres et al. performed numerical and experimental study of Rayleigh–Bénard–Kelvin Convection [5]. Pyrda has provided an analysis of unsteady thermal convection of paramagnetic fluid in cubical enclosure under strong magnetic field gradient [6] and application of Fast Fourier Transform [7] in thermo-magnetic convection analysis.
Although magnetic convection and combined thermo-magnetic convection was investigated in last two decades, there is still lack of experimental data including effect of an aspect ratio in considered geometries on thermo-magnetic convection. Present paper will address this matter.

2. Experiment

2.1. Experimental setup

Experimental setup consisted of: considered experimental enclosures with different aspect ratios, superconducting magnet generating strong magnetic field gradients, a power supply, a thermo-stating bath and a temperature signal acquisition system. Experimental enclosure centre was placed 0.095 [m] above centre of utilized superconducting magnet, where maximum gradient of magnetic field is present and enhance natural convection due to its orientation and chosen fluid type occur. Schematic diagram of superconducting magnet and position of enclosure is shown in Figure 1. To calculate magnetic field distribution and its gradient, which is responsible for magnetic convection of non-electrically conducting fluid, besides real dimensions of system presented on Figure 1, current density in internal and external coil should be taken into account. For maximum value of magnetic field inside centre of magnet: $|b_0|_{max}$=10 [T], maximum current was equal to 140.9·10^{-6} [A/m^2] for internal coil and 167.3·10^{-6} [A/m^2] for external coil.

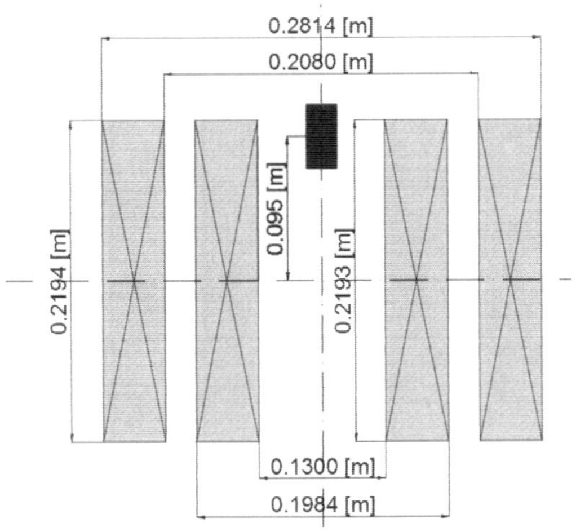

Figure 1. Schematic view of superconducting magnet solenoid used in presented research with marked position of the enclosure B.

Considered experimental enclosures, presented on Figure 2, possessed aspect ratio AR=0.5 (Enclosure A) and AR=2.0 (Enclosure B), which correspond to height H=0.016 [m] and H=0.064 [m] with same square base with side length equal to D=0.032 [m]. Both where build with same materials and manner: four sidewall were made from acrylic glass. In one of them, an additional holes where made to place five (Enclosure A) or six (Enclosure B) thermocouples at 6 [mm] deep from sidewall for insight of temperature variation in time. Holes position in both enclosures sidewall are presented on Figure 3. Bottom wall was made from cooper plate with additional three holes for thermocouples, necessary to calculate heat transfer ratio. Below bottom wall an electrically insulated nichrome wire was placed as a heater supplied from DC adapter. Top wall was made as well from cooper and above it a cooling chamber was placed, which temperature was maintained at constant value by coolant from thermo-stated water bath.

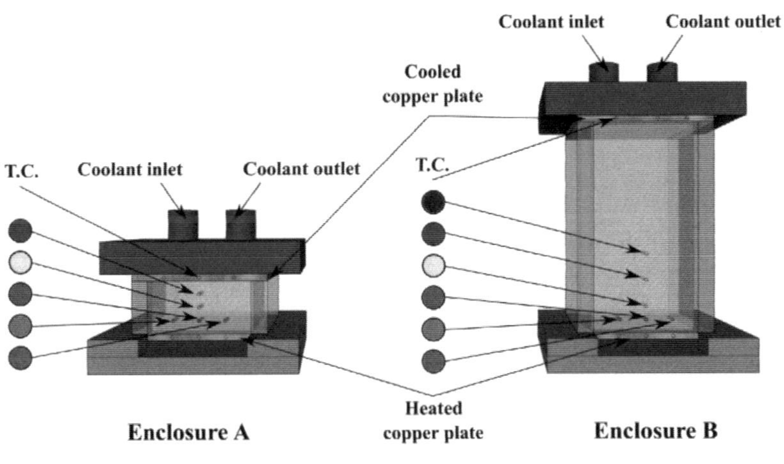

Figure 2. Schematic view of experimental enclosures.

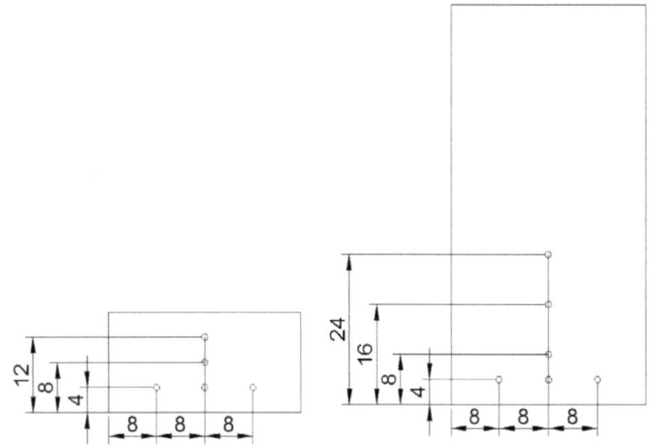

Figure 3. Arrangement of thermocouples positions in experimental enclosures in millimetre scale.

2.2. Experimental fluid

Chosen paramagnetic non-electrically conducting fluid was made from 50% volume aqueous solution of glycerol with 0.8 [mol/(kg of solution)] concentration of gadolinium nitrate hexahydrate ($Gd(NO_3)_3 \cdot 6H_2O$). Experimental fluid was made twice due to year apart between experiments, therefore properties are slightly different – fluid A for Enclosure A and fluid B for Enclosure B. All important, for presented studies, properties of the fluids were measured experimentally and are listed in Table 1.

Table 1. Properties of the working fluid at 298 K.

Property	Symbol	Value		Unit
		A	B	
Heat capacity	c_p	$2.92 \cdot 10^3$	$2.92 \cdot 10^3$	[J/kg·K]
Thermal diffusivity	α	$9.13 \cdot 10^{-8}$	$7.55 \cdot 10^{-8}$	[m^2/s]
Thermal expansion coefficient	β	$1.21 \cdot 10^{-5}$	$4.13 \cdot 10^{-6}$	[1/K]
Dynamic viscosity	μ	$1.30 \cdot 10^{-2}$	$1.56 \cdot 10^{-2}$	[kg/m·s]
Thermal conductivity	λ	0.376	0.376	[W/m·K]
Kinematic viscosity	ν	$9.25 \cdot 10^{-6}$	$1.10 \cdot 10^{-5}$	[m^2/s]
Density	ρ	1411	1418.1	[kg/m^3]
Mass magnetic susceptibility	χ_m	$2.39 \cdot 10^{-7}$	$2.58 \cdot 10^{-7}$	[m^3/kg]

Due to the narrow temperature range studied, it could be assumed that almost all needed properties were constant, except viscosity. Therefore the dependence of viscosity on the temperature was checked experimentally and approximated function could be written as follows:

for fluid A:

$$\mu_0 = 5.600 \cdot 10^{-6} \theta^2 - 7.354 \cdot 10^{-4} \theta + 2.702 \cdot 10^{-2} \quad (1)$$

and for fluid B:

$$\mu_0 = 2.078 \cdot 10^{-5} \theta^2 - 1.678 \cdot 10^{-3} \theta + 4.451 \cdot 10^{-2} \quad (2)$$

2.3. Heat loss estimation

To estimate heat loss at first conduction experiment was performed. Both enclosure, one after another were place upside down in same position as during main experiment. During that stage chosen power level was supplied to electrical heater. After thermal stabilization of system, applied power was measured and temperature on hot and cold wall. This experiment was repeated for various values of power supplied and together with heat flux calculated from Fourier law allow to estimate total heat loss of system for chosen temperature difference between walls:

$$Q_{\mathrm{loss}} = Q_{\mathrm{cond}} - Q_{\mathrm{theor\_cond}} \tag{3}$$

$$Q_{\mathrm{theor\_cond}} = \mathrm{D}^2 \lambda (\theta_{\mathrm{h}} - \theta_{\mathrm{c}}) \mathrm{H}^{-1} \tag{4}$$

where: θ_h, θ_c – are temperature of hot and cold walls respectively, D – enclosure basis side 0.032 [m], H – enclosure height 0.016 [m] for Enclosure A and 0.064 [m] for Enclosure B.
Heat loss was calculated according to eq. (3) and could be linearly approximated for various heating rates ($\Delta\theta=\theta_h-\theta_c=2.5\div15$ [K], where temperature of cold wall was kept at ambient temperature inside magnet), for Enclosure A:

$$Q_{\mathrm{loss}} = 0.0907\Delta\theta \tag{5}$$

and for Enclosure B:

$$Q_{\mathrm{loss}} = 0.0831\Delta\theta \tag{6}$$

2.4. Nusselt number calculation

Nusselt number was calculated as a ratio between the net convective heat transfer rate and the net pure conduction contribution with Ozoe and Churchill [8] method to estimate net heat fluxes:

$$\mathrm{Nu} = \frac{Q_{\mathrm{net\_conv}}}{Q_{\mathrm{net\_cond}}} = \frac{Q_{\mathrm{conv}} - Q_{\mathrm{loss}}}{Q_{\mathrm{cond}} - Q_{\mathrm{loss}}} \tag{7}$$

Assuming that convection heat flux was equal to heater heat flux ($Q_{\mathrm{net\_conv}}=Q_{\mathrm{heater}}$) and applying eq. (3) and (4) to eq. (7) it could be rewritten in following form:

$$\mathrm{Nu} = \frac{Q_{\mathrm{heater}} - Q_{\mathrm{loss}}}{Q_{\mathrm{theor\_cond}}} = \frac{IU - Q_{\mathrm{loss}}}{\mathrm{D}^2 \lambda (\theta_{\mathrm{h}} - \theta_{\mathrm{c}}) \mathrm{H}^{-1}} \tag{8}$$

The Prandtl number definition was:

$$\mathrm{Pr} = \frac{\nu}{\alpha} \tag{9}$$

The thermal Rayleigh number was defined as:

$$\mathrm{Ra_T} = \frac{\beta \mathbf{g} \Delta\theta \mathrm{H}^3}{\nu\alpha} \tag{10}$$

where β is the thermal expansion coefficient, $\mathbf{g}$ is the gravitational vector, $\Delta\theta$ is the temperature difference between horizontal thermally active walls, H is the enclosure height, ν is the kinematic viscosity coefficient and α is the thermal diffusivity.
The magnetic Rayleigh number was defined as:

$$\mathrm{Ra_M} = \left(1 + \frac{1}{\beta\theta_0}\right)\left(\frac{\gamma\beta\mathbf{g}\Delta\theta\mathrm{H}^3}{2\nu\alpha}\right) \tag{11}$$

The magnetization number:

$$\gamma = \frac{\chi_{\mathrm{m}} |\mathrm{b}_0|^2_{\mathrm{max}}}{\mu_0 \mathbf{g} \mathrm{H}} \tag{12}$$

where: $\mu_0=4\cdot\pi\cdot10^{-7}$ [H/m].
The thermo-magnetic Rayleigh number was defined as:

$$\mathrm{Ra_{TM}}=\mathrm{Ra_T}+\mathrm{Ra_M} = \left(\frac{\beta\mathbf{g}\Delta\theta\mathrm{H}^3}{\nu\alpha}\right)\left(1+\frac{\gamma}{2}+\frac{\gamma}{2\beta\theta_0}\right) \tag{13}$$

2.5. Signal analysis

Temperature signals, recorded during experiments, enabled investigation of the flow structure. This was obtained through Fast Fourier Transform (FFT), which computes discrete Fourier transform (DFT) of a sequence:

$$F_n = \sum_{i=0}^{N-1} x_i e^{-\frac{2\pi i}{N} ni} \tag{14}$$

where x_i is a sequence of length N, and F_n is its discrete Fourier transform.
Calculated results, in the form of amplitude:

$$\text{Amplitude} = \sqrt{\text{Re}^2 + \text{Im}^2} \,/\, n \tag{15}$$

versus frequency, were used to analyze the fluid behaviour in thermo-magnetic convection in enclosures with two different aspect ratios.

3. Results

Figures 4-6 present results of FFT in the form of amplitude versus frequency diagrams. All of the thermocouple signals (see Figures 2 and 3 for placement of the thermocouples in experimental enclosures) were analyzed.

Figure 4 shows results of flow analysis for both experimental enclosures for 3[K] temperature difference between thermally active walls. For natural convection case ($|b_0|_{max}$=0 [T]) fluid flow in Enclosure A is steady and no fluctuations of the temperature can be observed, while in Enclosure B small fluctuations of amplitude 0.00578 [-], 0.00496 [-] and 0.00287 [-] for characteristic frequency 0.6766 [Hz] and 0.00396 [-] and 0.12422 [Hz] appear for two thermocouples placed in bottom left and bottom central positions. After applying magnetic induction to both systems, flow becomes faster and temperature oscillations are more frequent. For $|b_0|_{max}$=5 [T] in enclosure A oscillation of temperature signal has characteristic frequency equal to 0.04218 [Hz]. In enclosure B, at the same magnetic induction value, all but one thermocouples have characteristic oscillations as for natural convection case. One thermocouple, placed in bottom left corner of the enclosure shows different behaviour – many peaks of relatively high amplitude can be observed. This character of fluid behaviour could indicate that a vertical structure appeared in experimental enclosure and was not a part of general fluid flow. With further increase of magnetic induction up to maximum value of 10 [T] oscillation amplitudes slightly increased and became more intensive than results for natural convection case, which confirms intensification impact of magnetic field to paramagnetic fluid flow.

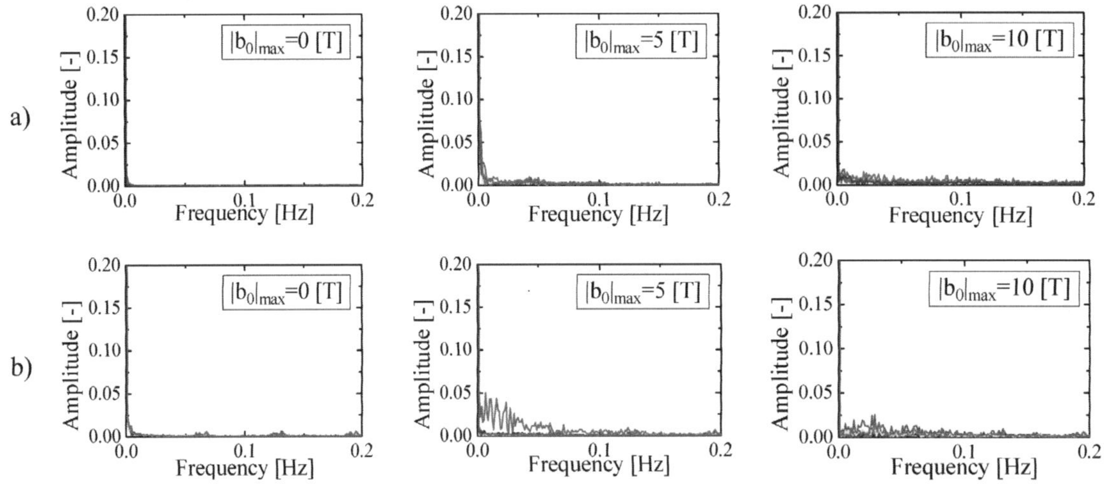

Figure 4. FFT amplitude versus frequency for a) Enclosure A with AR=0.5 and b) Enclosure B with AR=2.0 for $\Delta\theta$ =3 [K].

Figure 5 presents results for temperature difference of 5 [K] between bottom and top walls. As before in Enclosure A fluid is stable and no temperature oscillation can be observed. Different phenomena occurs in Enclosure B – many temperature fluctuations appear on amplitude vs. frequency diagram and the maximum amplitude has a value of 0.10239 for characteristic frequency 0.00888 [Hz]. After applying magnetic induction at $|b_0|_{max}$=5 [T] in smaller enclosure many small fluctuations appear with maximal amplitude values of 0.05466 and 0.05109 for frequencies 0.00333 [Hz] and 0.00666 [Hz] respectively. For larger enclosure fluid flow changes significantly. The thermocouple placed in bottom left corner shows different behaviour from the rest – it fluctuates with high amplitudes, while every other thermocouple show periodic fluctuations of small amplitudes. Further increase of magnetic induction up to 10 [T] causes unification of temperature field in Enclosure B with aspect ratio 2.0 – signal from thermocouple in bottom left corner changes and shows behaviour similar to other thermocouples. Probably high magnetic field caused dissipation of vertical structure that appeared in experimental enclosure for lower magnetic induction values.

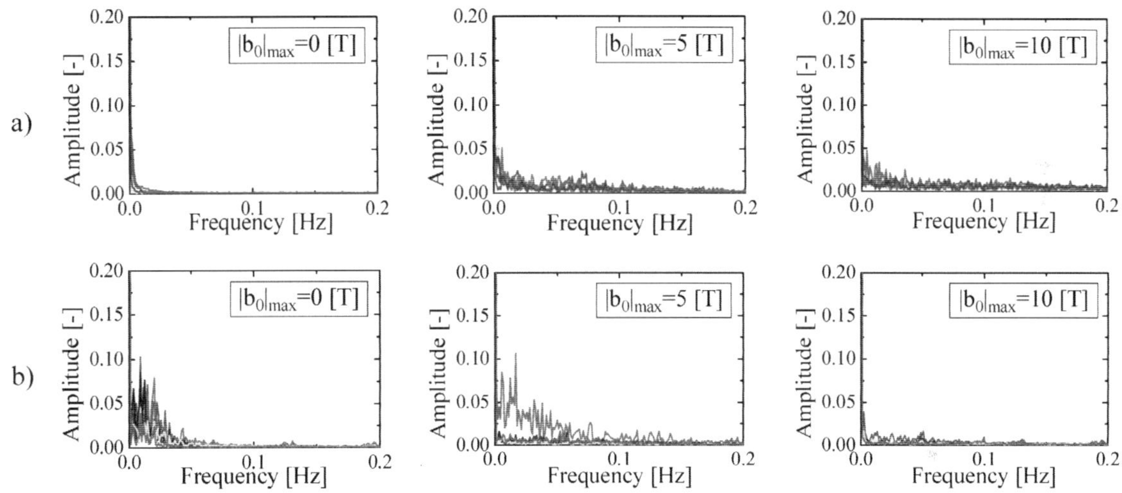

Figure 5. FFT amplitude versus frequency for a) Enclosure A with AR=0.5 and b) Enclosure B with AR=2.0 for $\Delta\theta$ =5 [K].

Amplitude vs. frequency diagrams for temperature difference between thermally active walls of 11 [K] are shown in Figure 6. Natural convection case shows results similar to $\Delta\theta$ =5 [K]. In Enclosure A flow is stable and temperature signals show no fluctuations. In Enclosure B flow field oscillates significantly – the maximum amplitude 0.18652 appears for frequency 0.00111 and more high amplitudes occur for a series of characteristic frequencies. Increasing magnetic induction to 5 [T] significantly changes the character of the flow in Enclosure A with aspect ratio 0.5. Many periodic oscillations appear on the spectrum for all thermocouples but one. Thermocouple placed in bottom right corner shows different behaviour – oscillations with high amplitude indicating that a structure separate from the main flow appeared in the enclosure. For the same magnetic induction in higher enclosure flow field is more uniform but higher temperature oscillations appear for a thermocouple placed in right bottom corner, as in Enclosure A. Further increasing the magnetic induction causes unification of the temperature field. For both enclosures temperature oscillations decreased, and while in Enclosure B fluid structure which appeared in bottom right corner dissipated, in Enclosure A periodic temperature oscillations in this region suggest that the structure is still present.

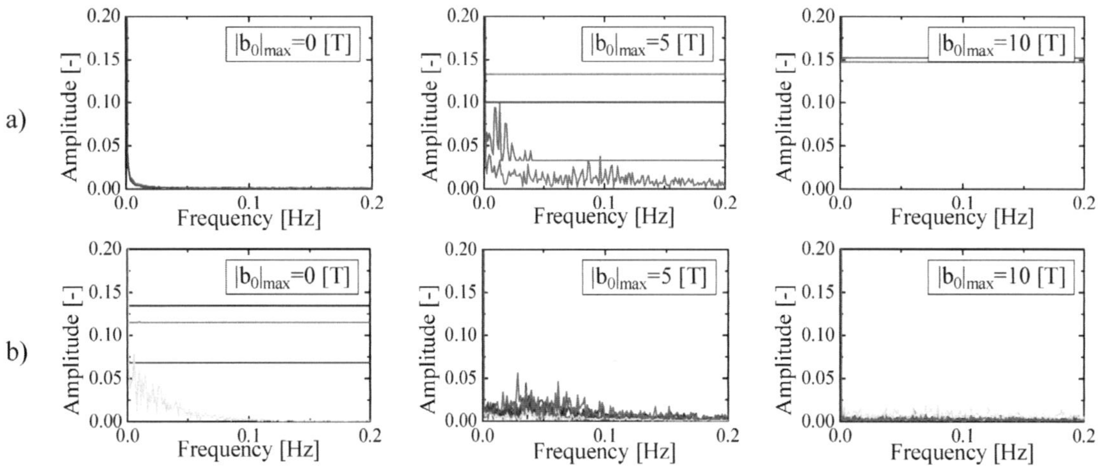

Figure 6. FFT amplitude versus frequency for a) Enclosure A with AR=0.5 and b) Enclosure B with AR=2.0 for $\Delta\theta$ =11 [K].

Results of heat transfer analysis are presented in Figure 7 as a ratio Nu/Nu_0 versus thermo-magnetic Rayleigh number, where Nu_0 is Nusselt number value for natural convection case ($|b_0|_{max}$=0 [T]) enlisted in Table 2.

Table 2. Nusselt number values for natural convection measurements

Case	$\Delta\theta$ [K]	Value
	3	3.458
AR=0.5	5	2.954
	11	3.618
	3	13.447
AR=2.0	5	14.498
	11	18.820

With increasing value of magnetic induction in the centre of the magnet enhancement of heat transfer was observed for every case. The largest growth in Nusselt number occurred for a small enclosure (AR=0.5) for $\Delta\theta$=5 [K], where the relative value of Nu/Nu_0 is about 3.73. Slightly lower rise appeared for tall enclosure, where the increase of heat transfer in the form of Nu/Nu_0 for $|b_0|_{max}$=10 [T] and $\Delta\theta$=5 [K] equals 3.37.

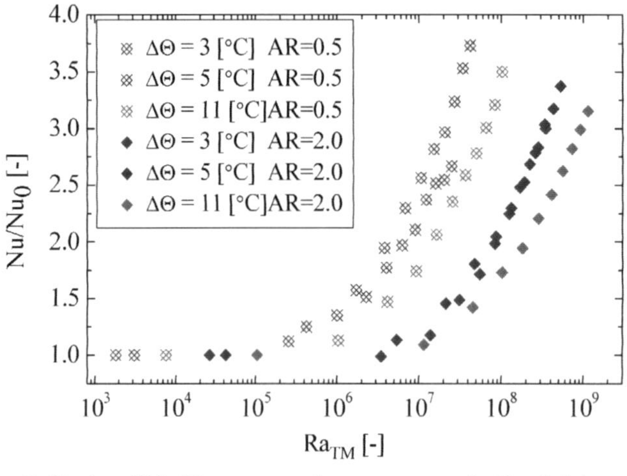

Figure 7. Ratio of Nu/Nu_0 versus thermo-magnetic Rayleigh number.

4. Conclusions

An experimental analysis of thermo-magnetic convection of paramagnetic fluid in two enclosures with different aspect ratios was conducted. Flow behaviour was investigated based on temperature variations and Fourier analysis. Application of strong magnetic field gradients strongly influence fluid flow which could be observed as grow of frequency of temperature signal, its amplitude and finally thermal heat flux presented as Nusselt number. Heat transfer enhancement with application of strong magnetic field gradient was significant for both enclosures. Presented approach could be applied in similar system with lack of classical visualization methods.

Acknowledgements
This work was supported by the National Science Centre (Project No. 12/07/B/ST8/03109).

References

[1] Braithwaite D, Beaugnon E and Tournier R 1991 *Nature* **354** 134–6
[2] Bednarz T, Fornalik E, Tagawa T, Ozoe H and Szmyd J S 2005 *Int. J. Therm. Sci.* **44** 933–43
[3] Kenjereš S, Wrobel W, Pyrda L, Fornalik-Wajs E and Szmyd J S 2011 *J. Phys. Conf. Ser.* **318** 072028
[4] Kenjereš S, Pyrda L, Wrobel W, Fornalik-Wajs E and Szmyd J 2012 *Phys. Rev. E* **85** 1–8
[5] Kenjereš S, Pyrda L, Fornalik-Wajs E and Szmyd J S 2013 *Flow, Turbul. Combust.* **92** 371–93
[6] Pyrda L, Kenjeres S, Fornalik-Wajs E and Szmyd J S 2012 A *J. Phys. Conf. Ser.* **395** 012125
[7] Pyrda L 2014 *J. Phys. Conf. Ser.* **530** 012060
[8] Ozoe H and Churchill S W 1973 *AIChE Symp. Ser. Heat Transf.* **69 (131)** 126–33

Analysis of water management at a closed cooling system of a power plant

P Regucki[1], B Engler[1] and Z Szeliga[2]

[1] Wrocław University of Technology, Faculty of Mechanical and Power Engineering, Wybrzeże Wyspiańskiego 27, 50-370 Wrocław, Poland
[2] VSB - Technical University of Ostrava, Faculty of Mechanical Engineering, Department of Energy Engineering, 17. listopadu 15/2172, 708 33 Ostrava-Poruba, Czech Republic

E-mail: pawel.regucki@pwr.edu.pl

Abstract. The paper presents a mathematical model describing the changes in SO_4^{2-} ions concentration in a closed system of cooling water circulation in a professional power plant due to the new restrictions posed on industrial emissions by EU and Polish government regulations [1]. Analyzing the behavior of the numerical solutions of the model Authors could determine the optimal volumetric flow rate of wastewater depending on the current configuration of operating power units in a professional power plant.

1. Introduction

In professional power plants, water is commonly used in many industrial processes including, among others: water-steam circulation in power boiler, cooling systems supporting boiler auxiliary equipment (e.g. heat exchangers, compressors) or condensers' cooling system. Each of these processes requires to ensure appropriate physical and chemical properties of water in order to maintain the continuity of the production of electricity and heat [2]. Due to the fact that daily demand for fresh water in power plants often reaches tens of thousands of m^3, these units have to be located near rivers, lakes or other natural reservoirs, where are also released industrial wastewater. Increase production of wastewater and its influence on environment was recently reported e.g. by Blog [3] what indicates on the need for new restrictive regulations for water environment protection. Taking into account its significant impact on natural environment, conventional power plants operated in Poland must satisfy the last industrial waste regulations of The Minister of Environment dated 18 November 2014 and the EU directive [1] which impose, with the beginning of 2016, new limits on the physical and chemical parameters of wastewater released to natural reservoirs. On the other hand, the increase of sewage costs enforces the companies to look for the optimal management of industrial wastes.

Particularly important component of power plant, from the water and wastewater management point of view, is connected with a condenser - cooling tower closed cooling system. Circulating in this system water is heated in the condenser by the condensing steam from the last stage of the turbine, and then cools down in the cooling tower. The main mechanism of the temperature decrease of water inside the cooling tower is connected with a partial evaporation of the flowing water. This phenomenon is associated with an increase of concentration of chemical compounds in the circulating water. This is disadvantageous effect, for instance because an excessive increase in the concentration of sulphate ions SO_4^{2-} in the water may cause corrosion of concrete parts of channels and cooling tower and the increase in the concentration of calcium salts can accelerate the process of their deposition on the exchanges' surfaces inside the condenser thereby impairing the heat exchange processes [4, 5].

2. Concentration of sulphate ions in cooling system

The scheme of closed cooling water system is presented at figure 1.

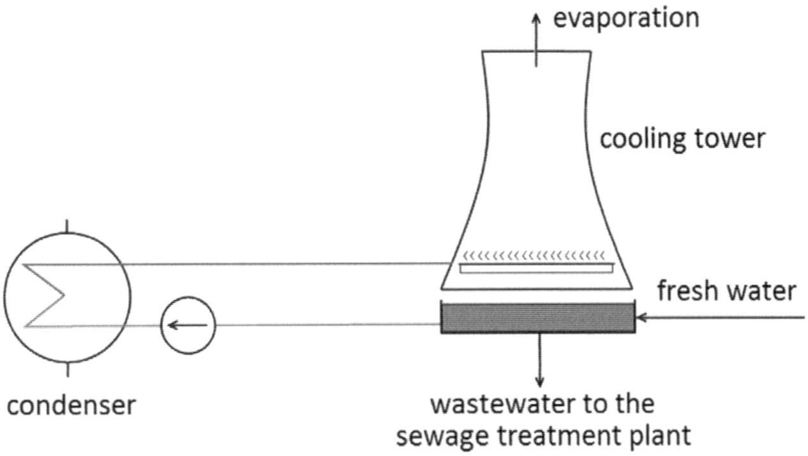

Figure 1. Scheme of closed system of cooling water in power plant

As it was mentioned above, the main mechanism leading to decrease of temperature of circulating water is associated with partial evaporation of the water inside the cooling tower. Due to the fact of continuous evaporation, the total mass of water circulated in the closed system decreases. This leads to the continuous increase of concentration of chemical components in the circulating water. The physical and chemical requirements for cooling water are not as restrictive as in the case of water prepared for steam-water circulation in power boiler, because water does not have a direct contact with boiler or turbine. Despite this, however, the special attention should be paid to its chemical composition, particularly in terms of sulphate (VI) ions. These ions are dangerous because of its corrosive influence on concrete components of the cooling system such as channels or cooling towers reservoirs. Their destructive action on the concrete is connected with ($3CaO \cdot Al_2O_3 \cdot 3CaSO_4 \cdot 31H_2O$) crystals which precipitate from water in the presence of SO_4^{2-} ions. Formation of these crystals in deposits is connected with considerable increase in its volume what may lead to disintegration of the concrete surface.

A simple mechanism leading to decrease of undesirable chemicals concentration in cooling water is associated with periodic discharge of its part to water treatment plants. The missing mass of water is replenished from the natural reservoirs like rivers or lakes. Usually the fresh water has lower concentration of SO_4^{2-} ions in comparison to its actual concentration in closed cooling system what causes that the total concentration of SO_4^{2-} ions in the closed system decreases. Studies carried out in the professional power plant indicate the monthly changes in concentration of SO_4^{2-} ions (denoted by x, [mg/l]). An example results are presented in table 1.

Table 1. Mean concentration of sulphate ions x and mean temperature T of cooling water during the year

month	I	II	III	IV	V	VI	VII	VIII	IX	X	XI	XII
T, °C	15,7	14,0	14,5	17,4	19,1	19,9	21,2	22,0	20,6	20,4	16,7	16,9
x, mg/l	175,0	160,0	190,2	206,4	252,0	303,7	313,0	316,2	253,6	276,4	190,7	195,4

Presented data indicate that mean monthly concentration of SO_4^{2-} ions changes during the year. Two reasons have the significant influence on this situation: high intensity of water evaporation in cooling tower during the summer time and reduction of operation parameters of cooling tower in winter. High ambient temperature during summer in Poland (usually between 20 - 25 °C) causes that approximate 1% of mass of circulated water evaporates in cooling towers. This intense water loss must be constantly supplemented with natural resources. On the other hand in winter, operators must keep the temperature of cooling water above at the level of 12 °C due to the technological reasons. Moreover, at this time the intensity of water evaporation is not so high because only a part of the circulating water is cooled down in cooling towers. Monthly changes in intensity of cooling water evaporation are reflected in mean concentration of sulphate (VI) ions. The highest concentration SO_4^{2-} of is observed in period June-August (more than 300 mg/l), when the lowest concentration is in February (approx. 160 mg/l). Keeping in mind that legal regulations allow on much higher SO_4^{2-} ions concentration in waste water (maximally x_{limit} = 500 mg/l [1]) it means that operator of the closed cooling system releases more sewage than it is required. This situation gives an opportunity to optimize water management in order to reduce operating costs of the system.

3. Mathematical model

Presented mathematical model enables the prediction of the daily increase in the concentration of sulphate (VI) ions in the closed cooling water system and determines the optimal volumetric flow rate of wastewater due to the current configuration of operating power units in a professional power plant. Balancing the SO_4^{2-} ions concentration changes in the cooling system the following processes are taking into account:

- evaporation of water in cooling towers, $q_{v,ct}$,
- discharge of waste water to the sewage treatment plant, $q_{v,dw}$
- replenish of fresh water, $q_{v,fw}$.

Assuming that the total volume of cooling water V circulating in the system is constant, the change of the current SO_4^{2-} ions concentration $x(t)$ with time t is described by the differential equation (where x_{ct} - SO_4^{2-} ions concentration in vapour released in cooling tower; x_{fw} - SO_4^{2-} ions concentration in fresh water delivered to the system):

$$V \cdot \frac{d\,x(t)}{dt} = -q_{v,dw} \cdot x(t) - q_{v,ct} \cdot x_{ct} + q_{v,fw} \cdot x_{fw} \tag{1}$$

If the total volume of water in the closed system V is constant, the volumetric flow rates must satisfy balance equation:

$$q_{v,fw} = q_{v,dw} + q_{v,ct} \tag{2}$$

Expressing volumetric flow rate of evaporated water in cooling towers in (1) by $q_{v,dw}$ and $q_{v,fw}$ from (2) one can get:

$$V \cdot \frac{d\,x(t)}{dt} = -q_{v,dw} \cdot (x(t) - x_{ct}) + q_{v,fw} \cdot (x_{fw} - x_{ct}) \tag{3}$$

Further simplification of (3) is possible if one assumes that $x_{ct} = 0$ and x_{fw} is constant. Then (3) has a form of linear differential equation of first order:

$$\frac{d\,x(t)}{dt} + a \cdot x(t) - b = 0 \tag{4}$$

where: $a = q_{v,dw}/V$ and $b = q_{v,fw} \cdot x_{fw}/V$. Analytical solution of (4) has a form:

$$x(t) = \frac{b}{a} + \left(x_0 - \frac{b}{a}\right) \cdot e^{-at} \quad (5)$$

where x_0 refers to initial SO_4^{2-} ions concentration in the closed system.
The SO_4^{2-} ions concentration in fresh water varies between 40 - 60 mg/l. Then in the model Authors used x_{fw} = 60 mg/l. Figure 2 presents an example of solutions obtained for $q_{v,ct}$ = 48000 m^3/d and total volume of the circulating water equals V = 76750 m^3.

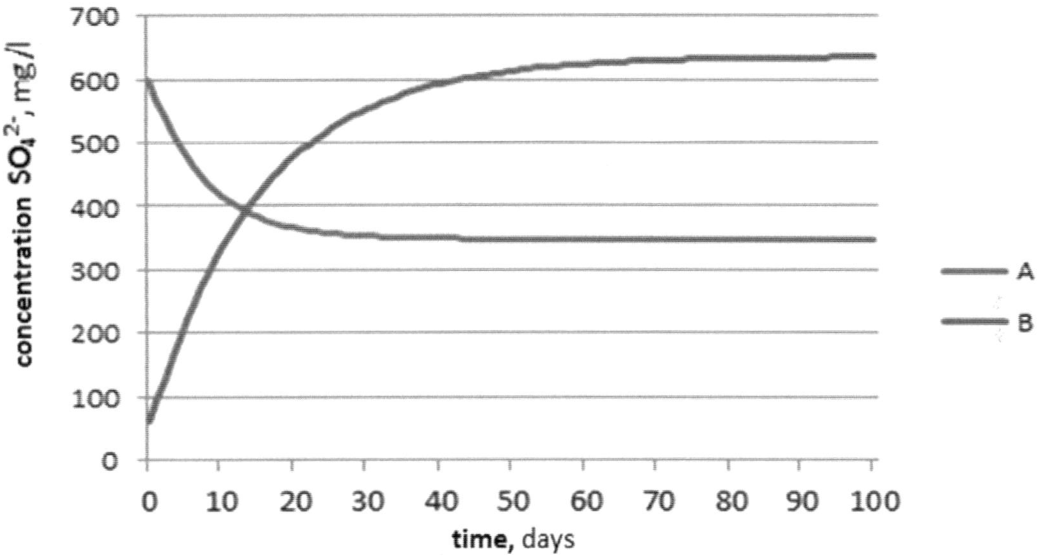

Figure 2. Example of analytical solutions of (4) for different initial conditions: A - x_0 = 600 mg/l and $q_{v,dw}$ = 10000 m^3/d; B - x_0 = 60 mg/l, $q_{v,dw}$ = 5000 m^3/d

Analyzing presented at figure 2 solutions one can note that in case A the volumetric flow rate $q_{v,dw}$ is relatively large and causes that sulphate ions concentration in closed system $x(t)$ decreases when in case B volumetric flow rate $q_{v,dw}$ is insufficient to prevent an increase in ion concentration with time. From these two examples one can deduce two important conclusions: controlling the volumetric flow rate of discharge of waste water $q_{v,dw}$ operator can control the actual concentration of SO_4^{2-} ions in the closed system and presented functions have asymptotic solutions which depend on initial values of a and b parameters. In the presented cases, the asymptotic values of the sulphate ions concentrations are x_{asym} = 348 mg/l and 636 mg/l, for A and B curve respectively.

4. Water waste management

Results obtained from the mathematical model have an important influence on water management in power plant. Firstly, it is not possible to reduce a flow rate of waste water too much because it will result in increase of sulphate ions concentration in circulating water. On the other hand, too large waste water discharge will generate increase of financial costs because of the high prices for industrial sewage. It is worth to mention that actual limit for concentration of sulphate ions in waste water (for a cooled surface having a temperature below 60 °C) in Poland is x_{limit} = 500 mg/l. Moreover, the legal regulations are provided for additional penalties for exceeding the permissible concentration limits in sewage. All these aspects cause that power plants must lead a very economical water policy if they want to get the minimum operating costs in this sector.

Analyzing the analytical solution (5) one can determine asymptotic value of SO_4^{2-} ions concentration x_{asym} in circulating water if the constant value of waste water flow rate $q_{v,dw}$ is preserved:

$$x_{asym} = \lim_{t\to\infty} x(t) = \lim_{t\to\infty} \frac{b}{a} + \left(x_0 - \frac{b}{a}\right) e^{-at} = \frac{b}{a} \tag{6}$$

If there is assumed that $x_{ct} = 0$ expression (b/a) simplifies to the form:

$$x_{asym} = \frac{b}{a} = \frac{q_{v,fw} \cdot x_{fw}}{q_{v,dw}} = \left(1 + \frac{q_{v,ct}}{q_{v,dw}}\right) \cdot x_{fw} \tag{7}$$

From the formula (7), one can notice that asymptotic value x_{asym} depends only on three parameters: SO_4^{2-} ions concentration in fresh water x_{fw}, volumetric flow rate of discharge of waste water $q_{v,dw}$ and volumetric flow rate of evaporated water in cooling tower $q_{v,ct}$. Power plant operator has an influence only on one of them - $q_{v,dw}$, because number of operating cooling towers and weather conditions determine $q_{v,ct}$. Also chemical parameters of fresh water (among them x_{fw}) are rather fixed. Asymptotic value x_{asym} can reach level of SO_4^{2-} ions concentration available at fresh water x_{fw} if volumetric flow rate of waste water tends to infinity. From waste water management point of view this solution is costly and thus less economical for power plant. On the other hand there are no operating reasons to maintain the SO_4^{2-} ions concentration as low. Example values of asymptotic ions concentrations x_{asym} for fixed value of $q_{v,ct} = 48000$ m^3/d are presented in table 2.

Table 2. Example values of asymptotic ion concentrations x_{asym} for selected values of volumetric flow rates $q_{v,dw}$ and SO_4^{2-} ions concentrations in fresh water x_{fw} (for $q_{v,ct} = 48000$ m^3/d)

SO_4^{2-} ions concentration x_{fw}, [mg/l] / flow rate $q_{v,dw}$,[m^3/d]	40	45	50	55	60
5000	424	477	530	583	636
7500	296	333	370	407	444
10000	232	261	290	319	348
12500	194	218	242	266	290
15000	168	189	210	231	252

Analyzing data from table 2 one can notice that there exists pairs of $q_{v,dw}$ and x_{fw} for which asymptotic value of x_{asym} is below legally acceptable limit x_{limit}= 500 mg/l. It is worth to mention that value of $q_{v,ct}$ is determined by operating parameters of cooling towers as well as its amount and configuration in closed cooling water system.

On the other hand one can use the mathematical model to determine the minimum volumetric flow rate of discharge waste water $q_{v,dw}$ requires to satisfy the legal requirements for SO_4^{2-} ions concentration in waste water releases to natural water reservoirs. If in formula (7) asymptotic ions concentration x_{asym} will be replaced by x_{limit} then one can transform (7) in order to obtain the formula for minimum volumetric flow rate of discharge waste water $q_{v,dw}$:

$$q_{v,dw} = \frac{q_{v,ct} \cdot x_{fw}}{(x_{limit} - x_{fw})} \tag{8}$$

Assuming that SO_4^{2-} ions concentration in fresh water x_{fw} is fixed as well as volumetric flow rate of evaporated water in cooling towers $q_{v,ct}$ for given number of working cooling tower, an operator responsible for waste water management can calculate for (8) the minimum volumetric flow rate of waste water $q_{v,dw}$. Example values of minimal volumetric flow rates $q_{v,dw}$, parameterized by sulphate (VI) ions concentration in fresh water x_{fw} and volumetric flow rate $q_{v,ct}$, for fixed x_{limit} = 500 mg/l are presented in table 3.

Table 3. Example values of minimal volumetric flow rates $q_{v,dw}$ parameterized by sulphate ion concentration in fresh water x_{fw} and volumetric flow rate $q_{v,ct}$ for fixed x_{limit} = 500 mg/l

sulphate ion concentration x_{fw}, [mg/l] / volumetric flow rate $q_{v,ct}$, [m^3/d]	40	45	50	55	60
9600	835	949	1067	1187	1309
19200	1670	1899	2133	2373	2618
28800	2504	2848	3200	3560	3927
38400	3339	3798	4267	4746	5236
48000	4174	4747	5333	5933	6545

Keeping the minimal volumetric flow rate $q_{v,dw}$ which preserves legal regulations for SO_4^{2-} ions concentration in waste water operator can minimize operating costs of closed cooling water system both saving fresh water resources as well as minimizing the costs for industrial waste water.

Presented mathematical model allows not only estimate the minimal requirements for water management in closed cooling water system but also can be useful for testing different waste water strategies. As an example of such possibilities, figure 3 presents two different approaches to the problem of waste water management during the period of 52 days.

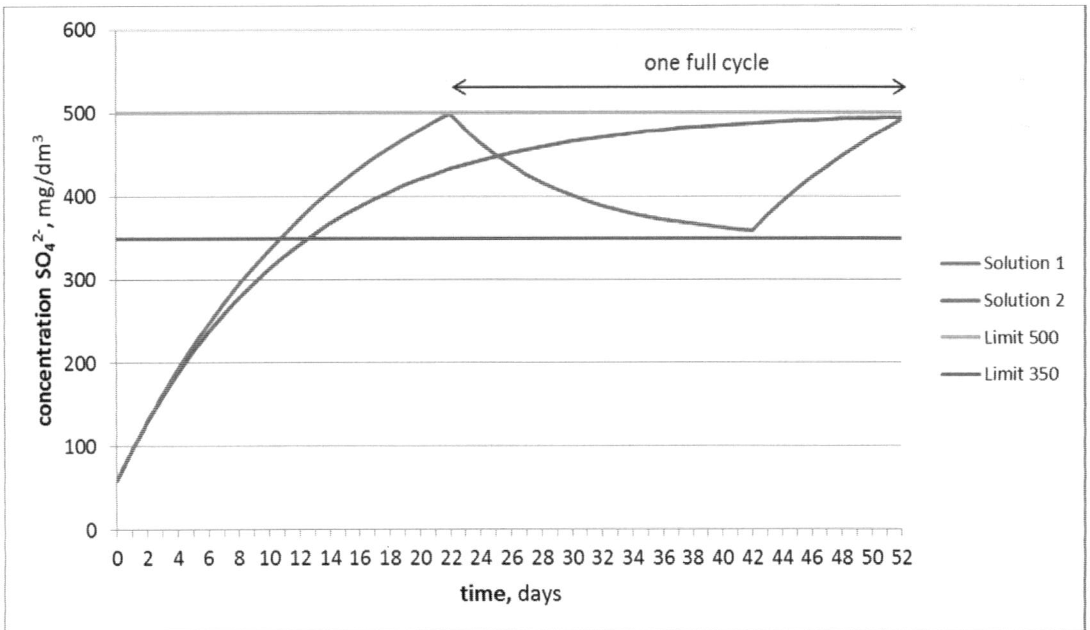

Figure 3. Example of the numerical calculations presented different solutions of the mathematical model (1) with initial condition x_0 = 60 mg/l and $q_{v,ct}$ = 48000 m^3/d. Red curve: $q_{v,dw}$ = 6545 m^3/d; blue curve: $q_{v,dw}$ = 5000 m^3/d and 10000 m^3/d respectively.

Both curves (red and blue one) are calculated for the same initial condition x_0 = 60 mg/l and fixed volumetric flow rate of evaporated water in cooling towers $q_{v,ct}$ = 48000 m^3/d. In case of first solution (red curve) the volumetric flow rate of waste water is fixed during the whole period of 52 days at the level of $q_{v,dw}$ = 6545 m^3/d (according to the table 3). The blue curve consists of two solutions calculated for $q_{v,dw}$ = 5000 m^3/d (first 22 days and period between 42 and 52 day) and 10000 m^3/d (period between 22 and 42 day). Changes in volumetric flow rate $q_{v,dw}$ for blue curve are forced by legal limit of sulphate (VI) ions concentration x_{limit} = 500 mg/l. Operator have to increase initial value of the volumetric flow rate of waste water up to $q_{v,dw}$ = 10000 m^3/d in order to avoid exceeding of the legal limit x_{limit}. When the ion concentration decreases to approx. 350 mg/l, operator again reduces volumetric flow rate of waste water to $q_{v,dw}$ = 5000 m^3/d. So the waste water management works in the cycle of 30 days with ion concentration between 350 and 500 mg/l and $q_{v,dw}$ between 5000 and 10000 m^3/d.
Both presented curves preserve legal limits but in first case operator released only 340340 m^3 of waste water during 52 days when in the other case 360000 m^3. The difference between these two approaches is 19660 m^3. It means that optimal waste water management uses approx. 5.8% less fresh water than in second case.

5. Conclusions

Presented mathematical model of changes of SO_4^{2-} ions concentration in closed cooling water system shows possibility of optimization of waste water management under different operating conditions of cooling towers. Basing on information about chemical parameters of fresh water and flow rate of evaporated water inside cooling towers one can estimate the minimum flow rate of waste water in order to preserve legal requirements. This knowledge helps operators and decision makers to lead a planned and cost-effective waste water management in power plants.

References

[1] Council Directive 91/271 / EEC of 21 May 1991. Urban Waste Water Treatment (Acts. Office. EC L 135, 05.30.1991, p. 40, as amended. d.; Acts. Office. Polish special edition, ch. 15, v. 2, p. 26), the Directive of the European Parliament and of the Council 2010/75/EU of 24 November 2010. on industrial emissions (integrated pollution prevention and control) (recast) (OJ. office . EC L 334, 17.12.2010, p. 17, as amended. d.
[2] Laudyn D., Pawlik M., Strzelczyk F., *Elektrownie*, Wydawnictwo WNT, Warszawa 1997.
[3] Blog, S. F., Message to EPA: Time to Modernize America's Power Plants - Cooling Systems Included, http://switchboard.nrdc.org/blogs/sfleischli/message_to_epa_it_is_time_to_m.html Last cited 4January, 2016.
[4] Hermanowicz W. et al., *Fizyczno – chemiczne badanie wody i ścieków*, Wydawnictwo Arkady, Warszawa 2010
[5] Stańda J., *Woda do kotłów parowych i obiegów chłodzących siłowni cieplnych*, Wydanie trzecie rozszerzone, Wydawnictwa WNT, Warszawa 1992, 1999.

Impact of numerical method on auto-ignition in a temporally evolving mixing layer at various initial conditions

A Rosiak[1] **and A Tyliszczak**[1]

[1] Czestochowa University of Technology, Faculty of Mechanical Engineering and Computer Science, Al.Armii Krajowej 21, 42-201 Czestochowa, Poland

E-mail:`rosiaka@imc.pcz.czest.pl`

Abstract. Numerical analysis of the auto-ignition of turbulent mixing layer between the cold fuel (hydrogen) and hot oxidizer (air) is presented. The research were performed using an Implicit-Large Eddy Simulation (ILES) method with attention on auto-ignition time, flame kernel localisation and propagation. We focused on an impact of discretization method on auto-ignition scenario and flame development. The results obtained showed that numerical approach plays an important role and to some extent may falsify the results, especially for low oxidiser temperatures.

1. Introduction

Turbulent combustion process, being a result of the two-way interaction of chemical reactions and turbulent structures, is characterized by significant oscillations of all flow variables. A flame interacting with turbulent flow leads to its modification caused by strong accelerations through the flame front and because of large changes of density and viscosity associated with temperature variations. An opposite impact is equally important, i.e. the turbulence affects the combustion process by enhancing the chemical reactions or inhibiting them, which may lead to a spontaneous ignition or extinction. The fact that accuracy of modelling of turbulent flows is directly related to an applied model and numerical method indicates that simulations of reacting flow are very complex problem and are regarded as one of the most difficult tasks in the contemporary CFD (Computational Fluid Dynamics). They need an involvement of computational methods capable to accurately represent a wide range of strongly unsteady phenomena occurring in the combustion process.

During the last decades numerous works devoted to DNS (Direct Numerical Simulation) of auto-ignition were presented. One of the first numerical analysis of auto-ignition (DNS calculations with one-step chemistry) was performed for two dimensional (2D) mixing layer (methane/air) by Mastorakos et al.[1] in 90s. This fundamental work constituted the basis for subsequent research in the field of numerical analyses of the auto-ignition phenomenon. Recently, the auto-ignition in 3D case was analysed using DNS by Doom and Mahesh [2]. They studied a configuration similar as in [1] and found that extending the mixing layer over the third dimension three possible auto-ignition scenarios can be observed, not reported for 2D cases.

The present work concerns Implicit Large Eddy Simulation (ILES) of non-premixed combustion in a temporally evolving mixing layer between cold fuel (hydrogen) and hot oxidiser (air). Similarly to [2], on the initial velocity field we impose an isotropic turbulence. The analysis is focused on the auto-ignition and flame propagation phenomena and their dependence on discretization method, which in ILES approach plays a crucial role.

2. Numerical methods

2.1. Governing equations

We consider low Mach number reactive flows governed by the Navier-Stokes equations, the transport equations for each of reactive species and energy equations complemented with the equation of state. In the context of LES method the set of the governing equations is given as:

$$\begin{aligned}
&\partial_t \bar{\rho} + \nabla \cdot (\bar{\rho}\widetilde{\mathbf{u}}) = 0 \\
&\bar{\rho}\partial_t \widetilde{\mathbf{u}} + (\bar{\rho}\widetilde{\mathbf{u}} \cdot \nabla)\widetilde{\mathbf{u}} + \nabla \bar{p}\mathbf{I} = \nabla \cdot \left(\boldsymbol{\tau} + \boldsymbol{\tau}^{\mathrm{SGS}}\right) \\
&\bar{\rho}\partial_t \widetilde{Y}_i + (\bar{\rho}\widetilde{\mathbf{u}} \cdot \nabla)\widetilde{Y}_i = \nabla \cdot \left(\bar{\rho}\left(D_i + D_i^{\mathrm{SGS}}\right)\nabla \widetilde{Y}_i\right) + \bar{\dot{\omega}}_i \quad \text{for } i = 1,\dots,N \text{ - species} \qquad (1) \\
&\bar{\rho}\partial_t \widetilde{h} + (\bar{\rho}\widetilde{\mathbf{u}} \cdot \nabla)\widetilde{h} = \nabla \cdot \left(\bar{\rho}\left(D + D^{\mathrm{SGS}}\right)\nabla \widetilde{h}\right) \\
&p_0 = \bar{\rho}R\widetilde{T}
\end{aligned}$$

The bar and tilde symbols denote filtered quantities [3]. The variables $\mathbf{u}, \rho, p, Y_i, h$ represent the velocity vector, density, hydrodynamic pressure, species mass fraction and enthalpy. The symbols $\boldsymbol{\tau}, D_i, D$ represent viscous stress tensor, mass and heat diffusivities. The SGS index refers to unresolved subgrid terms, where the subgrid viscosity $\boldsymbol{\tau}^{\mathrm{SGS}}$ is computed accordingly to the model proposed by Vreman [4] and the subgrid diffusivities D_i^{SGS} and D^{SGS} are computed based on the turbulent Schmidt or Prandtl number. The sources $\bar{\dot{\omega}}_i$ are the filtered reaction rates of species i which are non-linear functions of species mass fraction and temperature. The ILES approach is an attractive form of combustion closure wherein the reaction rates are obtained from the Arrhenius formula and the computational costs are reasonable. The ILES closure assumes [5] $\overline{\dot{\omega}_i(\mathbf{Y}, h)} = \dot{\omega}_i\left(\widetilde{\mathbf{Y}}, \widetilde{h}\right)$ and for a grid resolution sufficiently close to the DNS requirements the above assumption can be considered as appropriate.

2.2. Solution algorithm

The computations are carried out using an in-house numerical code (SAILOR) based on the low Mach number approach which was previously well verified in gaseous shear layer flows with/without chemical reactions (see [6-9]). The spatial discretization is performed on half-staggered meshes by high-order (6th) compact difference approximation for the Navier-Stokes and continuity equations while for the chemical transport equations two different schemes are examined: 2nd order TVD (Total Variation Diminishing) with van Leer's limiters and the 5th order WENO (Weighted Essentially Non-Oscillatory). The governing equations are integrated in time using the 2nd order predictor-corrector approach combined with the projection method for pressure-velocity coupling. In the half-staggered approach the velocity components and scalar quantities are stored in the same spatial locations (nodes) while the pressure nodes are shifted to the cell centres. Compared to the fully staggered grid arrangement, where the pressure and all scalar variables are shifted to the cell centres and the velocity components are placed on different cell corners, the use of half-staggered meshes facilitates the solution algorithm as the velocity needs to be interpolated only within the projection method [6,7].

2.3. Simulation details

The calculations are performed to study the auto-ignition in the temporally evolving mixing layer between two separate streams (cold fuel/hot air) moving with opposite velocities. The initial isotropic turbulence prescribed by a three dimensional kinetic energy spectrum [2]:

$$E(k) = 16\sqrt{\frac{2}{\pi}}\frac{u'^2}{k_0}\left(\frac{k}{k_0}\right)^4 \exp\left[-2\left(\frac{k}{k_0}\right)^2\right] \tag{2}$$

was combined with the Gaussian function $\phi(y) = \exp\left(-20(y/\mathrm{L}_y)^2\right)$ and superimposed on the initial flow field providing the largest velocity fluctuations in a region of fuel/oxidizer contact zone (see figure 1). In the equation (2), $u' = \left\langle \sqrt{\mathbf{u'u'}} \right\rangle$ is the RMS of initial velocity fluctuations and k_0 is used to adjust the turbulent length scale $\lambda = \langle \mathbf{u'u'} \rangle / \langle \partial(\mathbf{u'u'})/\partial x \rangle$. We consider the cases with the turbulence intensity $T_i = 0.01U$ and with the Reynolds numbers calculated based on the initial shear layer thickness $\delta = 0.5\times10^{-3}\mathrm{m}$ equal to $\mathrm{Re}_\delta = 50, 100$. The turbulent length scales are equal to $\lambda = 0.4\delta, 0.2\delta$, respectively, which gives $\mathrm{Re}_\lambda = 0.2$. A three dimensional view of the geometry used with dimensions: $\mathrm{L}_x \times \mathrm{L}_y \times \mathrm{L}_z = 0.025\mathrm{m}\times 0.04\mathrm{m}\times 0.01\mathrm{m}$, is presented in the figure 2. The yellow geometry edges indicate the size of whole flow region while the central part of the mixing layer shows the initial temperature distribution.. The periodic boundary conditions in the x and z directions and moving walls at $y = \pm 0.02$ were assumed. The calculations were performed with the constant time step $\Delta t = 1\times10^{-7}\mathrm{s}$ on the mesh with $\mathrm{N}_x \times \mathrm{N}_y \times \mathrm{N}_z = 512\times256\times128$ nodes, stretched in transversal (y) direction. The hydrogen diluted with nitrogen ($Y_{\mathrm{H}_2} = 0.1$) at 300K was chosen as a fuel and the analysis was performed for two oxidiser (air) temperatures $T_{\mathrm{air}} = 970\mathrm{K}, 980\mathrm{K}$. The chemical reactions were modelled using detailed mechanism (9species, 21 reactions) proposed by Mueller et al. [10].

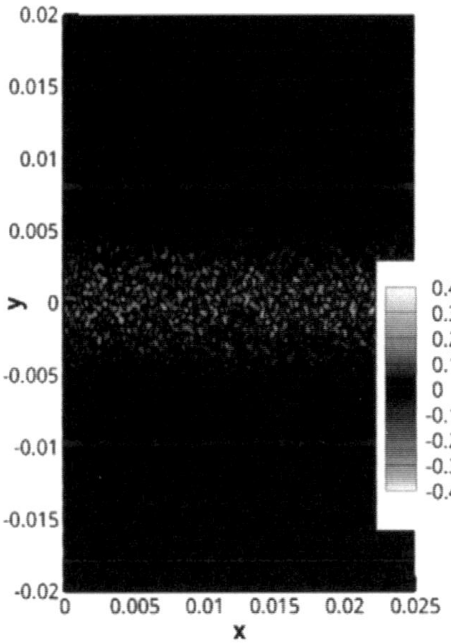

Figure 1. 2D contours of the initial isotropic turbulence field for the case with $\mathrm{Re}_\delta = 50$.

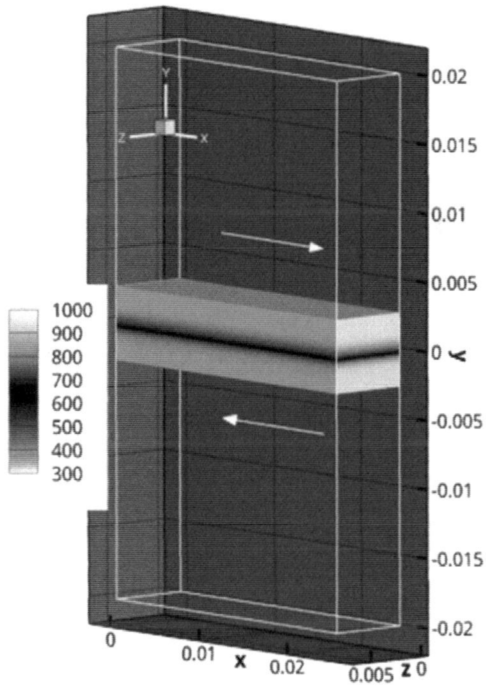

Figure 2. View of the computational domain.

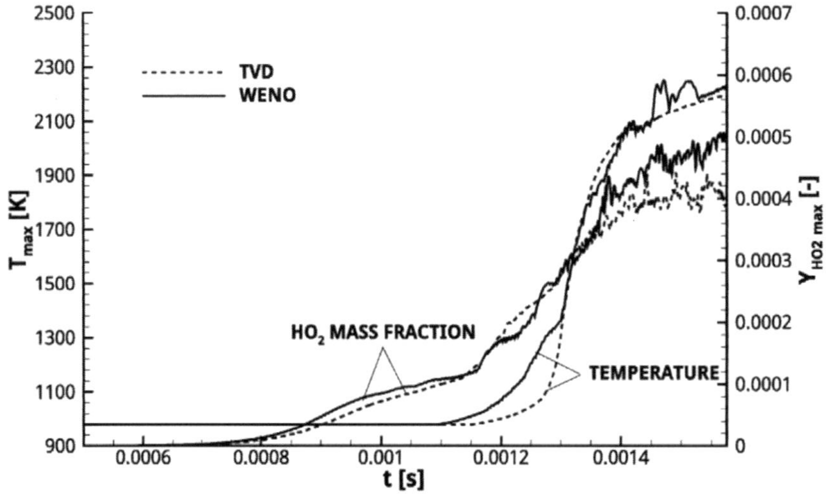

Figure 3. Time evolution of maximum temperature and HO_2 mass fraction. Results for $T_{air} = 980K$, $Re_\delta = 50$ and discretization schemes: TVD, WENO.

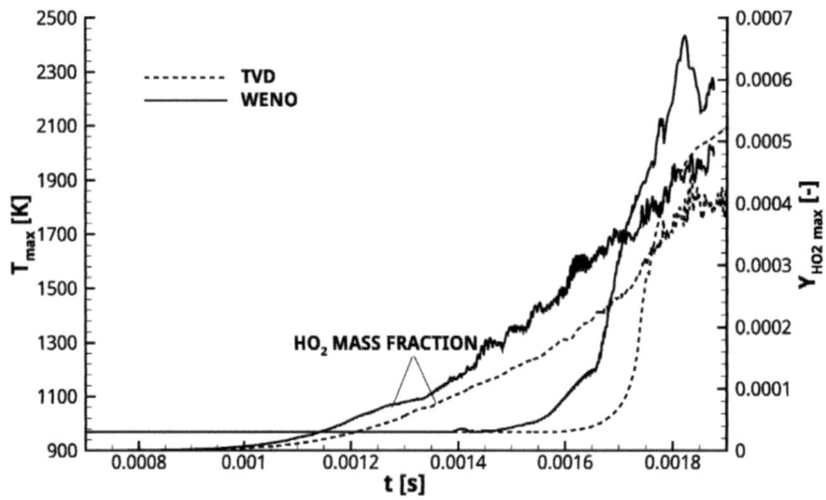

Figure 4. Time evolution of maximum temperature and HO2 mass fraction. Results for $T_{air} = 970K$, $Re_\delta = 50$ and discretization schemes: TVD, WENO.

3. Results and discussion

In all analyzed cases the simulations started from the initial conditions and continued until the auto-ignition occurred and the flame started to propagate. The obtained results clearly reveal influence of discretization schemes on the auto-ignition process and flame propagation phase. The results presented in the figures 3-9 show the time evolutions of maximum quantities in the domain (HO_2 mass fraction and temperature), 2D contours of instantaneous temperature distributions as well as 3D contours with iso-surfaces inside the flame. Figures 5-8 show the enlarged regions of the shear layer at selected time moments.

During the calculations the maximum values of temperature and species mass fractions were registered and their sudden increase indicated the auto-ignition. Figures 3 and 4 show the time evolution of T_{max} and $Y_{HO_2\,max}$ for TVD and WENO schemes, for the initial conditions corresponding to $\mathrm{Re}_\delta = 50$ and for both analyzed oxidizer temperatures. It is seen that the simulation time needed for the occurrence of auto-ignition, in the case of $T_{air} = 970\mathrm{K}$ (figure 4) is noticeably longer (≈0.4ms) compared to the case with the higher oxidizer temperature (figure 3). The profiles of mass fraction of OH radical, not presented here, starts to grow at the time instant close to the auto-ignition time indicated based on the temperature growth, which is consistent with literature [11] studies of auto-ignition processes. On the other hand the radical HO_2, the so-called pre-ignition species, appears soon before the temperature growth, which is well seen in both the figures. The results presented show that for the higher oxidiser temperature the profiles slightly depend on the discretization method and for the TVD scheme (dashed lines) their time evolution is smoother. Comparing the T_{max} profiles, the main differences are observed shortly after the auto-ignition when the flame starts to grow. The evolution of maximum HO_2 mass fraction is similar at the beginning of auto-ignition, however the profiles start to diverge at t = 1.35 ms. Figure 4 shows that in the case when the oxidizer temperature is reduced to 970K, a significant impact of numerical approach becomes more pronounced both during the auto-ignition and propagation phases. The maximum values of HO_2 and temperature obtained using the WENO scheme grow faster, achieve very high level and reveal some oscillations, contrary to rather smooth profiles obtained using TVD scheme.

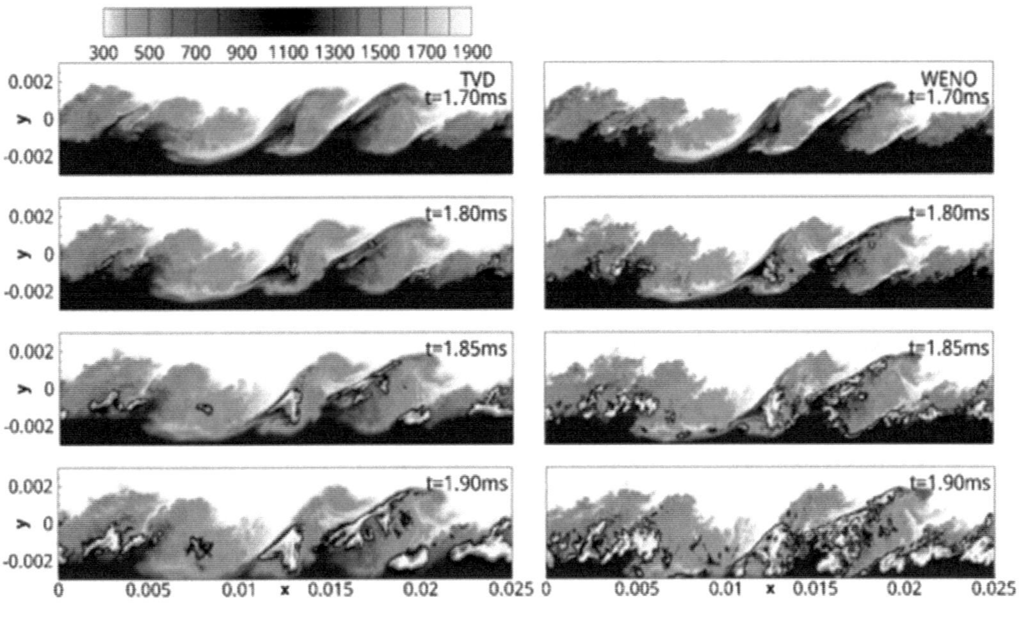

Figure 5. Instantaneous temperature contours for $T_{air} = 970\mathrm{K}$, $\mathrm{Re}_\delta = 50$ and discretization schemes: TVD (left column), WENO (right column).

Similar observations can be made analyzing the evolution of temperature in enlarged regions of 2D cross-sections located in the centre of 3D computational domain, shown in the figure 5. The time instances selected correspond to the subsequent flame development phases. It is seen that the flame develops starting from many separated ignition kernels distributed along the mixing layer. At the time instant t = 1.8 ms, corresponding to the rise of the T_{max} profiles in the figure 4, the differences observed in the temperature distributions, for both schemes are evident. Analysis of the contour plots reveals that the numerics affects both the appearance, distribution and evolution of the flame kernels. It seems that for lower oxidizer temperature, which corresponds to longer simulation time needed to the auto-ignition occurrence, which in turn enhances the mixing of fuel and oxidizer, the convective and diffusive mechanisms play an important role.

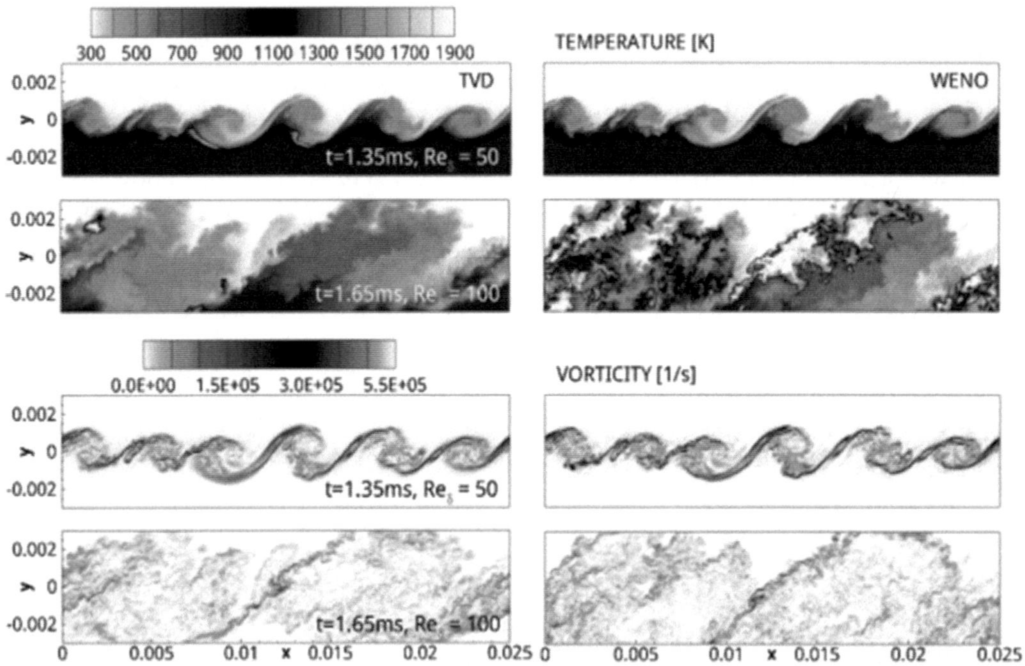

Figure 6. Instantaneous temperature and vorticity contours for $T_{\mathrm{air}} = 980\mathrm{K}$, $\mathrm{Re}_{\delta} = 50, 100$ and discretization schemes: TVD (left column), WENO (right column).

The contours of instantaneous temperature and vorticity for $T_{\mathrm{air}} = 980\mathrm{K}$, $\mathrm{Re}_{\delta} = 50, 100$ and both discretization schemes are shown in the figure 6. At first the analysis is focused on the case with $\mathrm{Re}_{\delta} = 50$. The time instances selected correspond to the flame development phases where the maximum temperature in the domain is higher, i.e. when the TVD scheme was used (see figure 3). It can be observed that in this case the use of TVD or WENO schemes leads to only minor discrepancies and the flames grow similarly. An evident rise of the temperature is observed in a thin braid connecting the shear layer and the vortex structure (cf. figure 5). The flame overlaps with the region of large values of the velocity and vorticity, where the intense mixing enhances the combustion process. At subsequent time instances the flame grows very fast and at t = 1.5 ms it is seen along the entire region of the shear layer, as can be observed in the figures 7-8 presenting the temperature contours and iso-surfaces of the temperature inside the flame.

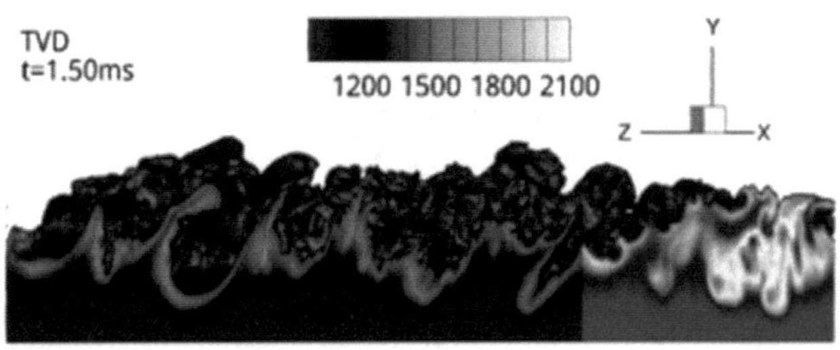

Figure 7. Instantaneous temperature contours and iso-surfaces for $T_{\mathrm{air}} = 970\mathrm{K}$, $\mathrm{Re}_{\delta} = 50$ and discretization scheme TVD.

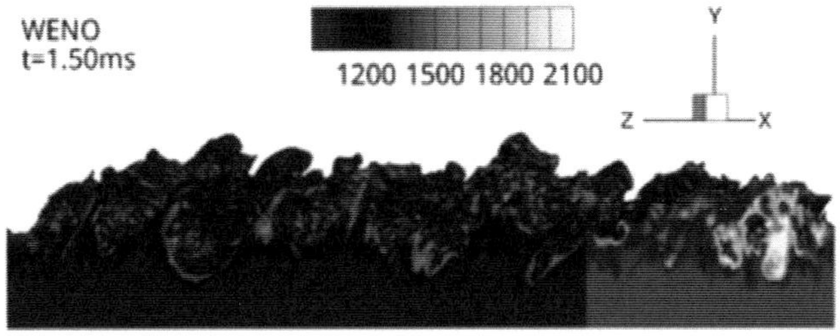

Figure 8. Instantaneous temperature contours and iso-surfaces for $T_{air} = 970K$, $Re_{\delta} = 50$ and discretization scheme WENO.

Comparing the figures 7-8 one may conclude that the main differences between the obtained solutions using TVD and WENO schemes lie in the smoother resolution of the flow field obtained using the TVD approach. Concerning the results for the flow conditions at $Re_{\delta} = 100$, the flame kernels localization and maximum values of variables describing the flow field seem to be also significantly dependent on the applied discretization scheme. In the figure 6 it is seen that the flame develops from single sites for TVD, whereas when the WENO is used, the flame appears in a large region of the mixing layer. The profiles of maximum temperature and HO_2 mass fraction, presented in the figure 9, confirm assumptions that the use of various numerical approaches may lead to large discrepancies and falsification of the results. Future studies will include a mesh sensitivity analysis. One could expect that when the ILES mesh resolution approaches a mesh resolution required for DNS an impact of the numerical method becomes definitively smaller.

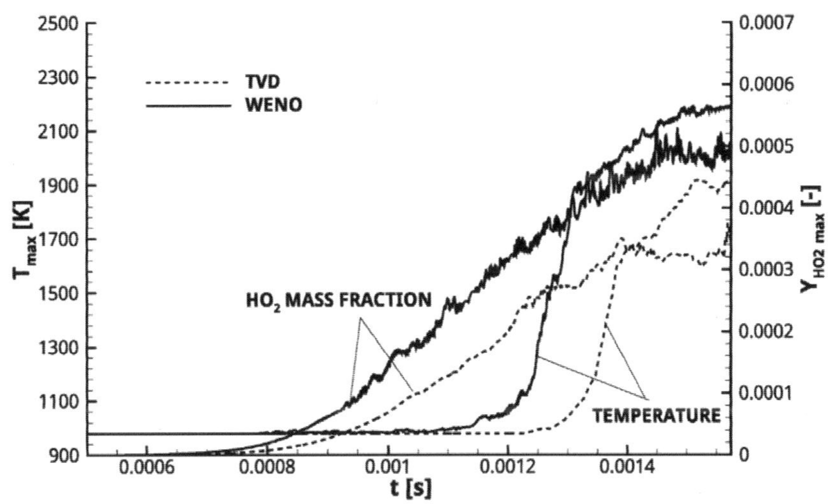

Figure 9. Time evolution of maximum temperature and HO_2 mass fraction. Results for $T_{air} = 980K$, $Re_{\delta} = 100$ and discretization schemes: TVD, WENO.

4. Conclusions

The paper presented the numerical analysis of temporally evolving mixing layer with the particular attention on the numerical impact of discretization method on auto-ignition time, localization of ignition kernels and flame propagation. The results were obtained using the TVD and WENO schemes. We analyzed the mixing layer between two separate streams (cold hydrogen/hot air) moving in opposite directions. As a initial fluctuation an isotropic turbulence was prescribed in the region of the shear layer. It was found that in the cases when the oxidizer temperature was low (970K) the numerical approach significantly affected the ignition scenario and growth of the flame. On the other hand, when the oxidiser temperature was higher (980K) the numerical method did not play such an important role and it seems that in this case the chemical kinetics was a crucial element of the simulation. However, this observations are valid only for specified initial conditions.

Acknowledgment
This work was supported by grant 2014/13/B/ST8/04246 (National Science Center, Poland) and statutory funds BS/PB-1-103-3010/11/P. PL-Grid infrastructure was used to carry out the computations.

References

[1] Mastorakos E, Barituad T A and Poinsot T J 1997 *Combust. Flame* **109** 198-223
[2] Doom J and Mahesh K 2009 *Proc. 47th AIAA Aerospace Sciences Meeting* AIAA2009-240
[3] Geurts B J 2004 *Elements of direct and large-eddy simulation* (Edwards Publishing Philadelphia)
[4] Vreman A W 2004 *Phys. Fluids* **16** 3670-81
[5] Duwig C, Nogenmyr K J, Chan C K and Dunn M J 2011 *Combust. Theor. Model.* **15** 537-68
[6] Tyliszczak A 2014 *J. Comp. Phys.* **276** 438-67
[7] Tyliszczak A 2016 *Comput. Fluids* **127** 131-45
[8] Tyliszczak A and Geurts BJ 2014 *Flow Turbul. Combust.* **93** 221-247
[9] Rosiak A and Tyliszczak A 2016 *Int. J. Hydrogen Energ.* **41** 9705-17
[10] Mueller M, Kim T and Dryer F 1999 *Int. J. Chem. Kinet.* **31** 113-25
[11] Mastorakos E 2009 *Prog. Energ. Combust.* **35** 57-97

Application of a single-fluid model for the steam condensing flow prediction

K Smołka[1], S Dykas[1], M Majkut[1] and M Strozik[1]

[1] Silesian University of Technology, Institute of Power Engineering and Turbomachinery, Konarskiego 18, 44-100 Gliwice, Poland

E-mail: Krystian.Smolka@polsl.pl

Abstract One of the results of many years of research conducted in the Institute of Power Engineering and Turbomachinery of the Silesian University of Technology are computational algorithms for modelling steam flows with a non-equilibrium condensation process. In parallel with theoretical and numerical research, works were also started on experimental testing of the steam condensing flow. This paper presents a comparison of calculations of a flow field modelled by means of a single-fluid model using both an in-house CFD code and the commercial Ansys CFX v16.2 software package. The calculation results are compared to in-house experimental testing.

1. Introduction

It is a long-known fact that a considerable part of losses arising in the steam flow through the turbine low-pressure stages is due to the two-phase character of the process. Despite the impressive progress made in the development of the computational fluid dynamics (CFD) methods, modelling the two-phase flow in the turbine blade-to-blade cascade remains a complex task. First attempts to perform two-dimensional numerical modelling of steam flows with homogeneous condensation were made by Bakhtar and Tochai [1]. In the last few years a number of new publications concerning the flow of wet steam have appeared, which proves how topical the problem is worldwide. The progress made in theoretical studies of steam flows is reflected in a series of papers concerning the extension of the computational algorithm by the viscous gas model [2] and by the polydispersed model which can be applied in the Euler-Lagrange description of the liquid phase flow [3]. An interesting survey of selected works concerning the research on wet steam was presented by Bakhtar [4]. He pointed to the confirmed substantial share of thermodynamic losses resulting from water vapour condensation in the total loss value. He also indicated the impact of droplets and their size on generation of losses due to primary and secondary condensation. A survey of issues related to the formation of a water film on the surfaces of the steam turbine blades was made by Crane [5]. Apart from offering an explanation of the phenomenon physical aspects, he turned his attention to the impact of the droplet size on the film formation. Because the film on the blade moves due to the steam-related resisting force and the centrifugal force (on the rotor), the liquid phase on the trailing edge returns to the main flow in the form of large-sized droplets. These droplets are essential for the blade erosion process.

Many works on modelling the steam flow have been devoted lately to the problem of including the droplet size distribution in the analysis, for example the studies of Gerber [6, 7] or White [8]. Describing the droplet distribution, they all used the transport equation for the dispersed phase size according to the concept presented by Hill [9]. In the calculations, all these approaches draw on the assumption of no slip between the phases. A proposal was put forward to extend this method for aerosol flows, taking the slip between the phases into account [10]. The possibility of analysing two-phase systems with different velocities of the phases and different temperatures is discussed in the work of Zywotiagin et al. [11], where conditions for the Godunov method application are defined and a one-dimensional analysis of a sudden outflow of water with high parameters is presented.

The commercial CFD tools which are now in common use, such as the Ansys CFX package, create an opportunity to perform numerical calculations of the steam condensing flow using two-phase flow models with a different degree of complexity.

One of the results of many years of research conducted in the Institute of Power Engineering and Turbomachinery of the Silesian University of Technology are computational algorithms for modelling steam flows with a non-equilibrium condensation process [12, 13]. The considered computational models either assume that droplets are small enough to omit the slip between the phases (single-fluid model) or take the slip between the gaseous and the liquid phase into account by solving all conservation equations for the two phases separately (two-fluid model) [14]. In parallel with theoretical and numerical studies, works were also started in the Institute on experimental research on the steam condensing flow [15].

The numerical investigations into the two-fluid model are still in the early stage of development and its numerical algorithm is much more complex compared to the single-fluid model. Due to that, it is difficult to carry out calculations for the two-fluid model for complex geometries and even the application of the Ansys CFX commercial code is not effective enough. Generally, the single-fluid model has been adopted for engineering applications of the steam flow with homogeneous condensation. Such calculations may be performed using in-house CFD codes or the Ansys CFX program.

This paper presents a comparison of calculations of a flow field modelled by means of a single-fluid model using both an in-house code and the commercial Ansys CFX v16.2 software package. Both codes use the same flow model, i.e. the model based on the Reynolds-averaged Navier-Stokes equations (the RANS model) supplemented with steam properties as specified in the IAPWS'97 standard, the same viscous turbulence model, the same condensation model based on the classical models of nucleation and droplet growth as described in [16] and [17], respectively, and the same numerical grids. Additionally, the calculation results are compared with those obtained from in-house experimental testing performed for the de Laval nozzle and the blade channel.

2. Single-fluid, no-slip model

Steam expansion may involve the occurrence of homogeneous condensation. One of the ways to describe such a flow process is to use a no-slip model, where the two-phase system is treated as a homogeneous mixture. The no-slip two-phase flow model describing the flow of steam with phase transitions is composed of:

- the mass, momentum and energy conservation equations for the mixture,
- the turbulence model equations,
- the transport equations for the liquid phase arising due to homogeneous condensation,
- the relations modelling the condensation process,
- the real gas equation of state.

The presented model of a no-slip flow of steam as a two-phase medium assumes that the droplet-steam relative velocity is omitted, i.e. a droplet moves at the same velocity as steam. Another assumption is that, compared to the volume of the gaseous phase, the liquid phase volume is small and may therefore be ignored. The steam pressure is equal to the liquid phase pressure, $p = p_v = p_l$. This makes it possible to determine the mixture specific volume from the following relation:

$$v = v_v(1-y) + v_l y\big|_{v_l \to 0} \cong v_v(1-y) \tag{1}$$

where y denotes the non-equilibrium moisture degree (liquid phase mass content).

Moreover, in the single-fluid model the interaction between the liquid phase and the walls limiting the flow is not taken into account, which makes modelling the water film on the walls impossible.

The flow of steam as a compressible and viscous medium is described using the Reynolds-averaged Navier-Stokes (RANS) equations. In the no-slip model, conservation equations are formulated for parameters of a fictitious fluid – a steam-water mixture, for which the following relations between water and steam quantities are valid:

$$\begin{aligned} \alpha &= \frac{V_l}{V_m} \\ \rho_m &= (1-\alpha)\rho_v + \alpha\rho_l \\ h_m &= (1-y)h_v + yh_l \\ y &= \alpha\frac{\rho_l}{\rho_m} \end{aligned} \tag{2}$$

where α is the volume content. The mixture density ρ_m is a function of steam density ρ_v, water density ρ_l and α. The mixture enthalpy h_m is defined in a similar manner. The mass content (the moisture degree) y depends on the volume content and on the water-to-mixture density ratio. In the case under analysis, the mass content value is about 10^3 times higher than the volume content.

A detailed description of the single-fluid model applied in the in-house code can be found in [13] and [14]. In the case of the Ansys CFX code, the model is described in the Ansys CFX-solver Theory Guide.

3. Flow analysis

3.1. Experiment

The experimental analysis of the flow field in the blade channel presented in this paper was conducted using the steam facility installed in the Institute of Power Engineering and Turbomachinery of the Silesian University of Technology (cf. Fig. 2).

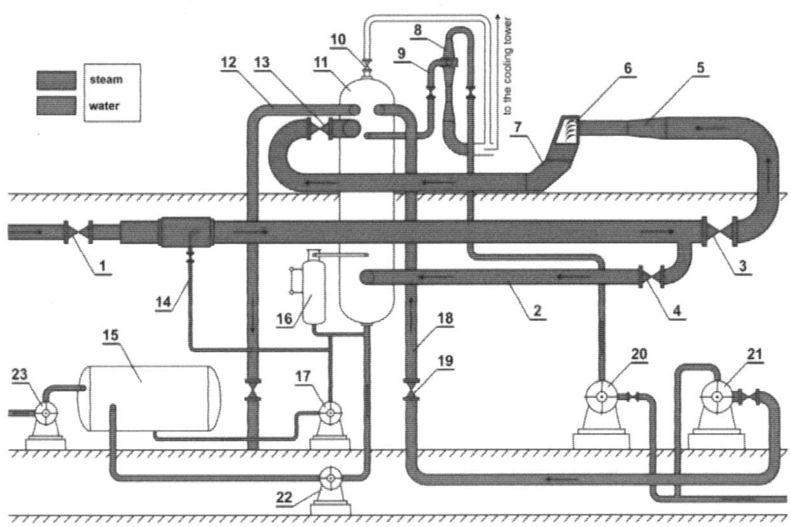

1) Control valve 2) By-pass 3) Cut-off valve 4) Cut-off valve at by-pass 5) Inlet nozzle 6) Test section 7) Outlet elbow 8) Water injector 9) Pipe 10) Safety valve 11) Condenser 12) Suction line 13) Throttle valve 14) Desuperheater 15) Condensate tank 16) Control system of condensate level 17) Condensate pump 18) Discharge line 19) Stop valve 20) Water injector pump 21) Cooling water pump 22) Condensate pump 23) Pump

Figure 1. Steam experimental facility with fittings

The wide range of changes in the channel inlet parameters (total inlet absolute pressure may vary in the range of 70–150 kPa(a), total inlet temperature – in the range of 70–150°C, steam maximum mass flow rate is about 3 kg/s) makes it possible to carry out tests of transonic steam condensing flows in different types of nozzles and/or flat blade cascades. The experimental facility and the measurement system used during the experiments are described in detail in [15] and [18].

3.2. De Laval nozzle

The nozzle geometry is described using a fourth degree polynomial $y = \sum_{i=0}^{4} a_i x^i$, where for the nozzle inlet part and for the interval of $-0.05m<x<-0.01m$, the polynomial coefficients are $a_0=1.9E\text{-}4$, $a_1=1.4E\text{-}3$, $a_2=3.67E0$, $a_3=0$, $a_4=0$, respectively. For the nozzle critical area, i.e. for $-0.01m<x<0.02m$, the values of the polynomial coefficients are $a_0=0$, $a_1=-1.365E\text{-}2$, $a_2= 3.665E0$, $a_3=-4.688E1$, $a_4=-7.268E2$, whereas for the divergent part, for $x>0.02m$, the values are: $a_0=-3.88E\text{-}4$, $a_1=5.45E\text{-}2$, $a_2=0$, $a_3=0$, $a_4=0$. The distance between the nozzle bottom and top walls in the nozzle critical cross-section is $y^*=0.04m$.

The experiment results of inlet and outlet parameters measured for two selected measuring series in the nozzle are presented in Table 1. The parameters were directly implemented in the numerical TraCoFlow and Ansys CFX codes.

Table 1. De Laval nozzle inlet and outlet parameters

Case	p_0, Pa(a)	t_0, °C	Δt, °C	p_1, Pa(a)	$Ma_{1,s}$
NS1 (nozzle, superheated steam)	$107000^{\pm 250}$	$113^{\pm 0.25}$	11.5	$21000^{\pm 250}$	1.78
NS2 (nozzle, superheated steam)				$81700^{\pm 250}$	0.72

Fig. 2 presents an example comparison of pressure distributions along the nozzle centre obtained from numerical calculations performed using the in-house (TraCoFlow) code and the commercial (Ansys CFX v16.2) program. It should be noted that due to difficulties that arise in numerical analyses using the Ansys CFX code, many correction factors modifying the nucleation rate are proposed in reference literature. One of them is the nucleation bulk tension factor β, which may be applied to limit the nucleation rate value. The numerical testing results indicate that if factor β is assumed within the limits of $1.0–1.2$, better convergence between the Ansys CFX code results and the experimental data can be achieved. Because no correction factors are used in the in-house (TraCoFlow) code, the value of factor β assumed for calculations performed by means of the Ansys CFX program is 1.

Fig. 3 presents Schlieren images obtained from the experiment and from numerical testing. It can be noticed that, like in the static pressure distribution, both the intensity and the location of the condensation wave right downstream the nozzle throat are modelled well by the applied codes, except that the in-house TraCoFlow results are closer to those obtained experimentally. Better convergence of the TraCoFlow numerical data can also be seen for the NS2 measuring series at the place where the shock wave is generated at the nozzle outlet.

In the figures below, presenting the moisture degree and the average size of the liquid phase droplets (Fig. 4) and the gaseous phase temperature (Fig. 5), it can be noticed that for the NS1 numerical test the values of both y and T_{vap} are similar along the nozzle entire length, no matter which code is applied. Discrepancies arise, however, in the values of r – the average value of r modelled in the Ansys CFX v16.2 program is by about $2{\cdot}10^{-8}m$ smaller compared to the TraCoFlow code. For the NS2 measuring series, good convergence of T_{vap} can be observed for both numerical codes only at the place where the shock wave is generated at the nozzle outlet. Additionally, a violent rise occurs on this wave in the gaseous phase temperature, which involves evaporation of a part of the droplets and a reduction in the moisture degree (Fig. 4, right).

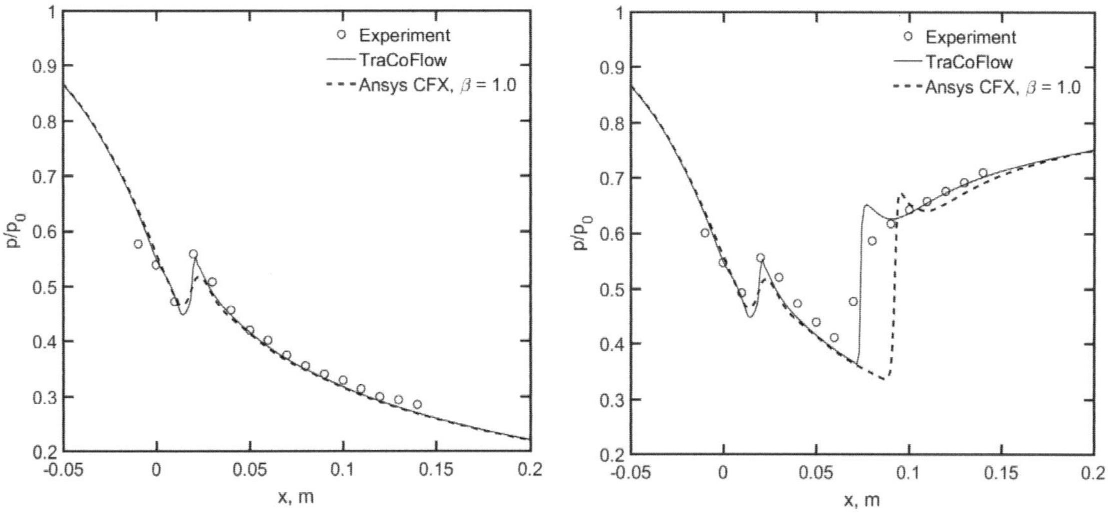

Figure 2. Static pressure distribution along the nozzle – Case NS1 (left) and Case NS2 (right)

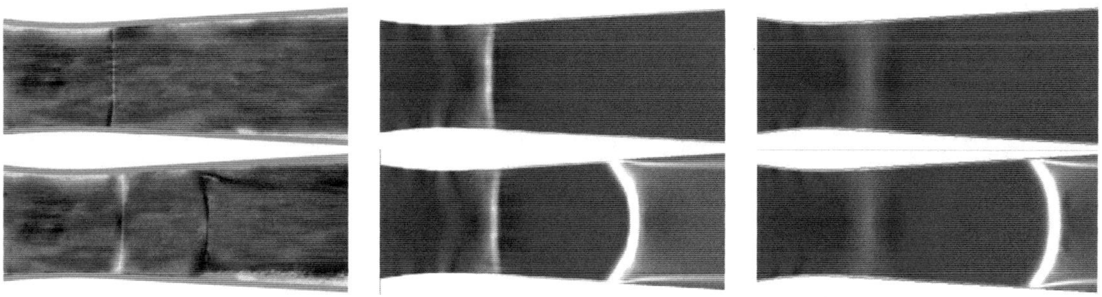

Figure 3. Schlieren distribution – Case NS1 (top), Case NS2 (bottom) – Experiment (left), TraCoFlow (centre) and Ansys CFX (right)

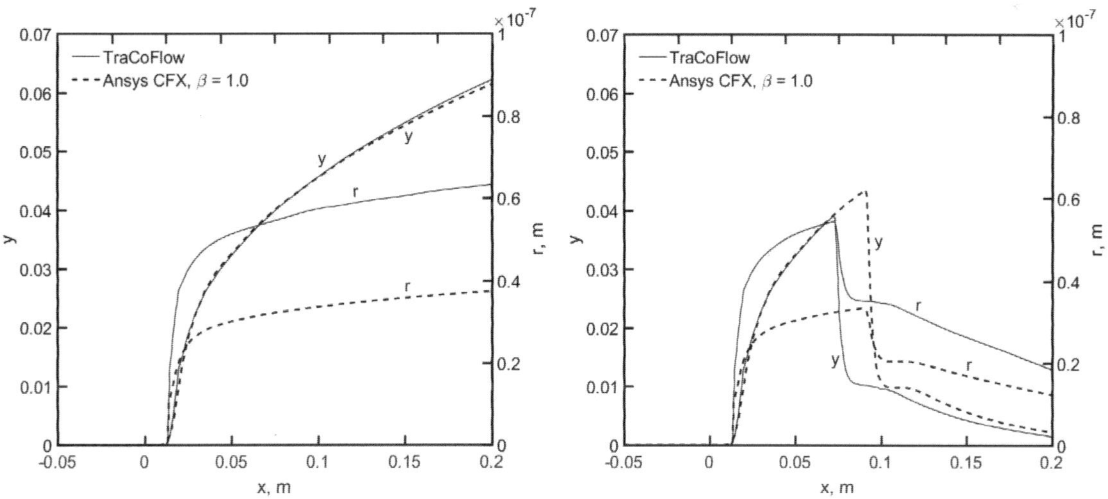

Figure 4. Moisture degree and droplet radius distribution along the nozzle – Case NS1 (left) and Case NS2 (right)

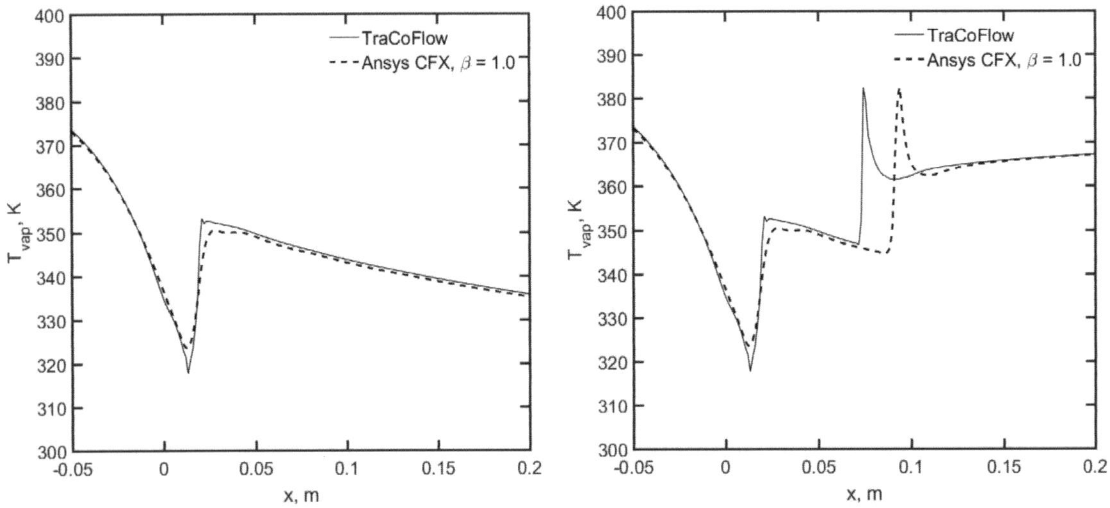

Figure 5. Vapour temperature distribution along the nozzle – Case NS1 (left) and Case NS2 (right)

3.3. Blade channel

The test section intended for analyses of the flow field in blade-to-blade channels was designed as a flat blade cascade. The geometry of the blades corresponds to that of the low-pressure part last stage stator of a real turbine. The test section is made of 4 blades, which gives 3 complete blade-to-blade channels. The blade axial chord is 173.97 mm, the blade pitch – 91.74 mm and the test section width – 110 mm. The geometrical details and arrangement of the static pressure measuring points on the surface of the blades are presented in [18].

The inlet parameters, i.e. pressure and total temperature, for the presented measuring series (cf. Table 2) correspond to about 60 ÷ 70% of the nominal load of a 200 MWe steam turbine, i.e. to 120 ÷ 140 MWe. Under such conditions, the condensation process in this turbine type begins in the stator of the low-pressure part last stage, whereas for the nominal load – the process usually commences in the penultimate stage.

Table 2. Blade channel inlet and outlet parameters

Case	p_0, Pa(a)	t_0, °C	Δt, °C	p_1, Pa(a)	$Ma_{1,s}$
CS1 (cascade, superheated steam)	$103000^{\pm 250}$	$106^{\pm 0.25}$	5.6	$42000^{\pm 250}$	1.29
CW1 (cascade, wet steam)	$114000^{\pm 250}$	$99^{\pm 0.25}$	-4.3	$60000^{\pm 250}$	1.09

In the static pressure distribution (Fig. 6), at the distance of about 0.12 m, a slight increase in pressure can be noticed on the blade suction side, which corresponds to the separation area. The next rise in pressure is related to the reflection of the shock wave formed downstream the blade trailing edge. The shock wave location in the wet steam flow is strongly dependent on the condensation process, which is triggered a little further downstream the sonic flow region. Good agreement between numerical and experimental data can be seen in Fig. 6 in the separation area for both measuring series, but it is only in the in-house TraCoFlow code that the shock wave reflection is modelled correctly. The shock waves generated past the blade trailing edge can also be seen very well in the flow field visualization using the Schlieren image (Fig. 7). It can be seen here that the shock waves have a similar nature, both for the experiment and for the numerical computations. However, in the case of the Ansys CFX v16.2 analysis they are too weak to get reflected off the neighbouring blade suction surface.

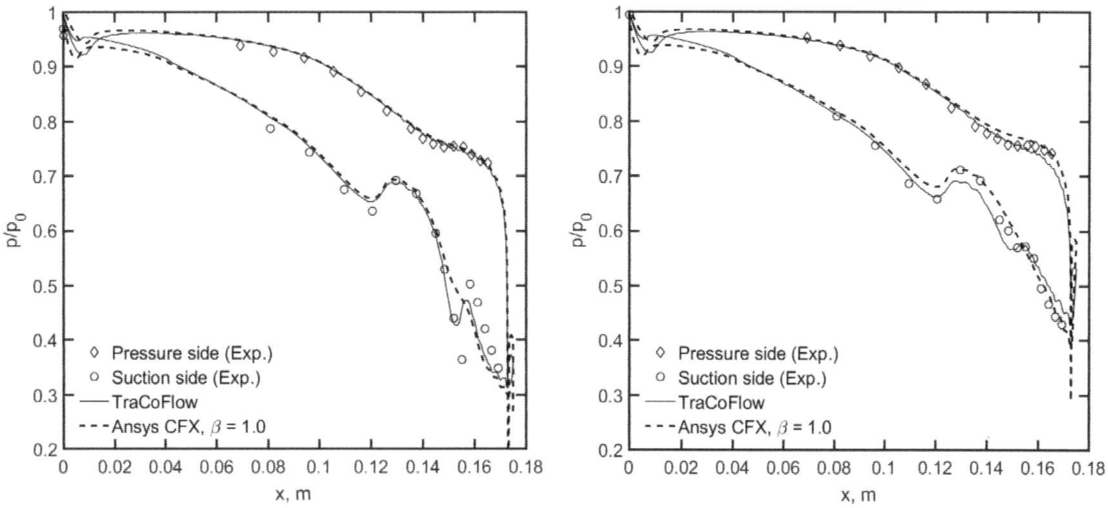

Figure 6. Static pressure distribution on the blade surface – Case CS1 (left) and Case CW1 (right)

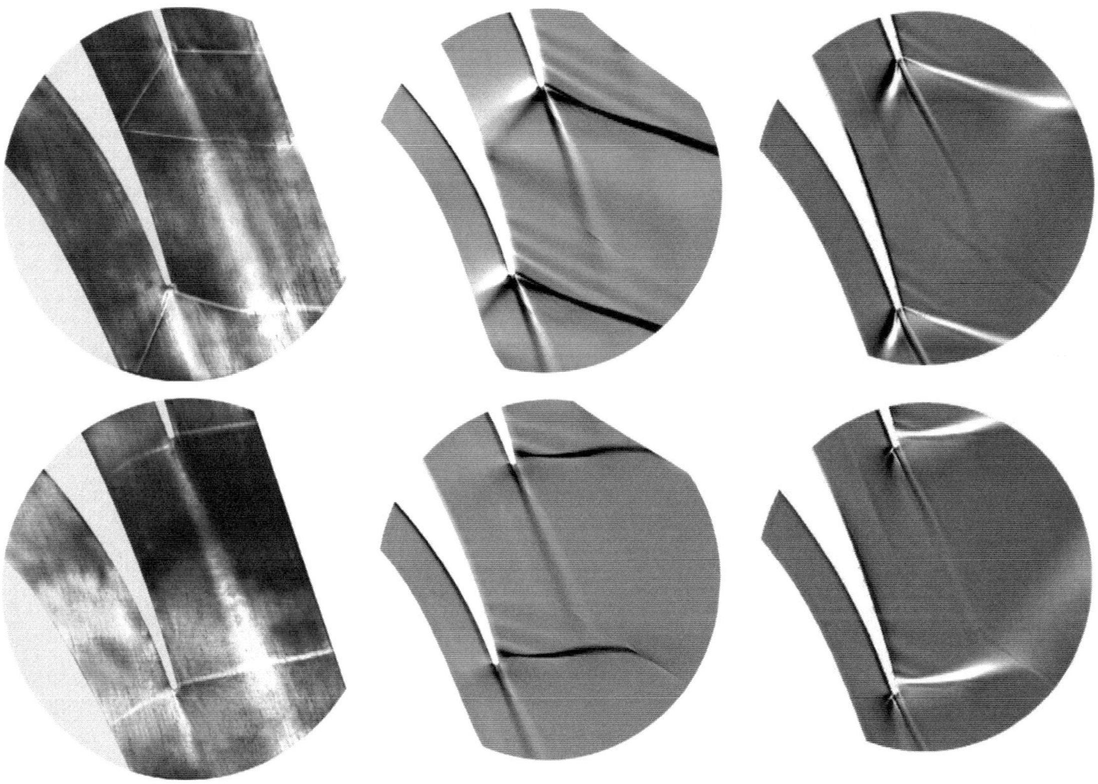

Figure 7. Schlieren distribution – Case CS1 (top), Case CW1 (bottom) – Experiment (left), TraCoFlow (centre) and Ansys CFX (right)

The following figures present the moisture degree and the liquid phase average diameter distributions (Fig. 8 and Fig. 9, respectively) obtained from numerical calculations for the blade cascade outlet part. For the CS1 measuring series, the distributions of both y and r have a similar nature in the two codes, but – like in the case of the nozzle – the values of r are smaller. By contrast, for the CW1 measuring series the results presented below are completely different for the two numerical codes. The differences are probably due to the way in which parameters are assumed at the blade channel inlet below the saturation line (Δt_{CW1} = -4.3°C). In the TraCoFlow code, the "local" equation of state is used to determine steam parameters, which enables extrapolation of the surface approximating the steam properties into the wet steam region [14]. In the Ansys CFX v16.2 code, on the other hand, steam parameters are taken as table-format values of the IAPWS'IF97 standard.

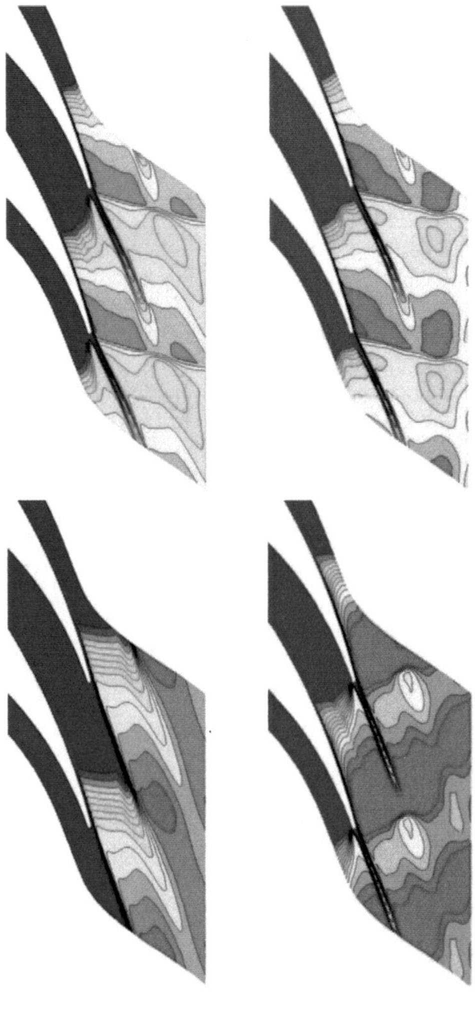

Figure 8. Moisture degree distribution – Case CS1 (top), Case CW1 (bottom) – TraCoFlow (left) and Ansys CFX (right)

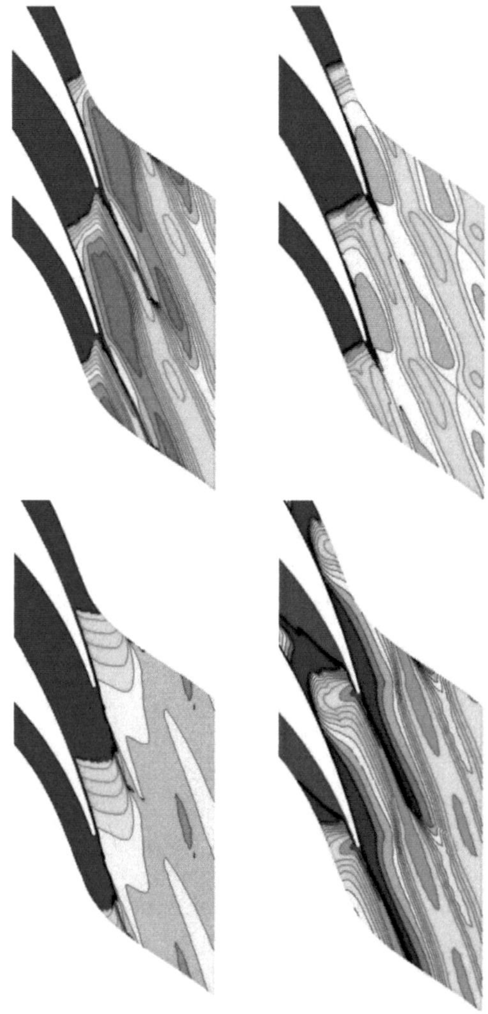

Figure 9. Droplet radius distribution – Case CS1 (top), Case CW1 (bottom) – TraCoFlow (left) and Ansys CFX (right)

3.4. Determination of losses

The quantitative assessment of losses arising in nozzles and blade channels is made herein using the entropy loss coefficient. The form of the entropy loss coefficient is derived directly from the definition of isentropic efficiency, which is defined for static parameters upstream and downstream the blade ring, where the difference between the values of real and isentropic enthalpy at the end of the expansion process is replaced with an increment in entropy, according to the second law of thermodynamics.

Journal of Physics: Conference Series **760** (2016) 012028 doi:10.1088/1742-6596/760/1/012028

$$\zeta_{S\,0-1} = 1 - \eta_{S\,0-1} = 1 - \frac{h_0 - h_1}{h_0 - h_1 + T_1(s_1 - s_0)} = \frac{T_1(s_1 - s_0)}{h_0 - h_1 + T_1(s_1 - s_0)} \tag{3}$$

The coefficient is based on the theorem that the only reliable representation of losses occurring in the flow where an adiabatic process takes place is the increase in entropy. This increase may have several sources. One of them is internal friction resulting from the flow of any viscous fluid. Apart from that, an increment in entropy may be caused by heat transfer at a finite temperature difference and also by the unbalanced nature of some processes.

The loss coefficients calculated in this manner for individual measuring series in the de Laval nozzle are listed in Table 3 together with numerically determined enthalpy and entropy values and the gaseous phase temperature in characteristic points. Additionally, Fig. 10 presents numerically determined expansion lines in the nozzle centre for the NS1 measuring series. Although the expansion lines differ slightly from each other, especially in the nozzle outlet part, the differences are so small that they have a very small impact on losses generated in the flow. Therefore, the values of ζ_S presented in Table 3 do not differ for the same measuring series. A strange thing, however, is the increment in entropy beyond the condensation region for the results obtained using the Ansys CFX code, which has no physical justification.

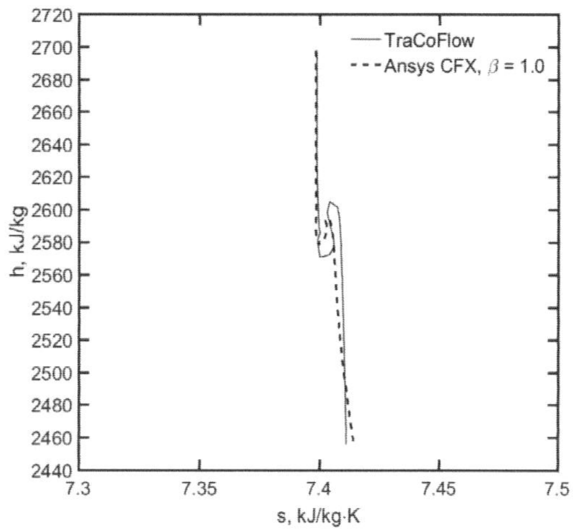

Figure 10. Expansion line along the nozzle centre – Case NS1

Table 3. Numerically calculated thermodynamic properties and the entropy loss coefficient for measuring series in the de Laval nozzle

Case	Numerical code	h_0, kJ/kg	s_0, kJ/kg·K	$h_{1,v}$, kJ/kg	$s_{1,v}$, kJ/kg·K	$t_{1,v}$, °C	y, %	ζ_S
NS1	TraCoFlow	2697.6	7.40	2455.9	7.41	61.0	6.5	0.01
	Ansys CFX	2697.5	7.40	2457.9	7.41	60.4	6.4	0.01
NS2	TraCoFlow	2697.6	7.40	2664.8	7.42	94.6	0.1	0.18
	Ansys CFX	2697.5	7.40	2664.6	7.42	94.6	0.1	0.18

Table 4 presents loss coefficients and the values of h, s and t_v needed to calculate them for individual measuring series in the blade channel. It can be noticed that the losses calculated by means of the TraCoFlow code are slightly higher compared to the results of the Ansys CFX program. This is the effect of the differences in the values of entropy (Fig. 11 and Fig. 12) and of enthalpy and the gaseous phase temperature calculated for individual numerical codes (Table 4).

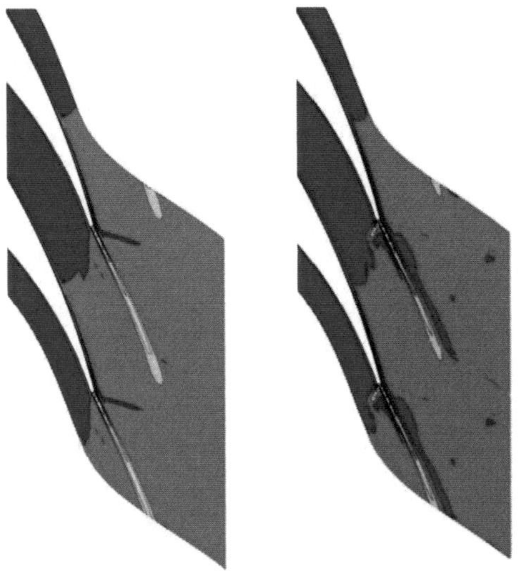

Figure 11. Entropy distribution – Case CS1 – TraCoFlow (left) and Ansys CFX (right)

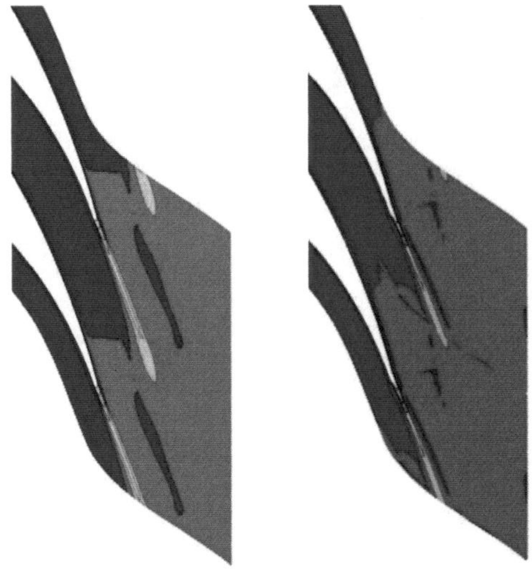

Figure 12. Entropy distribution – Case CW1 – TraCoFlow (left) and Ansys CFX (right)

Table 4. Numerically calculated thermodynamic properties and the entropy loss coefficient for measuring series in the linear blade cascade

Case	Numerical code	h_0, kJ/kg	s_0, kJ/kg·K	$h_{1,\mathrm{v}}$, kJ/kg	$s_{1,\mathrm{v}}$, kJ/kg·K	$t_{1,\mathrm{v}}$, °C	y, %	ζ_S
CS1	TraCoFlow	2687.7	7.37	2557.5	7.42	76.4	3.4	0.12
	Ansys CFX	2683.5	7.38	2552.6	7.41	76.7	3.7	0.07
CW1	TraCoFlow	2671.6	7.29	2576.2	7.32	85.8	3.3	0.10
	Ansys CFX	2667.6	7.29	2573.2	7.31	85.6	3.4	0.07

4. Summary and conclusions

Comparing the calculation results obtained using the in-house code and the Ansys CFX software with the results of the experiment, it can be noticed that both TracCoFlow and Ansys CFX v16.2 give qualitatively similar results. Among others, they model the location of the condensation wave occurrence correctly. The biggest differences can be observed, both for the de Laval nozzle and for the blade channel, in the distributions of the moisture degree, the liquid phase droplet average diameter and the gaseous phase temperature. However, quantitative differences can also be noticed between the results despite the fact that for both codes identical numerical meshes were used and the same flow and condensation models were applied. Moreover, the same IAPWS'IF97 standard was used for both numerical codes to describe the properties of steam. In order to make a quantitative assessment of the calculation results, distributions of the wet steam flow main parameters were compared and a comparison was also made between the results obtained from the experiments and the calculations. The good agreement between the numerical and experimental data justifies the conclusion that both numerical codes model the steam transonic condensing flow correctly. It should be emphasized, however, that the results of the in-house TraCoFlow code are closer to those obtained from the experiment.

Acknowledgements
The presented work was supported by the Polish National Science Centre with funds from Project UMO-2014/15/B/ST8/00203.

References

[1] Bakhtar F and Mohammadi Tochai M T 1980 An Investigation of Two-Dimensional Flows of Nucleating and Wet Steam by the Time-Marching Method *International Journal of Heat and Fluid Flow* **2(1)** pp. 5-18
[2] Simpson D A and White A J 2005 Viscous and unsteady flow calculations of condensing steam in nozzles *International Journal of Heat and Fluid Flow* **26(1)** pp. 71-79
[3] White A J and Hounslow M J 2000 Modelling polydispersed droplet spectra in nucleating steam flows *AIP Conf. Proc.* **534** (Rolla, Missouri)
[4] Bakhtar F 2006 Recent Advances in Steam Turbine Research *Proceedings of the International Conference on Energy and Environment* (Malaysia)
[5] Crane R I 2004 Droplet Deposition in Steam Turbines, *Proceedings of the Institution of Mechanical Engineers, Part C: Journal of Mechanical Engineering Science* **218(8)** pp. 859-870
[6] Gerber A G, Sigg R, Völker L, Casey M V and Sürken N 2007 Predictions of non-equilibrium phase transition in a model low-pressure steam turbine, *Proceedings of the Institution of Mechanical Engineers, Part A: Journal of Power and Energy* **221(6)** pp. 825-835
[7] Gerber A G and Mousavi A 2007 Application of quadrature method of moments to the polydispersed droplet spectrum in transonic steam flows with primary and secondary nucleation *Applied Mathematical Modelling* **31(8)** pp. 1518-1533.
[8] White A J 2003 A comparison of modelling methods for polydispersed wet-steam flow *International Journal for Numerical Methods in Engineering* **57(6)** pp. 819-834
[9] Hill P G 1966 Condensation of water vapour during supersonic expansion in nozzles *Journal of Fluid Mechanics* **25(3)** pp. 593-620
[10] Beck J C and Watkins A P 2003 The droplet number moments approach to spray modelling: The development of heat and mass transfer sub-models *International Journal of Heat and Fluid Flow* **24(2)** pp. 242-259
[11] Zywotiagin A F, Kaliniczenko S D, Kroszilin A E and Kroszilin W E 2000 Wysokotocnyj metod cisliennowo modelirowania wolnowowo istiecenia dwuchfaznych sried, *Teploenergetika* **5**
[12] Wróblewski W, Dykas S, Gardzilewicz A and Kolovratnik M 2009 Numerical and Experimental Investigations of Steam Condensation in LP Part of a Large Power Turbine *Journal of Fluids Engineering* **131(4)**
[13] Wróblewski W, Dykas S and Gepert A 2009 Steam Condensing Flow Modelling in Turbine Channels *International Journal of Multiphase Flow* **35(6)** pp. 498-506
[14] Dykas S and Wróblewski W 2011 Single- and two-fluid models for steam condensing flow modelling *International Journal of Multiphase Flow* **37(9)** pp. 1245-1253
[15] Dykas S, Majkut M, Strozik M and Smołka K 2015 Experimental study of condensing steam flow in nozzles and linear blade cascade *International Journal of Heat and Mass Transfer* **80** pp. 50-57
[16] Frenkel J 1955 *Kinetic Theory of Liquids* (New York: Dover Publications Inc)
[17] Gyarmathy G 1960 *Grunglagen einer Theorie der Nassdampfturbine* (Zürich: Juris Verlag)
[18] Dykas S, Majkut M, Strozik M and Smołka K 2015 Losses Estimation in Transonic Wet Steam Flow through Linear Blade Cascade *Journal of Thermal Science* **24(2)** pp. 109-116

Properties of velocity field in the vicinity of synthetic jet generator

P Strzelczyk[1] **and P Gil**[1]

[1] Rzeszow University of Technology, al. Powstańców Warszawy 12, Rzeszów, 35-959, Poland

E-mail: `piotstrz@prz.edu.pl`

Abstract. The paper presents the results of experimental investigation of velocity field in the vicinity of synthetic jet actuator as a function of Stokes number and for constant Reynolds number. A constant temperature hot-wire anemometer with tungsten–platinum coated single wire probe used for the velocity measurements. Synthetic jet flow visualization was presented, especially process of vortex ring development. Synthetic jet velocity profiles were compared with a solution to fully-developed pipe flow with an oscillating pressure gradient.

1. Introduction

Since the end of the last century, the synthetic jet has been the subject of both experimental and numerical investigations, usually under the name "SJ" synthetic jet [4][5] or "ZNMF" zero net mass flux [8][7]. A synthetic jet actuator consists of an oscillating driver and cavity that contains one or more orifice. This driver may be for example a loudspeaker, piezoelectric diaphragm or a mechanical piston. The device is called ZNMF because the integration of the mass flow rate across the orifice over an integer number of cycles is equal to zero. Although there is no net mass transfer to ambient fluid, the ZNMF device has the interesting property of causing a finite amount of momentum transfer to ambient fluid [2].

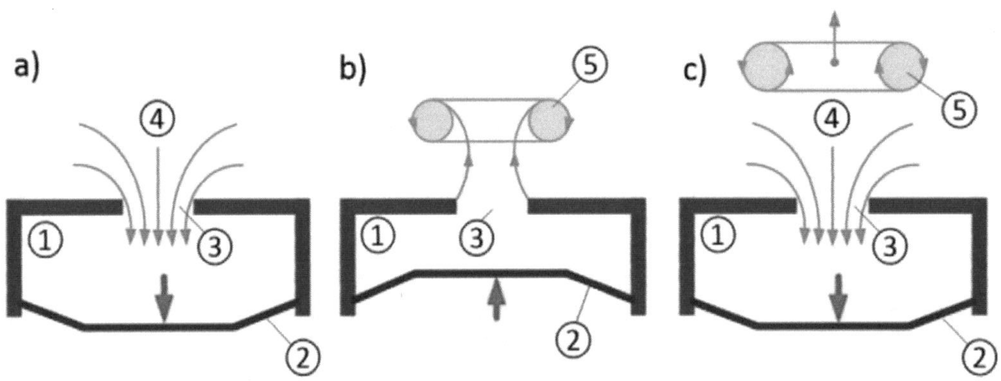

Figure 1. Schematic of a device producing a synthetic jet, a) suction stroke, b) production a vortex ring during ejection stroke c) vortex ring propagates away from the orifice: ① - cavity, ② - diaphragm, ③ - orifice, ④ - ingested fluid, ⑤ - vortex ring

Figure 1 illustrates a typical synthetic jet actuator being operated to producing a synthetic jet. If the diaphragm amplitude ② is high enough, as fluid is expelled through the orifice ③ the boundary layer separates from the wall and, at edge of the orifice, it rolls up to produce a vortex ring ⑤. This vortex ring propagates away from the orifice under its own self-inducted velocity. During the subsequent suction stroke, fluid ④ is drawn into the cavity ① from the surrounding, but the vortex ring has moved sufficiently far from the orifice so as to be relatively unaffected. A new vortex ring is then ejected and the cycle continues, producing a train of vortex rings (fig. 2a). In a time-averaged sense, this behaviour creates a jet-like flow which is synthesized from the ambient fluid (fig. 2b).

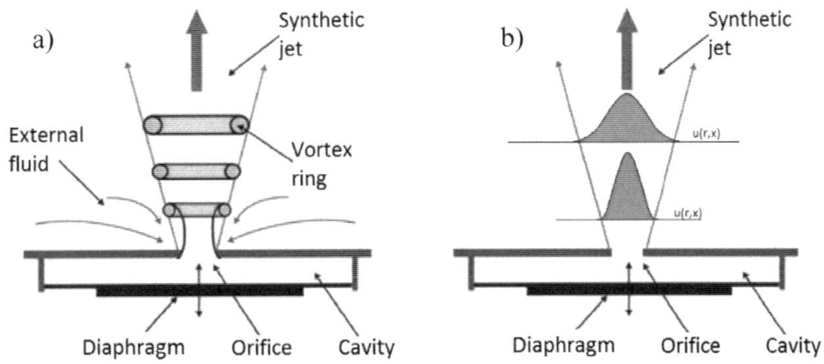

Figure 2. Synthetic jet: a) instantaneous b) time-averaged

Synthetic jet can be used to reduce aerodynamic drag. Reducing aerodynamic drag of bodies was the subject of studies presented in [9]. Synthetic jet actuator mounted in the rear bumper of the Renault Altica has been used to reduction of frontal drag, which at a speed of 130km/h has about 15% lower drag coefficient using the synthetic jet as compared to the case when the synthetic jet generator was switched off [10]. Underwater engines using synthetic jet recently attracted the attention of researchers because they solve the limitations of the current techniques of manoeuvring underwater vehicles [11]. The most promising application of synthetic jet is heat transfer enhancement, both impinging jet as well as in combination with cross-flow [12][13][14]. Synthetic jet actuator have also been used for controlling airflow in aircraft to enhance lift, increase manoeuvrability, control stalls, and reduce noise [15].

2. Experimental setup

a) Synthetic jet actuator

The cavity designed for the purpose of the present study is shown in (fig. 3). It consists of an actuator (loudspeaker STX W.18.200.8.FGX) of 150 mm diameter as the vibrating element fitted to a plexiglas sheet having a centered bore of 150 mm. The loudspeaker nominal impedance is 8Ω, and nominal resonance frequency is 37Hz. The sheets are fastened using four 8 mm bolts. The depth of the cavity H can be adjusted by adding (or removing) plexiglas plates with 150 mm bore in between. Experiments are carried out for three orifice diameters d=9, 15, 24 mm, for one orifice lengths t=5 mm, and for one cavity depths H=20 mm. Working fluid is air. We used a sharp edge orifice. The coordinate system is presented in the (fig. 3). Note that x is axial and r is radial coordinate. All connections between sheets and the loudspeaker have been sealed with silicone paste for the purpose of the leakages elimination. The actual measured resonance frequency of the loudspeaker is 38Hz.

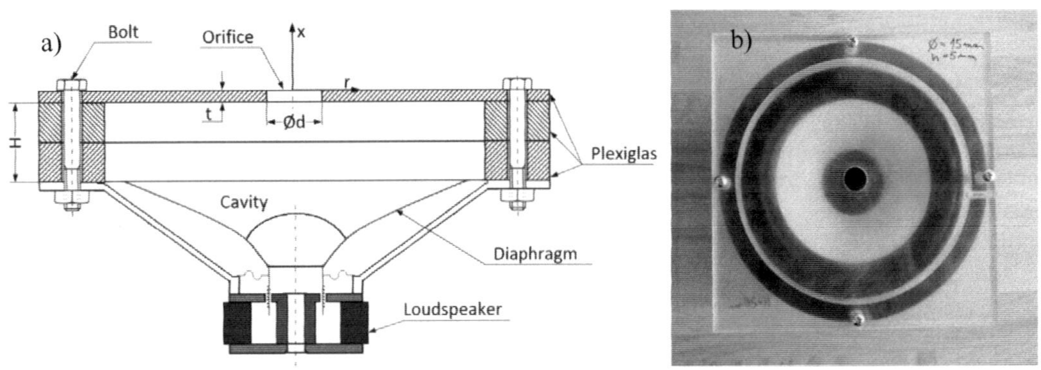

Figure 3. Schematic b) photo of the synthetic jet actuator. Note that *t* is the orifice thickness, *d* is the orifice diameter, *H* is the cavity depth, *r* – radial coordinate, *x* – axial coordinate.

A digital sinusoidal signal generated from LabView application is converted to analog signal with 16-bit 48kHz Realtek ALC889 DAC (Digital to Analog Converter). The LM3886 Texas Instruments amplifier is used to amplify analog signal from DAC and to excite the actuator. Multifunction board (Keithley KPCI-3116A) is used for the purpose of continuous monitoring of the actuator input. The measurement of current in conjunction with voltage gives electrical power supplied to the loudspeaker. The output voltage from the actuator is maintained constant for particular set of experiments. The sampling frequencies of current and voltage measurements are 32 times the actuator excitation frequency.

b) Velocity measurement

A constant temperature hot-wire anemometer (HPA 98 The Strata Mechanics Research Institute) with tungsten–platinum coated single wire probe of sensing element length 1mm is used for the velocity measurements. During calibration conducted in the low-speed wind tunnel of Rzeszow University of Technology, the reference velocity was measured with a Pitot tube connected to the FirstSensor HCLA differential pressure transducer. Measurement points are fitted with a 6th order polynomial curve with a maximum error of 2%. The hot-wire probe is mounted on a two-dimensional manipulator, which allow positioning the probe with an accuracy of 0.01 mm. The sampling frequency was at least 32 times the excitation frequency for all velocity measurements. The Constant temperature anemometer (CTA) probe was 1mm offset from orifice, such procedure turned out to be infeasible due to the hazard of the destruction of hot-wire probe (fig. 4).

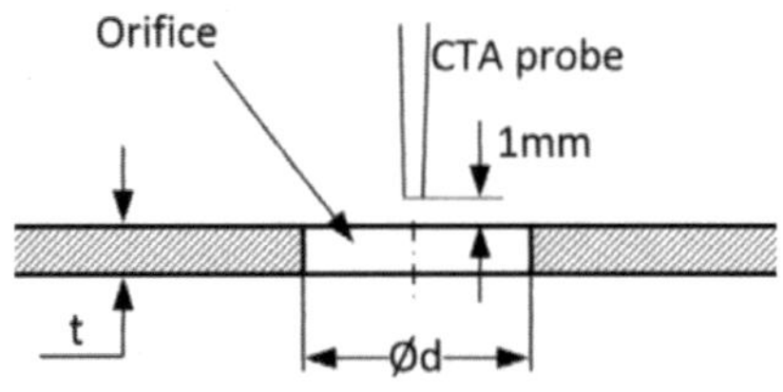

Figure 4. Constant temperature anemometer (CTA) probe position with respect to synthetic jet generator orifice exit

c) Flow visualization

Synthetic jet flow visualisation was realized utilizing light sheet and glycerine vapour (fig. 5). Flow visualization setup consists of: green laser, cylindrical lens, synthetic jet actuator, heaters soaked with glycerine and HD digital camera. Glycerine vapour was illuminated by light sheet.

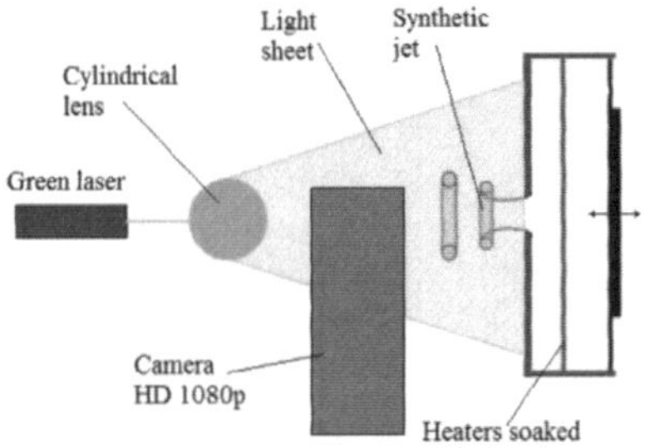

Figure 5. Synthetic jet flow visualization setup: a) schematic, b) photo

The images were captured with a digital camera with a resolution of 1920x1080 HD and 24 frames per second. Due to limitations in the recording speed, a synthetic jet was generated for a constant Stokes number Stk=22, a variable Reynolds number in range, Re = 0 ~ 3340. The uncertainty of the start of the cycle for each series is +/- 1 frame.

3. Data reduction

The Reynolds number is calculated using the procedure given by Holman et al. [1] based on spatial and time-averaged exit velocity during the ejection stroke which is the most common definition:

$$Re = \frac{U \cdot d}{\nu} \tag{1}$$

d- Orifice diameter, *ν*- kinematic viscosity.
Where characteristic velocity *U* was defined as:

$$U = \frac{2}{T} \cdot \frac{1}{A} \int_{A} \int_{0}^{T/2} (u) d\tau dA \tag{2}$$

T – Period, *A* – orifice area, *u* – instantaneous velocity
Root mean square velocity at orifice centreline:

$$U_{RMS.C} = \sqrt{\frac{1}{T} \cdot \int_{0}^{T} u^2 d\tau} \tag{3}$$

Average velocity at orifice centreline:

$$U_{AVG.C} = \frac{1}{T} \cdot \int_{0}^{T} u d\tau \tag{4}$$

The Stokes number is calculated according to the definition given by Holman et al. [1]:

$$Stk = \sqrt{\frac{2\pi f d^2}{\nu}} \tag{5}$$

f – sinusoidal signal frequency
The Strouhal number:

$$Sr = \frac{St^2}{Re} \tag{6}$$

4. Results

a) Flow visualization

In this paper synthetic jet flow visualisation was reduced to one vortex ring development in one period.

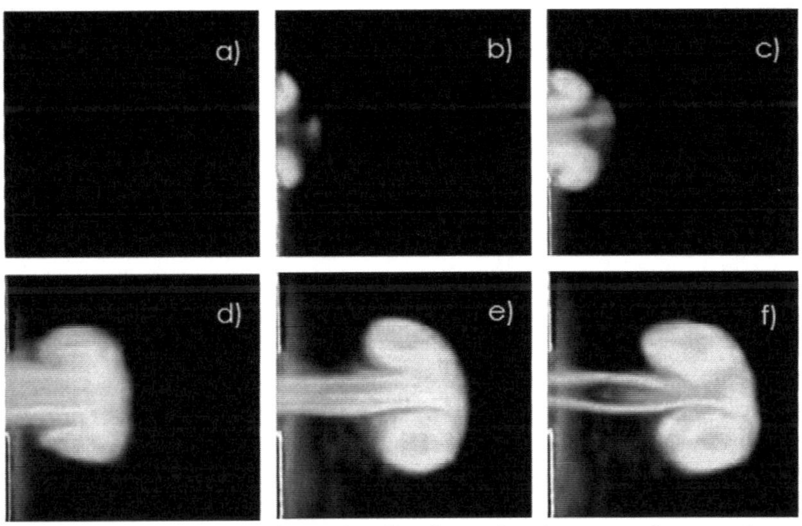

Figure 6. Synthetic jet flow visualization – development of vortex ring

Vortex ring formed in synthetic jet actuator orifice are presented in (fig. 6). Visualization of the formation of the synthetic jet shows the steps to create a coherent structure of the vortex ring. Fluid is expelled through the orifice (fig. 6a). The discharging liquid from an orifice begins to curl on the edge of the orifice (fig. 6b). Initial thin boundary layer ejected from the orifice, it begins to curl into a vortex ring (fig. 6c, d). Ejected fluid from the cavity pushes vortex ring out of the orifice (fig. 6e). On the last picture we can observe fully developed vortex ring (fig. 6f).

b) Axial velocity profile

Synthetic jet close to orifice is dominated by vortex rings which along axial distance increasing their mass and then break down to form a turbulent flow. Instantaneous velocity close to orifice consists of ejection and suction stroke, so time dependent velocity look like sinusoidal wave. Therefore average centreline velocity tends to zero for x/d→0. Average velocity reach maximum at about x/d=2, which indicate that suction stroke has area of influence up there. Root mean square velocity reaches maximum close x/d→0, after which there is appear a plateau.

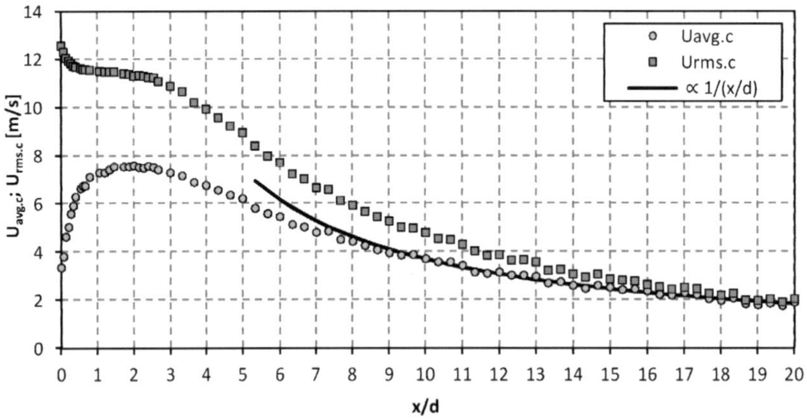

Figure 7. Streamwise variation of time average velocity and momentum synthetic jet velocity at centreline d=15mm; t=5mm; H=40mm; x/d=0÷20; r/d=0, Re=22950, 1/Sr=12.2

Chaudhari et al. [17] has shown that self-similarity of the velocity profile occurs after a distance of 7 times the orifice diameter. Mallison et al. [18] reports a value of 10d and Cater and Soria [19] report a value of 15d based on the mean velocity. The experimental results show that beyond x/d of 9 the velocity indeed decays as x^{-1}, which indicates that the jet is self-similar beyond this distance.

c) Analytical model

The problem of fully-developed pipe flow with an oscillating pressure gradient is the most reasonable exact solution available to model flows emanating from the orifice of synthetic jet actuator.
The nondimensional solution for the radial velocity profile [1]:

$$\frac{u}{u_{max}} = -i \cdot \frac{16}{Stk^2}\left[1 - \frac{J_0\left(\frac{r}{d} \cdot Stk \cdot \sqrt{-i}\right)}{J_0\left(0{,}5 \cdot Stk \cdot \sqrt{-i}\right)}\right] \cdot e^{i\omega\tau} \tag{7}$$

Where: u – velocity, Stk – Stokes number, τ – time, r – radial coordinate, d – orifice diameter, J_0 – zero order Bessel function.
This solution of Navier-Stokes equations shows the velocity profiles during the ejection stroke (fig. 7) for several Stokes number in range Stk=1÷100 obtained from the solution (7). For small values of the Stokes number Stk=1÷8 radial velocity profile is similar to a parabola (fig. 7), representing Poiseuille'a pipe solution. Then, with the increase of the Stokes number Stk>8 it can be seen that the maximum velocity does not occur in the centreline, but shifts to the edges it is known, as Richardson effect [16].

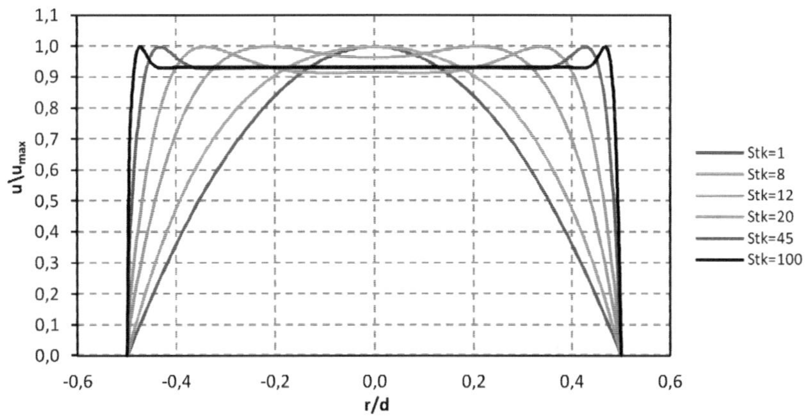

Figure 8. Normalized radial velocity profile vs. normalized radius for several Stokes number

d) Radial velocity profile

For comparison the profiles obtained from the solution of the Navier-Stokes equations (7) for an infinite tube with oscillating pressure gradient, radial velocity profiles of synthetic jet was experimentally studies for three Stokes numbers Stk=8.2; 45.1; 84.8 and constant Reynolds number Re≈4450. The sampling frequency was increased up to 360S/(s·Hz) samples per cycle, measuring and averaging 100 cycles. The sampling frequency has been chosen due to the uncertainty of angular velocity, which in this case is ± 1° for 360° range.
In (fig. 9) shows the synthetic jet radial velocity profile with the velocity profile obtained from the analytical solution (black line). As can be seen, there are discrepancies in the shape of the profile of synthetic jet for maximum velocity during the ejection stroke Φ=300°, and the mathematical model.

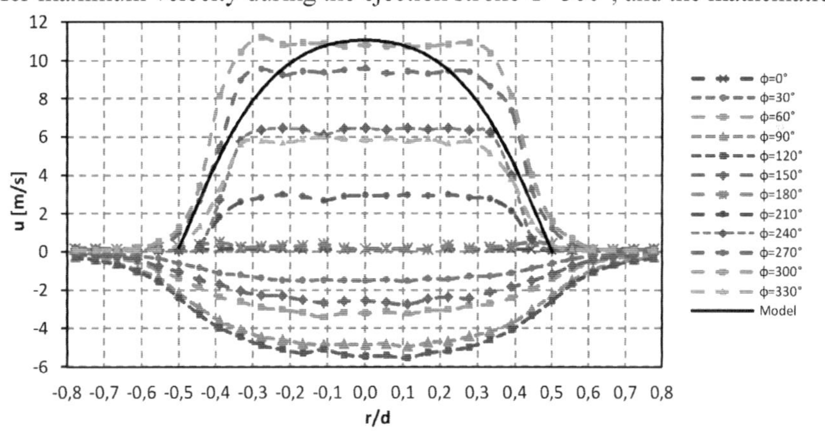

Figure 9. Synthetic jet radial velocity profile vs. normalized radius for Stokes number: Stk=8.2; Re=4334; t=5mm, d=9mm, H=40mm

Differences are due to several reasons; the first is that the radial velocity profiles were measured at the distance 1mm from orifice. Another reason is the finite length of the nozzle which in this case is 5mm (t/d=0.56), compared with an infinite length in the mathematical model may be another cause of the discrepancy.

For the Stokes number Stk=8.2 radial velocity profile obtained from a solution of the mathematical model has a flattened shape of a parabola, with no visible Richardson effect. Synthetic jet velocity profile for Stokes number Stk=8.2 has a much uniform profile with a visible Richardson effect Φ=300 °. Current velocity reaches a maximum of approximately 11m/s in the ejection stroke, while the minimum velocity in the suction stroke reaches about -5.5m/s.

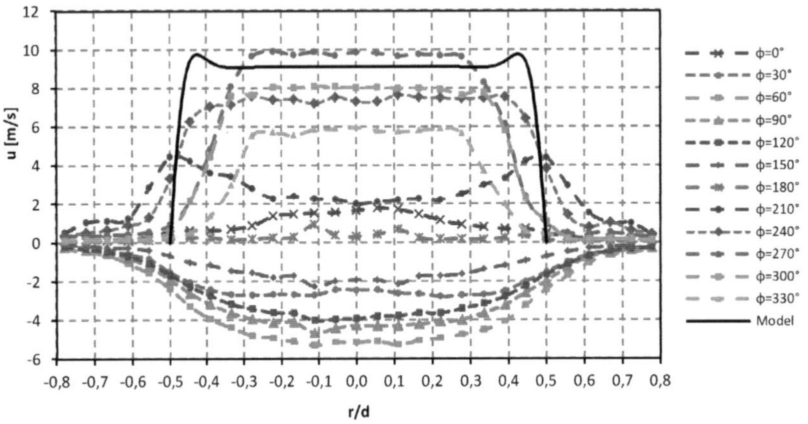

Figure 10. Synthetic jet radial velocity profile vs. normalized radius for Stokes number: Stk=45.1; Re=4687; t=5mm, d=9mm, H=40mm,

In the (fig. 10) velocity profile for the synthetic for Stokes number Stk=45.1 and Reynolds number Re=4687 with the analytical solution for the same number of Stokes number are presented. It can be seen that the profile of the synthetic jet is significantly narrower in comparison to the mathematical model solution. Furthermore there are no Richardson effects for synthetic jet velocity profile for the maximum ejection stroke. It is however clearly visible for Φ=210°. The instantaneous velocity reaches up to 10m/s in a ejection part, while the minimum velocity reaches about -5,3m/s.
In the (fig. 11) velocity profile for the synthetic jet for the Stokes number Stk=84.4 and Reynolds number Re=4458 with the analytical solution for the same Stokes number are presented.

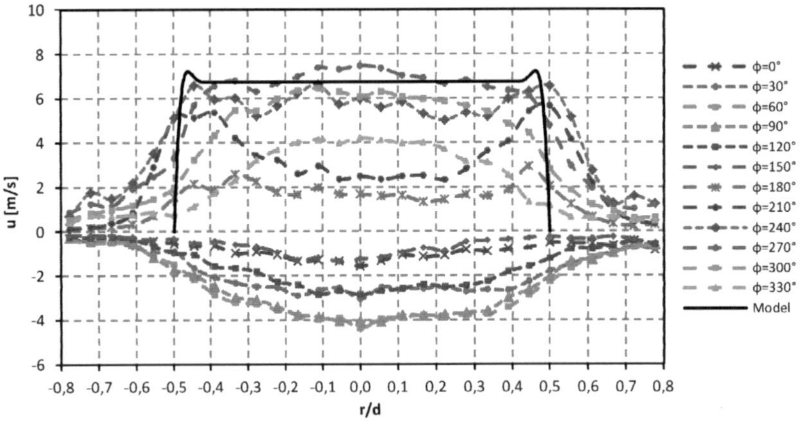

Figure 11. Synthetic jet radial velocity profile vs. normalized radius for Stokes number: Stk=84.8; Re=4458; t=5mm, d=9mm, H=40mm,

It can be seen that the Richardson effect appears at profiles Φ=180°÷240° but does not exist for the maximum velocity at ejection stroke. In addition, the Richardson's effect of synthetic jet is much broader than that implied by the analytical solution. It can be seen that the red line Φ=270° is similar in shape to the profile obtained from a mathematical model. Both models are flat. In contrast, showing significant differences in the width of the profiles during the ejection stroke which is much wider compared to the profile 1, for a low Stokes number. Velocity reaches a maximum of about 7.5 m/s, while the minimum velocity reaches about -4,4m/s.

5. Discussion
Tang et al. [3] made experimental and numerical investigation of flows result from the orifice of synthetic jet actuator. A good agreement between the prediction and the measurement was achieved. For low Stokes number radial velocity profile is remains parabolic, while for high Stokes number it is

more uniform for described solution. Experimental investigated velocity profile behave similarly but only for low Reynolds number, for high Reynolds number are less qualitative compliance.

References

[1] Holman, R., Utturkar, Y., Mittal, R., Smith, B. L., & Cattafesta, L. Formation criterion for synthetic jets. AIAA journal, 43(10), 2005, 2110-2116.

[2] Smith, B. L., and Glezer, A., The Formation and Evolution of Synthetic Jets, Physics of Fluids, Vol. 10, No. 9, 1998, 2281–2297.

[3] Tang H., Zhong S.: Incompressible flow model for synthetic jets actuators, AIAA Journal 44 (4), 2006, 908-912

[4] Jain, M., Puranik, B., & Agrawal, A. (2011). A numerical investigation of effects of cavity and orifice parameters on the characteristics of a synthetic jet flow. Sensors and Actuators A: Physical, 165(2), 351-366.

[5] James, R. D., Jacobs, J. W., and Glezer, A., "A Round Turbulent Jet Produced by an Oscillating Diaphragm," Physics of Fluids, Vol. 8, No. 9, 1996, pp. 2484–249

[6] Mallinson, S. G., Reizes, J. A., and Hong, G., "An Experimental and Numerical Study of Synthetic Jet Flow," The Aeronautical Journal, Vol. 105, No. 1043, 2001, pp. 41–49.

[7] Pack, L. G., and Seifert, A., "Periodic Excitation for Jet Vectoring and Enhanced Spreading," Journal of Aircraft, Vol. 38, No. 3, 2001, pp. 486–495.

[8] Cater, J. E., & Soria, J., "The Evolution of Round Zero-Net-Mass-Flux Jets," Journal of Fluid Mechanics, Vol. 472, 2002, pp. 167–200.

[9] Jeon S., Choi J., Jeon W. P., Choi H., & Park J. (2004). Active control of flow over a sphere for drag reduction at a subcritical Reynolds number. Journal of Fluid Mechanics, 517, 113-129

[10] Renault, Renault Altica: 44MPG Diesel Concept with Active Airflow Management, 2006, http://www.greencarcongress.com/2006/02/renault altica.html.

[11] Mohseni K., (2006). Pulsatile vortex generators for low-speed maneuvering of small underwater vehicles. Ocean Engineering, 33(16), 2209-2223

[12] Gillespie M., Black W., Rinehart C., Glezer A.: Local convective heat transfer from a constant heat flux flat plate cooled by synthetic air jets. ASME Journal of Heat Transfer 128, 2006, s. 990-1000

[13] Gil P., Strzelczyk P.: Porównanie właściwości chłodzących strugi syntetycznej i strugi swobodnej. Zeszyty Naukowe Politechniki Rzeszowskiej 291, Mechanika 87 (2/15), 2015 s.105-117

[14] Timchendo V., Reizes J.A., Leonardi E., Stella F.: Synthetic jet forced convection heat transfer enhancement in micro-channels, International Journal of Numerical Methods for Heat & Fluid Flow 17, 2007, s. 263-283

[15] Glezer A., Amitay M.: Synthetic Jets, Annual Review of Fluid Mechanics 34, 2002, s. 503-529

[16] Richardson E.G., Tyler, E.: The transverse velocity gradients near the mouths of pipes in which an alternating or continuous flow of air is established. The Proceedings of the Physical Society 42 (231), 1929, s. 1-15

[17] Chaudhari, M., Verma, G., Puranik, B., & Agrawal, A. (2009). Frequency response of a synthetic jet cavity. Experimental Thermal and Fluid Science, 33(3), 439-448.

[18] Mallinson, S. G., Reizes, J. A., and Hong, G., "An Experimental and Numerical Study of Synthetic Jet Flow," The Aeronautical Journal, Vol. 105, No. 1043, 2001, pp. 41–49.

[19] Cater, J. E., & Soria, J., "The Evolution of Round Zero-Net-Mass-Flux Jets," Journal of Fluid Mechanics, Vol. 472, 2002, pp. 167–200.

Boundary layer eruption behind the bridge pillar model

K Strzelecka[1] and H Kudela[1]

[1] Wrocław University of Technology, Faculty of Mechanical and Power Engineering, Wybrzeże Wyspiańskiego 27, 50-370 Wrocław, Poland

E-mail: katarzyna.strzelecka@pwr.edu.pl

Abstract. Experimental quantitative (local velocity measurements by Laser Doppler Anemometry) and qualitative researches (visualization by dye marker) of flow around a bridge pillar model for the Reynolds number Re_D in the range of 40 up to 350 (for laminar, transitional and turbulent flow) were conducted. Re_D was the Reynolds number referred to the diameter of the model (cylinder), D = 14,65 mm.

1. Introduction

Experimental quantitative (local velocity measurements by Laser Doppler Anemometry) and qualitative researches (visualization by dye marker) of flow around a bridge pillar model for the Reynolds number Re_D of laminar, transitional and turbulent flow were conducted. Re_D was the Reynolds number referred to the diameter of the model D = 14,65 mm (vertical cylinder placed in shallow water).

It is known from the literature (e.g. [1]) that vortex structure above the plate can lead to the boundary layer eruption. It seems that similar phenomenon was observed behind a model of bridge pillar placed in shallow water. Previous investigations of flow past a vertical cylinder placed in shallow water were carried out for low values of h_w/D (h_w is the water depth in the test section), e.g. $h_w/D \ll 1$ [2] or h_w/D = 0,5; 1; 2 [3]. Chen and Jirka conducted extensive experimental researches for the case where the cylinder diameter D greatly exceeded the water depth. Akilli and Rockwell carried out investigations using a combination of visualization by dye marker and a PIV technique. They presented visualization results of vortex formation behind a cylinder for relatively low velocities in test section resulting in the Reynolds number (based on the water depth h_w: $Re_{hw} = U \cdot h_w/\nu$, where: U – mean velocity in a tunnel, ν –kinematic viscosity of water) from about Re_{hwmin} = 500 up to Re_{hwmax} = 2500.

Both authors mentioned above and also the others referred to the problem of flow past cylinder in shallow water (e.g. [4–6]) didn't investigate the vortex interaction with boundary layer and the boundary layer eruption. The phenomenon of nonstationary separation of a boundary layer leading to the eruption of fluid elements from the wall neighbourhood to the main flow was a subject matter of other researchers (e.g. [7–9]).

To put more light into the processes behind a model of bridge pillar for the Reynolds number Re_D of laminar, transitional and turbulent flow, authors of this paper decided to widen experiments on that subject.

2. Experimental setup

Experiments were conducted in water tunnel. Schematic of experimental setup working in closed circle is shown in Figure 1. In order to clean water in a system, part or whole mass of the water can be exchanged. It is significant when carrying out visualization by dye marker researches, as colouring water can lead to gradual loss of contrast in recorded images.

Water is pumped from bottom reservoir (*BR*) by one, two or three pumps (P_1, P_2, P_3) – depending on the required flow rate, than it flows through the filter (*F*), rotameter (system is equipped with four rotameters of measuring ranges: R_1: 2–19 l/h, R_2: 18–180 l/h, R_3: 170–1700 l/h, R_4: 1500–15000 l/h), pressure vessel (*PV*) to damp pressure fluctuation (coming from pumps) and flow regulating valve (*FRV*) to inlet collector (*IC*). End of the pipe supplying the inlet collector is perforated and placed inside the collector to avoid excessive disturbances.

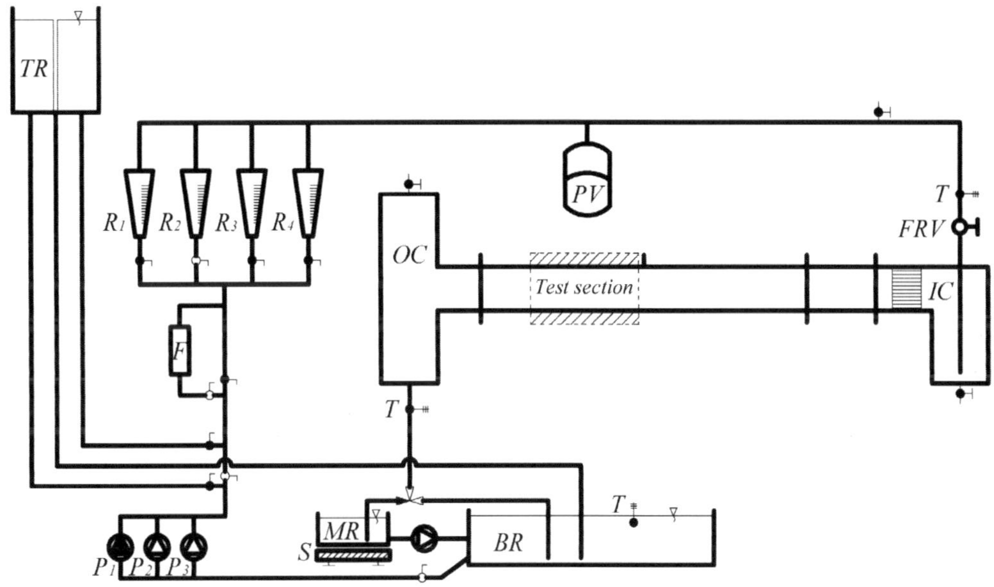

Figure 1. Schematic of experimental setup.

Rotameters fulfills the role of indicators. The accurate volumetric flow rate is done by mass measurement method. For low flow rates water is directed from the bottom reservoir (*BR*) to the top reservoir (*TR*), where the liquid level is fixed by overfall, than it flows through the rotameter (*R*) and flow regulating valve (*FRV*) to the inlet collector (*IC*). Next water inflows to the section consisting of two packets of drinking straws (5 mm in diameter) where disturbances are reduced. The length of each packet was determined in conformity with [10]. Then water flows through the Witoszynski nozzle behind which uniform velocity profile is formed. Next the liquid inflows to test section. The main test section built of plexi is H = 100 mm in depth, B = 100 mm in width and L = 2000 mm in length. Next water outflows to the outlet collector (*OC*), measuring reservoir (*MR*) placed on scales (*S*) – to determine the flow rate by mass method – and finally to bottom reservoir (BR). Temperature of the liquid is measured in three points: at inflow to the inlet collector (*IC*), at outflow from the outlet collector (*OC*) and in bottom reservoir (*BR*).

In order to determine the structure of flow in the area of the bridge pillar model (cylinder) impact, measurements of local averaged velocity (average in time) in the range of Reynolds numbers $Re_D \in \langle 40;350 \rangle$ were conducted. Re_D values characterize flow in the tunnel with respect to the cylinder diameter D and, therefore laminar flow, laminar-turbulent transition as well as turbulent flow occur in test section [11]. Quantities characterising tested flows during anemometric researches are presented in Table 1. All the anemometric measurements were carried out for constant water depth in a tunnel h_w = 65 mm. Figure 2 shows schematic of test section for anemometric measurements.

Table 1. Fluid flow's hydrodynamic parameters for anemometric researches, water temp. $T = 22°C$.

Re_D	U	h_w	h_w/D	q_V
-	mm/s	mm	-	dm^3/h
40	2,61	65	4,44	61
100	6,53	65	4,44	153
150	9,79	65	4,44	229
250	16,32	65	4,44	382
350	22,84	65	4,44	535

Instantaneous local velocities were measured in 7 cross sections placed in a tunnel behind the bridge pillar model in the distances: 1*D*, 2*D*, 4*D*, 8*D*, 12*D*, 16*D* and 20*D* (Figure 2(b)). The number of measuring points spaced across the width of the tunnel *B* was 39. Additionally, measurements of instantaneous local velocities in a cross section designated as "cylinder axis" were performed. In this case, measurements were made at 17 points placed between the tunnel wall and the cylinder wall.
Laser Doppler Anemometry was the method applied for measuring local velocities (Figure 2(a)) in subsequent measuring points [12–14]. Data obtained from measurements (number of measurements $n = 500$ values of seeds particles velocity in each measuring point), were recorded in the form of text files and further processed. The obtained results enabled to determine the time-averaged local velocities (u), and then their profiles (relative values applied to mean velocity U, where $U = q_V/(h_w \cdot B)$) in the subsequent cross sections.

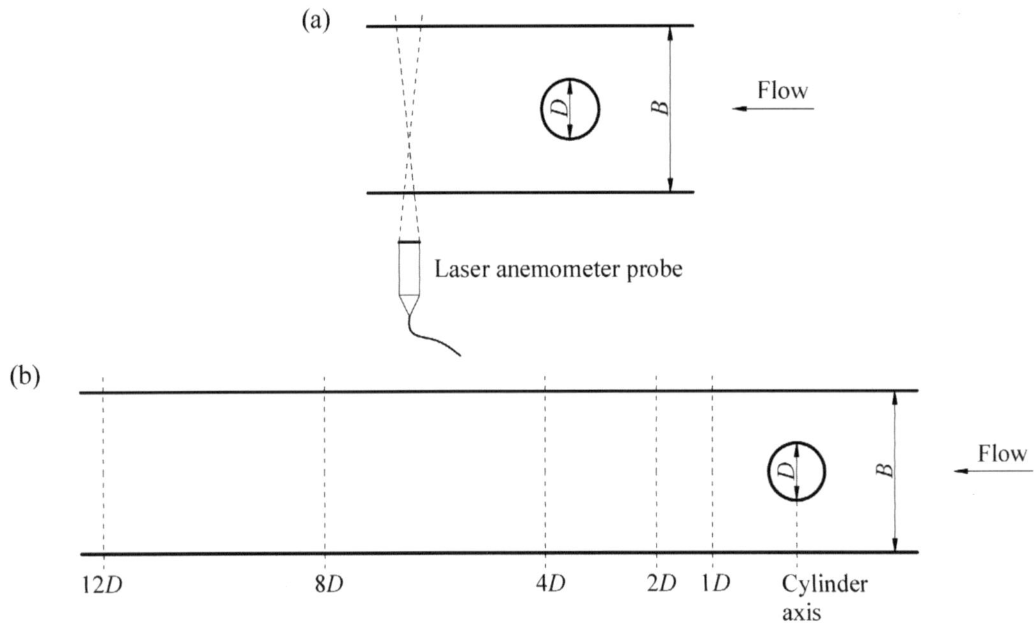

Figure 2. Schematic of test section for anemometric measurements: (a) plan view, (b) chosen cross sections in which researches were provided.

Vortex formation from a vertical cylinder in shallow water was investigated using visualization by dye marker. The vertical cylinder $D = 14,65$ mm in diameter was separated from the entrance to the test section and located at a distance of 1000 mm from it to ensure the fully formed velocity profile. Schematic of test section for visualisation experiments is shown in Figure 3. The dye injection was placed in the base of cylinder and carried out by means of the infusion pump.
Dye marker – through a tube ended with a needle – is placed inside the cylinder. The dye flows out from the cylinder at its bottom and is introduced into the boundary layer area (Figure 3 (c)). One

should be careful of the dye marker flow rate, not to have it very high and consequently not to disturb the formation of the vortex structure that is going to be observed behind the obstacle.
Visualization researches were carried out also for five values of the Reynolds number Re_D, but in contrast to anemometric measurements, they were extended to various water depth in the tunnel. Two visualisations were made for each Reynolds number as a consequent. In the first case the water depth was h_w = 65 mm, then it was reduced to h_w = 30 mm. The purpose of reducing the tunnel filling is to demonstrate the difference in the phenomena occurring in the flow around the cylinder without changing the Re_D, that is, assuming the diameter of the cylinder as a characteristic quantity. Flow characteristic values during visualization researches are summarized in Table 1.

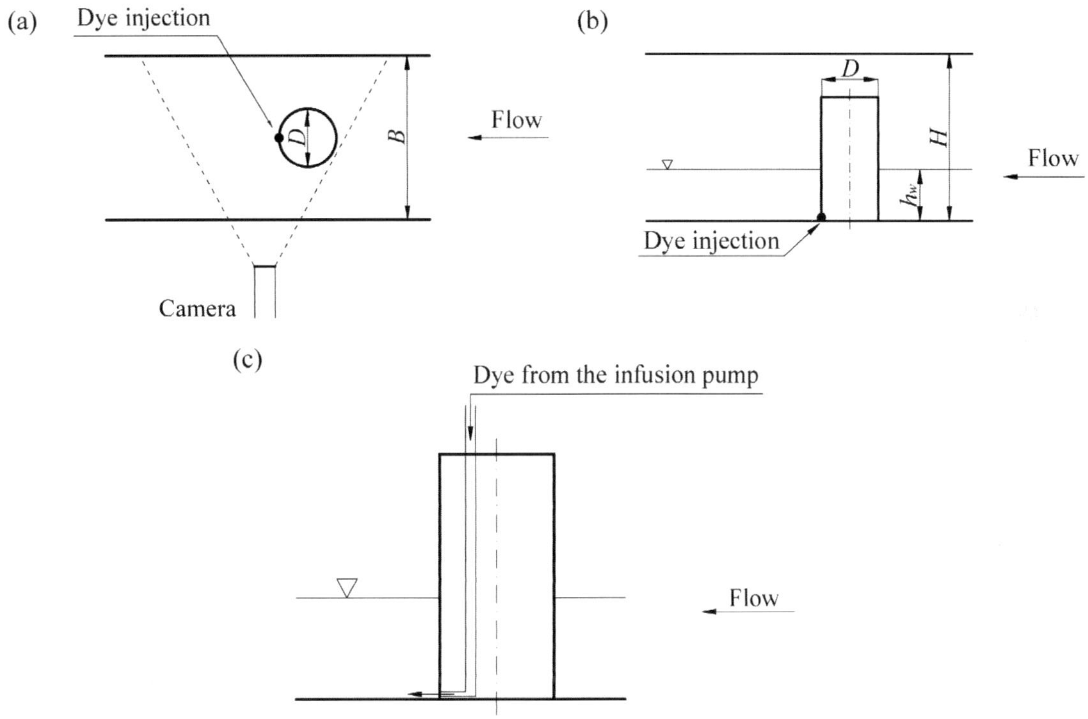

Figure 3. Schematic of test section for visualization experiments: (a) plan view, (b) side view, (c) dye marker letting in.

Table 2. Fluid flow's hydrodynamic parameters for visualisation researches, water temp. T = 22°C.

Re_D	U	h_w	h_w/D	Re_{hw}	q_V
-	mm/s	mm	-	-	dm^3/h
40	2,61	65	4,44	177	61
		30	2,05	82	28
100	6,53	65	4,44	444	153
		30	2,05	205	70
150	9,79	65	4,44	666	229
		30	2,05	307	106
250	16,32	65	4,44	1109	382
		30	2,05	512	176
350	22,84	65	4,44	1553	535
		30	2,05	717	247

3. Results of quantitative researches

Average local velocity measurements made by Laser Doppler Anemometry allowed to determine the velocity profiles in 7 cross sections behind the cylinder (1*D*, 2*D*, 4*D*, 8*D*, 12*D*, 16*D* i 20*D*) as well as velocity distribution in the cross section signed as "cylinder axis" – at the distance from the wall of the tunnel to the cylinder wall. The analysis of quantitative measurements was carried out for five values of the Reynolds number Re_D: 40, 100, 150, 250 and 350. Figures 5, 7, 9, 11, 13 illustrate dimensionless velocity distribution obtained for examined Re_D in successive cross sections.

Average velocity distribution in the cross section signed as "cylinder axis" for subsequent Re_D are presented in Figures 4, 6, 8, 10, 12. It should be noted that in this cross section the dimensionless velocity distribution is the most disturbed for Re_D = 40, 100 and 150. For higher values of Re_D disturbances are noticeable only at the measuring points in the immediate vicinity of the tunnel wall and the wall of the cylinder.

For the Reynolds number of 40 (Figure 5), a recirculation zone was observed in the area of wake just behind the obstacle, in cross section 1*D* (Figure 5 (a)). This zone decreases with the distance from the obstacle and disappears in the distance of 8*D* behind the body flown around. In addition, in each of the cross sections till the distance of 16*D* the reverse flow zones near the tunnel walls were observed. In a distance 20*D* velocity profile is approximately formed.

For Re_D = 100 (Figure 7) significantly greater disturbances are visible than it was in the case of Re_D = 40. The recirculation zone in the wake of the cylinder has a slightly smaller width than that one observed for Re_D = 40, but here it reaches further – till the cross–section 2*D* (Figure 7 (a)–(b)). The reverse flow near the wall of the tunnel are observed till the distance of 8*D* (Figure 7 (a)–(d)). In the distance of 12*D* behind the cylinder disturbances of the velocity profile are still visible, velocity profile stabilizes at a distance of 16*D* and it looks similar in cross section 20*D*.

Figure 9 shows the dimensionless velocity distribution for Re_D = 150. Both the shape of the dimensionless velocity profiles in successive measuring cross sections, and the range of the recirculation zones are here similar to those in the case of Re_D = 100.

For Re_D = 250 negative velocity values were measured in the area of the cylinder wake at a distance of 1*D*, 2*D* and 4*D*, thus the recirculation zone reaches further distance than it was for smaller values of the Reynolds number. There are no reverse flows observed near the walls of the tunnel. Velocity distribution looks very similar for the next examined Reynolds number, Re_D = 350. Here the recirculation zone is also observed only in the wake of the cylinder and reaches the cross–section 4*D*. For both values of the Reynolds number the zone of reverse flows is much wider than for smaller values of Re_D. For Re_D = 250 and 350 velocity profile stabilizes at a distance of 12*D*. In the cross sections 16*D* and 20*D* shapes of the velocity profiles are already very similar to each other.

The recirculation zone in the wake of the cylinder was observed for each examined Re_D Reynolds number. This zone is the more extensive the higher the Re_D is. That is connected with the regimes of flow around a cylinder, where laminar vortex street is observed for Re_D between 40 and 200, than transition to turbulence in the wake occurs for Re_D between 200 and 300 and finally the wake becomes completely turbulent for Re_D higher than 300 [11].

4. Results of qualitative researches

Qualitative researches performed with the use of visualization by dye maker were conducted for the same values of the Reynolds number Re_D as quantitative (LDA) measurement, but for two water levels (h_w = 65 mm, h_w = 30 mm) – in order to demonstrate the differences occurring in the flow with a Re_D Reynolds number (referred to the diameter *D* of the model) immutability.

Exemplary images of flow behind the bridge pillar model for subsequent Re_D and both water depths are presented in Figures 14–18.

For Re_D = 40 and h_w = 65 mm (Figure 14 (a)) the dye marker spreads along the bottom surface of the test section, in the area of the boundary layer. Then a thin thread of the dye marker is slowly displaced from the boundary layer area. This thread of dye is taken away and dissipated by the main stream of water flowing in a tunnel at a height about 1/3 of the water depth.

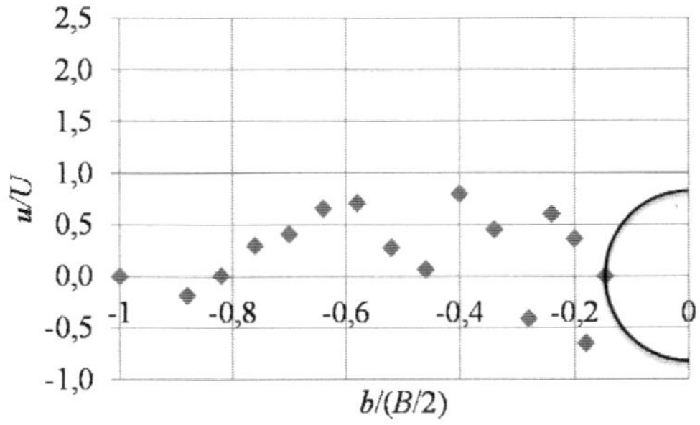

Figure 4. Average velocity distribution in the cross section of the cylinder axis for $Re_D = 40$

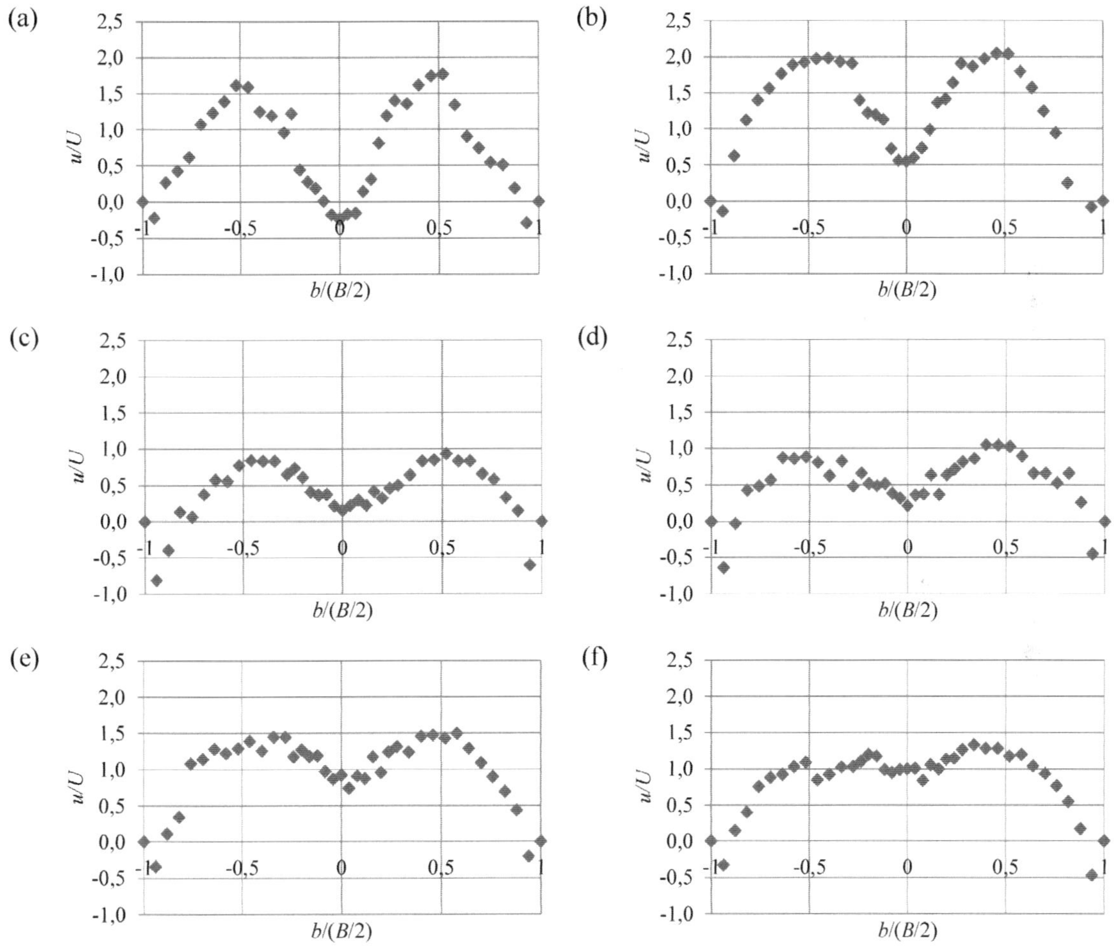

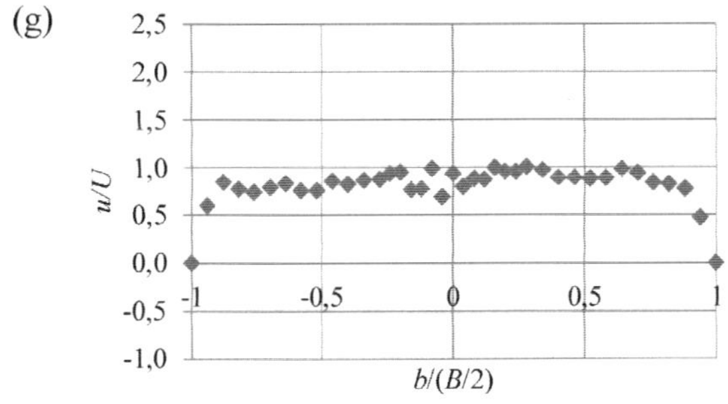

Figure 5. Average velocity profile evolution behind the cylinder for $Re_D = 40$ in cross sections: (a) 1*D*, (b) 2*D*, (c) 4*D*, (d) 8*D*, (e) 12*D*, (f) 16*D*, (g) 20*D*

Journal of Physics: Conference Series **760** (2016) 012030 doi:10.1088/1742-6596/760/1/012030

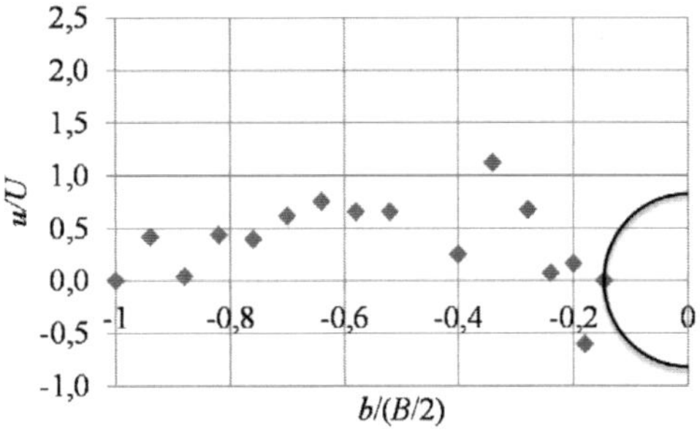

Figure 6. Average velocity distribution in the cross section of the cylinder axis for $Re_D = 100$

(a)

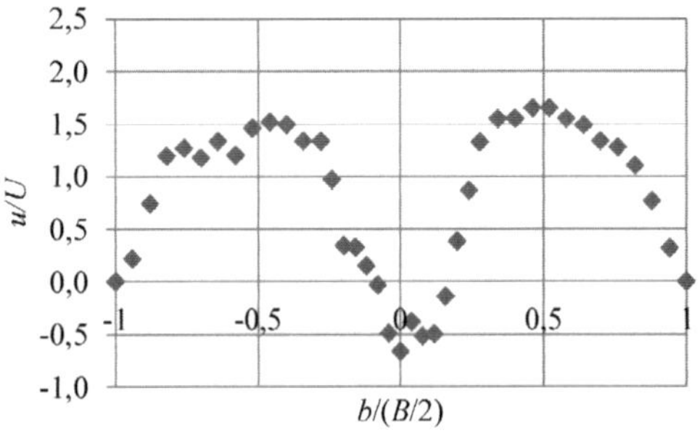

(b)

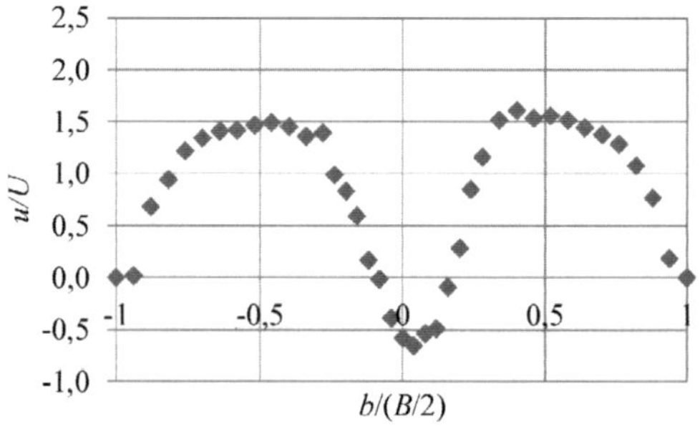

(c)

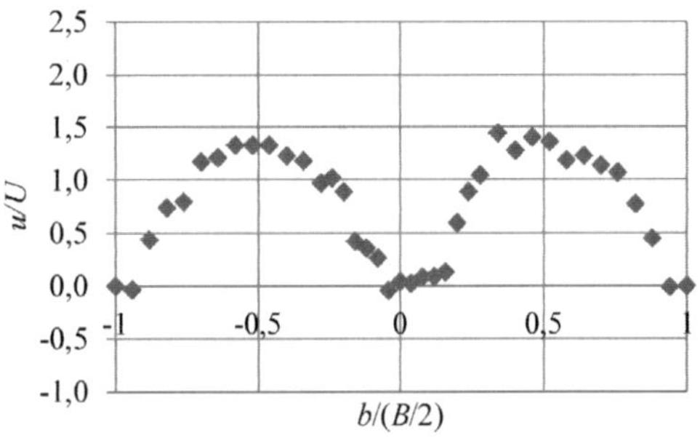

(d)

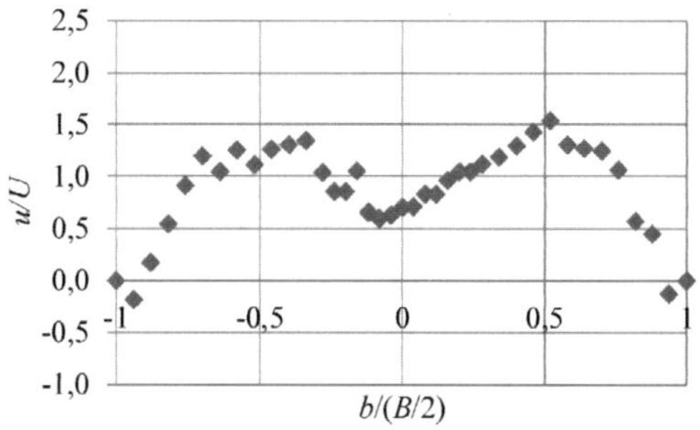

(e)

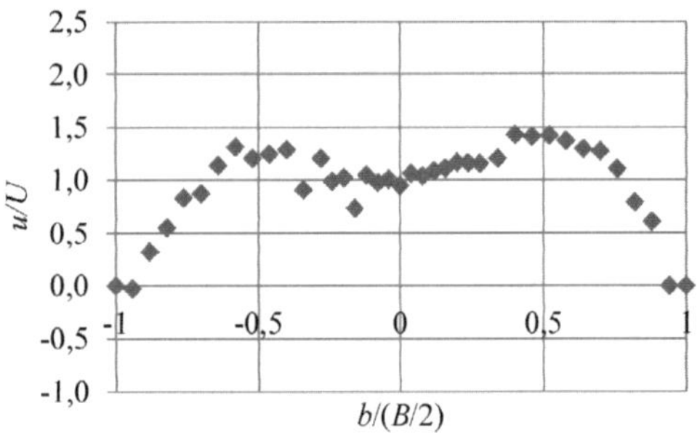

(f)

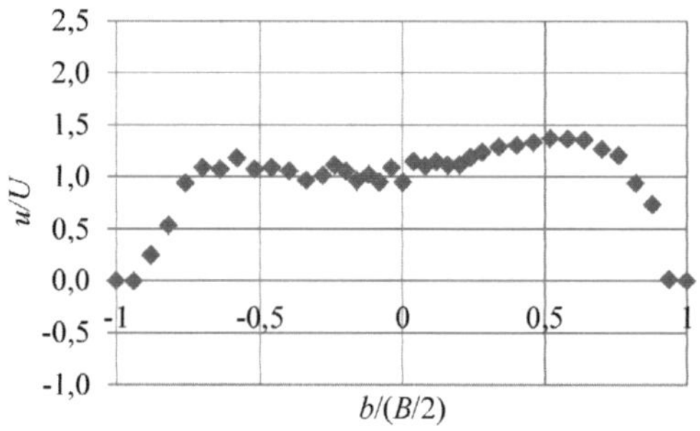

(g)

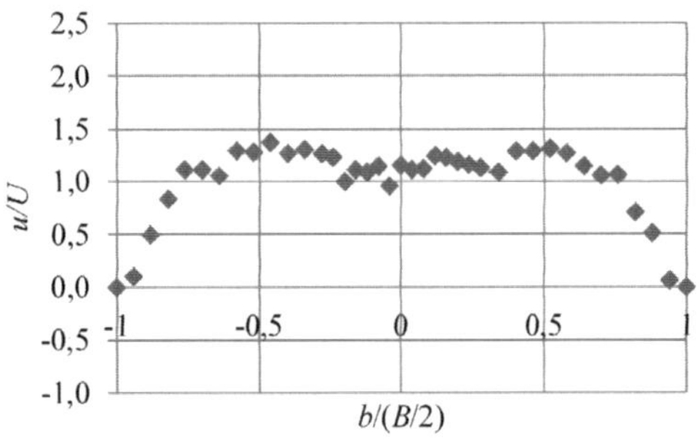

Figure 7. Average velocity profile evolution behind the cylinder for $Re_D = 100$ in cross sections: (a) 1*D*, (b) 2*D*, (c) 4*D*, (d) 8*D*, (e) 12*D*, (f) 16*D*, (g) 20*D*

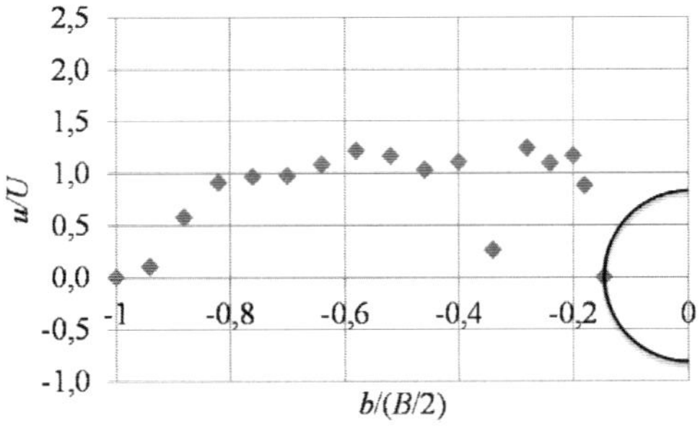

Figure 8. Average velocity distribution in the cross section of the cylinder axis for $Re_D = 150$

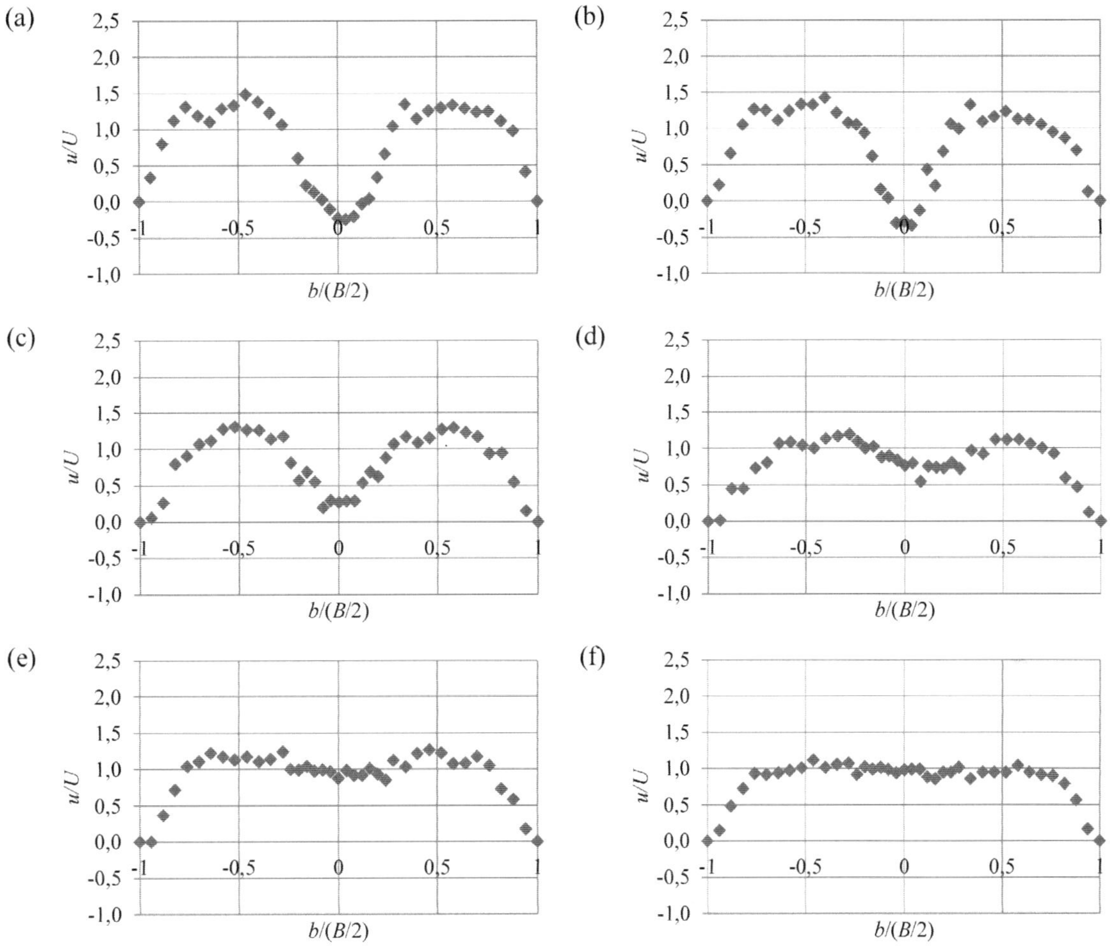

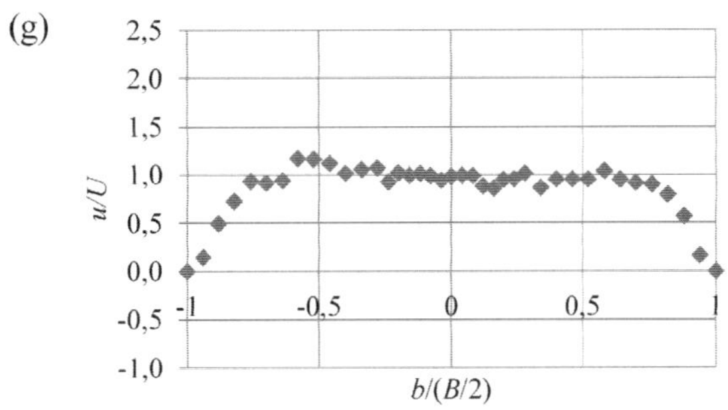

Figure 9. Average velocity profile evolution behind the cylinder for $Re_D = 150$ in cross sections: (a) 1D, (b) 2D, (c) 4D, (d) 8D, (e) 12D, (f) 16D, (g) 20D

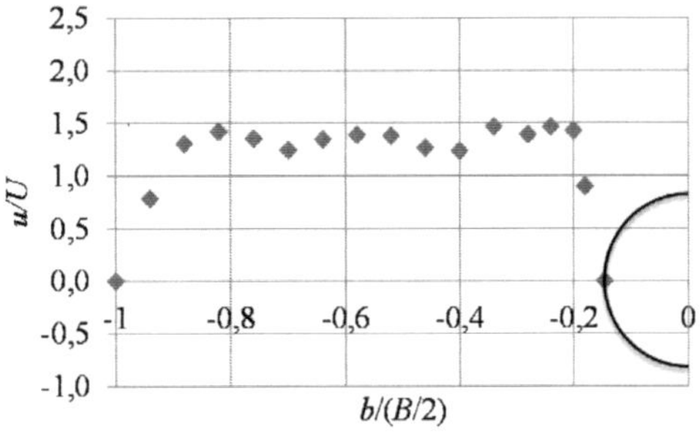

Figure 10. Average velocity distribution in the cross section of the cylinder axis for $Re_D = 250$

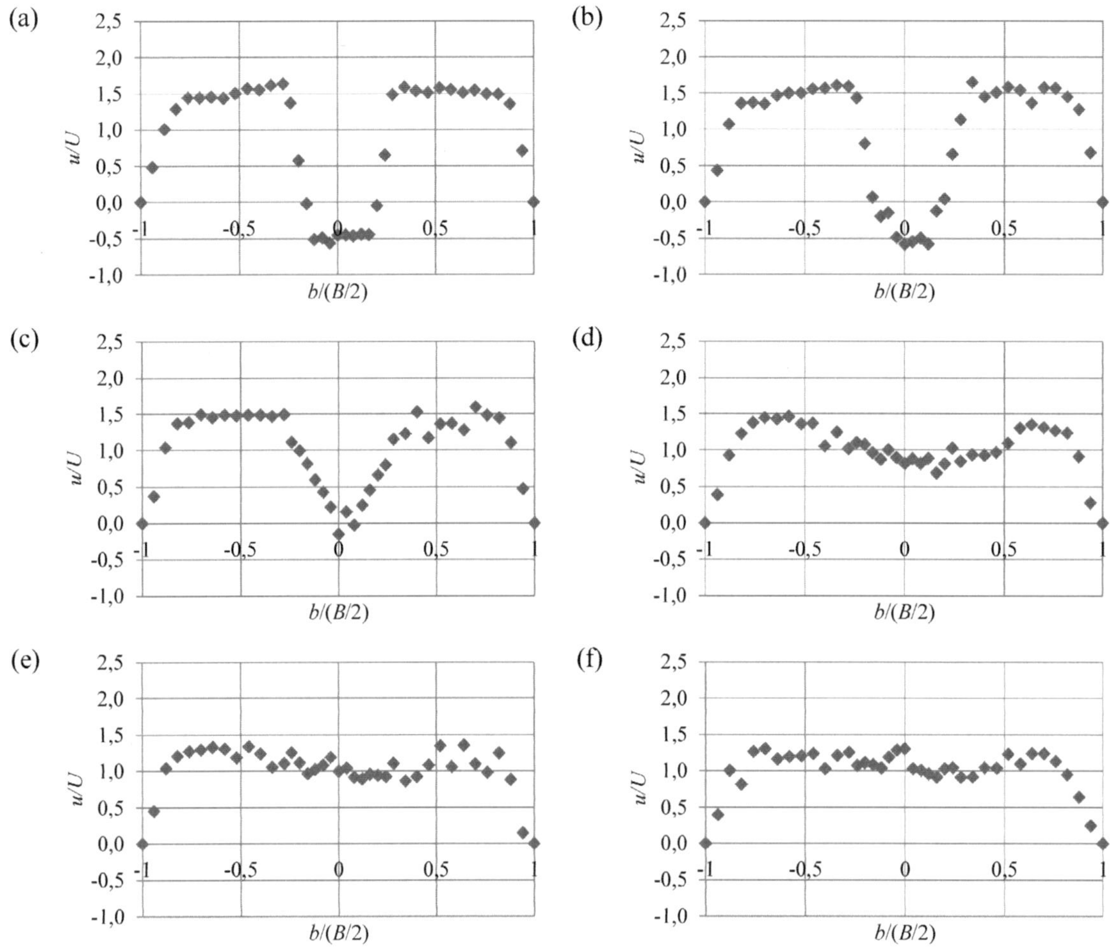

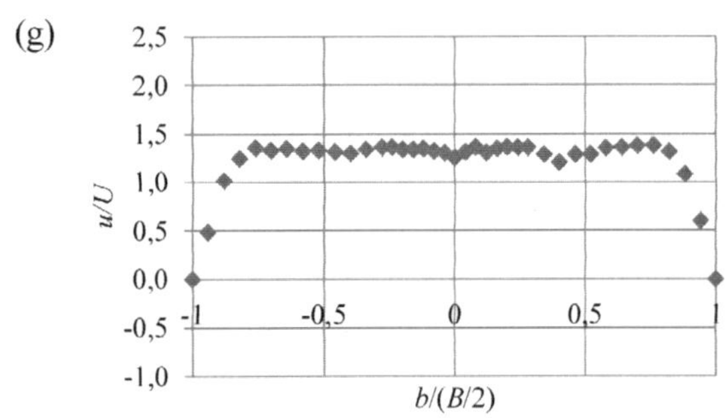

Figure 11. Average velocity profile evolution behind the cylinder for $Re_D = 250$ in cross sections: (a) 1*D*, (b) 2*D*, (c) 4*D*, (d) 8*D*, (e) 12*D*, (f) 16*D*, (g) 20*D*

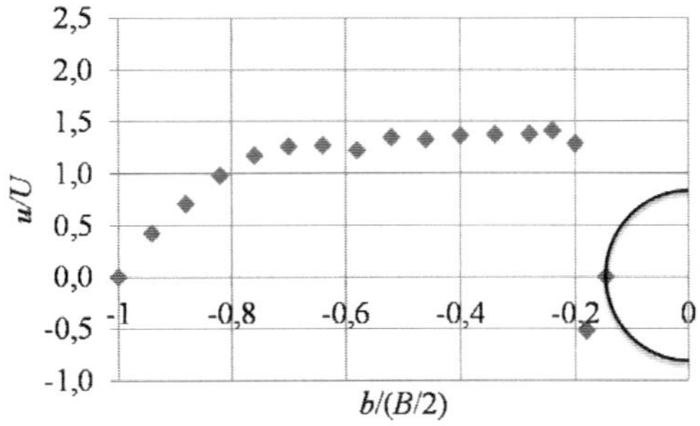

Figure 12. Average velocity distribution in the cross section of the cylinder axis for $Re_D = 350$

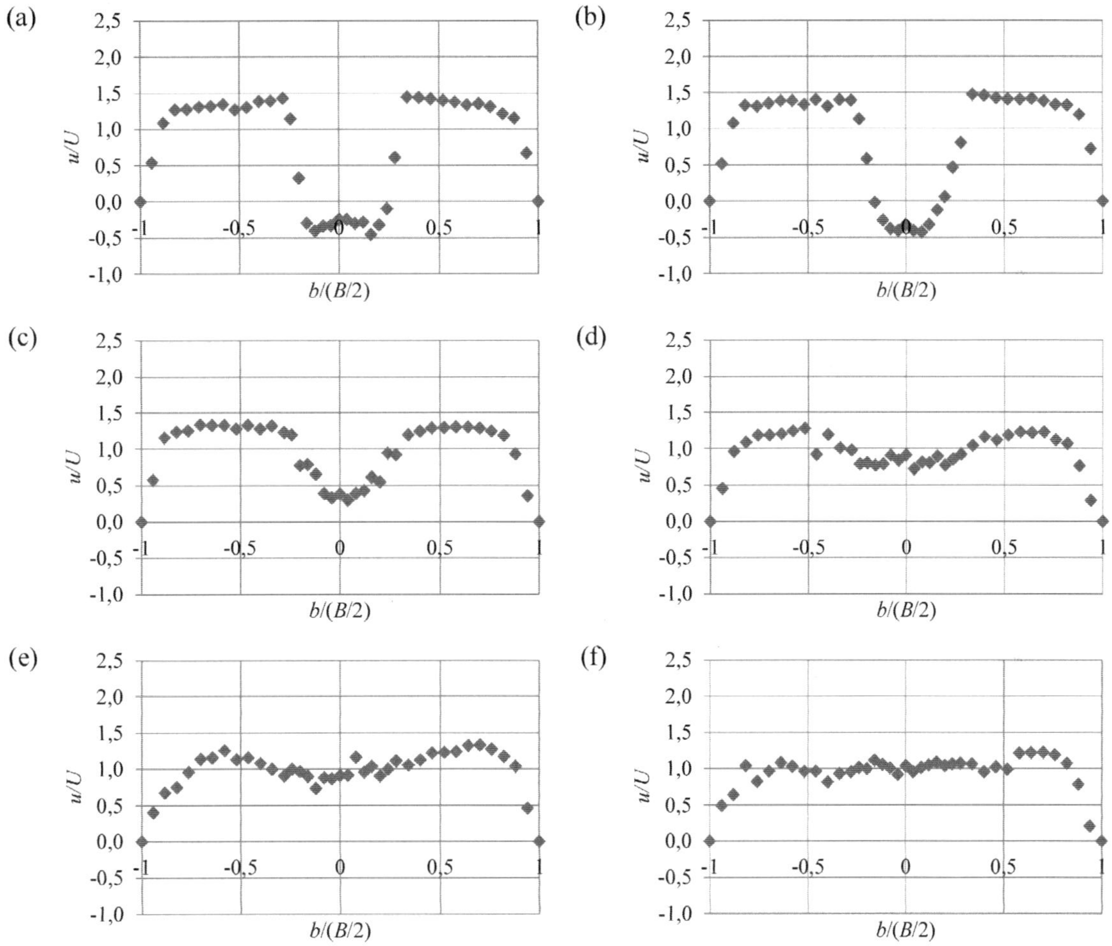

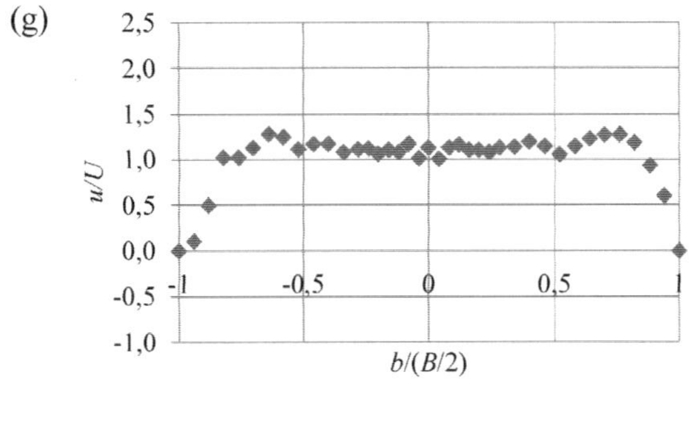

Figure 13. Average velocity profile evolution behind the cylinder for Re_D = 350 in cross sections: (a) 1D, (b) 2D, (c) 4D, (d) 8D, (e) 12D, (f) 16D, (g) 20D

Changing the depth of water in a tunnel changes the nature of the flow (Figure 14 (b)). Although the Re_D does not change, the Re_{hw} ($Re_{hw} = U \cdot h_w / \nu$) decreases twice (Table 2). For h_w = 30 mm a part of the dye spreads along the tunnel bottom, a part of it is however pulled upwards. At a height of approximately 1/4 of the water depth the dye marker is again taken away and dissipated by the main flow. For Re_D = 40 no vortex structure is observed in the wake of the cylinder regardless of the water depth.

For Re_D = 100 and h_w = 65 mm (Figure 15 (a)) it can be easily seen that the thread of the dye marker is displaced from the bottom surface of the tunnel in the direction of the water level. Pulling out the thread of dye from the bottom of the tunnel may be caused by negative pressure, which is the result of an increase in flow velocity in the area of the free water surface. At the height of approximately 1/2 of the water depth, the dye is rolled up in the direction of the cylinder wall, then dye marker rotates in the direction of cylinder and flows down along the back side of the cylinder. A clear vortex structure increasing with time can be observed. The entire phenomenon takes place only in the lower half of the filling height of the water tunnel.

For h_w = 30 mm and the same Re_D (Figure 15 (b)) the thread of dye taken away from the boundary layer and placed in in the direction of the water level is rolled up in the direction of cylinder and creates a vortex structure. Initially it is seen only in the lower half of the tunnel, but with time it increases in volume, expands up to the free water surface and finally is extended between the bottom and the free surface. Then the vortex structure “sticks” to the back side of the cylinder.

Re_D = 150 and h_w = 65 mm (Figure 16 (a)), the dye marker is rapidly pulled off from the bottom of the tunnel in the direction of the free surface. Then the dye starts to vibrate violently showing the disorder. Due to the fact that the velocity of the flow is high, at the height of approximately 3/4 of the water depth, the dye marker is taken away and dissipated by the main stream instead of being rolled up towards the cylinder wall.

After lowering the height of the tunnel to h_w = 30 mm (Figure 16 (b)) a gentle thread of dye taken from the boundary layer is observed. It is rolled up in the direction of cylinder wall just under the water level and yields a clear vortex structure. The dye marker "sticks" to the cylinder wall in time, crawling on it and directing to the bottom of the tunnel. Just before the bottom the dye marker is picked up again. The observed vortex structure occupies almost the entire space between the bottom and the water surface.

For both analysed water depths for Re_D = 250 a clear vortex structure in the cylinder wake, rotating in the direction of cylinder wall was observed to be formed. In the case of h_w = 65 mm (Figure 17 (a)), this structure reaches a height of about 2/3 of the water depth. It should be noted that a considerable amount of dye is placed from the bottom of the tunnel to the interior of the vortex structure. The dye does not spread along the bottom of the tunnel in the direction of the main flow, it is accumulated just behind the obstacle. In the case of h_w = 30 mm (Figure 17 (b)) the vortex structure occupies the entire space between the bottom of the tunnel and the water level. It seems to be slightly flattened by the free surface. The dye marker spreads along the tunnel bottom behind the cylinder at a much greater distance than it was for h_w = 65 mm.

The increase in the Reynolds number to Re_D = 350 causes that regardless of the water depth in the tunnel, a vortex structure is no longer observed in the wake of the bridge pillar model. For h_w = 65 mm (Figure 18 (a)), the dye is placed up and dispersed by the main stream of water before it reaches the water level. Due to the very intensive transport upwards, it practically does not spread in the area of the boundary layer. For h_w = 30 mm (Figure 18 (b)) the thread of the dye marker reaches the free surface but is immediately taken away by the main flow. A clear vortex structure rotating in the direction of cylinder is not observed.

(a)

(b)

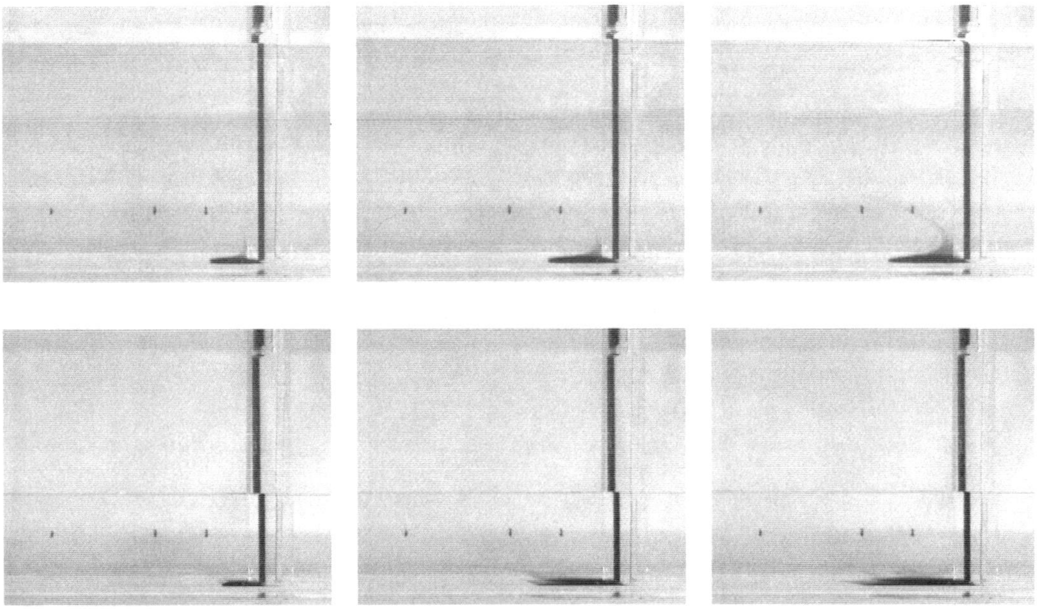

Figure 14. Exemplary images of flow behind the bridge pillar model for Re_D = 40, side view: (a) h_w = 65 mm, h_w/D = 4,44, (b) h_w = 30 mm, h_w/D = 2,05

(a)

(b)

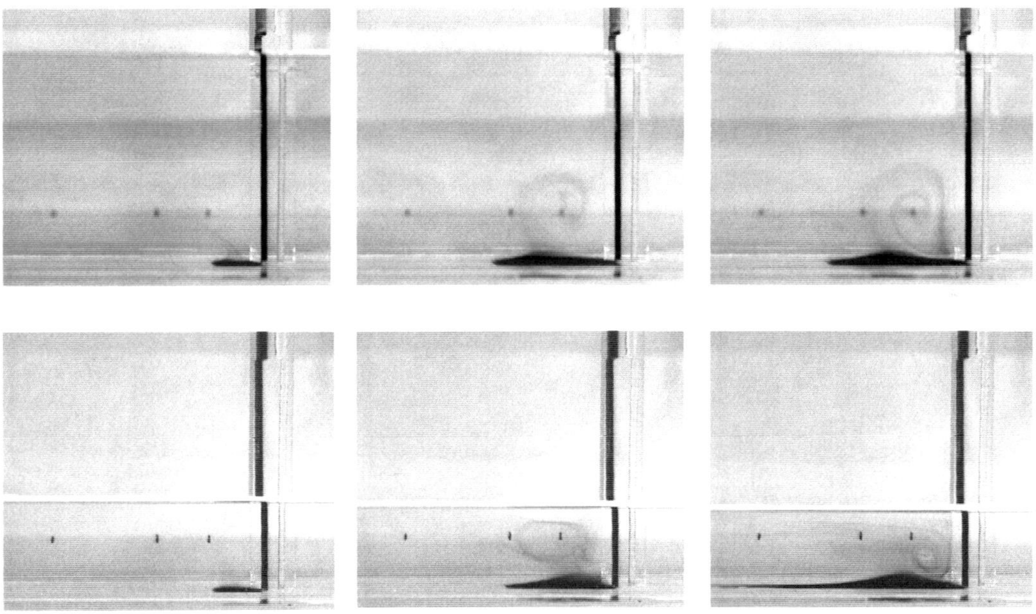

Figure 15. Exemplary images of flow behind the bridge pillar model for Re_D = 100, side view: (a) h_w = 65 mm, h_w/D = 4,44, (b) h_w = 30 mm, h_w/D = 2,05

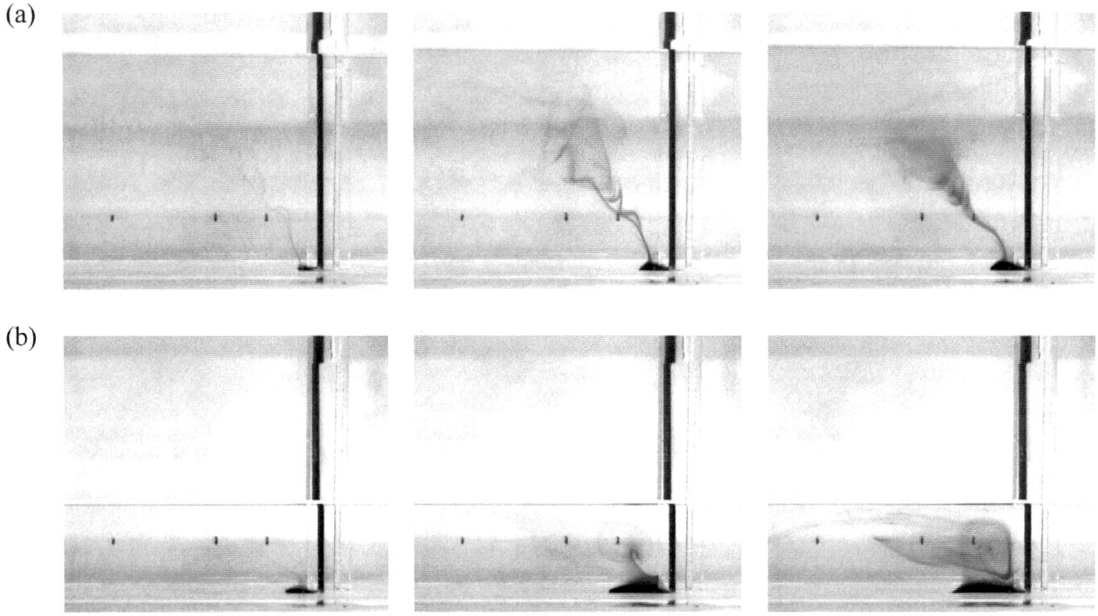

Figure 16. Exemplary images of flow behind the bridge pillar model for Re_D = 150, side view: (a) h_w = 65 mm, h_w/D = 4,44, (b) h_w = 30 mm, h_w/D = 2,05

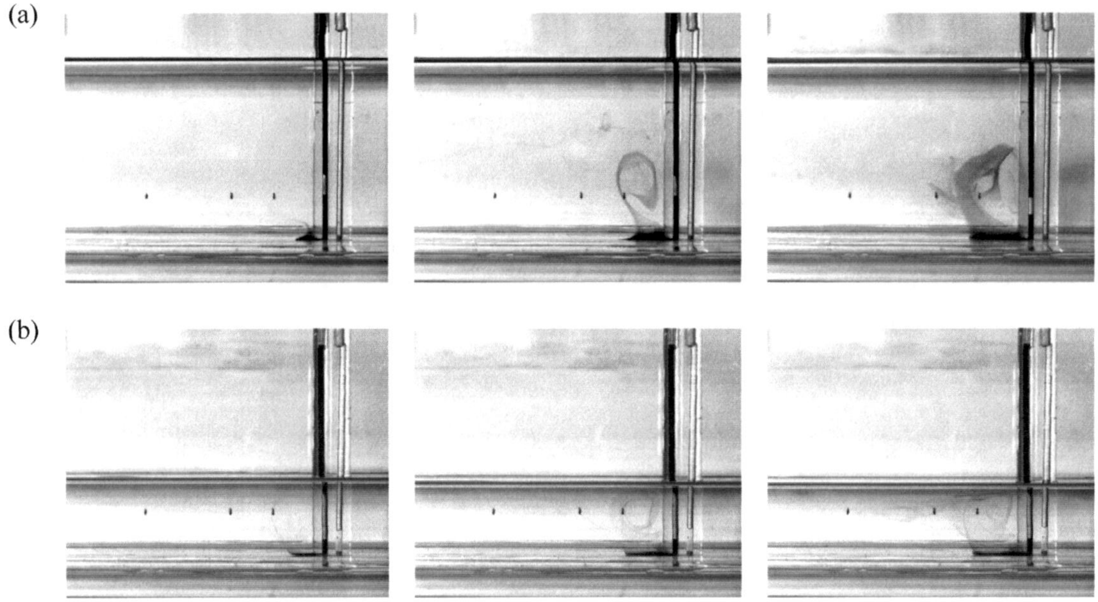

Figure 17. Exemplary images of flow behind the bridge pillar model for Re_D = 250, side view: (a) h_w = 65 mm, h_w/D = 4,44, (b) h_w = 30 mm, h_w/D = 2,05

(a)

(b)

Figure 18. Exemplary images of flow behind the bridge pillar model for Re_D = 350, side view: (a) h_w = 65 mm, h_w/D = 4,44, (b) h_w = 30 mm, h_w/D = 2,05

5. Summary

Quantitative and qualitative researches of flow around a bridge pillar model (cylinder) for the Reynolds number Re_D ($Re_D = U{\cdot}D/\nu$) were conducted. All researches were carried out for five values of the Reynolds number Re_D = 40, 100, 150, 250 and 300, therefore laminar flow, laminar–turbulent transition as well as turbulent flow occurred in test section.

Measurements of local average velocity (average in time) using Laser Doppler Anemometry were conducted in order to determine the structure of flow in the area of the bridge pillar model impact. All the anemometric measurement were carried out for constant water depth in a tunnel h_w = 65 mm.

Deformation of velocity profiles were observed for each examined Re_D. These deformation was decreasing with the distance from the model and in cross–section 16*D* all profiles were already stable. Recirculation zone behind such an obstacle was observed for each Re_D. For Re_D = 40 (the smallest examined Re_D) this recirculation zo a) $\alpha_0 = \pi/6$ne was reaching the 1*D* cross–section, for Re_D = 100 and 150 – 2*D* cross–section, for Re_D = 250 and 300 it reaches further distance (4*D*) and is much wider than it was for smaller values of Re_D.

Vortex formation from a vertical cylinder in shallow water was investigated using visualisation by dye marker. Visualization researches were carried out also for five values of the Reynolds number Re_D, but in contrast to anemometric measurements, they were extended to various water depth in the tunnel. Two visualisations were made for each Reynolds number as a consequent. In the first case the water depth was h_w = 65 mm, then it was reduced to h_w = 30 mm. The purpose of reducing the tunnel filling is to demonstrate the difference in the phenomena occurring in the flow around the cylinder without changing the Re_D, that is, assuming the diameter of the cylinder as a characteristic quantity.

A distinguishing feature of the shallow water wakes is the three–dimensional roll–up of the dye marker between the bottom and the free surface which yields a spiral vortex structure that is possible to be observed when too large disturbances does not occur in a flow. It was observed that the dyed fluid particles spread along the bottom surface of test section are forced to climb up in form of the narrow thread of vorticity. This phenomenon was called as the eruption of the boundary layer.

The mechanism of the dye marker displacing form the tunnel bottom to the main flow remains the same independently of the Reynolds number. However size, shape of the vortex structure and intensity of the dye marker relocation from the area near the bottom to the main flow do vary.

References

[1] Seal C V and Smith C R 1999 Visualization of a mechanism for three-dimensional interaction and near-wall eruption *J. Fluid Mech.* **394** 193

[2] Chen D and Jirka G H 1995 Experiential study of plane turbulent wakes in a shallow water layer *Fluid Dyn. Res.* **16** 11

[3] Akilli H and Rockwell D 2000 Vortex formation from a cylinder in shallow water *Phys. Fluids* **14** 2957

[4] CHEN D and JIRKA G H 1997 Absolute and convective instabilities of plane turbulent wakes in a shallow water layer *J. Fluid Mech.* **338** 157

[5] FU H and ROCKWELL D 2005 Shallow flow past a cylinder: control of the near wake *J. Fluid Mech.* **539** 1

[6] KAHRAMAN A, SAHIN B and ROCKWELL D 2002 Control of vortex formation from a vertical cylinder in shallow water: Effect of localized roughness elements *Exp. Fluids.* **33** 54

[7] KUDELA H and MALECHA Z M 2006 Badanie zjawiska erupcji warstwy przyściennej metodą cząstek wirowych *Inż. Chem. Proc.* **27** 833

[8] KUDELA H and MALECHA Z M 2007 Investigation of unsteady vorticity layer eruption induced by vortex patch using vortex particles method *JTAM* **45** 785

[9] KUDELA H and MALECHA Z M 2009 Eruption of a boundary layer induced by a 2D vortex patch *Fluid Dyn. Res.* **41**

[10] PN–EN 1205:25

[11] Sumer B M and Fredsoe J 1999 *Hydrodynamics around cylindrical structures* (Singapore: World Scientific Publishing Co. Pte. Ltd.)

[12] ELSNER J and DROBNIAK S 1995 *Metrologia turbulencji przepływów* (Wrocław, Warszawa, Kraków: Ossolineum)

[13] *FlowLite Installation & User's guide* 1995 (Dantec Measurement Technology A/S)

[14] GOLDSTEIN R 1983 *Fluid mechanics measurements* (Washington, New York, London: Hemisphere Publishing Corporation)

Journal of Physics: Conference Series **760** (2016) 012031 doi:10.1088/1742-6596/760/1/012031

Passive control of rotorcraft high-speed impulsive noise

O Szulc, P Doerffer and F Tejero

Institute of Fluid-Flow Machinery, Polish Academy of Sciences,
Fiszera 14, 80-231 Gdańsk, Poland

E-mail: Oskar.Szulc@imp.gda.pl

Abstract. A strong, normal shock wave, terminating a local supersonic area located at the tip of a helicopter blade, not only limits the aerodynamic performance, but also constitutes an origin of the High-Speed Impulsive (HSI) noise. The application of a passive control device (a shallow cavity covered by a perforated plate) just beneath the interaction region weakens the compression level, thus reducing the main source of the HSI noise. The numerical investigation based on the URANS approach and Bohning/Doerffer (BD) transpiration law (SPARC code) confirms a large potential of the new method. Two exemplary implementations, adapted to model helicopter rotors tested at NASA Ames facility in transonic conditions: Caradonna-Tung (lifting, transonic hover) and Caradonna-Laub-Tung (non-lifting, high-speed forward flight), demonstrate the possible gains in terms of the reduction of acoustic pressure fluctuations in the near-field of the blade tip. The CFD results are validated against the experimental data obtained for the reference configurations (no control), while the analysis of the passive control arrangement is based on a purely numerical research. The normal shock wave is effectively eliminated by the wall ventilation exerting a positive impact on the generated level of the HSI noise.

1. Introduction

Impulsive noise is believed to be one of the most annoying and loud sounds that can be generated by a helicopter rotor. It is described as a series of intense, low-frequency impulses radiated in the direction of forward flight. Two main types of impulsive noise (often called "blade slap") are attributed to helicopters: Blade Vortex Interaction (BVI) and High-Speed Impulsive (HSI) [1]. The HSI noise develops when the advancing tip Mach number Ma_{AT} reaches values above 0.75, giving rise to transonic conditions in high-speed forward flight. For $Ma_{AT} > 0.9$ shock wave delocalization may arise, emitting intense thumping, harsh sounds heard for many kilometres, compromising the detection distance

of an approaching helicopter and increasing the community annoyance. Delaying the appearance of HSI noise and shock wave delocalization is still one of the fundamental considerations in the design and development process of rotor systems. The target is often met by a passive tip shape alteration, relying on combined sweep, taper and thinning [2]. Such a major modification of the blade's most important part effectively damps the HSI noise. Unfortunately, the achieved benefits in the acoustic signature may be outbalanced by a decreased performance. On the contrary, the proposed type of a passive control device does not require any alteration of the outer shape (section) of the blade, assuring a beneficial effect on the radiated noise without degradation of the retreating side aerodynamic performance [3].

Journal of Physics: Conference Series **760** (2016) 012031 doi:10.1088/1742-6596/760/1/012031

At the beginning of the 90's W. Braun (University of Karlsruhe, Germany) investigated experimentally passive control of the shock wave terminating a local supersonic area [4] – a flow configuration commonly found on airfoils or helicopter rotor blades in transonic conditions. A shallow cavity covered by various perforated plates was embedded into the convex wall of the test section, just below the interaction, to permit air ventilation. It was shown for the first time that when the cavity covered by a perforated plate is sufficiently prolonged (called here "extended" passive control) a strong, normal shock wave may be interchanged with a system of oblique waves, reflecting between the wall and the edge of the supersonic region (see the experimental interferograms in figure 1, obtained at shock upstream Mach number Ma = 1.32). Due to the expected high loss of the aerodynamic performance this configuration has been abandoned for many years in favour of the "classical" passive control. In contrast to the fixed wing applications, the "extended" passive control may constitute a basis of a new method designed for the HSI noise reduction. Fortunately, in high-speed forward flight the advancing blade is overproducing the lift, which necessitates a very low (or even negative) pitch angle setting close to the tip (almost non-lifting conditions) in order to maintain the lateral stability of a helicopter. In this case the reduction of lift due to the effect of "extended" passive control seems not to be a relevant factor, but other questions immediately arise. What benefits are expected in terms of the acoustic pressure attenuation and what impact on the aerodynamic performance of the rotor is anticipated?

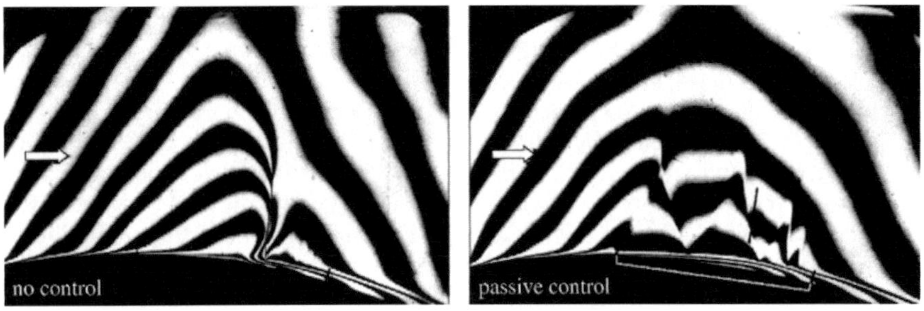

Figure 1. "Extended" passive control of shock wave at convex wall (Ma = 1.32) [4].

2. Mechanism of passive control of shock wave by wall perforation

A shallow cavity covered by a perforated surface, located just beneath the shock wave-boundary layer interaction region, significantly alters the flow topology due to the occurrence of the transpiration through the plate (figure 2). The depicted transformation of the shock system on the NACA 0012 profile was obtained at exemplary inflow conditions: Mach number $Ma_\infty = 0.8$, Reynolds number $Re_\infty = 9 \cdot 10^6$ and incidence $\alpha = 1°$. The method of control is passive in nature and does not require any external supply of energy. The cavity connects zones of low and high pressure, located just upstream and downstream of the shock wave. Upstream of the shock wave pressure in the cavity is higher than in the flow-field, leading to blowing into the main stream. An opposite situation arises downstream of the shock wave where lower pressure in the cavity induces suction. The blowing is perceived by the main, supersonic stream as a ramp (or a bump) resulting in a generation of front, oblique compression which in case of the "classical" passive control intersects the normal shock wave, creating a large λ-foot structure. The suction of the boundary layer downstream of the interaction is not sufficient to counteract the negative influence of blowing. Therefore, the final effect is a balance between two opposing trends: growth of viscous losses (on account of boundary layer disturbance) and reduction of wave losses (through a replacement of a normal shock by a λ-foot). It was proven during the Euroshock project that for specific configurations the method of passive control in its "classical" form is capable of marginally improving the aerodynamic performance of an airfoil in transonic conditions [5].

Another attitude (undertaken in our research group) towards airfoil performance has to be considered in respect to the rotor blade of a helicopter. In high-speed forward flight the most negative effect is a generation of shock wave which is responsible for the HSI noise. In contrast to the "classical" passive control it is proposed to limit the shock intensity by a prolonged cavity covered by a perforated plate (i.e. "extended" passive control in figure 2). The induced flow recirculation through the cavity leads to a substitution of the strong, normal compression by a system of oblique waves, reflecting between the surface and the edge of the supersonic region. As a result the static pressure variation in the volume above the suction side of the profile is reduced – counteracting the main source of the HSI noise (figure 3). Without any modification to the blade section (outer shape) the extent of the supersonic area is significantly restricted together with a maximum Mach number in the flow-field.

An obvious consequence of the passive ventilation is flattening of the pressure distribution at the wall, but the most relevant effect becomes apparent right above the surface of the airfoil (figure 3). At a constant distance $h = 0.2\ c$ an intense and sudden increase of pressure coefficient c_p, related to a normal shock wave (black colour), is interchanged with a much more gradual compression (red colour), almost entirely eliminating total pressure wave losses at a vertical cross-section of $x/c = 0.65$, just downstream of the interaction (both locations marked in figure 2). Unfortunately, the viscous effects are magnified due to the disturbing blowing from the upstream part of the cavity, thickening boundary layer and usually intensifying flow separation, therefore augmenting total pressure losses in the region close to the wall. For the "classical" passive control similar trends are present, but barely noticeable (blue colour). A significant drawback of the "extended" passive control is a serious limitation of the normal force coefficient c_n (accompanied by a slight reduction of the drag coefficient c_d), making the method unsuitable for fixed-wing applications, where the production of lift is of a primary importance. Favourably, for a helicopter rotor in high-speed forward flight the most intense acoustic signal of the HSI noise is generated when the advancing tip Mach number Ma_{AT} is approaching maximum. At the corresponding blade azimuthal position of $\psi = 90^o$ the pitch angle is often limited to approx. $\theta = 0^o$, creating nearly non-lifting conditions and a double shock pattern.

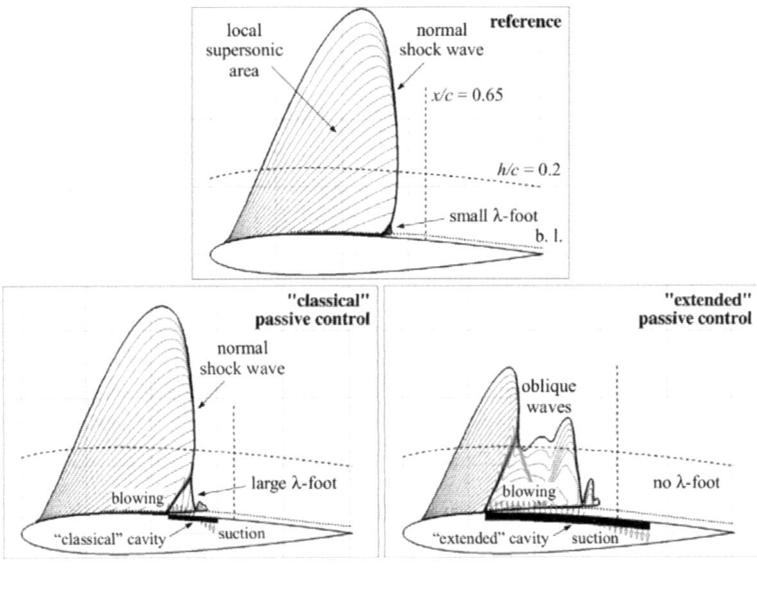

Figure 2. "Classical" and "extended" passive control of shock wave, NACA 0012 airfoil ($Ma_\infty = 0.8$, $Re_\infty = 9{\cdot}10^6$ and $\alpha = 1^o$).

3. Numerical implementation of the Bohning/Doerffer (BD) transpiration law

One of the achievements of the Euroshock project is a development of the BD transpiration law [6]. The empirical model describes a relation between the pressure difference over the perforated plate and the induced mass flow rate. The experiments proved that the pressure in the cavity p_c may be considered as constant. The determination of the pressure difference $\Delta p = p(x) - p_c$ requires the wall distribution in the main stream $p(x)$. According to the BD formulation for passive control of the shock wave the effective Mach number in a single perforation hole Ma_h is calculated locally based on the expression:

$$Ma_h = 1.2\left(\frac{|\Delta p|}{p_0}\right)^{0.55}, \tag{1}$$

where p_0 denotes the stagnation pressure value at the inlet side of the orifice, different for blowing ($\Delta p < 0$) and suction ($\Delta p > 0$). Knowing the aerodynamic porosity of the perforated plate p_{aero} (often depending on the flow direction due to low manufacturing quality of the holes), the corresponding mass flow rate is estimated. Passive character of the interaction is reflected in an instantaneous zero net mass flux through the ventilated surface area. As a result an effective transpiration velocity V_t (normal to the wall) distribution is determined and applied in the numerical simulation as a boundary condition.

The present investigation has been carried out with a cell-centred, block-structured, parallel code SPARC, developed by F. Magagnato at the University of Karlsruhe [7]. It solves numerically the compressible, mass-weighted, Favre-averaged Navier-Stokes equations with several turbulence models. The shock wave-boundary layer interaction phenomenon under the influence of the transpiration is fairly predicted by a low-Reynolds closure of Spalart-Allmaras (SA) [8]. The numerical algorithm is based on a semi-discrete scheme, utilizing a finite-volume formulation for the spatial discretisation (central, 2nd order) and an implicit dual-time-stepping approach for the temporal discretisation (2nd order), supplemented with an explicit Runge-Kutta method. To damp numerical oscillations the SLIP artificial dissipation model is implemented. In order to increase the convergence rate the local time-stepping, implicit residual averaging and full-multigrid techniques are incorporated in the solver.

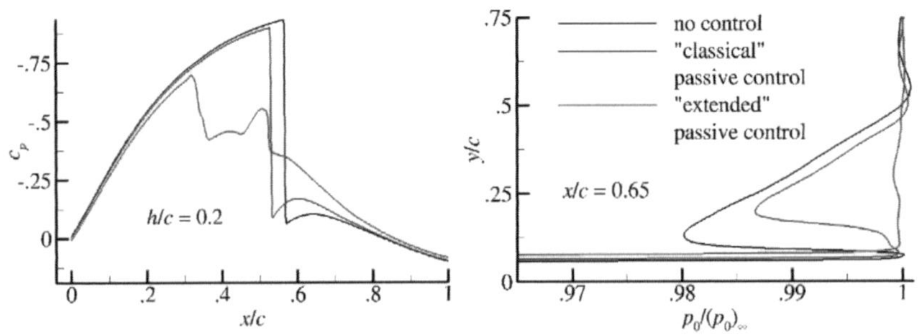

Figure 3. Pressure coefficient c_p and total pressure $p_0/(p_0)_\infty$ variation above the surface, NACA 0012 airfoil ($Ma_\infty = 0.8$, $Re_\infty = 9\cdot10^6$ and $\alpha = 1°$).

A new, perforated wall boundary condition, designed towards the modelling of passive control of shock wave, has been implemented into the SPARC code. The validation process, based on a comparison with the experimental data obtained for a range of flow configurations: a transonic nozzle with a flat wall (Onera), curved duct with local supersonic area (University of Karlsruhe) and NACA 0012 profile (NASA Langley), has verified the extended functionality of the tool [3, 9]. In case of a classical no-slip wall boundary condition the tangential and normal velocity components (mean and fluctuating), as well as the eddy viscosity, are set to zero at the surface as a consequence of the adhesion of fluid. In an adiabatic flow (no heat-flux) pressure and density are extrapolated from the interior of the computational domain. A construction of the ventilated wall boundary condition is analogous, taking into account a non-zero normal (transpiration) velocity V_t, estimated from the BD law in accordance with the local properties, i.e. pressure $p(x)$ and temperature $T(x)$. The pressure in the cavity volume p_c is assumed uniform, adapting itself naturally to ensure vanishing of the total mass flow rate through the perforated plate (passive environment). The temperature in the control system T_c

is fixed to the stagnation value of the free-stream. If $V_t = 0$, then a standard no-slip wall boundary condition is restored, allowing for modelling of the perforated and solid surfaces in a common framework.

For each section of a rotating helicopter blade the stagnation pressure p_0 and temperature T_0 of the incoming air depend on the radial location r/R. In such conditions the application of a passive control system would induce an undesirable spanwise flow inside the cavity towards the tip. To counteract this effect it is proposed to divide the cavity volume into smaller, independent and sealed sub-domains, experiencing a quasi-2d, chordwise recirculation (figure 4). An additional adaptation process of pressure within each sub-cavity p_c separately assures passive ventilation. Similarly, the temperature inside every sub-cavity T_c has to be computed as an average value of the temperature of the sucked-in air. Frequently, the mesh density is increasing in the direction of the tip, the area subjected to large flow gradients and shocks. Hence, the resulting radial arrangement of sub-cavities is uneven with decreasing spanwise extent as the position is shifted towards the outer sections of the blade. The estimated transpiration intensity, based on the BD law, is relying on accurate prediction of the shock location and wall pressure distribution in the interaction area. As a consequence, a suitable correlation of the numerical results with the measurements for the reference rotors (no control) is of an utmost importance.

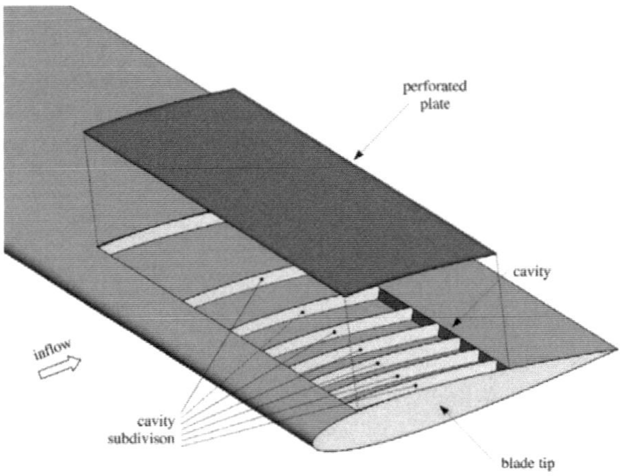

Figure 4. Passive cavity arrangement at the tip of a helicopter rotor blade.

4. Caradonna-Tung (1981) model helicopter rotor in lifting, transonic hover

To present the effectiveness of the proposed HSI noise reducing method the Caradonna-Tung (C-T) rotor was chosen, operating in severe high-speed, transonic hover conditions (tip Mach number $Ma_T = 0.877$, tip Reynolds number $Re_T = 3.931 \cdot 10^6$ and collective $\theta_0 = 8^o$) [10]. The model consists of 2 rectangular, untwisted and untapered NACA 0012 rigid blades, mounted on a tall column (figure 5). The rotor aspect ratio $AR = 6$, chord length $c = 0.1905$ m and diameter $2R = 2.286$ m. The pressure coefficient c_p distributions at 5 cross-sections were measured for the reference case (no control).

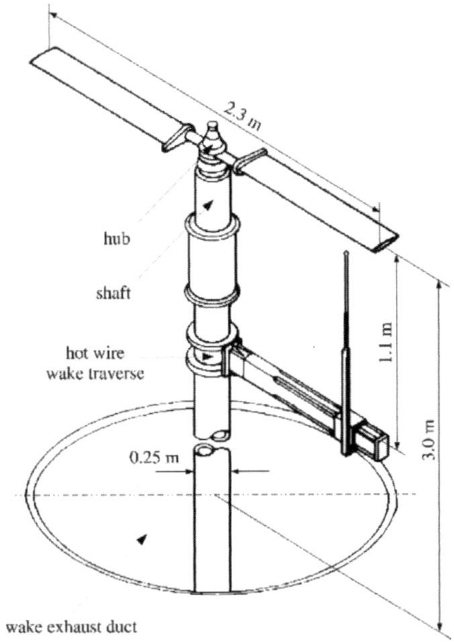

Figure 5. Experimental set-up (NASA) [10].

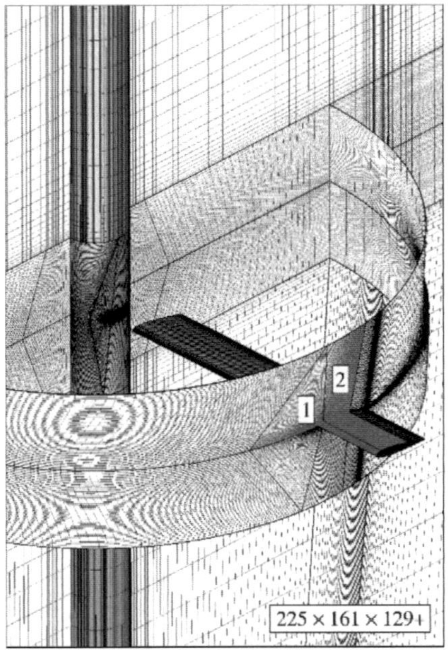

Figure 6. Multi-block computational grid.

The computational domain was subdivided into 80 blocks (figure 6). The flow-field of a hovering rotor is quasi-steady in respect to the blade and periodic in nature – only a single blade needs to be considered. The base computational grid ($5.7 \cdot 10^6$ of volumes) was refined locally within two most important mesh blocks (marked by red colour in figure 6), containing the region of transonic flow and shock wave-boundary layer interaction. It increased the mesh size up to $7.0 \cdot 10^6$ of volumes. The cavity covered with a perforated plate was placed at the outer 20% of the span, on the suction side, between $x/c = 0.1$ and $x/c = 0.65$ (within block 2). The aerodynamic porosity value was taken as $p_{aero} = 5\%$. The computational domain was surrounded by boundary conditions of 4 types: viscous and inviscid walls, rotational periodicity and pressure inlet/outlet. The unsteady simulation (necessary for the acoustic analysis) was progressed with a time step $\Delta t = 0.25^\circ$ of azimuth (1440 time steps per period of rotation).

A fully turbulent unsteady simulation of the hovering rotor flow-field requires 40 revolutions to develop the complex wake, tip vortices and their interactions to reach a quasi-steady state. The experimental thrust coefficient $c_t = 0.00473$ is overpredicted by 20% (inviscid model, $c_t = 0.00565$) and 10% (turbulent SA, $c_t = 0.00522$). The rotor wake (vertical inflow) affects mainly the inner sections of the blade where only low velocities are induced by the rotation. At the outer sections rotational velocity dominates. At $r/R = 0.5$ and 0.68 flow is subsonic (see comparison of c_p distributions in figure 7). The shock system is building up at the outer 20% of the span ($r/R = 0.8$, 0.89 and 0.96). The agreement with the test data is satisfactory at all cross-sections, validating the computational model.

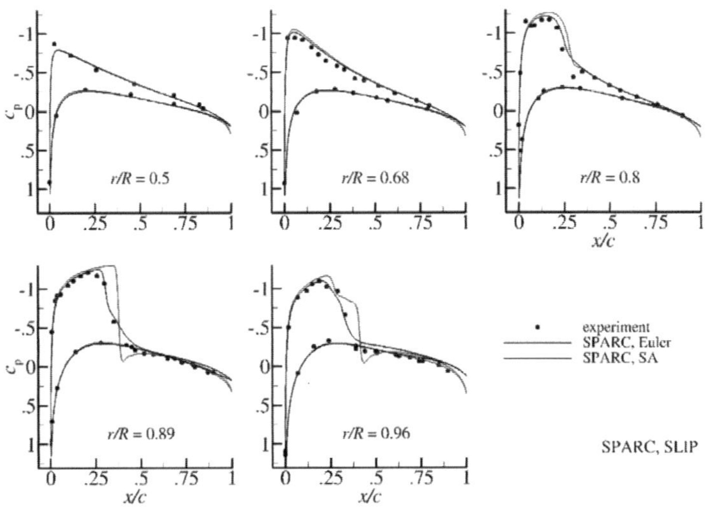

Figure 7. Pressure coefficient c_p distributions at r/R = 0.5, 0.68, 0.8, 0.89 and 0.96.

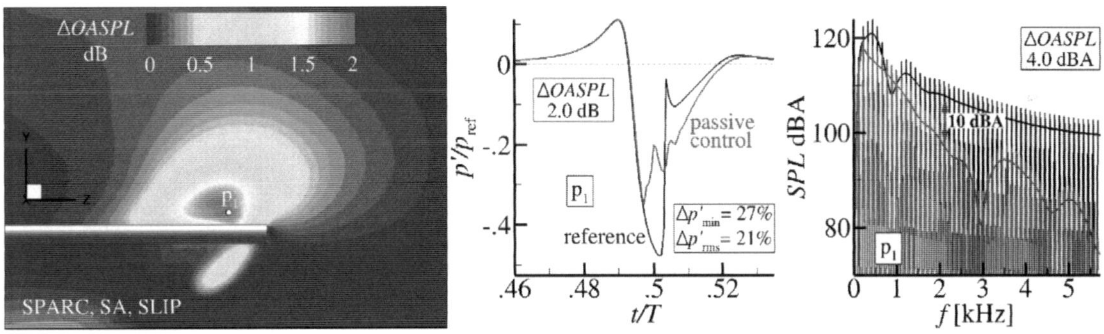

Figure 8. Effect of "extended" passive control on overall sound pressure level *OASPL*, acoustic pressure fluctuation p'/p_{ref} and A-weighted sound pressure level spectrum *SPL*.

The acoustic analysis is based on a large set of data files (containing 3d flow-field) recorded during 1 rotor revolution, at every time step (1440 samples per period T) and averaged in time t in the near-field of the rotor (figure 8). The overall sound pressure level *OASPL* at point p_1, located near the blade tip, is reduced by 2 dB. The corresponding acoustic pressure fluctuation p'/p_{ref} is indicating that not only the root mean square p'_{rms} is damped by 21%, but also the amplitude of the negative peak p'_{min} is decreased by 27%. Moreover, the duration time of the signal is extended, limiting the impulsive character of the fluctuation. At some frequency (f) bands of the A-weighted sound pressure level spectrum *SPL* the drop of more than 10 dBA is predicted ($\Delta OASPL$ = 4.0 dBA). The cost of passive control is a decrease of thrust coefficient c_t by 5% and an increase of torque coefficient c_q by 9%.

5. Caradonna-Laub-Tung (1984) model helicopter rotor in non-lifting, high-speed forward flight

To confirm the potential of the new HSI noise attenuating method the Caradonna-Laub-Tung (C-L-T) teetering rotor was chosen, operating in severe high-speed, non-lifting forward flight conditions (tip Mach number $Ma_T = 0.8$, tip Reynolds number $Re_T = 2.89 \cdot 10^6$ and advance ratio $\mu = 0.2$) [11]. The model consists of 2 rectangular, untwisted and untapered NACA 0012 rigid blades, mounted on a tall column. The rotor aspect ratio AR = 7, chord length c = 0.1524 m and diameter $2R$ = 2.134 m. The instantaneous pressure coefficient c_p distributions at the cross-section of r/R = 0.89 and azimuthal positions of $\psi = 30°$, 60°, 90°, 120°, 150° and 180° were recorded for the reference case (no control).

The computational domain was subdivided into 160 blocks. The base computational grid ($12.2 \cdot 10^6$ of volumes) was refined locally within eight most important mesh blocks, containing the region of transonic flow and shock wave-boundary layer interaction (located above and below the surface of the tip of both rotor blades, that is in high Mach number area). It increased the mesh size up to $18.2 \cdot 10^6$ of volumes. Two identical cavities covered with perforated plates were placed at the outer 20% of the span, on upper and lower sides, between $x/c = 0.3$ and $x/c = 0.9$. The aerodynamic porosity value was higher, i.e. $p_{aero} = 10\%$. The transpiration was activated only in the range of ψ between 60° and 150°. The computational domain was surrounded by boundary conditions of 4 types: viscous and inviscid walls, rotational periodicity and pressure inlet/outlet. The forward flight state was realised through grid rotation and translation. The unsteady simulation was progressed with a time step Δt = 0.25° of azimuth (1440 time steps per period). The unsteady shock wave development is captured by the numerical method (inviscid Euler and turbulent SA), based on the comparison of c_p distributions of figure 9. The agreement with the test data is satisfactory at all azimuths, validating the computational model.

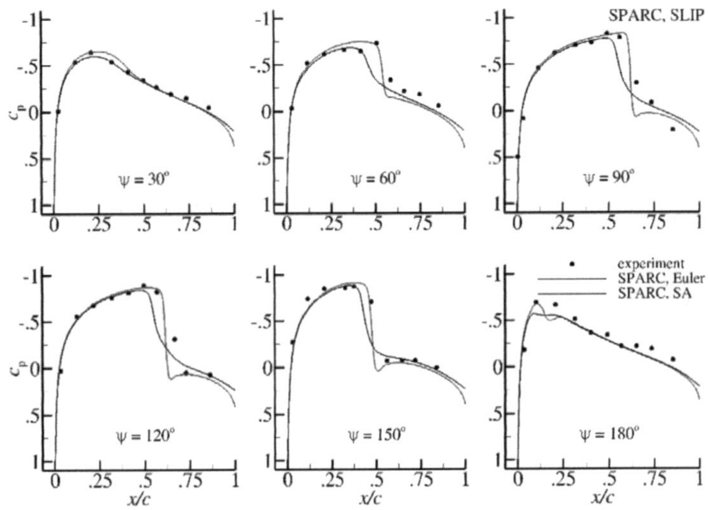

Figure 9. Pressure coefficient c_p distributions at $r/R = 0.89$.

The acoustic averaging of the data is similar to the process described for the C-T rotor in hover. The overall sound pressure level *OASPL* at point p_2, located at $r/R = 0.89$, $y/c = 0.3$ and $\psi = 129°$, is reduced by almost 2 dB (figure 10). The corresponding acoustic pressure fluctuation p'/p_{ref} is indicating that not only the root mean square p'_{rms} is damped by 20%, but also the amplitude of the negative peak p'_{min} is decreased by 18% (figure 11). Moreover, the duration time of the signal is prolonged, limiting the impulsive character of the fluctuation. At some frequency bands of the A-weighted sound pressure level *SPL* spectrum the drop of more than 10 dBA is predicted ($\Delta OASPL$ = 4.1 dBA). The cost of the application of passive control method is a substantial increase of torque coefficient c_q by 22%. Since the rotor operates in non-lifting conditions (symmetric flow) and two identical cavities (covered by perforated plates) are implemented on both sides of the blade tip, the thrust coefficient c_t is still equal 0.

6. Conclusions

The article describes the application of a new method of HSI noise reduction, designed for rotorcrafts, based on the application of perforated plates and passive wall ventilation. It is evident that the normal shock wave has been eliminated for presented C-T and C-L-T rotors in transonic hover and forward flight conditions, causing a positive impact on the generated level of the acoustic pressure fluctuations in the near-filed of the blade. Unfortunately, the cost of the operation cannot be neglected, requiring further investigations and optimisation of the device focused on the aerodynamic performance of the rotor. Another option is to develop and incorporate a cavity control system, allowing for an instantaneous activation and deactivation of the transpiration depending on the actual flight conditions.

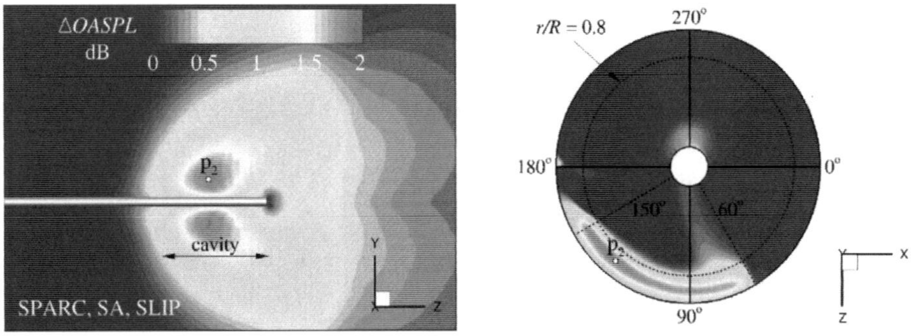

Figure 10. Effect of "extended" passive control on overall sound pressure level *OASPL*.

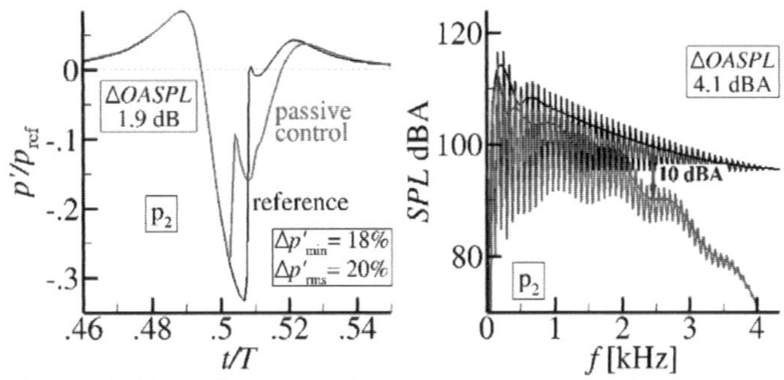

Figure 11. Effect of "extended" passive control on acoustic pressure fluctuation p'/p_{ref} and A-weighted sound pressure level spectrum *SPL*.

Acknowledgment
This research was supported by CI TASK and in part by PL-Grid Infrastructure.

References

[1] Schmitz F H and Yu Y H 1983 *Technical Memorandum* (NASA Ames) 84390
[2] Baeder J D 1998 *J. of the American Helicopter Society* **43**(3) pp. 222–34
[3] Doerffer P and Szulc O 2011 *Int. J. of Eng. Systems Modelling and Simulation* **3**(1-2) pp. 64–73
[4] Braun W 1990 *PhD Dissertation* (University of Karlsruhe)
[5] Stanewsky E, Délery J, Fulker J and Geißler W 1997 *EUROSHOCK – Drag Reduction by Passive Shock Control* (Braunschweig: Vieweg)
[6] Doerffer P and Bohning R 2000 *Aerospace Sci. and Technol.* **4**(8) pp. 525–34
[7] Magagnato F 1998 *TASK Quarterly* **2**(2) pp. 215–70
[8] Spalart P R and Allmaras S R 1992 *Proc. 30th Aerospace Sciences Meeting & Exhibit* (Reno) 92-0439
[9] Doerffer P and Szulc O 2006 *Arch. of Mechanics* **58**(6) pp. 543–73
[10] Caradonna F X and Tung C 1981 *Technical Memorandum* (NASA Ames) 81232
[11] Caradonna F X, Laub G H and Tung C 1984 *Technical Memorandum* (NASA Ames) 86005

Journal of Physics: Conference Series **760** (2016) 012032 doi:10.1088/1742-6596/760/1/012032

Transpiration effects in perforated plate aerodynamics

R Szwaba[1] **and T Ochrymiuk**[1]

[1] Institute of Fluid Flow Machinery Polish Academy of Sciences, Fiszera 14, 80-231 Gdansk, Poland

Abstract. Perforated walls find a wide use as a method of flow control and effusive cooling. Experimental investigations of the gas flow past perforated plate with microholes (110μm) were carried out. The wide range of pressure at the inlet were investigated. Two distinguishable flow regimes were obtained: laminar and turbulent regime.The results are in good agreement with theory, simulations and experiments on large scale perforated plates and compressible flows in microtubules. Formulation of the transpiration law was associated with the porous plate aerodynamics properties. Using a model of transpiration flow the "aerodynamic porosity" could be determined for microholes.

1. Introduction

The motivation behind this work was to find a more efficient cooling system for the first stages of gas turbine blades. In such configurations, the coolant is supplied by the set of small holes. The film cooling process has been widely adopted on the walls of a high temperature system, such as gas turbine blades, nozzles and the walls of the combustion chamber to protect the surfaces from being overheated. Turbine airfoils of aircraft engine are small and therefore require small flow passages and film holes. For example a typical aircraft engine needs film holes diameter as small as 0.4 mm, while a typical industrial power turbine can be limited to a minimum diameter of 1 mm. The concept of microchannel cooling for gas turbine blades [14] is the natural application of thermodynamics and heat transfer to accomplish two goals: first to spread out the cooling network in a series of smaller and highly distributed channels to provide much better uniformity of cooling and thermal gradients; second to bring the cooling fluid closer to the blade surface and create more efficient heat transfer. In the past, the film jet used in the film cooling process was so big thus the amount of cooling air used was so large that it could effectively reduce the performance of a gas turbine engine. In addition, the relatively thick film jet is expected to cause rapidly mix with the hot gas in the freestream and to reduce its protection effectiveness. The challenges of microchannel cooling applications include hole plugging, wall strength, film cooling, manufacturing and costs. Up to now no commercial use of such microchannel cooling solutions was applied.

Perforated plate aerodynamics has been extensively studied over many decades by experiments [1÷7] and theory [8÷11]. The application of interest has varied: from sound absorption [12], microfilters and separators [4] and heat exchangers [13] to the method of production of a uniform turbulent flow [2] and a new concept of a blade turbine cooling technique [14]. The concept of microchannel cooling (or effusive cooling) for the gas turbine blades is a natural application of thermodynamics and heat transfer. It allows us to accomplish two goals: first to spread out the coolant to provide a much better uniformity of the cooling and thermal gradients and second to bring the cooling fluid closer to the blade surface and create a more efficient heat transfer. In the past the blade cooling process was very intensive and could reduce the performance of a gas turbine engine [15]. This paper is focused on the investigation of the aerodynamic performance of a plate with microholes using the global macroscale experimental data. For that purpose the experiments were performed in air over a perforated plate with a perforation of 5.2% and holes 110 μm in diameter. Various ranges of pressure at the inlet were studied. This allowed us to obtain the characteristics of two distinguishable regimes: a laminar and turbulent one with the transition to turbulence at around *Re*=80. The studies were performed for various Mach numbers from *Ma*<0.1 up to the critical conditions (choking flow in the microholes). The theory of compressible gas flow in the channels has been studied by continuum and molecular approaches and can be found in various textbooks; for example, continuum [16], molecular [17] and simple analytical relations [18]. Beneath there is briefly described the basic theory required to extract the information from experimental data. The relation between the gas parameters is given as an ideal gas equation of state. Based on that equation the speed of sound a can be defined as:

$$a = \sqrt{\kappa RT} = \sqrt{\frac{\kappa p}{\rho}} \quad (1)$$

κ - specific heat constant, R - gas constant, T - temperature, p - pressure, ρ - density.
Another characteristic parameter of the gas relates its molecular properties to the continuum quantities. The mean free path represents an average distance between the molecules of the gas and is defined as

$$\lambda = \frac{k_\lambda \mu(T)\sqrt{2RT}}{p} \quad (2)$$

where $k_\lambda = \frac{\sqrt{\pi}}{2}$. The viscosity in this relation depends on the temperature and can be described as:

$$\mu(T) = \left(\frac{T}{T_{ref}}\right)^{\varpi} \quad (3)$$

μ - dynamic viscosity, k - thermal conductivity, ω - viscosity index, p - pressure, U – velocity.
The nondimensional parameters determining the relevant scales for the flow are the Mach number $Ma = \frac{U}{a}$ and the Reynolds number based on the hole's diameter and average velocity $\mathrm{Re} = \frac{\rho UD}{\mu}$.

2. Experimental setup

The experimental investigations were carried out in the test section shown in Fig. 1. The flow direction is from the left to the right as indicated by an arrow. The flow is a result of an imposed pressure difference between the ambient condition and pressure in the vacuum tanks, which was placed behind valve 7. The ambient condition corresponds to the air parameters at the laboratory space, e.g. temperature about 20–22 °C and atmospheric pressure. The pressure in the vacuum tanks was equal to 16 hPa. The capacity of the vacuum tanks was 120 m^3, since it allows a long enough time for measurements at very low pressures where the mass flow is only about 12 standard litters per minute (SLPM). The air goes from the ambient flow through the flow meter 1, through the control valve 2 and compensatory chamber 3 to frame 4, where the membrane with microholes is mounted. Downstream of the membrane the flow goes through another chamber 5, a second valve 6 (which controls the flow condition downstream of the microholes), and the cut-off valve 7 to the vacuum tanks. The change of the stagnation parameters (pressure and density) was obtained by means of control valve 2. However, the pressure downstream of the microholes was controlled by means of valve 6. The particular characteristics can be measured throughout an adequate manipulation of these two control valves. The scheme of a measurements arrangement is shown in Fig. 2. The measurement of the stagnation parameters was located in compensatory chamber 3. The pressure was measured by means of a Prandtl probe with a Kulite pressure transducer with the full scale (FS) accuracy of 0.1%, e.g., ±1 hPa. The temperature was measured by means of the thermocouple element with an accuracy of 0.1 K.

Journal of Physics: Conference Series **760** (2016) 012032 doi:10.1088/1742-6596/760/1/012032

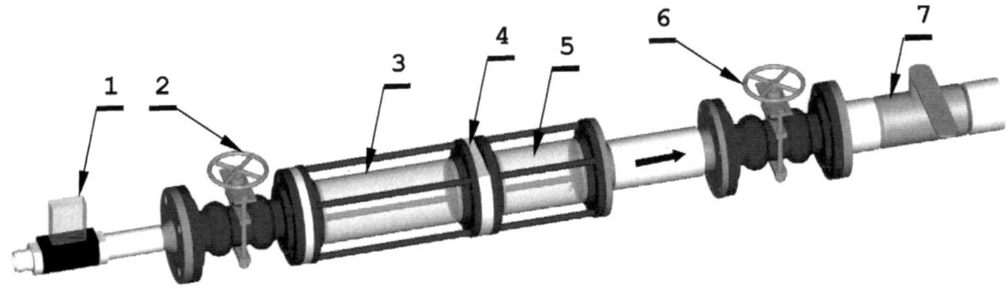

Figure 1. View of the test section.

The static pressure upstream and downstream of the membrane was measured by means of a pressure scanner with the FS accuracy of 0.05 %, e.g. ±0.5 hPa. The mass flow rate was measured by means of three different laminar flow meters, depending on the range of the mass flow (Alicate: 20, 100, and 1500 SLPM; at the ambient pressure equal to 1013 hPa and the temperature of 298 K, the air density for these conditions was equal to 1.184 kg/m^3). The accuracy of these flow meters was equal to ±0.01 SLPM of the measured value. The measurements were comprised of the barometric pressure; stagnation parameters upstream of the microholes, static pressure upstream and downstream of the microholes, and the mass flow rate. The pressure versus mass flow characteristics at constant pressure upstream of the measurement membrane was obtained. The following upstream microholes pressures were chosen: 25, 30, 40, 50, 100, 200, 400 and 550 hPa.

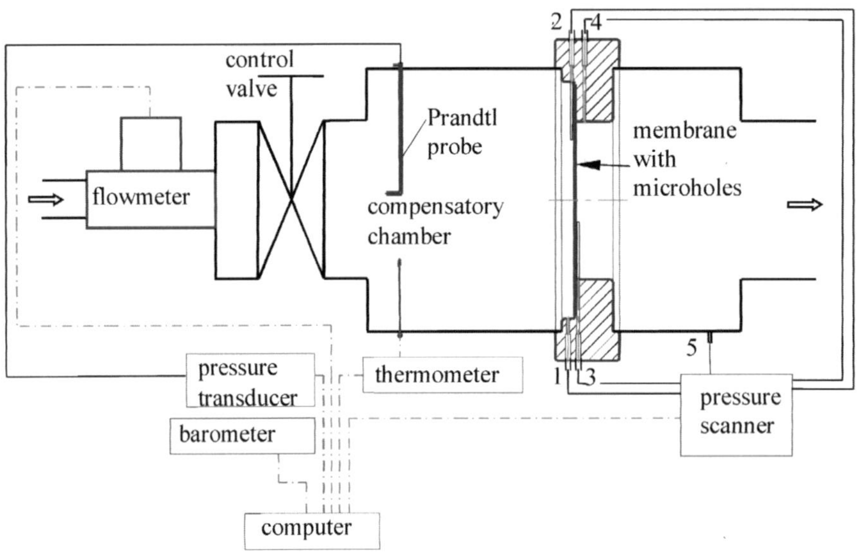

Figure 2. Scheme of measurement system.

The membrane diameter was 60 mm. The membrane was produced from steel. The diameter of a single microhole was equal to 110 μm and the length was equal to 500 μm. The membrane perforation, e.g., the ratio of the surface of all holes to the whole membrane surface was equal to 5.2%, which means that the membrane contained about 15 500 microholes. A detailed view of the membrane is shown in Fig. 3. The accuracy of the estimation of the microhole diameter and length is 10 μm and 2 μm respectively.

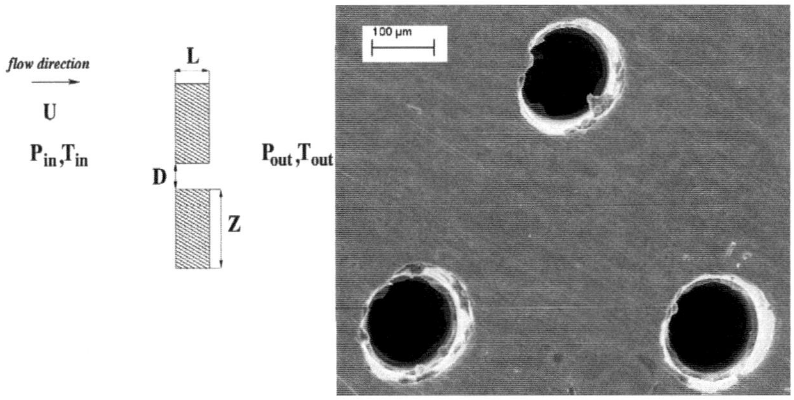

Figure 3. Schematic representation of the perforated plate (D =110 µm, L=500 µm) and the microscopic picture.

3. Experimental results

The Reynolds number is determined by the hole's diameter and the outlet parameters. The Mach number in the hole is a function of Reynolds numbers. As can be noted from Fig. 4, there are two distinguishable flow regimes: one for low Reynolds numbers (Re < 10) and one for high Reynolds numbers (Re > 50). The latter is related to the turbulence transition and the evidence for that is provided later. Curves in both of the regimes behave similarly and the changes are related to the adiabatic transition between the conditions of the gas. Figure 5 shows the mass flow rate as a function of the Reynolds number. The nature of the curves is not surprising since the velocity from the mass flow rate was taken into account to estimate the Reynolds number. Hence, the linear dependency between the mass flow rate Q_h and the Reynolds number Re is observed and plotted in Fig. 5

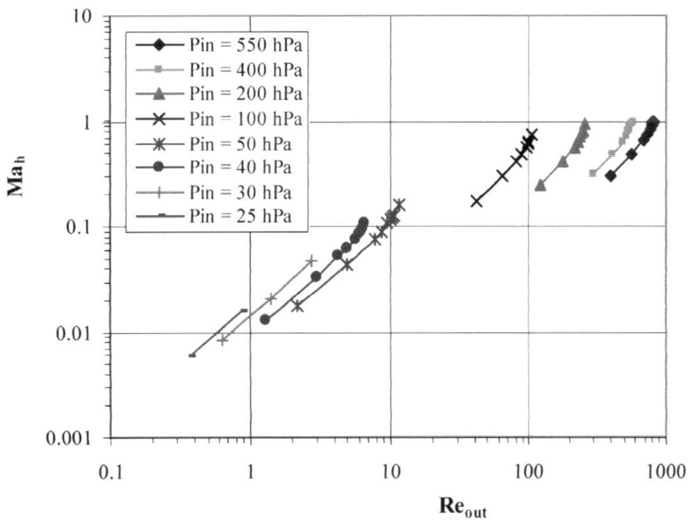

Figure 4. Nondimensional number dependencies, the Mach number as a function of the Reynolds number.

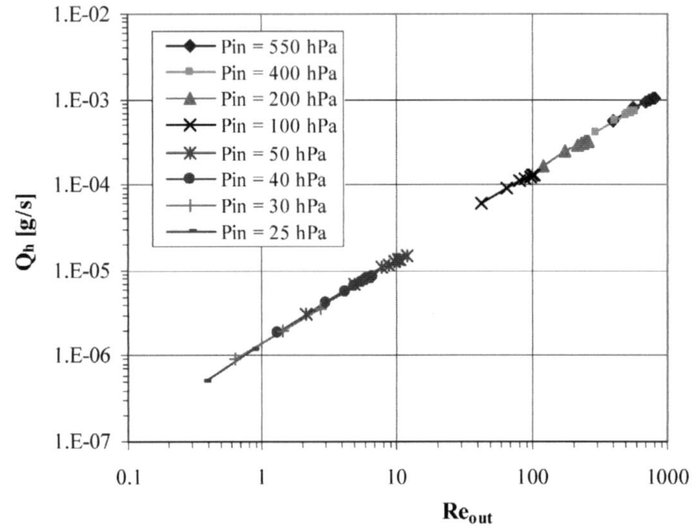

Figure 5. Mass flow rate as a function of the Reynolds number.

In Fig. 6 the velocity in the hole is plotted as a function of the pressure difference on both sides of the perforated plate. This standard plot was first proposed by Hagen [19] and shows most clearly the transition to turbulence which occurs for cases of the inlet pressure of P_{in}=100÷550 hPa. The velocity of the flow should obey the following law for the laminar flow – $\Delta P=P_{in}-P_{out}\sim U$ and for the transition to turbulence regime – $\Delta P=P_{in}-P_{out}\sim U^{1.75}$ (see Ref. [18]). Figure 7 shows the mass flow rate changes in relation to the pressure drop. The characteristics are similar to the velocity changes and show the dependence of the velocity changes from the flow regime. Table 1 shows the best fit values for the velocity where the function is as follows:

$$\Delta P = AU^{B} \tag{4}$$

where A, B are fitting coefficients

Table 1. Coefficients for the velocity pressure relation in equation (4)

P_{in} [hPa]	25÷50	100	200	400	550
A	0.62	0.04	0.02	0.02	0.04
B	0.99	1.35	1.54	1.71	0.577

It can be seen that the laminar flow (Figure 6b shows an enlargement of the laminar part) has indeed, a linear velocity dependency, while the transition to the turbulence regime is different showing the similarity to the exponential function. It can be noted that the results obtained in our experiments are in agreement with the literature. The point of this work was to study microscale effects and evidence of these effects is much more difficult to find in the turbulent regime due to the insufficient theoretical description. Hence our further investigations are mainly focused on the laminar flow. Figure 7 shows the mass flow rate as a function of the pressure difference. It can be noted that the compressibility effects are large in the transitional regime. Moreover it can be seen that the transition in our experiments occurs at very similar Reynolds number as in other investigations. We obtained a transitional Reynolds number around Re =80 and in the case concerning the pipe of similar diameter, it has occurred for Re = 81.5 [20].

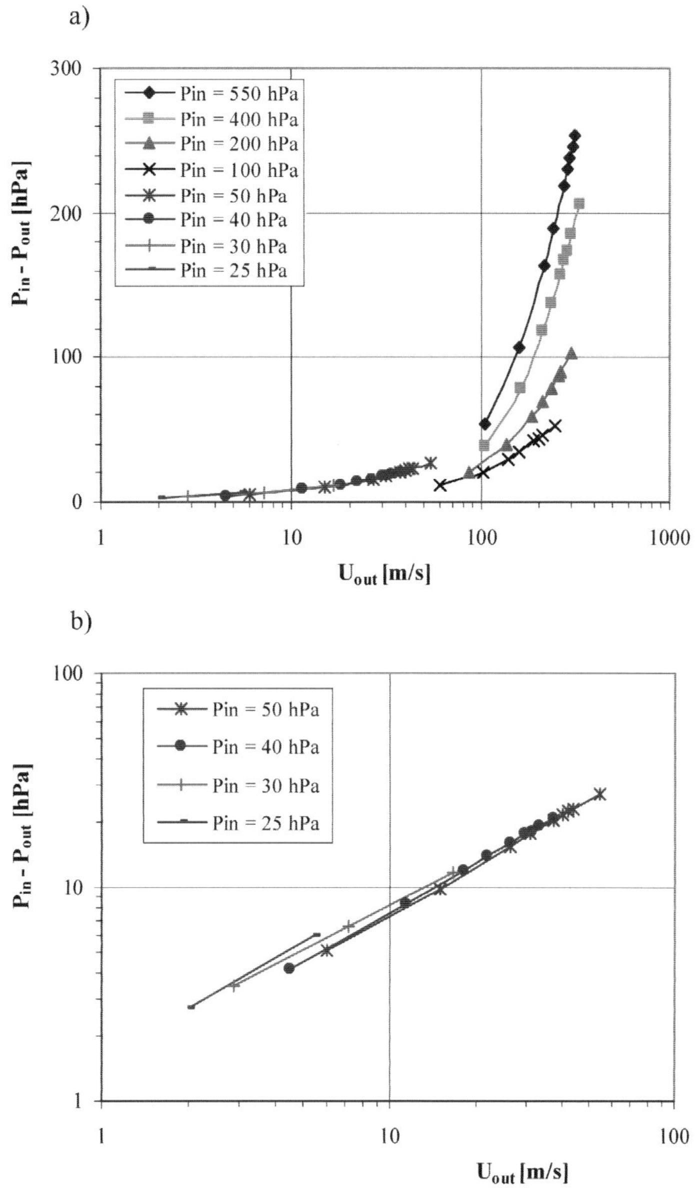

Figure 6. Transition to turbulence, a) laminar and turbulent regime, b) only laminar regime.

In Figure 8 the universal dimensionless pressure drop results defined as $\frac{\Delta P}{\rho U_0^2}$ are plotted and can be very good illustrations of the breakdown of the laminar, friction dominated flow regime [24].This figure shows clearly that transitional Reynolds number is around *Re* =80.

4. Perforated plate aerodynamics

Perforated plate aerodynamics has been studied for many years, mostly focusing on plates with large holes. Both the laminar and turbulent regimes have been considered. Some specific examples can be found in the literature; for example the pressure loss coefficient over plate thickness [21], the turbulent flow over a perforated plate [2] and a very high Reynolds number (*Re* > 10 000) [22]. In this section

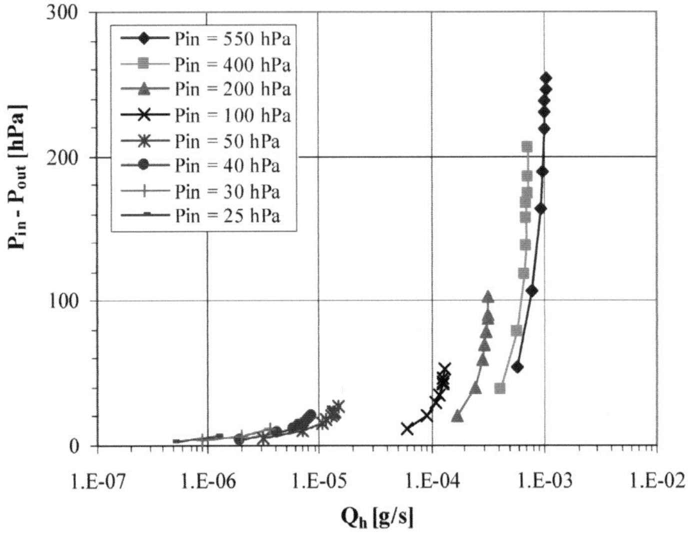

Figure 7. Transition to turbulence and compressibility effects shown as changes of the mass flow rate.

the obtained results will be compared with the ones of two specific papers. One considers various large scale holes in different types of plates [23] and the second is a study of various shapes of microholes on a large plate [4]. The mass flow rate Q is given by the equation from the empirical studies in Ref. [23]

$$Q=FS\varrho_h U_h \tag{5}$$

which by implementing the stagnation parameters transforms to:

$$Q = FS\frac{Ma_h}{\left(1+\frac{\kappa-1}{2}Ma_h\right)^{\frac{\kappa+1}{2(\kappa-1)}}}p_0\sqrt{\frac{\kappa}{RT_0}} \tag{6}$$

F-perforation, S-perforated plate surface, $P_0=P_{in}$, h-hole.
The experimental relation derived in Ref. [23] leads to:

$$Ma_h = 1.2\left(\frac{\Delta P}{P_0}\right)^{0.55} \tag{7}$$

Curve representing the equation (7) (B/D model) is plotted in Fig. 9 together with experimental data. Two regimes are clearly seen. The turbulent flow for the highest inlet pressures fits well to the relation derived from the experiments in Ref. [23]. The laminar flow and turbulent one at low Reynolds number behave differently. The exact fitting coefficients are presented in Table 2, where the function is given by the equation:

$$Ma_h = A\left(\frac{\Delta P}{P_0}\right)^{B} \tag{8}$$

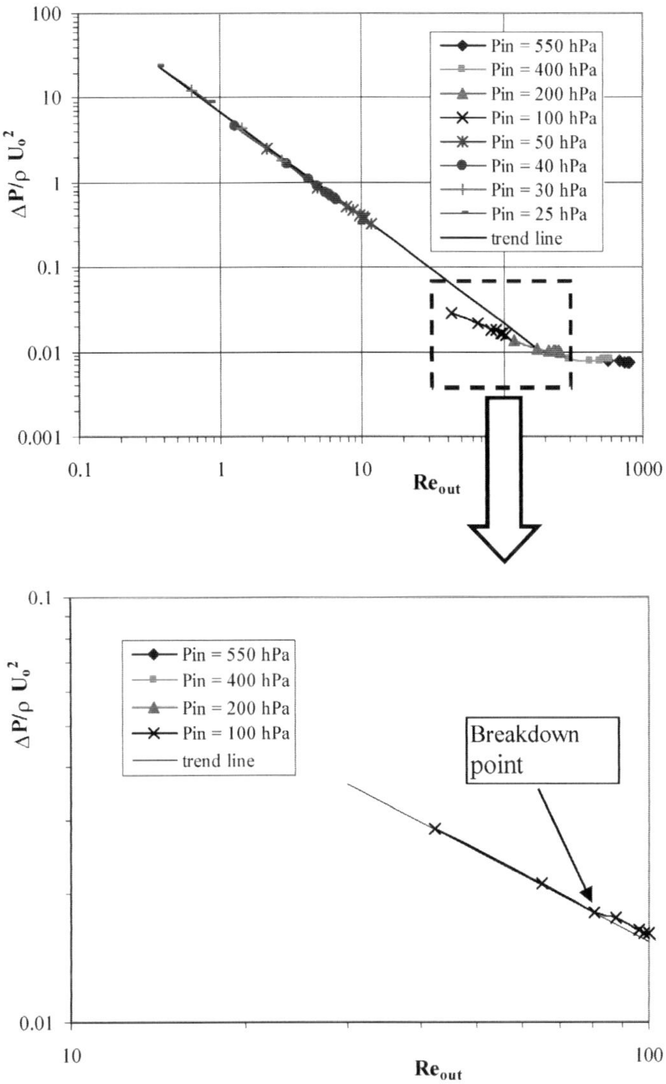

Figure. 8. Illustration of the breakdown of laminar friction dominated flow.

Table 2. Coefficients for the Mach number pressure drop dependency (see Eq. (**8**))

P_{in} [hP a]	A	B
30	0.18	1.01
40	0.25	0.96
50	0.32	0.93
100	1.16	0.65
200	1.3	0.53
400	1.43	0.5
550	1.95	0.49

It can be seen that the coefficient B for the laminar flow is twice as large as the one for the transition to turbulence. This means that at lower Reynolds numbers, for the same pressure drop, there is an increase in the importance of viscous effects. The results confirm that despite the small size of the holes in our experiments the global plate aerodynamics fits very well to the large scale investigation. However to extend B/D model it need to be introduced variables as function of P_{in} coefficient $A=A(P_{in})$ and $B=B(P_{in})$ in equation (8). These functions are a polynomials of the 4th order and probably they fit well only for our specific experiments. Therefore the B/D model needs further development and modification by using some detailed insightful theoretical aerodynamical models.

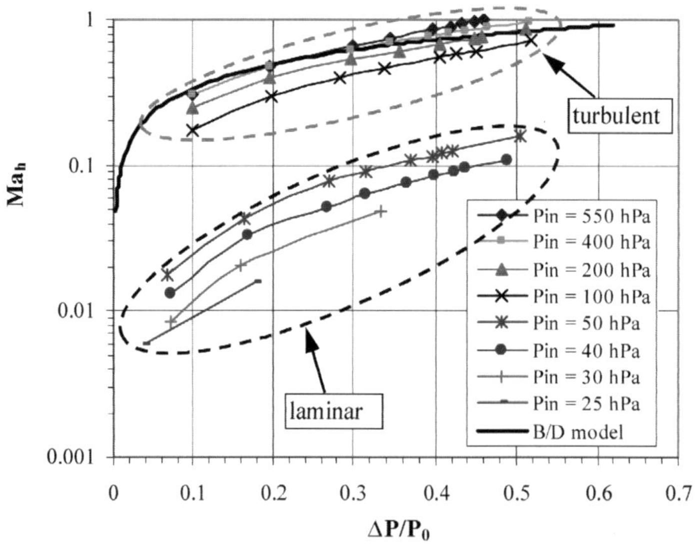

Figure 9. Comparison with the other experiments for the perforated plates, B/D model for perforated plates in parallel flow.

5. Conclusions

This paper focuses on the experimental investigation of the perorated plate aerodynamics for wide ranges of Reynolds numbers. It was shown that two Reynolds dependent regimes existed: the laminar one and the transition to turbulence occurring at Re=80. The observed Reynolds number value for the transition to turbulence is very low. It confirms observations performed by other researchers, where transition to turbulence in microchannel is present for Reynolds numbers much smaller than in the macroscale. The carried out experiment is the first one which shows the turbulent transition on the perforated plate with microscale influence. The perforated plate had the holes of a micrometre size (110 μm). The experimental results were compared with compressible experimental models for a large scale [23]. For all of these cases, the measured values agreed only for high Reynolds numbers regions. To confirm that a more extensive study with various plate perforation models and flow parameters would need to be performed.

It was observed during the experiments that small increase of mass flow through the perforated plate was accompanied by high pressure losses what can be an important parameter in potential application of effusive cooling technique in gas turbines. Therefore the future work should be also focused on finding the optimal coolant mass flow, adequate for blade thermal loading and simultaneously not introducing to much pressures drop and losses in turbine blades cooling system.

References

[1] Budoff M and Zorumski W 1971 *Flow Resistance of Perforated Plates in Tangential Flow* NASA Report No TM X-2361

[2] Liu R and Ting D 2007 *Turbulent Flow Downstream of a Perforated Plate Sharp-Edged Orifice Versus Finite-Thickness Holes* ASME Journal of Fluids Engineering 129 pp 1164–1171

[3] Van Rijn C Van der Wekken M Hijdam W and Elwenpoek M 1994 *Deflection and Maximum*

Load of Microfiltration Membrane Sieves Made With Silicon Micromachining Journal of Microelectromechanical Systems 6 pp 48–54

[4] Yang J Ho C Yang X and Tai Y 2001 *Micromachined Particle Filter With Low Power Dissipation* ASME Journal of Fluids Engineering 123 pp 899–908

[5] Zierep J Bohning R and Doerffer P 2003 *Experimental and Analytical Analysis of Perforated Plate Aerodynamics* Journal of Thermal Science 12(3) pp 193–197

[6] Zhang T Jia L Li C Yang L and Jaluria Y 2011 *Experimental Study on Single-Phase Gas Flow in Microtubes* ASME Journal of Heat Transfer 133

[7] Lahjomri J and Obarra A 2013 *Hydrodynamic and Thermal Characteris- tics of Laminar Slip Flow Over a Horizontal Isothermal Flat Plate* ASME Journal of Heat Transfer 135 p 021704

[8] Barashkin S Porodnov B and Syromyatnikov S 1989 *Density Distribution in Rarefied Gas Flow Behind a Perforated Plate* Fluid Dynamics 24(3) pp 482–484

[9] Edwards D Shapiro M Bar-Yoseph P and Shapira M 1990 *The Influence of Reynolds Number Upon the Apparent Permeability of Spatially Periodic Arrays of Cylinders* Physics of Fluids A 2(1) pp 45–55

[10] Tio K and Sadhal S 1994 *Boundary Conditions for Stokes Flows Near a Porous Membrane* Applied Science Research 52 pp 1–20

[11] Duan Z 2011 *Incompressible Criterion and Pressure Drop for Gaseous Slip Flow in Circular and Noncircular Microchannels* ASME Journal of Fluids Engineering 133 p 074501

[12] Maa D 1975 *Potentials of Micro Perforated Absorbers* Journal of the Acoustical Society of America 104(5) pp 2866–2868

[13] Babak V Babak T Kholpanov L and Malyusov V 1986 A *Procedure for Calculating Matrix Heat Exchangers Formed From Perforated Plates* Journal of Engineering Physics 50(3) pp 330–335

[14] Bunker R 2007 *Gas Turbine Heat Transfer Ten Remaining Hot Gas Path Challenges* ASME Journal of Turbomachinary 129(2) pp 193–201

[15] Gau C and Huang W 1990 *Effect of Weak Swirling Flow on Film Cooling Performance* ASME Journal of Turbomachinary 112(4) pp 786–791

[16] Howarth L 1953 *Modern Developments in Fluid Dynamics High Speed Flow* Vol. 2 Clarendon Oxford UK

[17] Bird G 1994 Molecular *Gas Dynamics and the Direct Simulation of Gas Flows* Oxford University Press New York

[18] White F 1998 *Fluid Mechanics* McGraw-Hill New York

[19] Hagen G 1839 *Uber den bewegung des wassers in engen cylindrischen rohren* Annual Review of Physical Chemistry 46 pp 423–442

[20] Eckhardt B 2008 *Turbulence Transition in Pipe Flow Some Open Questions* Nonlinearity 21 pp 1–11

[21] Gan G and Riffat S 1997 *Pressure Loss Characteristics of Orifice and Perforated Plates* Experimental Thermal and Fluid Science 14(2) pp 160–165

[22] Cuhadaroglu B Akansu Y and Turhal A 2007 *An Experimental Study on the Effects of Uniform Injection Through One Perforated Surface of a Square Cylinder on Some Aerodynamic Parameters* Experimental Thermal and Fluid Science 31 pp 909–915

[23] Doerffer P and Bohning R 2000 *Modelling of Perforated Plate Aerodynamics Performance* Aerospace Science and Technology 4 pp 525–534

[24] Bayazit Y Sparrow EM Joseph DD 2014 *Perforated plates for fluid management Plate geometry effects and flow regimes* International Journal of Thermal Sciences 85

Experimental and numerical study on the performance of the smooth-land labyrinth seal

A Szymanski[1], S Dykas[1], W Wróblewski[1], M Majkut[1] and M Strozik[1]

[1] Silesian University of Technology, Institute of Power Engineering and Turbomachinery, ul. Konarskiego 18, 44-100 Gliwice, Poland

E-mail: Artur.Szymanski@polsl.pl

Abstract. In turbomachinery the secondary flow system includes flow phenomena occurring outside the main channel, where the gaseous medium performs work on blades. Secondary air distribution constitutes a very complex and closely interrelated system that affects most of the gas turbine components. One of the most important examples of the secondary flow is leakage occurring in seals, e.g. at the rotor and stator tips, on the shaft or on the sides of the blade rim. Owing to its simplicity, low price, easy maintenance and high temperature capability, the labyrinth seal is a prime sealing solution that may be selected from numerous types of sealing structures applied in turbomachinery. For this reason, an experimental study of this particular structure has been carried out. The paper presents leakage performance of the smooth-land labyrinth seal.

1. Introduction

The continuous efforts to reduce emissions of pollutants and improve the reliability, comfort and safety of the operation of aircraft engines force designers and scientists to implement new structural solutions. One important way for enhancing energy conversion efficiency in turbine engines is to improve and perfect the sealing operating conditions. The sealing type selection affects not only the size of leakage but also the temperature distribution downstream the stage, the character of the main flow, the entire system fuel consumption, the dynamics of the system operation, as well as the risk of the creep phenomenon or of a decrease in the engine thrust. The results presented in [1] may serve as an example here. They indicate that a rise in the air flow through sealing from 3.5% to 4.5%, calculated as the main flow percentage (a relative 50% rise), involves an increase in the turbine outlet temperature by 15°C. This in turn decreases durability by approximately 30% due to the risk of creep occurrence [1].

Many different kinds of sealing are used in turbomachinery. Such seals comprise the following types: brush seals, finger seals, leaf seals, graphite seals (contact seals) as well as labyrinth, foil, gas and liquid seals (non-contact). Labyrinth seals are characterized by a very simple design and are the most common turbomachinery sealing type. However, the leakage size is big compared to other solutions. Due to that fact, turbomachinery labyrinth seals have been the focus of interest of numerous academic and research centres for many years. The works undertaken in them, both experimental and numerical [2], have concentrated mainly on determining the impact of geometrical parameters on leakage, distribution of pressure [3], temperature or propagation of aeroacoustic phenomena. In experiment, most researchers focus on stationary configurations, where the motion of the rotor with the labyrinth seal relative to the casing is ignored. In experimental works [4], [5] it is proved that such simplification may be assumed if the flow velocity over the labyrinth seal fin, c_{ax}, is higher than the rotor tangential velocity, u – in this case stationary test rigs simulate the flow behaviour correctly and the condition $u/c_{ax} < 1$ is satisfied. However, if the above-mentioned condition $u/c_{ax} < 1$ is not met and a substantial increase occurs in the mass flow difference between the stationary model and the rotating one. In most of the cases, assumption of non- rotating test rig gives satisfactory results. Researchers comparing behaviour of labyrinth seals test rigs with and without rotation pay attention to static pressure uniformity in circumferential direction. The works mentioned above also prove that the mass flow through the seal depends on many parameters, such as the inlet pressure, the inlet temperature, the Reynolds number, the angle of inflow onto the seal (initial preswirl), the pressure ratio, the seal relative motion and the architecture of the seal itself – the size of the clearance over the labyrinth fin in particular. In experiments carried out on stationary test rigs [6] [7] [8] the observed parameters are the dimensionless mass flow rate and the temperature and pressure distribution along the seal structure. These parameters make it possible to determine the loss coefficient distribution. Both high pressure feeding system (where test rig was supplied with air at pressure higher than ambient) and vacuum feeding system were used for test stands in the presented works. In all cases the authors emphasize the importance of the measurement of the clearance size as the parameter with the biggest impact on the measuring uncertainty. Under the work mentioned herein, a concept is used of feeding the measuring stand with air with ambient temperature at a stationary – axial architecture of the sealing. The air is sucked by the air blower, working as a vacuum pump, into a vessel where the set pressure value prevails. The sealing architecture maintains all dimensions of the fins and of the clearance, as well as the shape of the element over the fins. This paper presents the approach and results of experimental testing carried out on an in-house stationary (the labyrinth does not move) measuring test rig of the Institute of Power Engineering and Turbomachinery (IPET) of the Silesian University of Technology (SUT). The aim of the tests was to establish the impact of the clearance size on leakage and the static pressure distribution along the labyrinth structure. The experimental research was supported by CFD simulations by means of Ansys CFX commercial package.

2. Experimental facility

The experimental part of the testing was performed using the air installation in the Turbomachinery Laboratory of the Institute of Power Engineering and Turbomachinery of the Silesian University of Technology. The installation includes a Roots air blower with the output of 600 Nm^3/min (0.2 kg/s), a 3 m^3 pressure vessel and a pipeline system connecting the air installation to the research stand equipped with a measuring system. The system main components and the measuring instrumentation calibration procedures are described in detail in the next subsections. The minimum achievable pressure at the air blower inlet is 50 kPa(a). The total capacity of the installation (vessel and pipelines) exceeds 3.5 m3, which is relatively big compared to the expected volume flow rates, but such dimensions are indispensable to maintain stable and repeatable pressure distributions at the measurement stand inlet. Fig. 1 presents a simplified diagram of the installation. In Table 1 each part of the installation is explained. The diameter of the pipelines is DN100, which ensures a low flow velocity and small pressure losses. Secondary air (Fig. 1) is sucked in from the surroundings through the DN100 and DN 50 with throttling valve. The valve makes it possible to additionally regulate the pressure value in the vessel. The valve is controlled by an electric actuator, which allows a precise and repeatable setting of the opening angle which is remote-adjusted from the measuring system. On the stand side, the air is sucked in through an inlet preceded by a pipe which is 4 m long, which makes it possible to create appropriate conditions for the mass flow measurement upstream the test rig. After the air passes through the stand, it gets into the vessel and then – through the blower – into the environment. Mass flow evaluation is placed at inlet pipe, 3m downstream the pipe inlet.

2.1. Test rig

A stationary test rig was designed to carry out experimental testing of the flow through labyrinth sealing using a stationary linear model of the labyrinth seal. Details of the stand are presented in Figures 1 and 2. The stand is composed of 3 main elements: the inlet channel, the outlet channel and the main chamber where the tested labyrinth sealing structures are installed. As it can be seen, the inlet channel is much longer than the outlet one. This is dictated by the channel opening angle of 7°. This opening value ensures smooth transition from a small diameter, DN100, to a big one without the risk of the separation phenomenon arising in the boundary layer. The outlet channel is much shorter. On the side walls of the measuring chamber there is an inspection peephole with a window made of BK7 optical glass, which allows measurements by means of optical methods such as Laser Doppler Anemometry. The interior dimensions of the chamber are (L x H x W) 478x240x240mm. Mentioned width of test section is more than 200 times higher than examined clearances, which minimizes the effect of side walls on the main flow. Fig. 3 presents an analysed model of the seal, with placement of pressure taps, and main dimensions. Relative values of the most important dimensions are presented in Table 2. In presented study, pressure distribution is measured on upper wall only. In Fig. 3, place of pressure measurement is coloured red. Black arrow shows place of static pressure measurement, which is used for operating pressure ratio evaluation. Pressure ratio is defined as the ratio of inlet total pressure to static pressure measured in presented point. It is described by the following equation:

$$\pi = \frac{p_{0,in}}{p_{s,16}} \tag{1}$$

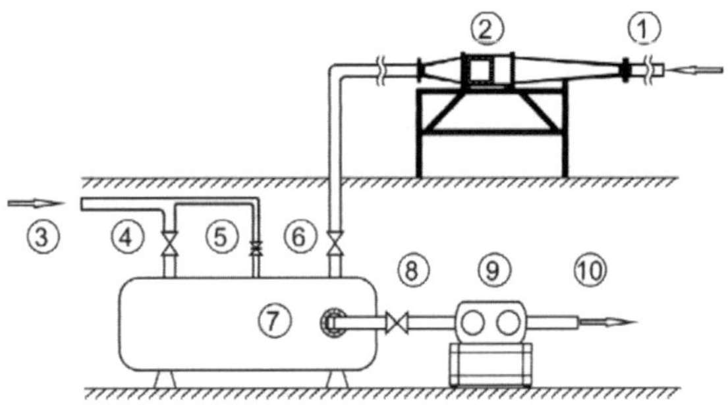

Figure 1. Experimental installation intended for the turbine seal testing in the IPET of the SUT.

Table 1. Installation description

Point	
1	Test rig inlet (mass flow measurement)
2	Test rig
3	Secondary air inlet
4	Cut-off valve
5	Remote controlled valve
6	Cut-off valve
7	3 m^3 vessel
8	Cut-off valve
9	Roots air blower
10	Outlet

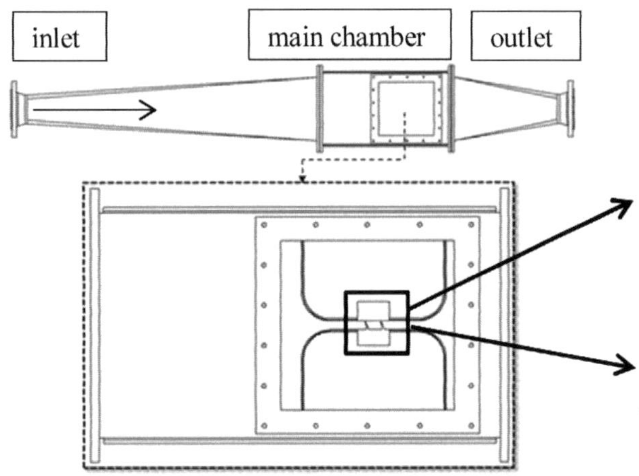

Figure 2. View of the stationary test rig.

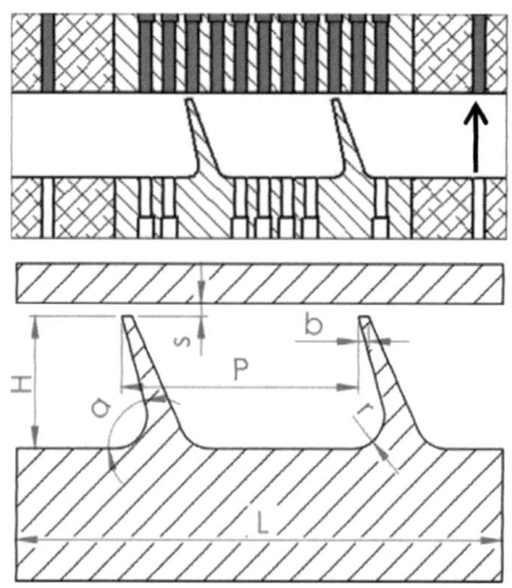

Figure 3. The geometry of examined labyrinth seal.

Table 2. Dimensions of the labyrinth seal

Dimension	Value	Dimension	Value
H/b	12.5	r/b	1.875
P/b	23.19	a	20°
L/b	47.5	s	0.6, 0.75, 1 mm

2.2. Data acquisition system

The measurement system may be divided into several subsystems. The first is the air installation measurement subsystem. It allows the assessment of the blower operation conditions and provides operating information, like bearing temperatures, outlet air temperature and pressure in the vessel and pipeline. The second is the measurement subsystem of the test rig. The quantities measured therein are strictly related to the sealing model operation. The test rig system allows to measure the inlet air velocity, pressure differences, static pressure and temperature distribution along the investigated structure. Accuracy of velocity measurement done by Schmidt 20.500 thermal anemometer does not exceed 2 % measured value, pressure is measured by PC-28 Aplisens pressure transducer with accuracy of ±256 Pa, temperature measurement accuracy, done by means of TTP-TKb, J thermocouple, stands at 0.5°C at 100°C. Investigated clearances has been evaluated using normalised blade feelers. The execution elements and the data acquisition process are managed by a LabView handler co-operating with a PC class computer. The handler was specially developed for the needs of this project. The measuring stand and the software make up an open system that may be developed further by connecting more measuring clusters and modules for a program-based analysis and visualization of data. For results evaluation, about 1000 samples is taken, which corresponds to 90 seconds of measurement.

2.3. Hot Wire Anemometry calibration against Laser Doppler Anemometry

The mass flow was measured using a technique based on the measurement of the stream velocity in a certain point of the flow cross-section. Using the information about the local velocity and knowing of the velocity profile and the flow cross-section size as well as the calibration coefficients determined experimentally, it is possible to find the volume flow of the medium.

Within this project, the stream local velocity was determined using a Schmidt 20.500 thermal anemometer installed in the stand inlet pipe. Pitot tubes were also used for control purposes. The velocity profiles in the pipeline were determined using an LDA anemometer made by Dantec. The LDA technique is characterized by high measuring accuracy and may be treated as a model tool for the stream velocity determination. The velocity profiles, obtained from the measurement (Fig. 4) were used to determine calibration coefficients, which made it possible to calibrate the thermal anemometry probe and find the volume flow rate. The calibration was verified by additional LDA-based determination of velocity profiles before the seal structure for a few selected pressure ratios. The error in the post-calibration volume flow determination does not exceed 2% of the measured value. The manufactures pre calibration correction factor is in range 0.72 – 0.73. Presented procedure made it possible to enhance the measurement accuracy.

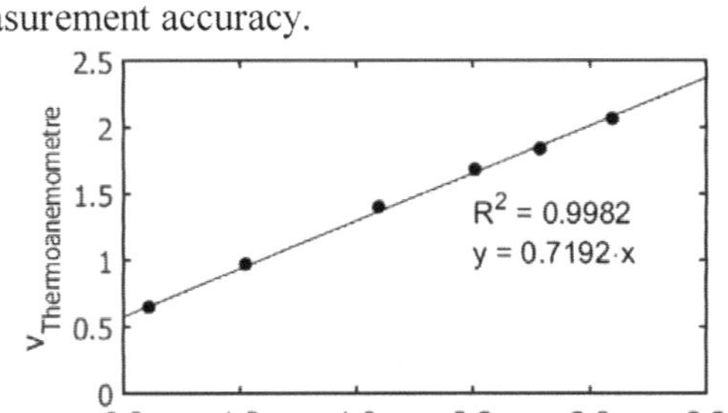

Figure 4. Results of Hot Wire Anemometer measurement against LDA.

3. CFD approach

This section presents the methodology of CFD computations of the flow through the seal model. The aim of the numerical analyses was to determine the air mass flow and the pressure distribution along the seal, especially in the area of the measuring hole location. A simple model of the seal was used for this purpose. Its view is shown in Fig. 5. The clearance between the fin tip and the casing is the same as in experiment, and the width – 5.5 mm, with periodicity boundary condition. These values make it possible to limit the number of the domain elements, ensuring a short computation time and keeping high convergence of residues. Consequently, the velocity components in the circumferential direction are practically eliminated and the model may be considered as a two-dimensional one. The model presented below was used in a further stage to generate the mesh. Computational mesh was generated in ICEM-CFD software, as fully structured. The correct solution to the flow patterns should be mesh-independent. To check this property, a numerical study with seven different mesh sizes with 105k, 167k, 554k, 1.1M, 8.9M, 17.5M and 38.8M grid nodes were used for the mesh dependence tests. The initial mesh with 105k nodes was not refined entirely, but only in regions with greater parameter gradients. The deviation between results for meshes 167k to 38.8M was less than 1% (C_D = 0.591 – 0.605). Although C_D value for initial mesh with 105k nodes was over estimated (C_D = 0.62). This indicates that acceptable tolerance was reached, and the 167k-node mesh can be used for a further analysis. Mesh employed for this analysis is shown in Fig. 6. The mesh has the maximum value of y^+ less than 2, but in the regions close to the rotor walls – $y^+ \ll 1$.

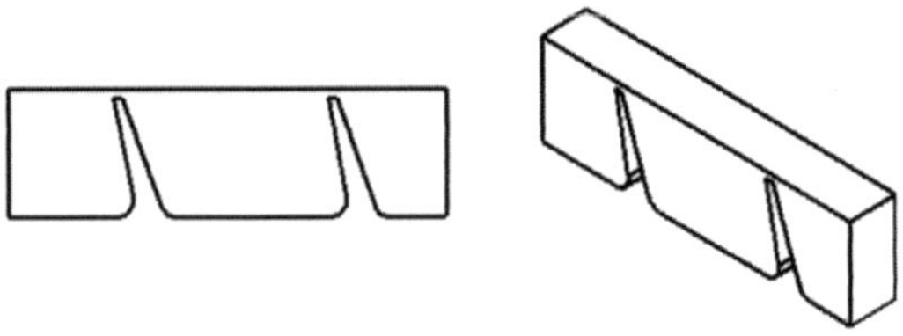

Figure 5. Geometry of CFD model employed for CFD analysis.

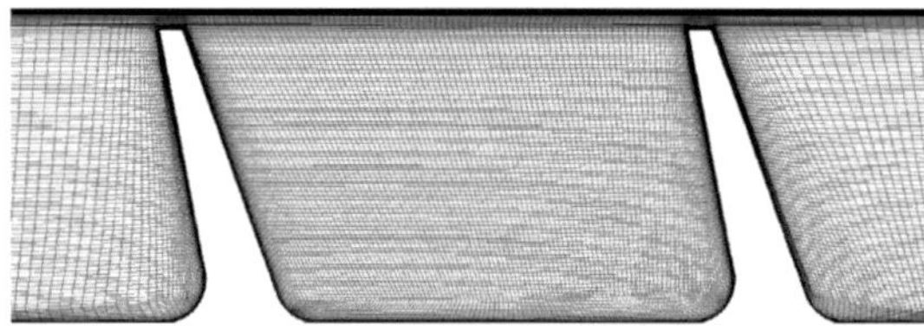

Figure 6. Final mesh of fluid domain above labyrinth structure.

The computations were performed using the commercial code Ansys CFX 16.2 using steady state scheme with Shear Stress Transport turbulence model. Boundary conditions corresponded to the measurements conditions. The working medium was dry air treated as perfect ideal gas. Moreover, the Sutherland formula was applied to determine dynamic viscosity and thermal conductivity as a function of temperature. Due to the fact that, especially after behind the seal second fin, the stream accelerated considerably causing turbulence in the outlet area, the outlet channel was lengthened. Such an approach eliminates the impact of boundary conditions on the flow structure in the sealing and on the possibility of reverse flow formation on the outlet cross-section. A similar procedure was used in the inlet area, where the swirl forming before in front of the seal first fin could distort the process of inflow onto the sealing.

4. Results evaluation

The operation of labyrinth seals is described in literature using dimensionless parameters. One of the most common ones is the discharge coefficient – C_D, which is the ratio of the current mass flow – measured or being the result of numerical calculations – to the mass flow resulting from ideal expansion in a nozzle with a surface area equal to the surface area of the clearance. It takes account of the inlet total pressure and total temperature at the inlet and the pressure ratio forcing the flow:

$$C_D = \frac{\dot{m}}{m_{id}} \tag{2}$$

$$m_{id} = \frac{p_{0,in} \cdot A}{\sqrt{T_{0,in}}} \left(\frac{2\kappa}{R(\kappa-1)} \left[\left(\frac{1}{\pi}\right)^{\frac{2}{\kappa}} - \left(\frac{1}{\pi}\right)^{\frac{\kappa+1}{\kappa}} \right] \right)^{1/2} \tag{3}$$

Moreover, there is flow function Ψ employed, which is defined as follows:

$$\psi = \frac{\dot{m}\cdot\sqrt{T_{0,in}}}{A\cdot p_{0,in}} \tag{4}$$

For comparison, the curve illustrating the history of the ideal flow function Ψ determined from analytical relationships is also presented:

$$\psi_{ideal} = \left(\frac{2\kappa}{R(\kappa-1)}\left[\left(\frac{1}{\pi}\right)^{\frac{2}{\kappa}} - \left(\frac{1}{\pi}\right)^{\frac{\kappa+1}{\kappa}}\right]\right)^{1/2} \tag{5}$$

4.1. Flow indicators evaluation

In this subsection, flow indicators – discharge coefficient C_D , flow function Ψ and flow function for ideal parameters Ψ_{ideal} described by equations (2-5) have been presented. Mentioned results show high repeatability of obtained results for mentioned values of clearance. Marked errorbars have been evaluated using the total differential method . It has to be noticed here that each measurement was preceded by disassembling and assembling of the sealing, what was associated with setting of the clearance. Average measurement uncertainty does not exceed 2.5% of C_D or Ψ for clearance s = 1mm, 3.5% for s = 0.75 mm and 4.5% for s = 0.6 mm. Discharge coefficient C_D, as well as Ψ values rise with pressure ratio. For higher clearances, the higher flow indicators were obtained. This is a result of higher mass flow rate in structure with higher clearance, as well as lower losses caused by flow through narrow gap. For comparison, characteristics of discharge coefficient and flow function, for labyrinth seal with smooth land, obtained by means of CFD simulations has also been presented (figures 7, 8). Results from numerical simulations were similar to experimental ones. Average difference between experiment and CFD stood below the uncertainty of measurement.

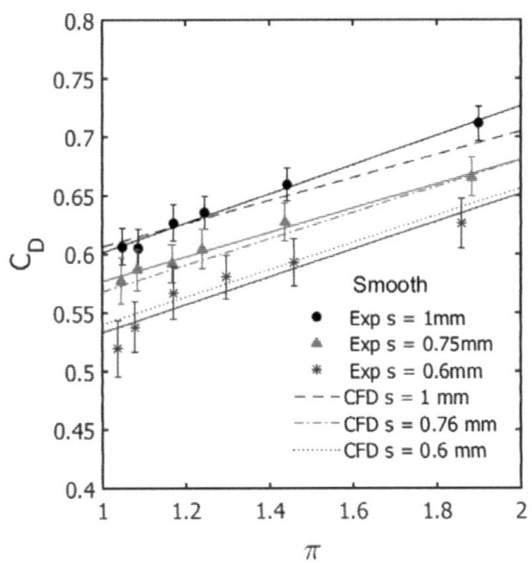

Figure 7. Discharge coefficient with pressure ratio – experiment and CFD.

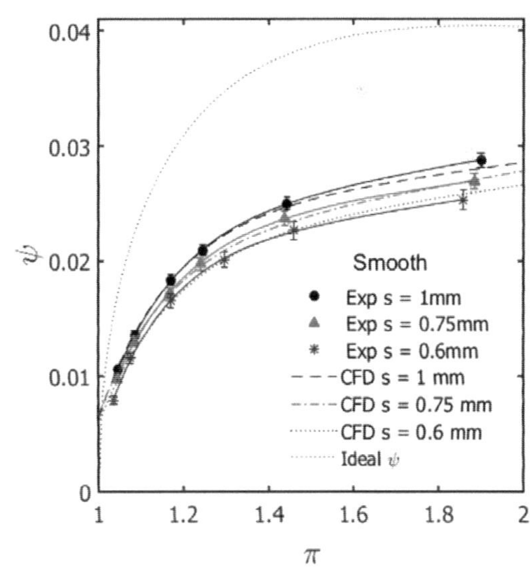

Figure 8. Flow function with pressure ratio – experiment, CFD and ideal value.

4.2. Relative pressure distribution

Figures 9-12 present distributions of static pressure to inlet total pressure ratio along the seal found experimentally for the upper wall. An accurate configuration of the applied sealing is also marked in the figures. Distances marked as x correspond to axial position from the beginning of the labyrinth seal element. Pressure drop measured at first fin is much higher than pressure drop at second fin. This phenomena is stronger for higher clearance – in case with s = 1 mm, pressure in cavity between fins, is almost the same as behind the second fin. This effect is visible for pressure ratios lower than π = 1.4. Furthermore, in front of second fin, at distance x = 25 mm some static pressure growth has been measured. This effect is mostly caused by velocity drop. Experimental data have been compared with CFD results. The comparison shows some disagreement between experimental and numerical results. Although the tendencies are similar, the CFD results under predict the static pressure value along the structure. The reason of this differences is still being investigated.

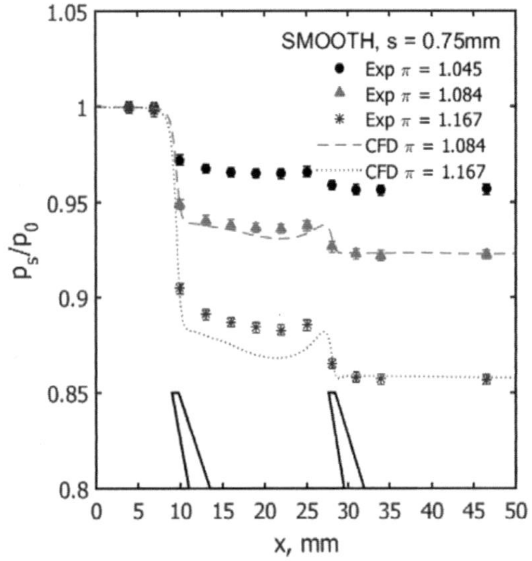

Figure 9. Static pressure distribution for smooth s = 0.75 mm, by low pressure ratios.

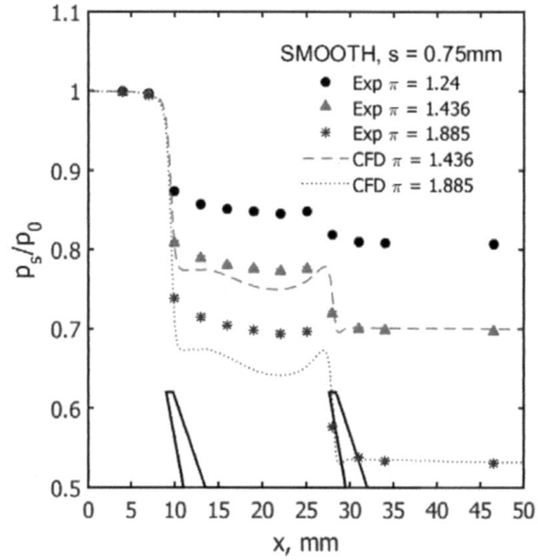

Figure 10. Static pressure distribution for smooth s = 0.75 mm, by high pressure ratios

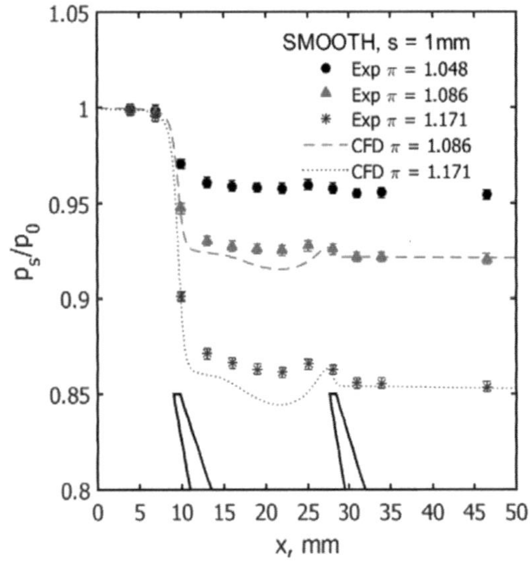

Figure 11. Static pressure distribution for smooth s = 1 mm, by low pressure ratios.

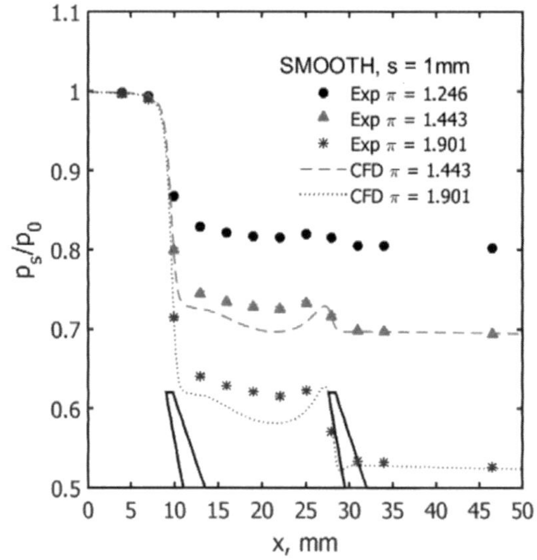

Figure 12. Static pressure distribution for smooth s = 1 mm, by high pressure ratios.

5. Summary and final remarks

In this paper description of experimental test rig for investigation of the flow through the sealing with different architecture has been presented. Together with a design of experimental facility a measurement system and CFD approach have been described. The results of the measurements for one sealing configuration with various clearances have been presented. On the basis of the performed measurements and CFD calculations one can draw the following main conclusions:

— Mass flow measurements by mean of thermo-anemometer have been calibrated against the LDA measurements of the velocity profile, what gives a very high accuracy, which is desirable in case of low mass flow rate measurement.

— Average measurement uncertainty stood below $2 - 4.5\%$, depending on the clearance size.

— CFD calculations met agreement in case of the flow indicators evaluation, although in case of the static pressure measurement, the CFD code underestimates the pressure values.

Further work will focus on more detailed CFD analysis, for explanation the static pressure behaviour, as well as experimental investigation exploiting different labyrinth seals structures.

Acknowledgments
This work was supported by the National Centre for Research and Development and Avio Polska within the Innolot Programme INNOLOT/I/11/NCBiR/2014, project Coopernik SEK-13/RIE-5/2014.

References

[1] H. L. Stocker, D. M. Cox and G. F. Holle, "Aerodynamic performance of conventional and advanced design labyrinth seals with solid-smooth abradable, and honeycomb lands," NASA-CR-135307, 1977.

[2] Kim T S Kang Y and Moon H K 2008 *Aerodynamic Performance of Double Sided Labyrinth Seals*, 4th International Symposium on Fluid Machinery and Fluid Engineering, p. 377

[3] Massini D Facchini B Micio M Bianchini C Ceccherini A and Innocenti L 2014 *Analysis of flat plate honeycomb seals aerodynamic losses: effects of clearance,* 68th CITMEA, p. 502

[4] V. Schramm, K. Willenborg, S. Kim and S. Wittig, "Influence of a Honeycomb Facing on the Flow Through a Stepped Labyrinth Seal," Journal of Engineering for Gas Turbines and Power, vol. 124, pp. 140-146, January 2002.

[5] J. Li, S. Kong, X. Yan, S. Obi and Z. Feng, "Numerical Investigations on Leakage Performance of the Rotating Labyrinth Honeycomb Seal," Journal of Engineering for Gas Turbines and Power, vol. 132, no. 062501, March 2010.

[6] J. Denecke, V. Schramm, S. Kim and W. S., "Influence of Rub-Grooves on Labyrinth Seal Leakage," Journal of Turbomachinery, vol. 123, pp. 387-393, April 2003.

[7] M.-O. Biester, L. Mueller, J. R. Seume and Y. Guendogdu, "GT2011-45883 Time-Resolved Numerical Investigation of the Interaction of Labyrinth Seal Leakage and Main-Flow in a 1.5-Stage LP Turbine," Proceedings of the ASME Turbo Expo 2013, vol. 5, pp. 1623-1632, June 2011.

[8] J. Denecke, K. Dullenkopf, W. S. and H.-J. Bauer, "GT2005-68677 Experimental Investigation Of The Total Temperature Increase And Swirl Development In Rotating Labyrinth Seals," Proceedings of ASME Turbo Expo 2005 -Power for Land, Sea, and Air, vol. 3, pp. 1161-1171, July 2005.

Unsteady conjugate heat transfer analysis for impinging jet cooling

F Tejero, P Flaszyński, R Szwaba and J Telega
The Szewalski Institute of Fluid-Flow Machinery (IMP PAN),
Fiszera 14, 80-231 Gdańsk, Poland

E-mail: fernando.tejero@imp.gda.pl

Abstract. The paper presents the numerical investigations of the heat transfer on a flat plate cooled by a single impinging jet. The thermal conductivity of the plate was modified from a high thermal case (steel - λ = 35 W/m/K) to a low one (steel alloy Inconel - λ = 9.8 W/m/K). The numerical simulations results are compared with the experimental data from the Institute of Fluid-Flow Machinery Polish Academy of Sciences, Gdansk (Poland). The numerical simulations are carried out by means of Ansys/Fluent and k-ω SST turbulence model and the temperature evolution on the target plate is investigated by conjugated heat transfer computations.

1. Introduction

Heat transfer and cooling effectiveness of impinging jets application is being investigated by several researchers recently. It was during the 50s when Freidman and Mueller [1] reported the first research conducted on this topic. Among other purposes, there has been a big effort on turbine cooling applications [2,3] where the impinging jets might cool several sections of the engine such as the combustor, the turbine case and/or the turbine blades. The effects of a crossflow in a jet array impingement on the heat transfer was investigated by Flourscheutz et al. [4]. The research of the use of the liquid crystal technique to study the local heat transfer coefficient was carried out by Metzger and Bunker [5] and it is employed in the present investigation as well.

The proper design of an impinging jet is challenging due to the large number of parameters which should be considered and play an important role in the heat transfer effectiveness. During last years, there have been several investigations such as the studies of the effect of the jet-hole size and distribution, target surface shape (bumps configuration) or cooling channel cross-section.

The influence of plate material (change of the thermal conductivity, λ) on the heat transfer by impinging jets is investigated in the present paper. Two different material properties were studied for a plate thickness of 2 mm. The thermal conductivity of the plate was modified from a high thermal case (steel - λ = 35 W/m/K) to a low one (steel alloy Inconel - λ = 9.8 W/m/K). The unsteady simulations allowed for a better understanding of the flow physics when an impinging jet is applied. Besides, the temperature evolution on the investigated plate is successfully compared with the available experimental data (Transient Liquid Crystal measurements).

It is expected that the heat transfer efficiency might be improved by the use of rough plates (this paper presents only the smooth plate configuration). The final objective of the Innolot-Coopernik project investigations is to study the impact of different type of roughness (structured and/or unstructured) on the thermal effect of the applied impinging jet. Nevertheless, the numerical results presented here are very important for validation of the CFD and measurement techniques as first step for more complex configurations.

2. Experimental setup

In Figure 1, a sketch of the test section employed during the experiments is presented. The distance between the impinging jet nozzle outlet and the plate was modified from 4mm to 7mm (CFD results presented below are for the 4mm configuration). To avoid conduction between ambient and the side of the plate that is not interacting with the jet, it was isolated by a thick PMMA plate of low thermal conductivity ($\lambda = 0.21$ W/m/K).

The camera employed was the 3CCD JAY CV-M9GE which is capable of taking 30fps films with 24 bit color depth (8 bits for each of the RGB components). The plate is thermally isolated from its holder by PTFE elements. Once the plate is fixed, it can be precisely adjusted with micrometric screws along the flow direction. In order to control the activation of the impinging jet, a thermal shield was machined of 3mm thick plywood. If it is set in the "inactive" position, the shield covers the plate, isolating it from the oncoming flow (no influence of the jet on the plate temperature can be noticed). On the other hand, if the shield is removed, the interaction of impinging jet and plate starts.

The cylindrical nozzle of 2mm diameter and a hot air soldering station PT852 has been adapted to supply the air for the jet. Mass flow rate control is realized by an ALICAT flowmeter. The temperature is measured by a four K type thermocouples. Every temperature measurement channel has been calibrated using the DRUCK DBC150 Temperature Calibrator with 0.7°C absolute precision. The calibration function (mapping the measured temperature to real temperature) has been calculated by means of linear regression. This has been done for every thermocouple, and the coefficients have been implemented in the measurement software.

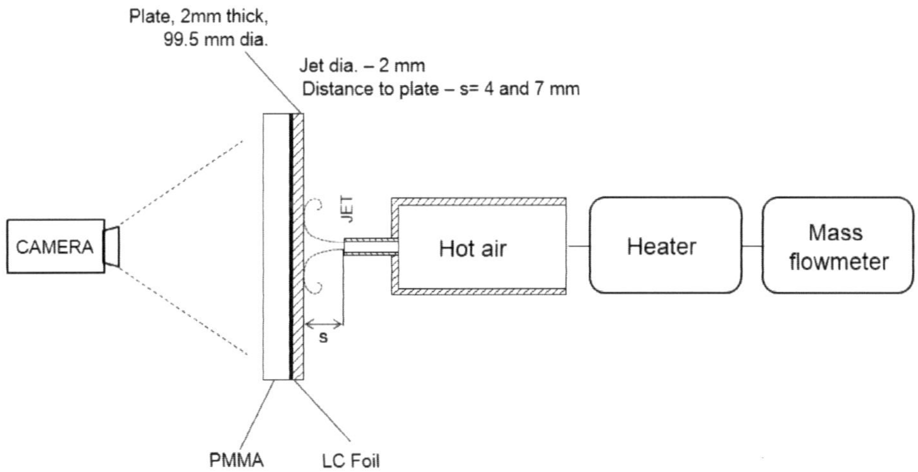

Figure 1. Sketch of the experimental setup

The investigated plate 2mm thickness and 99.5mm diameter are made of steel. One side of the plate is covered with TLC (Thermochromic Liquid Crystal) adhesive foil, while the hot air jet impinges the other side. Figure 2 presents both sides of the steel plates investigated. The red frame indicates the thermocouple wiring for plate temperature evolution in the experiment. The thermocouple sensor was located in the plate center and in the middle of plate thickness.

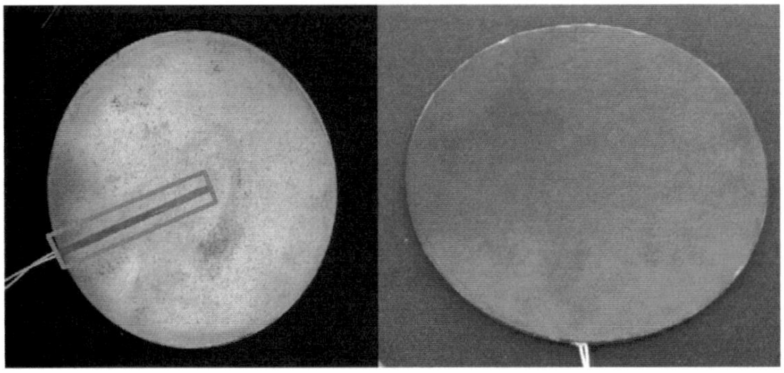

Figure 2. Investigated steel plates

3. CFD investigation

3.1. Numerical method

The numerical investigations were carried out by means of the Ansys/Fluent code which solves the Unsteady Reynolds-Averaged Navier-Stokes (URANS) equations with various turbulence models. From the different turbulent closures available, the two equation k-ω SST was chosen. The system of differential equations was closed by a perfect was model and the viscosity value was calculated according to the Sutherland's law. The numerical algorithm was based on a semi-discrete approach with finite-volume of 2nd order of accuracy for the spatial discretization as well as a 2nd order for the time-stepping scheme. The CFL number was set to 20 with the under-relaxation factor of density, body forces, turbulent kinetic energy, specific dissipation rate, turbulent viscosity and energy equal to 0.5. The unsteady simulations were performed in two stages: 1000 times steps with Δt = 5e-5s at the beginning of the computation to develop the flow in the domain and then the time step size was increased to Δt = 5e-4s for the remaining time of the simulation. It is important to mention that due to the large number of time steps needed (flow during more than 50 s is simulated which means more than 100,000 time steps), the computational domain was simplified to an axisymmetric approach, reducing considerably the number of volumes and the computational resources. The independence of the solution on the time resolution was successfully proven by the time step reduction by a factor of ten, in both stages of the simulations (i.e. Δt = 5e-6s for the first step 1000 iterations and Δt = 5e-5s for the second step of the simulation).

3.2. Computational domain – grid topology and boundary conditions

The computational domain of the impinging jet configuration is divided into 15 structured blocks (see Figure 3). The grid is generated semi-automatically using parameterized python scripts within the IGG (Numeca) software. In order to ensure a proper prediction of the boundary layer, the distance of the first grid point from the solid walls (steel and PMMA plates) is of the order if $y^+ = 1$. The proposed grid has approximately 40,000 cells. This low number of mesh cells is caused by the implementation of the axisymmetric approach in the numerical model. The number would grow up to 2,000,000 volumes for a fully 3d approach, which it is still acceptable, but due to the large number of time steps needed in the simulations, the reduction of the domain was applied.

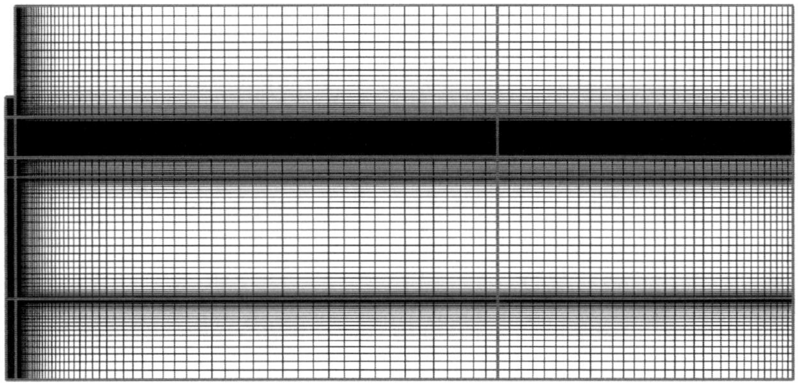

Figure 3. Structured grid and topology of the test section

The computational domain is closed by boundary conditions of 4 types: viscous and inviscid wall, mass flow inlet and pressure outlet (see Figure 4). Table 1 summarizes the conditions at the inlet and outlet of the domain. The outlet of the domain was set 30mm away in radial direction from the investigated plate (more than 30% of its diameter) which is proved to have no influence on the flow field in the near by the plate. On the other hand, the distance from the plate to the upper inviscid wall of the domain was set to 15mm (more than 7 times the thickness of the plate). It was observed how the interaction jet-surface created a wake which influenced the temperature propagation at the plane when the distance to the upper wall was not large enough. The faces of the domain colored by light blue and yellow were considered as solid (conjugated heat transfer is studied) and the heat conductivity of the materials were applied in the numerical model.

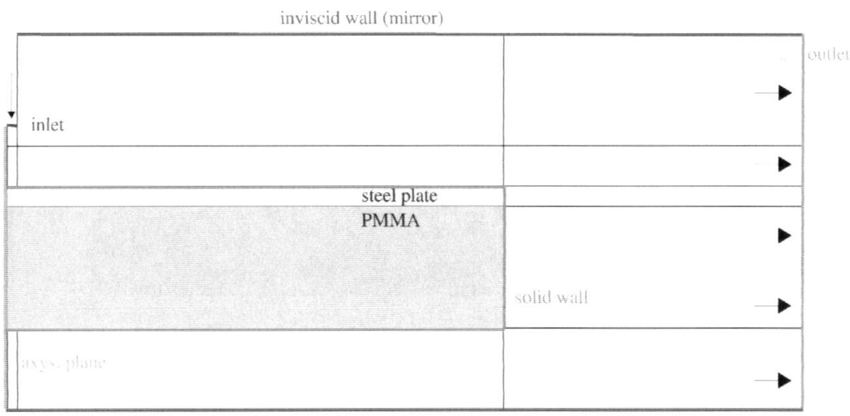

Figure 4. Boundary conditions of the numerical model

Table 1. Inlet and outlet boundary conditions

mass flow inlet	pressure outlet
Mass flow rate [kg/s]: 2.2e-4	Gauge pressure [Pa]: 101000
Turbulent intensity [%]: 0.5	Turbulent intensity [%]: 0.5
Turbulent viscosity ratio [-]: 10	Turbulent viscosity ratio [-]: 10
Total temperature [K]: 363	Total temperature [K]: 293

3.3. Numerical results for the single impinging jet

When a hot jet impinges on the plate, certain time is needed to propagate the temperature in radial direction. There are several parameters which determine the heat transfer speed such as: mass flow of the hot jet, distance jet/surface or temperature of the jet. In the present simulations, the full domain (fluid and solid) are initialized with the reference temperature of 293K. After certain time, the temperature starts to develop and the influence of the heat conduction of the target plate can be investigated. Figure 5 present three snapshots of the temperature contour map (t = 0.05, 20 and 40 s.) for the low conductivity case (λ = 9.8 W/m/K). At the beginning of the simulation (t = 0.05 s) the jet impacts the target surface and its influence is weakly visible. After certain time (e.g. t = 20s), the influence of the hot jet in all the domain is visible: above the plate, the fluid is heated with a propagation of the heat in radial direction inside the steel plate and with a small penetration in the PMMA. If the hot jet continues interaction with the target surface (t = 40s), the heat propagates in radial direction of the plate and penetrates more into the isolating material (PMMA).

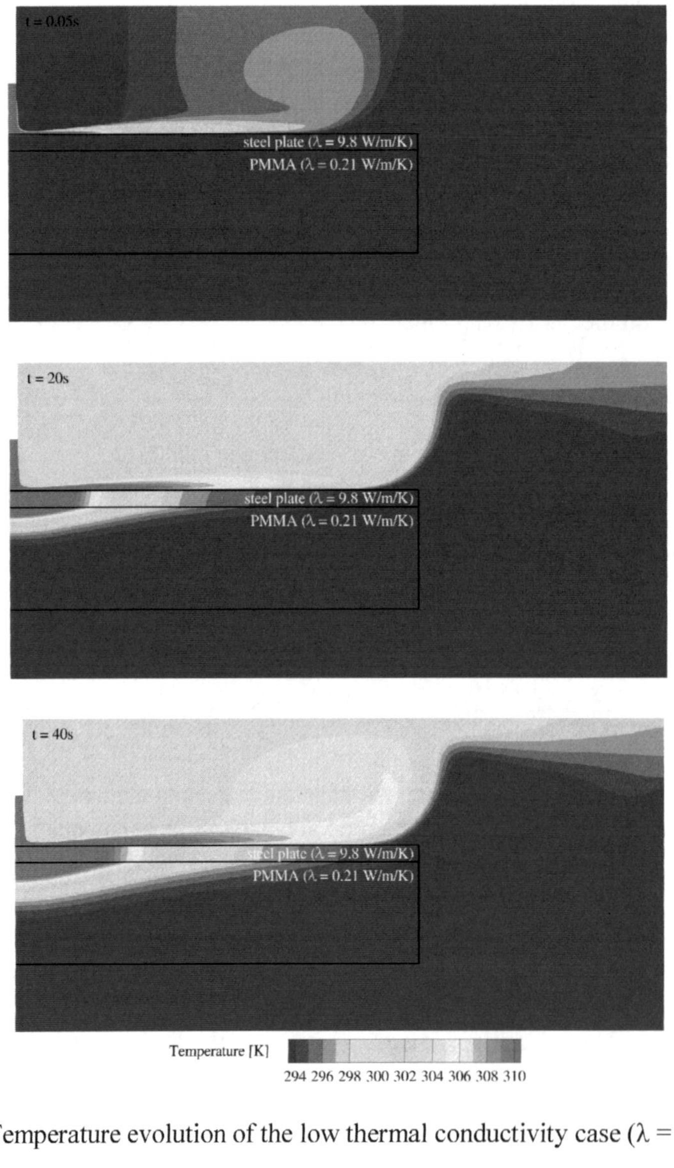

Figure 5. Temperature evolution of the low thermal conductivity case (λ = 9.8 W/m/K)

Figure 6 presents three snapshots of the temperature evolution for the high conductivity case (λ = 35 W/m/K). Since the time response for heating the plate is higher than 0.05s in both configurations investigated, no differences can be found in the first picture of Figures 5 and 6. After certain time, the heat starts to develop in radial direction of the target plate (see Figure 6, t = 20s). Although the thermal conductivity is higher in this case, it is required more time to propagate the heat in the plate. Finally, the last snapshot (t = 40s) shows how the conduction of the heat is still ongoing but also slower than for the low conductivity configuration.

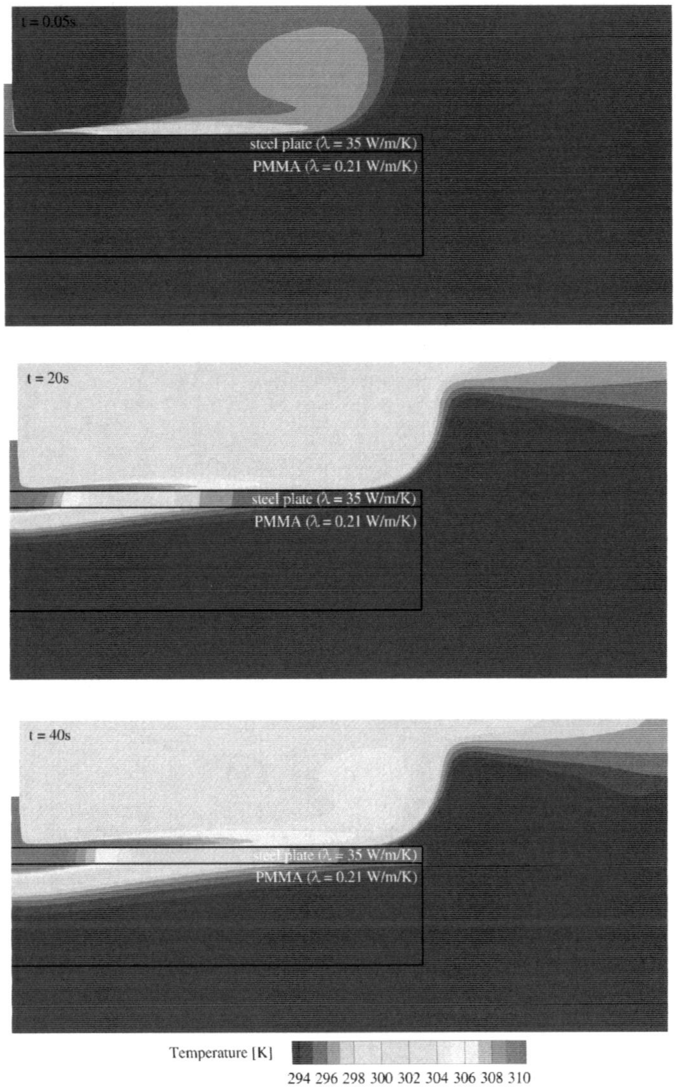

Figure 6. Temperature evolution of the low thermal conductivity case (λ = 35 W/m/K)

During the measurements, the temperature value of 308 K was chosen as reference and its propagation in the radial direction of the steel plate was tracked. Figure 6 presents the comparison of the temperature evolution in radial direction for two cases: high conductivity (λ = 35 W/m/K) and low conductivity (λ = 9.8 W/m/K). In the case of the low conductivity case, the time response of the investigated plate to heat up until the reference temperature near the stagnation point is well reproduced by CFD (approximately t = 2 s). With increased time, the target temperature moves in radial direction. Although, it is visible how the propagation is faster in the experiments than in the CFD, the slope in both investigations remains the same. In the high conductivity case, the response time for the propagation next to the stagnation point is satisfactory reproduced by the numerical simulations (approximately t = 9 s). Overall, it is visible how the CFD simulations presents a quicker propagation of the heat in the plate than the measurements. The differences arises from the change of slope caused, possibly, by the uncertainty of the exact value of the heat conductivity used during the experiments. It is also important to remark that the uncertainty of the measurements (0.7 °C) might influence the slope of the curve for high conductivity materials (due to the slow changes of the temperature in radial direction).

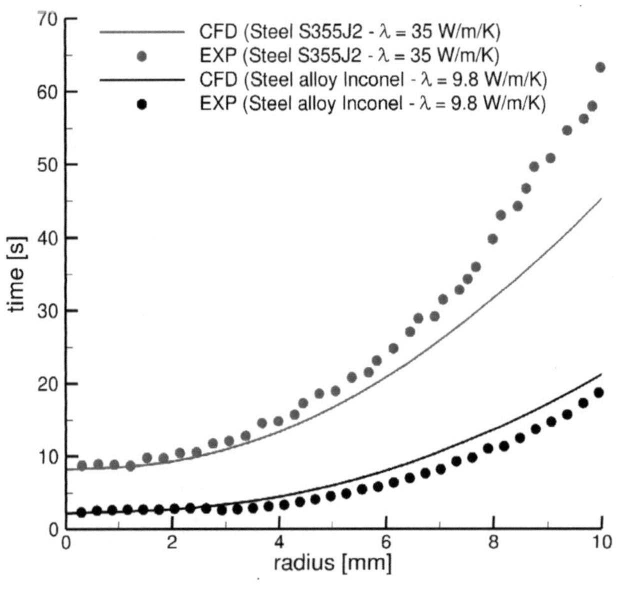

Figure 7. Comparison of the temperature propagation (T = 308K) in radial direction in the steel plate

Figure 8 presents a sensitivity investigation of the influence of the thermal conductivity on the radial propagation of a fixed temperature (308 K). It is shown how the time required to achieve the temperature threshold at the bottom side of the plate is shorter for low conductivity plates. Besides, if the certain temperature is reached close to the stagnation point, the evolution in radial direction is faster (shown by "flat" part of the curves). On the other hand, the flat behavior is followed by a curve of certain slope, which increases for higher conductivity cases.

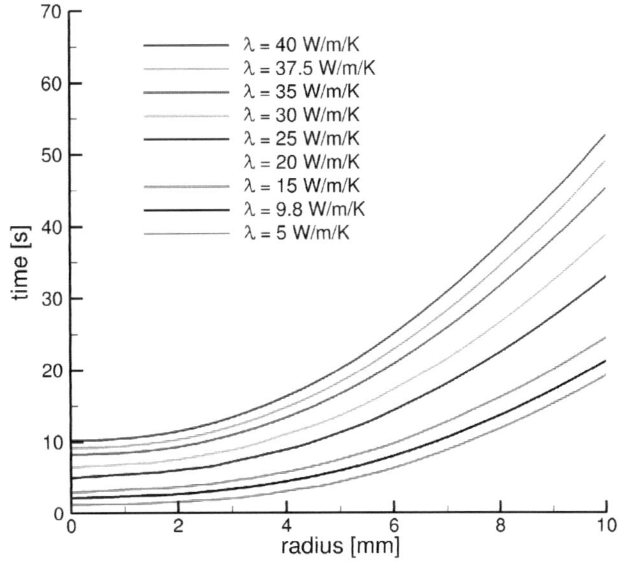

Figure 8. Sensitivity investigation of the plate thermal conductivity

Investigations of the surface modification effect on heat transfer can be assessed by heat transfer coefficient (HTC) comparison. In case of investigated configuration and applied Transient Liquid Crystal Technique, the influence of the conductivity and time dependence is an important feature. In Figure 9, the instantaneous pictures of heat transfer coefficient distribution in radial direction for investigated plates are shown. The HTC is comparable for both plates of different conductivity. Although the plates are heated up in time (as showed above), the HTC remains weakly dependent on time.

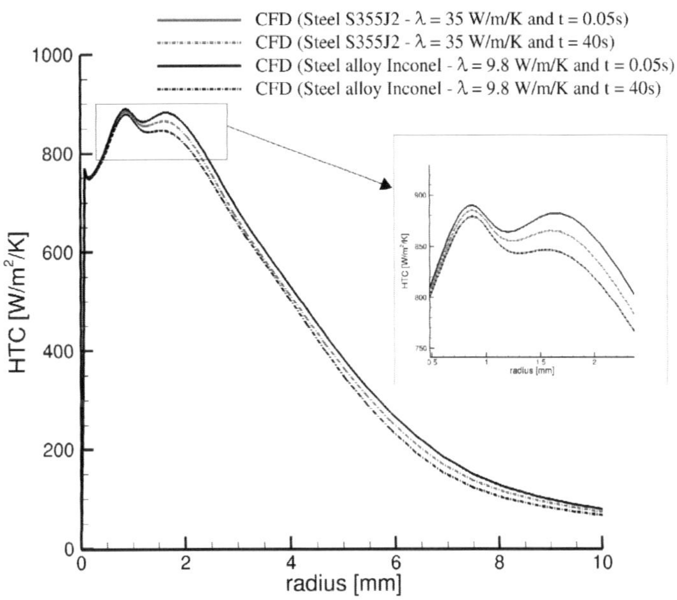

Figure 9. Comparison of heat transfer coefficient

4. Conclusions

The paper presents the details of the numerical simulations of heat transfer by a single impinging jet on a target surface. The effects on the thermal conductivity of the plate were successfully investigated. Numerical results and experimental data have satisfactory agreement for both configurations considered: low and high conductivity. The time needed to heat a plate to certain temperature and its evolution in radial direction was investigated. Heat transfer coefficient is weakly dependent on the plate conductivity and it is kept nearly constant in time during the heating up process. This validation is fundamental step before investigating the next, more complex configurations such as the influence of rough plates on the heat transfer.

Acknowledgment

This work was supported by the Innolot Programe - Coopernik project. The authors would like to thank the Academic Supercomputing Center TASK, Gdańsk (Poland) and PL-Grid Infrastructure for providing excellent HPC resources.

References

[1] Freidman S J and Mueller A C 1951 *Proc Gen Disc on Heat Transfer, Institution of Mech Engineers* pp 138-142

[2] Duckerman N and Lior N 2006 *Advances in Heat Transver* Vol 39, pp 565-631

[3] Chupp PRE, McFadden PW and Brown TR 1969 *Evaluation of internal heat-transfer coefficients for impingement-cooled turbine airfoils*, J Aircraft, 6, pp 203-208

[4] Florschuetz, L. W., Metzger, D. E., Su, C. C., Isod a, Y. and Tseng, H. H. (1984). *Heat transfer characteristics for jet array impingement with initial cross flow*, J of Heat Transfer, 106 (1), pp 34-41.

[5] Metzger, D. E. and Bunker, R. S. (1990). *Local heat transfer in internally cooled turbine airfoil leading edge regions: Part I – Impingement Cooling without Film Coolant Extraction*, J. of Turbo Machinery, 112 (3), pp 451-458.

The numerical simulation of Taylor-Couette flow with radial temperature gradient

E Tuliszka-Sznitko and K Kiełczewski

Institute of Thermal Engineering, Poznań University of Technology, Poland

E-mail: ewa.tuliszka-sznitko@put.poznan.pl

Abstract. The Taylor-Couette flow with radial temperature gradient is a canonical problem for the study of heat transfer in engineering issues. However, gaining insight into the transitional Taylor-Couette flow with temperature gradient still requires detailed experimental and numerical investigations. In the present paper we have performed computations for the cavity of aspect ratio $\Gamma = 3.76$ and radii ratios $\eta = 0.82$ and 0.375 with the heated rotating bottom disk and stationary outer cylinder. We analyse the influence of the end-wall boundary conditions and the thermal conditions on the flow structure, and on the distributions of the Nusselt number and torque along the inner and outer cylinders. The averaged values along the inner cylinder of the Nusselt number and torque obtained for different Re are analysed in the light of the results published in [2, 16, 17].

1. Introduction

Transitional flow driven by the combination of the rotation and the thermal gradients determines the dynamics which occur in complex industrial flows. The Taylor-Couette flow is one of paradigmatical systems in hydrodynamics very well suited for studying the primary instability, transitional flows and fully turbulent flows in the varying temperature fields. An overview of issues related to the Taylor-Couette flow with heat transfer can be found in [1, 2]. Regardless of the significance of the problem for basic research, the results obtained for the geometrically simple Taylor-Couette flow can be directly used in designing and optimizing many devices, such as: cooling systems in gas turbines and axial compressors, ventilation installations, desalination tanks and waste water tanks, nuclear reactor fuel rods [3, 4, 5, 6]. The Taylor-Couette flow is governed by the following parameters: radii ratio $\eta = R_1 / R_2$ (where R_1 and R_2 are the radii of the inner and outer cylinders respectively), by curvature parameter $Rm = (R_2 + R_1)/(R_2 - R_1)$ and by aspect ratio $\Gamma = H/(R_2 - R_1)$, where H is the axial dimension of the domain, figure 1. Reynolds number is defined in the following way: $\mathrm{Re} = \Omega(R_2 - R_1)R_1/\nu$ where Ω is the rotation of the inner cylinder and the bottom disk, ν is the kinematic viscosity of the fluid. The heat transfer is characterized by the thermal Rossby number $B = \beta(T_2 - T_1)$, where T_2 and T_1 are temperatures of the heated and cooled walls (β is thermal expansion coefficient).

The literature on the Taylor-Couette flow with heat transfer includes experimental, theoretical and numerical studies [7, 8, 9]. Instability and transitional process in the cavity of large aspect ratio $\Gamma > 100$ and narrow gap were investigated experimentally, among others, in the papers [10, 11]. The experimental results of the heat transfer obtained in the cavity of aspect ratio $\Gamma = 31.5$ ($\eta \sim 0.5$) were published in [12, 13]. The numerical simulations [14] obtained for the cavity of $\Gamma = 10$, $\eta = 0.5$ delivered detailed information on the flow structure, but the results revealed discrepancies between the experimental results and numerical ones. These discrepancies were attributed to the influence of the end-wall boundary conditions. The effect of the end-walls was intensively investigated for the isothermal fluid flow in [15], where the authors applied the asymmetric end-wall boundary conditions. In most of the Taylor-Couette flow numerical simulations the assumption of periodicity condition in axial direction has been used. This assumption significantly reduces the computational cost due to the fact that variables can be expanded as a Fourier series in the axial direction, which simplifies the numerical approach. Lopez et al. [2] investigated numerically the heat transfer in fluid flows between rotating cylinders ($5 \leq \Gamma \leq 80$) using no-slip boundary conditions at the end-walls and also, for comparison, using the periodicity condition in axial direction. They showed that the numerical results obtained with the axial periodicity condition agree with those obtained for closed cavities only for small Rayleigh numbers.

In the present paper we analyse the transition process in the Taylor-Couette flow in cavities of small aspect ratio $\Gamma = 3.76$, and radii ratios $\eta = 0.82$ and 0.375 with the heated rotating bottom disk and the heated stationary outer cylinder (the rotating inner cylinder and the stationary top disk are cooled). The Boussinesq approximation is used to take into account the buoyancy effect induced by the involved body forces i.e. the Coriolis force, the circumferential force resulting from the angular velocity of the rotor and the circumferential force, which is caused by the curvature of the particle track. In the paper we analyse the influence of the asymmetric end-wall boundary conditions and the thermal boundary conditions on the flow structure and on the axial and radial distributions of many physical parameters. We focus particularly on the dependence of the Nusselt number and torque [16, 17] averaged along the inner and outer cylinders of Reynolds number. The objective is also the examination how the influence of the end-wall boundary conditions on the flow structure depends on the curvature of cylinders, parameterized in the paper by η. These results are discussed in the light of the data obtained from correlation formulas proposed by Lopez et al. [2] for an infinitely long cavity and for the cavity of aspect ratio $\Gamma = 10$. The analysis is supposed to show the possibilities of using the existing data obtained for the infinitely long cylinders to predict processes in the cavities of small Γ closed by the end-walls.

The outline of the paper is as follows: the mathematical and numerical models are given in section 2. The flow structure, the radial profiles of the mean angular velocity and momentum, and the radial profiles of the dimensionless temperature obtained for cavity of $\Gamma = 3.76$, $\eta = 0.82$ are presented in section 3.1. The distributions of the Nusselt number and torque along the inner and outer cylinders, as well as, the dependence of the averaged values from Re are analyzed in section 3.2. In section 4 the flow structure obtained for the isothermal and the non-isothermal flow cases of $\Gamma = 3.76$, $\eta = 0.375$ are discussed. The conclusions are given in section 5.

2. The mathematical and numerical approaches

We consider the flow with heat transfer between two concentric cylinders of aspect ratio $\Gamma = 3.76$ and radii ratios 0.82 and 0.375 closed by end-walls. The inner cylinder of radius R_1 and the bottom disk rotates at a constant angular velocity Ω, while, the outer cylinder of radius R_2 and the top disk are at rest. The flow is described by the Navier-Stokes, continuity and energy equations written in a cylindrical coordinate system (R, φ, Z), with respect to rotating frame of reference:

$$\nabla \cdot \mathrm{V} = 0 \tag{1}$$

$$\rho \frac{\partial \mathrm{V}}{\partial t} + \rho(\mathrm{V} \cdot \nabla)\mathrm{V} + \rho\Omega \times (\Omega \times R) + 2\rho\Omega \times \mathrm{V} = -\nabla P + \mu \Delta \mathrm{V} \tag{2}$$

$$\frac{\partial T}{\partial t} + (\mathrm{V} \cdot \nabla)T = a\Delta T \tag{3}$$

where t is time, R is radius, P is pressure, ρ is density, **V** is the velocity vector, a is the thermal diffusivity and ν is the dynamic viscosity. The dimensionless axial and radial coordinates are: $z = Z/(H/2)$, $z \in [-1, 1]$, $r = [2R - (R_2 + R_1)]/(R_2 - R_1)$, $r \in [-1, 1]$. The velocity vector components in radial, azimuthal and axial directions are depicted by U, V and W, respectively, T is temperature. The

Boussinesq approximation is used to take into account the buoyancy effects induced by the involved body forces $\rho=\rho_r[1-\beta(T-T_1)]$, where $\beta=-1/\rho_r\cdot(\partial\rho/\partial T)_P$. For validity of Boussinesq approximation the thermal Rossby number $B=\beta(T_2-T_1)]$ is limited to $B\le 0.1$. The Prandtl number is equal to 0.71. The velocity components are normalized as follows: $u=U/\Omega R_2$, $v=V/\Omega R_2$, $w=W/\Omega R_2$. The dimensionless temperature is defined in the following way: $\vartheta=(T-T_1)/(T_2-T_1)$. The no-slip boundary conditions are applied to all rigid walls $(u=w=0)$. For the azimuthal velocity component the boundary conditions are: $v=0$ on the rotating inner cylinder and rotating bottom disk, and $v=-(Rm+r)/(Rm+1)$ on the stationary outer cylinder and stationary top disk. The bottom disk and the outer cylinder are heated $\vartheta=1$, and the top disk and the inner cylinder are cooled $\vartheta=0$. In order to eliminate singularities of the azimuthal velocity components at the junctions between the rotating and stationary walls, the azimuthal velocity is regularized by exponential profiles. The exponential profiles are used also for temperature at the junctions between the heated and cooled walls.

The numerical simulations (DNS/SVV) are based on a pseudo-spectral Chebyshev-Fourier-Galerkin collocation approximation. In time approximation we use a second-order semi-implicit scheme, which combines an implicit treatment of the diffusive term and an explicit Adams–Bashforth scheme for the non-linear convective terms. In the non-homogeneous radial and axial directions we use Chebyshev polynomials with the Gauss–Lobatto distributions to ensure high accuracy of the solution. Predictor/corrector method is used. All dependent variables i.e. predictors of three velocity components u, v and w, predictor of pressure and temperature, and corrector for pressure are obtained by solving Helmholtz equation [18, 19, 20].

For higher Reynolds numbers the SVV method is used. In this method an artificial viscous operator is added to Laplace operator to stabilize the computational process, [20]. The SVV operator is sufficiently effective to suppress Gibbs oscillations and simultaneously the SVV does not affect the solution accuracy. More detailed information about SVV algorithm can be found in [21, 22, 23]. For the visualization purpose we use the λ_2 criterion. The computations have been performed using mesh of 5-10 million collocation points. The time step is from the range $\delta t=0.01-0.0005$. The verification of the DNS/SVV algorithm has been done in [22, 23].

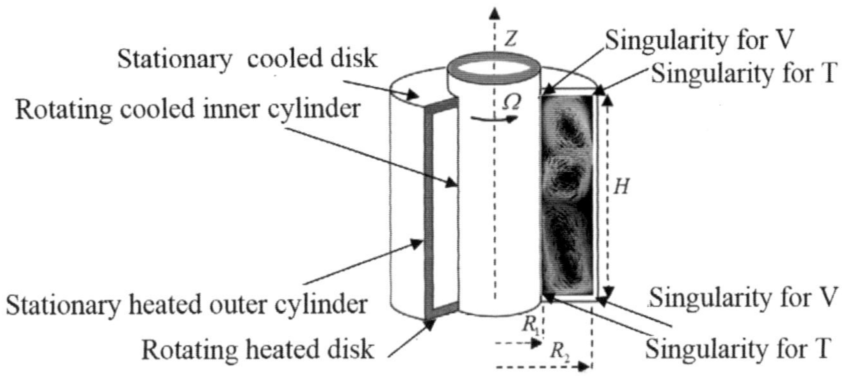

Figure 1. Schematic picture of the Taylor-Couette flow.

3. Selected results obtained for the flow case of Γ = 3.76, η = 0.82

3.1. The flow structure

In the cavity of aspect ratio Γ = 3.76 and radii ratio η = 0.82 with the heated bottom rotating disk and the stationary outer cylinder the Taylor-Couette vortices are formed at Re=80 above which three vortices in the meridian section are created, figure 3a. We observe the inflow jet along the top stationary end-wall and the outflow jet along the rotating bottom disk. The transition to unsteadiness takes place at Re = 916.7 (the critical Re of transition to unsteadiness for the isothermal flow case equals Re = 959). At Re = 916.7 we observe 10 spiral vortices (figure 2a) rising from the rotating bottom disk of the azimuthal wave lengths $\lambda_a/(H/2) = 1.7$ (in the isothermal flow case we observe 15 spirals of the azimuthal wave length $\lambda_a/(H/2) = 1.13$). The spirals occupy a large part of cavity and influence the total flow structure. We have traced the azimuthal velocity component in time in the middle section near the top disk, which allows us to characterize the growing disturbances, figure 2b.

Figures 3b, 3c, 3d and 3e show the axial profiles of the dimensionless temperature $<\vartheta>_{t,\varphi}$, the azimuthal velocity component $<v>_{t,\varphi}$, the turbulent heat flux components $<\vartheta'\vartheta'>_{t,\varphi}$ and $<\vartheta'v'>_{t,\varphi}$ obtained in the middle section of cavity ($<...>_{t,\varphi}$ stands for values averaged in time and in azimuthal direction). The meridian flow colored by the dimensionless temperature (figure 3a) allows for correlation between the flow structure and the presented axial profiles. The axial profiles of $<v>_{t,\varphi}$, and $<\vartheta>_{t,\varphi}$ clearly show the existence of two boundary layers at the bottom and top disks. We can see that all extreme values of the presented parameters are connected with the location of the inflow and outflow jets. We observe particularly large peaks in the axial profiles of the turbulent heat flux components $<\vartheta'\vartheta'>_{t,\varphi}$ and $<\vartheta'v'>_{t,\varphi}$.

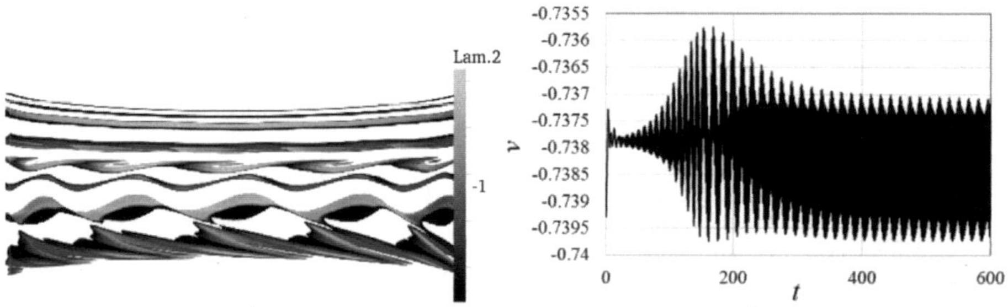

a) b)

Figure 2. The λ_2 iso-surfaces along the cylindrical surface near the inner cylinder (a), v = f(t) obtained in the middle section, r = 0. Re = 916.7, Γ = 3.76, η = 0.82, B = 0.1.

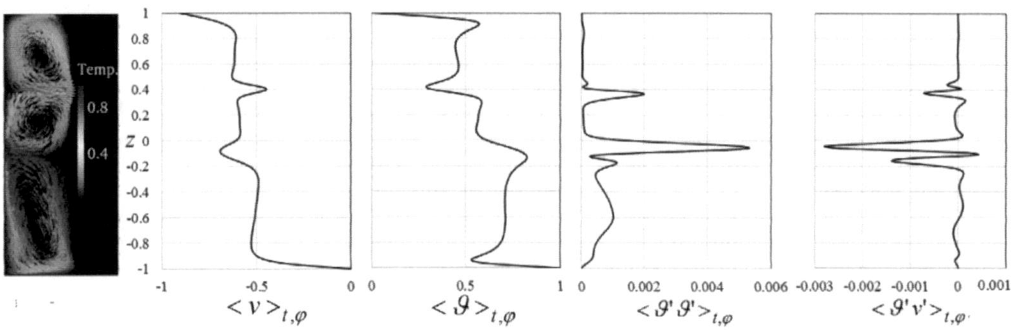

a) b) c) d) e)

Figure 3. The meridian flow structure (a), the axial profiles of: the azimuthal velocity component $<v>_{t,\varphi}$ (b), the dimensionless temperature $<\vartheta>_{t,\varphi}$ (c), the components of heat flux tensor $<\vartheta'\vartheta'>_{t,\varphi}$ (d) and $<\vartheta'v'>_{t,\varphi}$ (e). Γ = 3.76, η = 0.82, B = 0.1.

Figure 4 presents the radial profiles of the dimensionless temperature $<\vartheta>_{A(R),t}$, the angular velocity $<(V/R)/(V/R)_1>_{A(R),t}$ and the angular momentum $<(VR)/(VR)_1>_{A(R),t}$ normalized but its value at the inner cylinder ($<...>_{A(R),\varphi}$ stands for time and cylindrical surfaces average). The angular momentum (AM) and the angular velocity profiles (AV) are obtained for the isothermal (B = 0) and non-isothermal fluid (B = 0.1), Re = 1779. From figure 4b we can see that the heating of the rotating disk and the outer cylinder results in the decreasing of $<(V/R)/(V/R)_1>_{A(R),t}$ and of $<(VR)/(VR)_1>_{A(R),t}$ in the central part of the cavity, in comparison to the isothermal flow case (B =

0). This may be due to the fact that with increasing fluid temperature near the rotating disk the density of fluid decreases and so does the centrifugal force. The inverse effect is obtained by increasing Re.

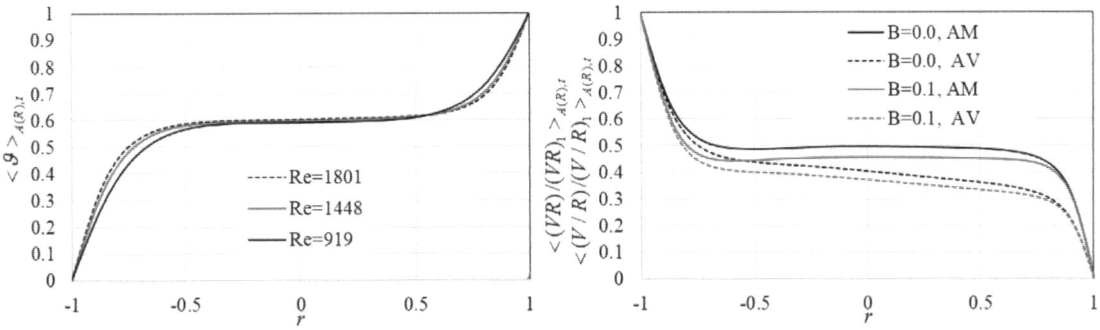

a) b)

Figure 4. The radial profiles: of the dimensionless temperature, B = 0.1 (a), of the angular velocity and the angular momentum, B = 0.1 and B = 0.0 (b). Re = 1779. Γ = 3.76, η = 0.82.

3.2. Distributions of Nusselt number and torque

The heat transfer across the gap is characterized by the distributions of Nusselt number along the inner cylinder and the outer one. The local Nusselt number is obtained in the following way $Nu_{local} = (R_2 - R_1)\alpha / \lambda$, where α ($\alpha = -\lambda(\partial T / \partial R)/(T_2 - T_1)$) is a heat transfer coefficient and λ is a thermal conductivity; the average value is depicted by Nu. Following [16, 17] we also analyze the distributions of the local transverse angular momentum current $j^{\omega} =< R^3[UV/R - \nu\partial(V/R)/\partial R] >_{t,\varphi}$ and, after averaging, in axial direction, its mean value $J^{\omega} =< R^3[UV/R - \nu\partial(V/R)/\partial R] >_{A(R),\varphi}$. The mean J^{ω} is normalized by its laminar value $Nu^{\omega} = J^{\omega} / J^{\omega}_{lam}$, [16]. In [16] the authors showed theoretically that the mean current Nu^{ω} is independent of radius in the infinitely long cavity.

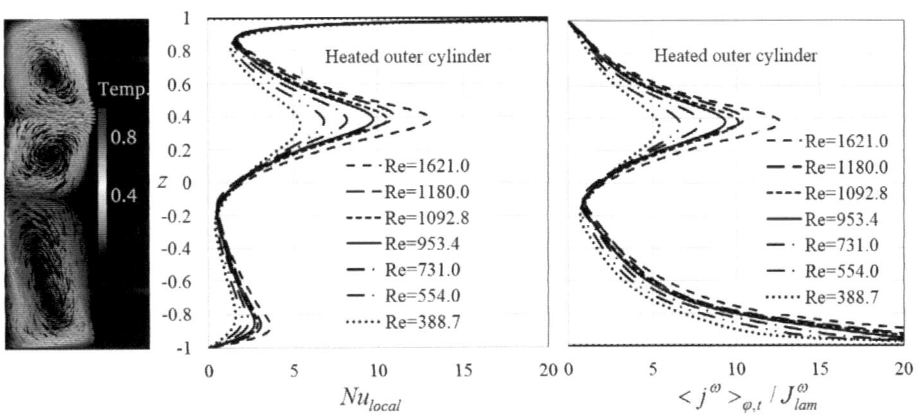

a) b) c)

Figure 5. a) The meridian flow, b) Nu_{local} as a function of z, c) $< j^{\omega} >_{\varphi,t} / J^{\omega}_{lam}$ as a function of z, the outer cylinder. B = 0.1, Γ = 3.76, η = 0.82.

In figure 5 we analyze the axial distributions of the local Nusselt number Nu_{local} (figure 5b) and $< j^{\omega} >_{\varphi,t} / J^{\omega}_{lam}$ (figure 5c) obtained along the outer cylinder for different Re. We can see that the peak values of Nu_{local} are connected with the outflow jets. The peak value at the top disk (figure 5b) is connected with the singularity of temperature at the junction between the heated outer cylinder and the cooled top disk. The extreme values of $< j^{\omega} >_{\varphi,t} / J^{\omega}_{lam}$ are observed at the bottom disk (such large value is partly due to the singularity of the azimuthal velocity component at the junction between the rotating bottom disk and the stationary outer cylinder). The remaining peak of $< j^{\omega} >_{\varphi,t} / J^{\omega}_{lam}$ is located at z = 0.4. Along the inner cylinder the peak values of Nu_{local} and $< j^{\omega} >_{\varphi,t} / J^{\omega}_{lam}$ are connected with the inflow jets. We observe the similarity between the axial profiles of Nu_{local} and $< j^{\omega} >_{\varphi,t} / J^{\omega}_{lam}$ in the central part of cavity (both in their shape and values).

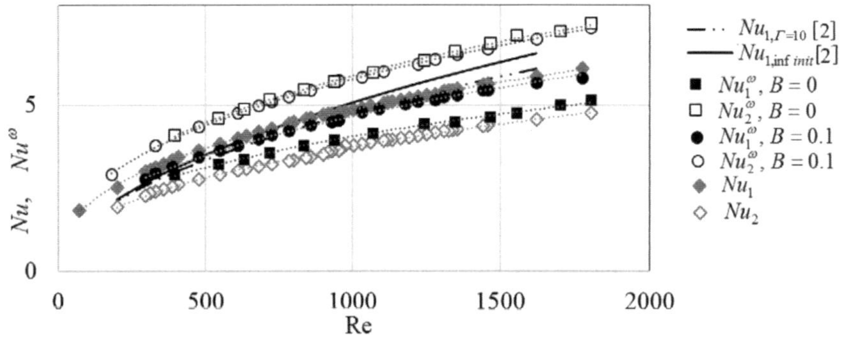

Figure 6. The Nusselt number Nu and Nu^{ω} as a function of Reynolds number with trend lines, ($\Gamma = 3.76$, $\eta = 0.82$), comparison with the results obtained in [2].

In figure 6 we summarize the results by presenting the averaged values Nu and Nu^{ω} as a function of Re. The averaged values are obtained at the inner cylinder (depicted by 1, $Nu_1, Nu^{\omega}{}_1$) and at the outer cylinder (depicted by 2, $Nu_2, Nu^{\omega}{}_2$). The results are compared with the data obtained from the correlation formula $Nu_1 = 0.13\,\mathrm{Re}^{0.53}$ proposed in the paper [2], where the authors investigated numerically (DNS) the flow in infinitely long cavity with the periodicity condition in axial direction (cavity with a heated rotating inner cylinder, a cooled stationary outer one, $\eta = 0.5$, Pr = 0.71, Ra = 1420, [2]). They also performed simulation in the cavity of $\Gamma = 10$ with the no-slip velocity condition at the end-walls and they proposed the following formula $Nu_1 = 0.14\,\mathrm{Re}^{0.51}$ for this flow case. From figure 6 we can see that our results agree with the data obtained from $Nu_1 = 0.14\,\mathrm{Re}^{0.51}$ ($\Gamma = 10$), in spite of differences in thermal boundary conditions. The differences between results are larger, for Re > 1000, when we compare our results with the data [2] obtained with the periodicity condition in axial direction. The Nusselt numbers at the outer cylinder Nu_2 are smaller in comparison to Nu_1, approximately $Nu_2 / Nu_1 = 0.78$. From figure 6 we can also see that our Nusselt number distribution Nu_1 agrees well with the distribution of $Nu^{\omega}{}_1$, and that $Nu^{\omega}{}_2$ is larger than $Nu^{\omega}{}_1$. For Re=1620, B=0.1 we estimate the difference between values $Nu^{\omega}{}_2$ and $Nu^{\omega}{}_1$ as follows: $(Nu_2^{\omega} - Nu_1^{\omega}) / Nu_1^{\omega} = 0.23$. The result shows that in the considered flow cases the influence of the end-wall boundary conditions is large (for infinitely long cylinders $Nu_2^{\omega} = Nu_1^{\omega}$).

4. Selected results obtained for the flow case of $\Gamma = 3.76$, $\eta = 0.375$

In this section we present a comparison of the flow structure obtained with the isothermal and non-isothermal boundary conditions in the cavity of radii ratio $\eta = 0.375$. For the isothermal fluid flow (B = 0) the three-cell Taylor-Couette flow structure is formed at Reynolds number Re = 80. At about Re = 270 the middle vortex is squeezed by the growth of the vortex adjacent to the bottom rotating disk (this process was previously described in [15]). Finally, the steady three-cell structure collapses to a one-cell structure at the critical Reynolds number Re = 281 via a saddle-node bifurcation. The transition to unsteadiness occurs at Re = 492. Above Re = 492 we observe six spirals of the dimensionless azimuthal wavelength $\lambda_a/(H/2) = 0.6128$, figure 7d.

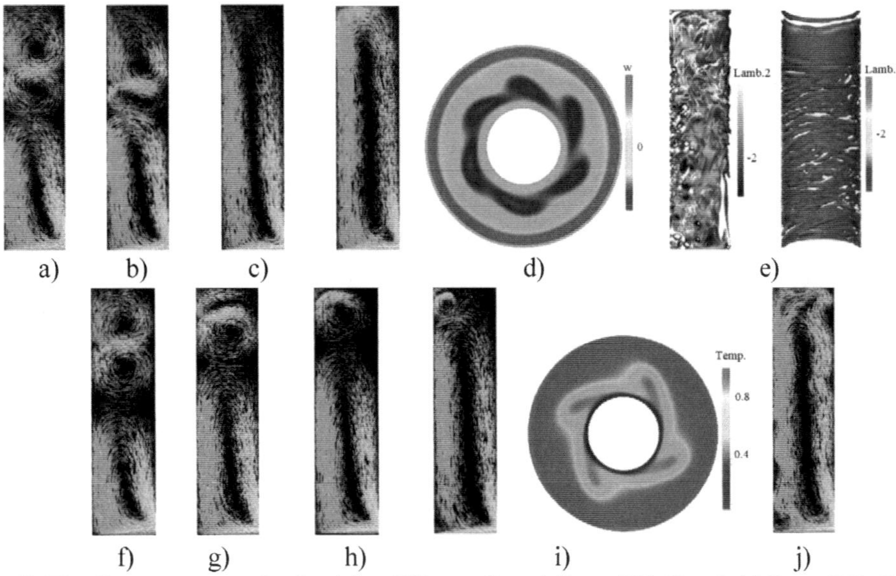

Figure 7. The flow structure obtained for different Re: a) Re = 176, B = 0, b) Re = 270, B = 0, c) Re = 281, B = 0, d) Re = 586, B = 0, e) Re = 4500, B = 0, f) Re = 176, B = 0.1, g) Re = 286, B = 0.1, h) Re = 288, B = 0.1, i) Re = 560, B = 0.1, j) Re = 890.6, B = 0.1. $\Gamma = 3.76$, $\eta = 0.375$.

The same computations have been performed for the thermal Rossby number B=0.1 with the heated outer stationary cylinder and the rotating bottom disk. The use of the non-izothermal boundary conditions changes the procedure of the transition from the three-cell structure to the one-cell structure. Again, the three-cell structure is formed at approximately Re = 80, then at approximately Re = 286 the top vortex is squeezed by the growth of the vortex adjacent to the bottom rotating disk, which results in the appearance of the two-cell structure at Re = 288, figure 7h. The flow is pumped outward along rotating bottom disk in accordance with the boundary condition, and it is pumped radially outward along the stationary top disk in opposite direction to what we can expect from the boundary condition. However, in the corner between the inner rotating cylinder and the stationary top disk there is a very small vortex (hardly visible in figure 7h) which rotates in accordance with the boundary condition. The transition to unsteadiness takes place at Re = 553, above which we observe four spiral vortices. From figures 7f-7i we can see that with the increasing Re the top vortex gradually shrinks and finally the one-cell structure is formed. We can conclude that the use of the non-isothermal boundary conditions delays the formation of the one cell-structure in comparison to the isothermal flow case (and generally slightly weakens the effect of the end-wall boundary conditions). The similar procedure of the transition from the three-cell structure to the one-cell structure we observe in the flow case of $\Gamma = 3.76$, $\eta = 0.524$, B = 0. For higher radii ratios $\eta = 0.756$ and 0.82 the classic Taylor-Couette laminar-turbulent transition takes place.

5. Conclusions

In the paper we investigate, with the use of DNS, the transitional Taylor-Couette flow between the rotating inner cylinder and bottom disk, and the stationary outer cylinder and top disk (the bottom disk and the outer cylinder are heated). The computations have been performed for the cavities of $\Gamma = 3.76$ and different radii ratios $\eta = 0.375$, 0.524, 0.615, 0.756 and 0.821 (in the paper attention is focused on the flow cases of $\eta = 0.375$ and 0.821; B = 0 and 0.1, Pr = 0.71).

For all flow cases we observe the strong effect of the Ekman boundary layers on the flow structure, however, this effect depends on the curvature of cylinders (η). For the flow case of $\eta = 0.375$ (B = 0), the effect of end-walls is so strong that it leads to a rapid transition from the three-cell structure to the one-cell structure. The use of the non-isothermal boundary conditions results in the appearance of an intermediate two-cell structure which delays the transition. The same scenario takes place in the flow case of $\eta = 0.524$. For other values $\eta = 0.756$ and 0.821 we observe the gradual development of the three-cell structure illustrated in section 3.1.

In order to better characterize the influence of the undertaken boundary conditions on the flow we examine the axial distributions of the local transverse angular momentum current $< j^{\omega} >_{\varphi,t} / J^{\omega}_{lam}$ and the distributions of the local Nussent number along the inner and outer cylinders. The comparison shows the similarity between the axial profiles of Nu_{local} and $< j^{\omega} >_{\varphi,t} / J^{\omega}_{lam}$ in the central part of the cavity but we observe large influence of singularities (of temperature and v) on the distributions near the disks. The distributions of the averaged values $Nu^{\omega}{}_2$, $Nu^{\omega}{}_1$, Nu_2, Nu_1 as a function of Re show that $Nu^{\omega}{}_1$ and Nu_1 take very similar values. Simultaneously, Nu_1 is larger than Nu_2; we estimate the correlation as $Nu_2 / Nu_1 = 0.78$. The comparison with the correlation formula $Nu_1 = 0.14\mathrm{Re}^{0.51}$ [2] obtained for $\Gamma = 10$ shows agreement. However, we observe discrepancies for Re > 1000 in comparison to correlation proposed in [2] for infinitely long cavity. The results obtained for $\Gamma = 3.76$ show that in the configuration of such low aspect ratio the influence of the end-walls is very large, particularly for smaller η.

Acknowledgements
The authors are grateful to the Poznań Supercomputing and Networking Center, where the computations have been performed.

References

[1] Fenot M, Bertin Y, Dorignac E and Lalizel G 2011 *Int. J. Thermal Sciences* **50** 7 1138
[2] Lopez J, Marques F and Avila M 2015 *Phys. Fluids* arXiv:1503.06816
[3] Owen J and Rogers R 1995 Research Studies Press Taunton, Somerset, England
[4] Singer P 1984 Semiconductor International 72
[5] Vives C 1988 *Int. J. of Heat and Mass Transfer* **31** 10
[6] Selimefendigil F 2014 *Int. J. of Thermal Sciences* **86** 258
[7] Lopez J, Marques F and Avila M 2013 *J. Fluid Mech.* **737** 56
[8] Poncet S, Haddadi S and Viazzo S 2011 *Int. J. Heat and Fluid Flow* **32** 128
[9] Yoshikawa H, Nagata M and Mutabazi I 2013 *Phys. Fluids* **25** 114104
[10] Sorour M and Coney J 1979 *J. Mech. Eng. Sci.* **21** 403
[11] Lepiller V, Goharzadeh A, Prigent A and Mutabazi I 2008 *The European Physical J.* B 71 445
[12] Ball K and Farouk B 1987 *Int. J. Numer. Meth. Fluids* **7** 857
[13] Ball K and Farouk B 1988 *J. Fluid Mech.* **197** 479
[14] Kuo D and Ball K 1997 *Phys. Fluids* **9** 2872
[15] Mullin T and Blohm C 2001 *Phys. Fluids* **13** 136
[16] Eckhardt B, Grossmann S and Lohse D 2007 *J. Fluid Mech.* **581** 221
[17] Brauckmann H and Eckhardt B 2013 *J. Fluid Mech.* **718**, 398
[18] Tuliszka-Sznitko E, Zieliński A and Majchrowski W 2009 *Int. J. Heat and Fluid Flow* **30** 534
[19] Tuliszka-Sznitko E, Majchrowski W and Kiełczewski K 2012 *Int. J. Heat and Fluid Flow* **35** 52
[20] Tadmor I 1989 *SIAM J. Numerical Analysis* 26 **30** 1989
[21] Severac E and Serre E 2007 *J. Comp. Phys.* **226** 2 1234
[22] Kielczewski K and Tuliszka-Sznitko E 2013 *Arch. Mech.* **65** 527
[23] Tuliszka-Sznitko E and Kielczewski K 2015 *Comp. Method in Science and Tech.* **21** 211

Comparing convolution-integral models with analytical pipe-flow solutions

K Urbanowicz[1], A S Tijsseling[2] and M Firkowski[1]

[1] West Pomeranian University of Technology Szczecin, Faculty of Mechanical Engineering and Mechatronics, Al Piastów 19, 70-310 Szczecin, Poland
[2] Eindhoven University of Technology, Department of Mathematics and Computer Science, PO Box 513, 5600 MB Eindhoven, The Netherlands

Abstract. This paper presents and discusses known analytical solutions for accelerated laminar pipe flow. On the basis of these solutions, formulas are given that enable the determination of shear stresses on the pipe walls together with a formula for coefficients of non-stationary friction losses. A small extension is introduced, which will make it possible to analyse flow in pipes inclined at any angle. Besides solutions related to acceleration, this paper also examines the analytical solution for decelerated laminar flow and the solution for flow changing its direction in both a stepwise and a linear manner. Calculations based on the analytical formulas are compared with numerical results obtained by Zielke's convolution-integral method. The convolution integral is treated both in the classical way and in a novel more efficient way. The classical approach requires the knowledge of the complete history of the flow, which makes it computationally inefficient and memory-wise burdensome. In the efficient calculation, only the velocity changes taking place in the last three time steps are needed.

1. Introduction

Over the last 150 years, many papers have been written on the flow of liquids in water supply systems, oil pipelines, blood vessels, and heating and cooling systems. Most of them concerned the analysis of steady states. A minority of authors dealt with transient states occurring in pipes during the first seconds after disturbing the established steady flow. The analytical solutions obtained in [1-4] required extensive knowledge of mathematics. These papers were written a long time ago, not always easy to obtain and not in English. Therefore they were largely forgotten by the scientific community. Only the progressing digitisation of many libraries, their free use through the global internet, and tools like Google Translate and Play, have made them accessible.

This year, 145 years have passed since the first publication of the analytical solution for the accelerated flow of an incompressible viscous fluid in a hydraulic pipeline or mammal blood vessel. The solution was presented in 1871 by Roiti at the University of Pisa in the university's first volume of their scientific journal [1]. Roiti developed a solution for unsteady flow in vertical pipes that was mainly based on earlier work in the field of unsteady heat transfer in cylinders by Poisson [5-7] and Betti [8,9]. Due to the local nature of the publication and the absence of English, French or German translations, this solution remained unknown for a long time after its publication. Even the Italian Fassò [10] did not refer to it sixty years ago and we think that it is still unknown to today's scientific community. Eleven years later, in 1882, Gromeka at the University of Kazan developed a similar solution but for horizontal pipes [2]. This solution stayed obscure for the same reasons as pointed out above, but finally became known to the scientific community [11].

Many years later, a similar problem was focused upon and studied by Szymański. While in 1928 staying in Paris on an internship, Szymański derived his solution which was based mainly on Fourier-Bessel series. The final works [3,12] contained a range of mathematical proofs on convergence and asymptotic behaviour of the analytical solutions. Yet another approach to find exactly the same solution was published in 1951 by Gerbes [4], who used the Laplace transform. Gerbes, besides showing the solution to accelerated flow, also proposed a solution for decelerated flow (the complete and rapid disappearance of the pressure gradient along the pipe was considered). The current study will also deal with other known analytical solutions, such as those presented in 1978 by Chambré et al [13], who analyzed flow reversal as a result of both a step change and a ramp change of the pressure gradient.

The main objective of this study is the comparison of the classic analytical solutions with solutions obtained by applying efficient numerical methods to the corresponding convolution integrals. Integral convolution models are widely used to introduce unsteady friction in numerical methods based on a mesh with fixed time steps (especially the method of characteristics). The results of the numerical simulations show that the conventional numerical solution [14] and a much more effective solution [15] are entirely consistent with the classic analytical results.

2. Analytical solutions for accelerating laminar flow

The oldest of the papers described below has been found quite by chance, while collecting materials on Gromeka. Attempts to find the papers of the Italian academic Betti, who, among other things, dealt with studies of transient heat flows [8,9], resulted in finding the forgotten work by Roiti, which will be discussed briefly below. Roiti worked at the same university (University of Pisa) as Betti, so he had a direct opportunity to study the work of his older colleague, which was especially important in an era where transfer of information, even written down, was limited.

2.1. Vertical accelerating pipe flow

The Roiti paper discussed below, entitled "Sul movimento dei liquidi" ("On the movement of liquid"), relates to transient flow in pipes and is probably the first work on this subject. Roiti, in order to find his solution, used the method for calculating transient heat flow in cylinders that was described three years earlier by Betti [8,9]. Roiti analysed accelerated flow of liquids in a vertical pipe. The hydrodynamic equation describing this type of incompressible flow reads:

$$g-\frac{1}{\rho}\frac{\partial p}{\partial x}=\frac{\partial v}{\partial t}-v\left(\frac{\partial^2 v}{\partial r^2}+\frac{1}{r}\frac{\partial v}{\partial r}\right)=\frac{\partial v}{\partial t}-\frac{v}{r}\frac{\partial}{\partial r}\left(r\frac{\partial v}{\partial r}\right), \tag{1}$$

where: r - radial coordinate [m], x - axial coordinate [x], t - time [s], v - flow velocity [m/s], p - pressure [Pa], g - acceleration due to gravity [m/s^2], ν - kinematic viscosity [m^2/s].

The solution of Eq. (1), for a suddenly applied pressure gradient and gravity acting vertically, was presented in the last chapter of Roiti's work in the form:

$$v(r,t)_{ver.}=\frac{2}{\nu R}\left(g+\frac{\Delta p}{\rho L}\right)\sum_{n=1}^{\infty}\frac{J_0(\chi_n r)}{\chi_n^3 J_1(\chi_n R)}\left(1-e^{-\nu\chi_n^2 t}\right)=\frac{2R^2}{\nu}\left(g+\frac{\Delta p}{\rho L}\right)\sum_{n=1}^{\infty}\frac{J_0\left(\lambda_n\frac{r}{R}\right)}{\lambda_n^3 J_1(\lambda_n)}\left(1-e^{-\lambda_n^2\hat{t}}\right), \tag{2}$$

where: $\lambda_n=\chi_n/R$ - the nth zero of the Bessel function of type J_0, $\hat{t}=\nu t/R^2$ - dimensionless time.

It is an expanded solution for the velocity profile in the horizontal cross-section of the vertical pipe. Roiti, with use of above formula, also determined a formula for the cross-sectional average value of the flow velocity, which is particularly useful in one-dimensional models:

$$v_{m,ver.}=\frac{2}{R^2}\int_0^R r\cdot v(r,\hat{t})dr=\frac{4R^2}{\nu}\left(g+\frac{\Delta p}{\rho L}\right)\sum_{n=1}^{\infty}\frac{1-e^{-\lambda_n^2\hat{t}}}{\lambda_n^4}=\left(\frac{R^2}{8\mu L}(g\rho L+\Delta p)\right)\left(1-32\sum_{n=1}^{\infty}\frac{e^{-\lambda_n^2\hat{t}}}{\lambda_n^4}\right). \tag{3}$$

Besides the theoretical part, Roiti's paper is distinguished by reporting a series of experimental tests carried out in a local laboratory which purpose was to confirm his newly derived formulas. But Roiti was not fully satisfied with the results, because he did not achieve full compliance of theory and experiment. The analytical results were characterised by a very good compatibility at low velocities, but significant deviations of measured results from theory was observed for higher velocities. It is difficult to clearly assess now, but perhaps these deviations were caused by the change in the nature of flow from laminar (for which the formulas were derived) to turbulent. At the time of publication of his paper, the volatile nature of turbulent flow had not been discovered yet; only twelve years later Reynolds published his most famous work [16].

For the purpose of conducting numerical studies in this paper, formulas for the shear stress on the pipe wall (4) and the temporary coefficient of resistance (5) for this type of flow have been determined:

$$\tau_{w,ver.} = \frac{R}{2}\left(\gamma + \frac{\Delta p}{L}\right)\left[1 - 4\sum_{n=1}^{\infty}\frac{e^{-\lambda_n^2 \hat{t}}}{\lambda_n^2}\right] = \frac{4\mu}{R} v_{\infty,ver.}\left[1 - 4\sum_{n=1}^{\infty}\frac{e^{-\lambda_n^2 \hat{t}}}{\lambda_n^2}\right], \tag{4}$$

$$f_{ver.} = \frac{8\tau_{w,ver.}}{\rho \cdot v_{m,ver.}^2} = \frac{64}{Re_{\infty,ver.}}\left[1 - 4\sum_{n=1}^{\infty}\frac{e^{-\lambda_n^2 \hat{t}}}{\lambda_n^2}\right] \cdot \left[1 - 32\sum_{n=1}^{\infty}\frac{e^{-\lambda_n^2 \cdot \hat{t}}}{\lambda_n^4}\right]^{-2}, \tag{5}$$

where: $\gamma = \rho g$ - specific weight of the liquid; $Re = (v_{\infty,ver.} D)/\nu$ and:

$$v_{\infty,ver.} = \left(\frac{R^2}{8\mu L}(g\rho L + \Delta p)\right). \tag{6}$$

2.2. Horizontal accelerating pipe flow

The Szymański paper [3] is widely known and repeatedly cited [17-26] and deals with accelerated flow in horizontal pipes. It contains a range of mathematical proofs on convergence and asymptotic behaviour of the derived solutions. In the 1930s that was most useful, since there were no computers to carry out quick calculational checks. More about Szymański's work and life can be read in the authors' recent paper [27]. However, it has been pointed out more and more often that an identical solution was also found by Gromeka [11,28,29]. He wrote in Russian and his paper was published in research communications of the university of Kazan [2]. This paper, however, still remains inaccessible to the wider public. Fortunately, a book published on the occasion of the hundred-year anniversary of Gromeka's birth contains his collected works [30]. It can be accessed via the internet [31]; the only problem might be the language barrier, since none of Gromeka's papers has been translated into English, French or German.

Both Gromeka and Szymański have sought a solution for the partial differential equation (1) assuming the absence of body forces (g=0). They independently presented an analogous final solution for the time-dependent velocity profile in the form:

$$v(r,\hat{t})_{hor.} = 2v_{\infty,hor.}\left[\left[1 - \left(\frac{r}{R}\right)^2\right] - 8\sum_{n=1}^{\infty}\frac{J_0\left(\lambda_n \cdot \frac{r}{R}\right)}{\lambda_n^3 \cdot J_1(\lambda_n)} \cdot e^{-\lambda^2 \cdot \hat{t}}\right], \text{ where: } v_{\infty,hor.} = \frac{R^2 \Delta p}{8\mu L}. \tag{7}$$

From the above formula, as previously from Eq. (3), we can determine the formula for the average velocity:

$$v_{m,hor.} = v_{\infty,hor.}\left[1 - 32\sum_{n=1}^{\infty}\frac{e^{-\lambda_n^2 \cdot \hat{t}}}{\lambda_n^4}\right], \tag{8}$$

and the shear stress:

$$\tau_{w,hor.} = \frac{4\mu}{R} v_{\infty,hor.} \left[1 - 4\sum_{n=1}^{\infty} \frac{e^{-\lambda_n^2 \hat{t}}}{\lambda_n^2}\right] = \frac{R\Delta p}{2L}\left[1 - 4\sum_{n=1}^{\infty} \frac{e^{-\lambda_n^2 \hat{t}}}{\lambda_n^2}\right]. \tag{9}$$

Just as in the previous section, the instantaneous coefficient of resistance has been designated:

$$f_{hor.} = \frac{8\tau_{w,hor.}}{\rho \cdot v_{m,hor.}^2} = \frac{64}{Re_{\infty,hor.}}\left[1 - 4\sum_{n=1}^{\infty} \frac{e^{-\lambda_n^2 \hat{t}}}{\lambda_n^2}\right] \cdot \left[1 - 32\sum_{n=1}^{\infty} \frac{e^{-\lambda_n^2 \cdot \hat{t}}}{\lambda_n^4}\right]^{-2}. \tag{10}$$

The latter will be useful in the calculations presented in Section 3.

2.3. General solution

Analysis of the above formulas, known from the literature, shows no immediate relation between the flow in horizontal (Gromeka, Szymański) and vertical (Roiti) pipes. In practice, hydraulic lines are often installed at different angles and are used in devices with variable use (for example excavators), in which the slope of the pipe depends on the actual situation. With this in mind, we suggest a universal form of the above formulas, in terms of piezometric heads as in steady flow [32]:

$$v(r,t) = \frac{2R^2}{\nu}\left(\frac{\Delta p}{\rho L} - g \sin\alpha\right) \sum_{n=1}^{\infty} \frac{J_0\left(\lambda_n \frac{r}{R}\right)}{\lambda_n^3 J_1(\lambda_n)}\left(1 - e^{-\lambda_n^2 \hat{t}}\right), \tag{11}$$

$$v_m = \left(\frac{R^2}{8\mu L}(\Delta p - g\rho L \sin\alpha)\right)\left(1 - 32\sum_{n=1}^{\infty} \frac{e^{-\lambda_n^2 \hat{t}}}{\lambda_n^4}\right), \tag{12}$$

$$\tau_w = \frac{4\mu}{R} v_\infty \left[1 - 4\sum_{n=1}^{\infty} \frac{e^{-\lambda_n^2 \cdot \hat{t}}}{\lambda_n^2}\right] \quad \text{where} \quad v_\infty = \left(\frac{R^2}{8\mu L}(\Delta p - g\rho L \sin\alpha)\right). \tag{13}$$

$$f = \frac{8\tau_w}{\rho \cdot v_m^2} = \frac{64}{Re_\infty}\left[1 - 4\sum_{n=1}^{\infty} \frac{e^{-\lambda_n^2 \hat{t}}}{\lambda_n^2}\right] \cdot \left[1 - 32\sum_{n=1}^{\infty} \frac{e^{-\lambda_n^2 \cdot \hat{t}}}{\lambda_n^4}\right]^{-2}. \tag{14}$$

From the above equations, it can be seen that the flow in a horizontal pipe (Gromeka-Szymański solution) is at an angle $\alpha = 0°$ (Eqs. 7-10). For angles in the range $0° < \alpha < 90°$, upward flow occurs if the pressure gradient overcomes gravity, while for angles in the range $-90° < \alpha < 0°$, there is a downward flow (Fig. 1), up to the extreme case of $\alpha = -90°$ for which the above formulas are reduced to the form derived by Roiti (Eqs. 2-5).

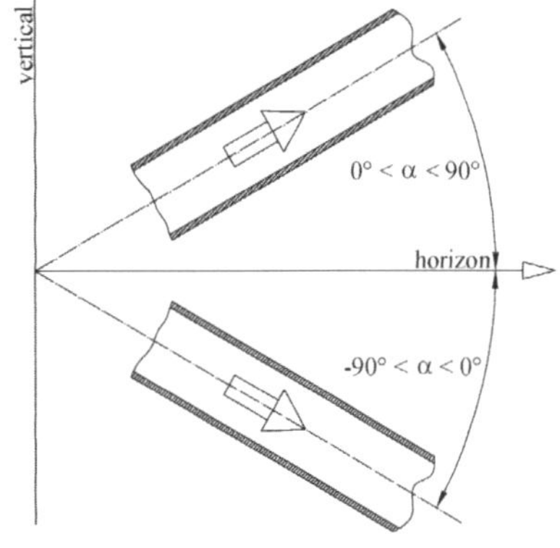

Figure 1. The tilt angle of the pipe and the positive direction of fluid flow.

To facilitate calculations, infinite sums appearing in the above formulas are saved in the selected range in the form of finite sums of exponential expressions. Thus, there is no need for the numerical determination of zeros of the Bessel function. For a single calculation of the coefficients, we can use the improved calculation algorithm presented in the first author's paper [33]. The improvement consists in the fact that both the current values of coefficients m_i and n_i can be calculated from precise mathematical formulas, such as:

$$\frac{C1-C2}{\exp(-n_{i+1}\cdot E1)}-\frac{C3-C4}{\exp(-n_{i+1}\cdot E2)}=0, \tag{15}$$

where: C1, ..., C4 and E1 and E2 are constants changing during the calculation of subsequent coefficients, which can be calculated analytically from

$$\ln\left(\frac{C2-C1}{C4-C3}\right)\cdot(E2-E1)^{-1}=n_{i+1}, \tag{16}$$

and not numerically as was mentioned in that particular paper (the author apologizes for this oversight). The resulting exponential functions presented below, which we will call weighting functions, will guarantee a high compatibility with the original infinite sums in the practical use of them, i.e. for $\hat{t}$ from 10^{-10} to ∞. The three new functions have the form:

$$\left(1-32\sum_{n=1}^{\infty}\frac{e^{-\lambda_n^2\hat{t}}}{\lambda_n^4}\right)=w_{vsr}^{-1}\left(\hat{t}\right)=\left[\sum_{i=1}^{22}a_i e^{-b_i\hat{t}}\right]^{-1}, \tag{17.1}$$

$$\left[1-4\sum_{n=1}^{\infty}\frac{e^{-\lambda_n^2\hat{t}}}{\lambda_n^2}\right]=w_{\tau}^{-1}\left(\hat{t}\right)=\left[\sum_{i=1}^{22}c_i e^{-d_i\hat{t}}\right]^{-1}, \tag{17.2}$$

$$\left[1-4\sum_{n=1}^{\infty}\frac{e^{-\lambda_n^2\hat{t}}}{\lambda_n^2}\right]\cdot\left[1-32\sum_{n=1}^{\infty}\frac{e^{-\lambda_n^2\cdot\hat{t}}}{\lambda_n^4}\right]^{-2}=w_{fc}\left(\hat{t}\right)=\sum_{i=1}^{22}e_i e^{-f_i\hat{t}}. \tag{17.3}$$

The values of the coefficients a_i, b_i, c_i, d_i, e_i and f_i obtained from the necessary calculations are summarised in Table 1 below.

Table 1. Calculated coefficients of the weighting functions.

i	a_i	b_i	c_i	d_i	e_i	f_i
1	1	0	1	0	1	0
2	1.1253	6.0053	0.7768	5.9421	1.5934	6.1776
3	2.7127	16.9400	1.0324	16.0510	6.0640	18.9757
4	8.0364	51.0539	1.9851	49.0212	27.9438	59.4110
5	24.3347	159.321	3.5383	153.327	140.1346	188.4939
6	74.931	502.258	6.3077	482.664	738.0994	598.58
7	233.297	1587.478	11.228	1524.031	4000.341	1898.58
8	731.22	5020.45	19.972	4817.25	22034.287	6015.23
9	2300.67	15878.53	35.520	15231.6	122474.44	19043.41
10	7254.65	50218	63.167	48165.3	684236.73	60260.3
11	22904.4	158817	112.33	152311	3833635.8	190631.9
12	72365	502249	199.75	481651	21513643	602960.7
13	228721	1588296	355.22	1523114	120839597	1906961
14	723072	5022718	631.67	4816511	679086243	6030748
15	2286156	15883308	1123.29	15231148	3817303983	19071499
16	7228764	50227233	1997.53	48165125	21462087350	60310528
17	22858248	158832515	3552.16	152311507	120668249757	190717542
18	72279258	502266929	6316.75	481651411	678652175201	603131809
19	228234286	1587473790	11232.9	1523116816	3812553675009	1906802180
20	721051335	5014638487	19975.3	4816508100	21394602492353	6024041424
21	2274403251	15835931280	35521.4	15231092712	120019731059804	19023982623
22	7327599236	50349124336	63165.7	48164347598	674994401528792	60136925028

Journal of Physics: Conference Series **760** (2016) 012036 doi:10.1088/1742-6596/760/1/012036

2.4. Horizontal decelerating pipe flow

Gerbes in his 1951 paper [4] used the Laplace transform. Starting from the Navier-Stokes equations for incompressible viscous liquid, he obtained the same formula (7) for the velocity profile in accelerated laminar flow as Gromeka and Szymański. In addition, Gerbes was the only one interested in the deceleration of the fluid. He presented the final formula for the velocity profile in a cross-section of the pipe during stopping in the following form:

$$v(r,\hat{t}) = \frac{R^2 \Delta p}{4\mu L} \sum_{n=1}^{\infty} \frac{8 J_0(\lambda_n \frac{r}{R})}{\lambda_n^3 J_1(\lambda_n)} \cdot e^{-\lambda_n^2 \hat{t}} . \tag{18}$$

By taking the average velocity of steady flow before the start of the deceleration as:

$$v_0 = \frac{R^2 \Delta p}{8\mu L}, \tag{19}$$

we obtain

$$v(r,\hat{t})_d = v_0 \sum_{n=1}^{\infty} \frac{16 J_0(\lambda_n \frac{r}{R})}{\lambda_n^3 J_1(\lambda_n)} \cdot e^{-\lambda_n^2 \hat{t}} . \tag{20}$$

Calculating the average velocity for decelerated flow gives:

$$v_{m,d} = \frac{2}{R^2} \int_0^R r \cdot v(r,\hat{t}) dr = v_0 \sum_{n=1}^{\infty} \frac{32}{\lambda_n^4} \cdot e^{-\lambda_n^2 \hat{t}} = \frac{4R^2 \Delta p}{\mu L} \sum_{n=1}^{\infty} \frac{e^{-\lambda_n^2 \hat{t}}}{\lambda_n^4} . \tag{21}$$

In contrast to acceleration, the shear stress on the pipe wall (r=R) during deceleration is:

$$\tau_{w,d} = -\mu \frac{\partial v(r,\hat{t})}{\partial r} = \frac{v_0 \cdot \mu}{R} \sum_{n=1}^{\infty} \frac{16}{\lambda_n^2} \cdot e^{-\lambda_n^2 \hat{t}} = \frac{2R\Delta p}{L} \sum_{n=1}^{\infty} \frac{e^{-\lambda_n^2 \hat{t}}}{\lambda_n^2} . \tag{22}$$

The coefficient of resistance in decelerated flow, on the basis of the above formulas, is calculated herein with the formula:

$$f_d = \frac{64}{Re_0} \left[4 \sum_{n=1}^{\infty} \frac{e^{-\lambda_n^2 \hat{t}}}{\lambda_n^2} \right] \cdot \left[32 \sum_{n=1}^{\infty} \frac{e^{-\lambda_n^2 \cdot \hat{t}}}{\lambda_n^4} \right]^{-2} \quad \text{where} \quad Re_0 = \frac{D \cdot v_0}{\nu} . \tag{23}$$

2.5. Reversed flow solution

Still another solution, which uses the finite Hankel transform, was derived in 1978 by a group of researchers under the leadership of Chambré [13]. It concerns reverse flow, which occurs after a sudden (stepwise) or a linear (ramp) change of the sign of the pressure gradient forcing the movement of the liquid. It is a kind of combination of accelerated flow with decelerated flow. This type of flow occurs, for example, in the cases of loss of coolant accidents in water reactors. In the case of an immediate change of the sign of the pressure gradient, the derived solutions were as follows:

$$v(r,t)_r = v_\infty \sum_{n=1}^{\infty} \frac{16 \cdot J_0\left(\lambda_n \frac{r}{R}\right)}{\lambda_n^3 J_1(\lambda_n)} \left(2e^{-\lambda_n^2 \hat{t}} - 1\right), \tag{24}$$

$$v_{m,r} = v_\infty \sum_{n=1}^{\infty} \frac{32}{\lambda_n^4}\left(2e^{-\lambda_n^2 \hat{t}} - 1\right), \tag{25}$$

$$\tau_{w,r} = \frac{v_\infty \cdot \mu}{R} \sum_{n=1}^{\infty} \frac{16}{\lambda_n^2}\left(2e^{-\lambda_n^2 \hat{t}} - 1\right), \tag{26}$$

$$f_r = \frac{64}{Re_\infty}\left[4\sum_{n=1}^{\infty} \frac{\left(2e^{-\lambda_n^2 \hat{t}} - 1\right)}{\lambda_n^2}\right] \cdot \left[32\sum_{n=1}^{\infty} \frac{\left(2e^{-\lambda_n^2 \hat{t}} - 1\right)}{\lambda_n^4}\right]^{-2} \quad \text{where} \quad Re_\infty = \frac{D \cdot v_\infty}{\nu}, \tag{27}$$

In contrast, when the pressure gradient is gradually changed in a linear manner, the final results are even more complicated:

$$v(r,t)_r = v_\infty \sum_{n=1}^{\infty} \frac{16 \cdot J_0\left(\lambda_n \frac{r}{R}\right)}{\lambda_n^3 J_1(\lambda_n)} \cdot F, \tag{28}$$

$$v_{m,r} = v_\infty \sum_{n=1}^{\infty} \frac{32}{\lambda_n^4} \cdot F, \tag{29}$$

$$\tau_{w,r} = \frac{v_\infty \cdot \mu}{R} \sum_{n=1}^{\infty} \frac{16}{\lambda_n^2} \cdot F, \tag{30}$$

$$f_r = \frac{64}{Re_\infty}\left[4\sum_{n=1}^{\infty} \frac{F}{\lambda_n^2}\right] \cdot \left[32\sum_{n=1}^{\infty} \frac{F}{\lambda_n^4}\right]^{-2} \quad \text{where} \quad Re_\infty = \frac{D \cdot v_\infty}{\nu}, \tag{31}$$

where for $0 \le \hat{t} \le \hat{t}_r$

$$F = F_1 = 1 - \frac{2}{\lambda_n^2 \hat{t}_r}\left(\lambda_n^2 \hat{t} + e^{-\lambda_n^2 \hat{t}} - 1\right), \tag{32}$$

and for $\hat{t} \ge \hat{t}_r$

$$F = F_2 = \frac{1}{\lambda_n^2 \hat{t}_r}\left(2e^{-\lambda_n^2 (\hat{t} - \hat{t}_r)} - 2e^{-\lambda_n^2 \hat{t}} - \lambda_n^2 \hat{t}_r\right). \tag{33}$$

3. Verification of the Zielke solution

The analytical solutions discussed in the previous section are distinguished by the assumed pressure gradient driving the flow and appearing on the left in Eq. (1). In the case of flow accelerating from rest, the gradient simply was not there for time $t \le 0$, and it is positive and constant for time $t > 0$ (Fig. 2a). In decelerated flow, the opposite is assumed, namely a positive value for time $t \le 0$ and a zero value for $t > 0$ (Fig. 2b). In reverse flow, there is a stepwise (Fig. 2c) or linear (Fig. 2d) change of its sign thereby maintaining the same absolute value (magnitude).

Apart from the discussed analytical solutions, there is a most clever solution which has been derived for any temporal change in the pressure gradient. Its inventor is Zielke. In 1966, in his dissertation [34] and then in an article [35], he used previously known solutions of Eq. (1) by Brown [36] and D'Souza and Oldenburger [37] in the Laplace domain, and he derived its inverse transform (for $g = 0$). From the two-dimensional solution in the Laplace domain he derived a linear (but frequency dependent) proportionality between the wall shear stress and the average (cross-sectional) velocity. In the time-domain this translates to the following formula relating wall shear stress to average flow acceleration:

$$\tau = \tau_q + \tau_u = \frac{1}{8}\lambda\rho v|v| + \frac{2\cdot\mu}{R}\cdot\int_0^t w(t-u)\cdot\frac{\partial v}{\partial t}(u)\,du. \tag{34}$$

This expression is the sum of the conventional quasi-steady shear stress τ_q and the new time-dependent shear stress τ_u, which is the convolution of a weight function w and the mean acceleration. It is now widely used for calculating non-stationary hydraulic resistance in pipelines during a pressure surge [38-41]. There are two ways of computing the above convolution integral: classic and efficient; in this paper, we will consider the results of both of them.

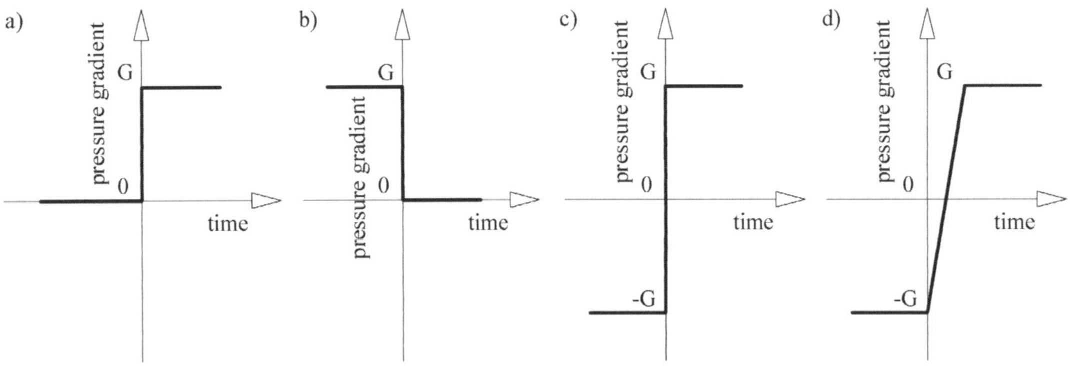

Figure 2. Pressure gradient variation: a) accelerated flow; b) decelerated flow; c) reverse flow (step change); d) reverse flow (linear ramp change).

The classical computation is inefficient, because dealing with all the stored values of changes in flow velocity. The impact of any temporal change of velocity is significant; therefore it is incorporated and weighted by the respective value of the weighting function:

$$\tau_u = \frac{2\mu}{R}\left(\sum_{j=1}^{n-1}\frac{\left(v_{i,n-j+1} - v_{i,n-j}\right)}{\Delta\hat{t}}\cdot\int_{(j-1)\Delta\hat{t}}^{j\Delta\hat{t}} w(\hat{t})d\hat{t}\right) = \frac{2\mu}{R}SUM_c. \tag{35}$$

This form is the classic approach, presented for example by Vardy and Brown [14]. The weight function in the above expression is determined for any dimensionless time from the infinite sum:

$$w_{classic}(t) = \sum_{n=1}^{\infty} e^{-\kappa_n^2\hat{t}}, \tag{35.1}$$

where: κ_n - n[th] zero of the Bessel function of type J_2.

Zielke approximated this function using the following two functions [34,35]:

$$w_{classic}(\hat{t}) = \sum_{i=1}^{6} m_i \hat{t}^{(i-2)/2}, \text{ for } \hat{t} \le 0.02, \tag{35.2}$$

$$w_{classic}(\hat{t}) = \sum_{i=1}^{6} m_i \hat{t}^{(i-2)/2}, \text{ for } \hat{t} \le 0.02, \tag{35.3}$$

where: m_1 = 0.282095; m_2 = −1.25; m_3 = 1.057855; m_4 = 0.9375; m_5 = 0.396696; m_6 = −0.351563; n_1 = 26.3744; n_2 = 70.8493; n_3 = 135.0198; n_4 = 218.9216; n_5 = 322.5544.

Unfortunately, this form is not suitable for use in the efficient solution of the convolution integral:

$$\tau_u(t+\Delta t)=\frac{2\mu}{R}\sum_{i=1}^{j}\underbrace{\left[y_{i(t)}A_i+\eta B_i\left[v_{(t+\Delta t)}-v_{(t)}\right]+\left[1-\eta\right]C_i\left[v_{(t)}-v_{(t-\Delta t)}\right]\right]}_{y_{i(t+\Delta t)}}=\frac{2\mu}{R}SUM_e, \tag{36}$$

where:

$$A_i=e^{-n_i\cdot\Delta\hat{t}};\quad B_i=\frac{m_i}{\Delta\hat{t}\cdot n_i}\cdot\left[1-A_i\right];\quad C_i=A_i\cdot B_i, \tag{36.1}$$

the idea of which was to reduce the time spent on computation The number of calculation in this approach depends only on the number of terms describing the effective weight function:

$$w_{\text{effective}}(\hat{t})=\sum_{i=1}^{k}m_i e^{-n_i\hat{t}}. \tag{36.2}$$

Formula (36) is a revised Schohl approximation [42]. For the case of water hammer, it was presented and verified in two papers by the first author [15,41]. However, further verification of this new numerical procedure is needed and this will be presented in this section for accelerated, decelerated and reverse laminar flows. The constants A_i, B_i, C_i and η:

$$\eta=\frac{\int_0^{\Delta\hat{t}}w_{\text{clasic}}(\hat{t})d\hat{t}}{\int_0^{\Delta\hat{t}}w_{\text{effective}}(\hat{t})d\hat{t}}, \tag{36.3}$$

are determined only once (for the pipe under consideration) at the start of the calculation. The coefficients m_i and n_i found in the formulas (36.1) for the constants A_i, B_i and C_i are suitable constants describing the effective form of the weight function w. Their values are specified in [39]. Additionally, the above procedure is characteristic for the fact that it does not introduce an error for the most recent calculation step. This is most important when using a reduced number of exponential expressions describing the weight function [15]. As seen from formulas (35) and (36), the wall shear stress is a function of temporal changes in the mean velocity and of the weighting function.

3.1. Calculation Example

In order to examine the compatibility of the calculation procedures presented in this section with the analytical solutions given in the previous section, the following examples have been chosen. Accelerated, decelerated and reverse flows of the liquid are considered in a horizontal ($\alpha=0$) pipe with length L = 250 m and inner diameter D = 0.025 m. The flowing liquid is an oil with density ρ = 900 kg/m^3 and kinematic viscosity $2.5\cdot10^{-5}$ m^2/s. The pressure difference Δp = 0.6624 MPa along the length of the pipe is calculated from the condition that the final (initial in decelerated and reverse flow) Reynolds number is 2300 so that the flow remains laminar. This difference in the case of accelerated flow exists for time t > 0, and in the case of decelerated flow for time $t \leq 0$. In reverse flow, a step change of the sign of the above difference is assumed at t = 0. Here $G = \Delta p/L$ = 2.65 kPa/m (see Fig. 2).

For the above conditions, analytical solutions have been plotted for **accelerated**, **decelerated** and **reverse** flow. They will be designated in detail for $t \geq 0$: **average velocity** from formulas (12), (21) (25) (change of mean velocity over time is illustrated in Fig. 3); **wall shear stress** (13), (22), (26) and the **unsteady coefficient of resistance** (14), (23), (27). The results obtained are compared with the results obtained from convolution-integral methods (35) and (36) discussed briefly in Section 3. Calculations have been performed for a fixed dimensionless time step in the interval from $\Delta\hat{t}=10^{-4}$ to $\Delta\hat{t}$ = 5 (dimensionless time is determined as: $\Delta\hat{t}=\nu\Delta t/R^2$).

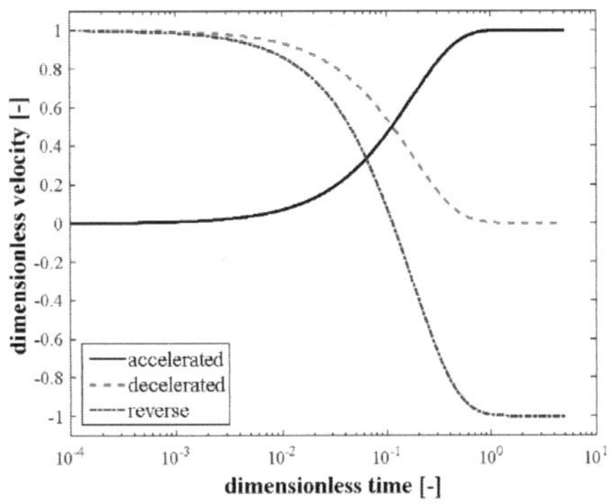

Figure 3. Velocity change over time.

The following ratios are of importance:

$$R_{u/q} = \frac{\tau_u}{\tau_q} = \frac{f_u}{f_q} \quad \text{and} \quad R_{tot./q} = \frac{f}{f_q} = \frac{\tau}{\tau_q}, \tag{37}$$

where: $\tau = \tau_u + \tau_q$ and $f = f_u + f_q$.

The final formulas for the above ratios for the different types of flows are summarised in Table 2.

Table 2. Formulas for the ratios (41).

Ratio	$R_{u/q}$	$R_{tot./q}$
Roiti-Gromeka-Szymański (accelerated flow)	$\frac{[1-S]}{[1-T]}-1$	$\frac{[1-S]}{[1-T]}$
Gerbes decelerated flow	$\frac{S}{T}-1$	$\frac{S}{T}$
Chambré et al. reverse flow	$\frac{X}{Y}-1$	$\frac{X}{Y}$
Zielke arbitrary flow	$\frac{SUM}{2v_{(t+\Delta t)}}$	$\frac{SUM}{2v_{(t+\Delta t)}}+1$

where: $S = 4\sum_{n=1}^{\infty}\frac{e^{-\lambda_n^2 \hat{t}}}{\lambda_n^2}$; $T = 32\sum_{n=1}^{\infty}\frac{e^{-\lambda_n^2 \cdot \hat{t}}}{\lambda_n^4}$; $X = 4\sum_{n=1}^{\infty}\frac{2e^{-\lambda_n^2 \hat{t}}-1}{\lambda_n^2}$; $Y = 32\sum_{n=1}^{\infty}\frac{2e^{-\lambda_n^2 \hat{t}}-1}{\lambda_n^4}$

$SUM = SUM_c$ *(in classic solution – (35));* $SUM = SUM_e$ *(in efficient solution – (36))*

The results obtained, showing the change in total shear stress over time, for the three analysed flows, are shown in the Figs 4a-4c. They indicate correctly that the values in the case of accelerated flow (Fig. 4a) increase from zero to a constant value equal τ_q, and that in decelerated flow (Fig. 4b) they decrease from the value τ_q to zero. The rapid decrease of the slope of the curve in Fig. 4a in the vicinity of the dimensionless time $\hat{t} = 1$ can be linked to the formation of a parabolic flow profile. Thus, the parabolic profile is formed only in the last phase of the acceleration. A different situation is observed in Fig. 4b, where there is a very slow relative change in shear stress, thus it can be assumed that the parabolic flow profile remains for quite a long time until the final stopping phase around time $\hat{t} = 0.22$. After that time curve have a rapid increase and flow profile cannot be assumed as parabolic anymore. In the example involving reverse flow (Fig. 4c), it can be seen that the stress correctly changes from a positive to a negative quasi-steady value. Complete reversal of flow, seen as a deceleration followed by an acceleration, is complete after $\hat{t} = 2$.

Journal of Physics: Conference Series **760** (2016) 012036 doi:10.1088/1742-6596/760/1/012036

a)

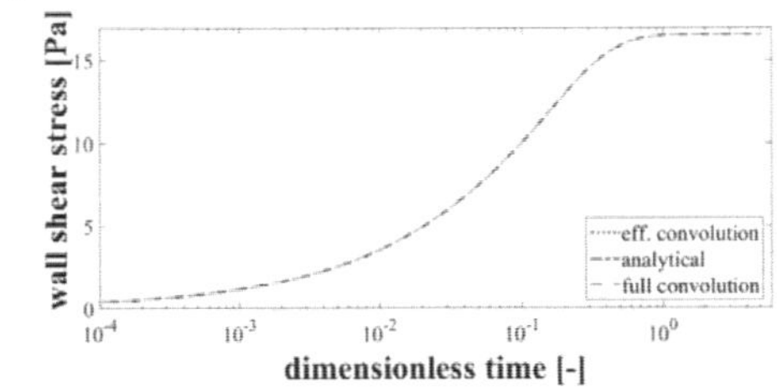

b)

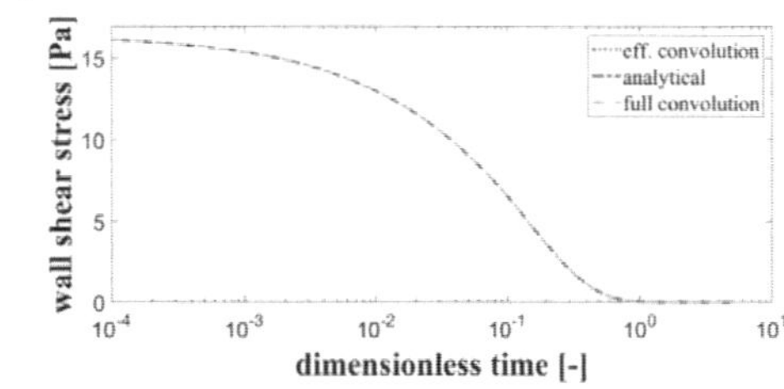

c)

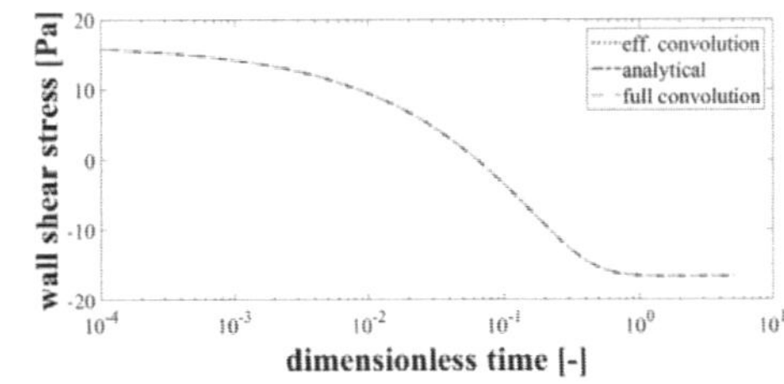

Figure 4. Total wall shear stress results: a) accelerated flow, b) decelerated flow, c) flow reversal

The results obtained for the change of the ratio $R_{u/q}$ of time dependent wall shear stress τ_u to quasi-steady wall shear stress τ_q and/or time dependent friction factor f_u to quasi-steady one f_q over time, for the three analysed flows, are shown in the Figs 5a-5c.

a)

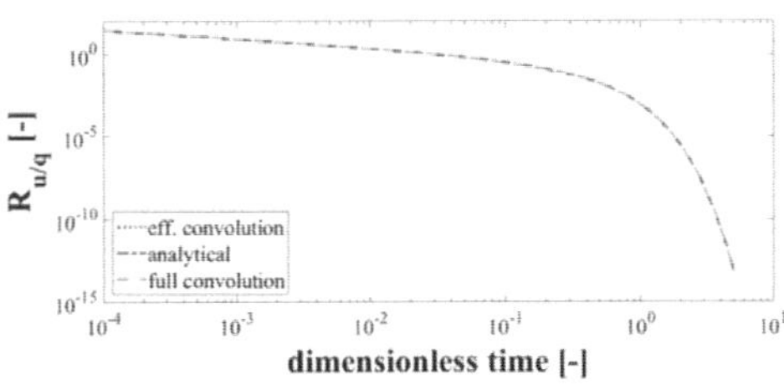

b)

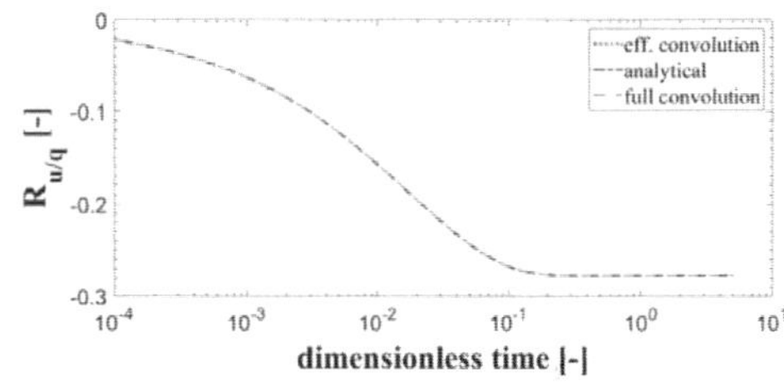

c)

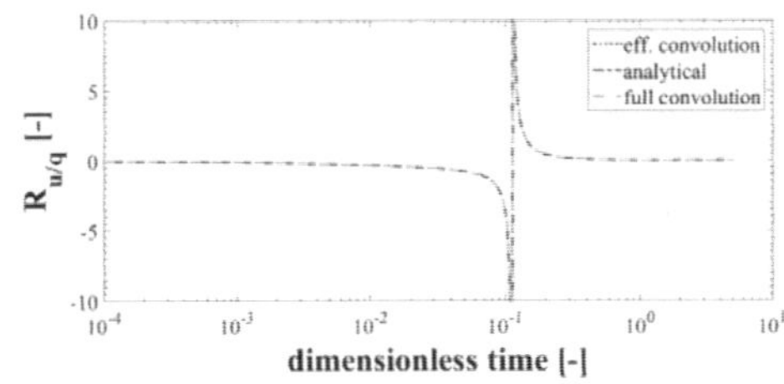

Figure 5. Calculated ratio $R_{u/q}$: a) accelerated flow, b) decelerated flow, c) flow reversal

The results of the $R_{u/q}$ ratio show that the initial phase of accelerated flow (Fig. 5a) is dominated by unsteady wall stress ($\tau_u > \tau_q$ for $\hat{t} < 0.03$). This situation is reversed for times $\hat{t} > 0.03$ (since henceforth $\tau_u < \tau_q$). In decelerated flow (Fig. 5b), the τ_u calculated is negative ($\tau_u < 0$), since along with the duration of this flow, the subsequent average flow velocity values decrease, and the quasi-steady component τ_q remains positive until the end. From time $\hat{t} > 0.45$, this ratio stabilises to the value −0.28, which is the final results for large dimensionless times (taking into account first terms of sums in S/T−1 one get $(4/32)\cdot \lambda_1^2 - 1 \approx -0.28$). This means that the unsteady wall stress τ_u affects significantly the total shear stress τ until the complete stop of fluid flow. Analysing the results for reverse flow (Fig. 5c), it can be seen that decelerating occurs in the initial phase. Just before $\hat{t} = 0.11$, the values of $R_{u/q}$ approach $-\infty$ because τ_q vanishes. However, after exceeding this reversal time and corresponding singularity, there is an acceleration of the liquid and $R_{u/q}$ ratio returns from $+\infty$ to reach the value of approximately 1 when $\hat{t} = 2$.

Figures 6a-6c show the ratio $R_{tot./q}$ of the total friction factor f to the quasi-steady one f_q and/or total wall shear stress τ to the quasi steady one τ_q, for the three analysed unsteady flows. In the case of accelerated flow (Fig. 6a), we can see that the resulting value moves from ∞ to 1 (steady flow). This means that the time dependent friction factor f_u in accelerated flow is greater than in quasi-steady flow f_q. The next graph (Fig. 6b) shows that in decelerated flows, the total friction factor f adopts values smaller than that calculated using the quasi-steady formula. But what is important time dependent friction factor f_u plays an important role until the final stop of flow as evidenced by the value $R_{tot./q}$ = 0.723 which this ratio tends to with increasing dimensionless time. In reverse flows (Fig. 6c), this ratio before changing the direction of flow reaches negative values (from $\hat{t} > 0.06$) and later approaches $-\infty$.

a)

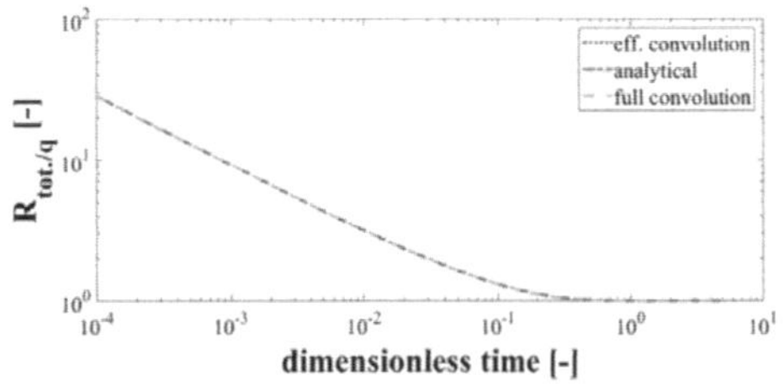

b)

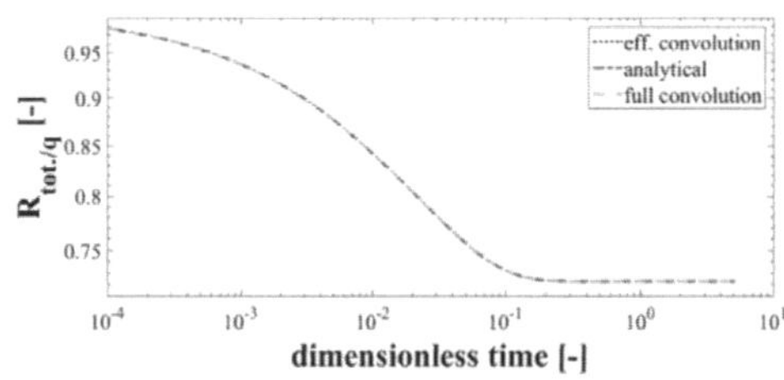

c)

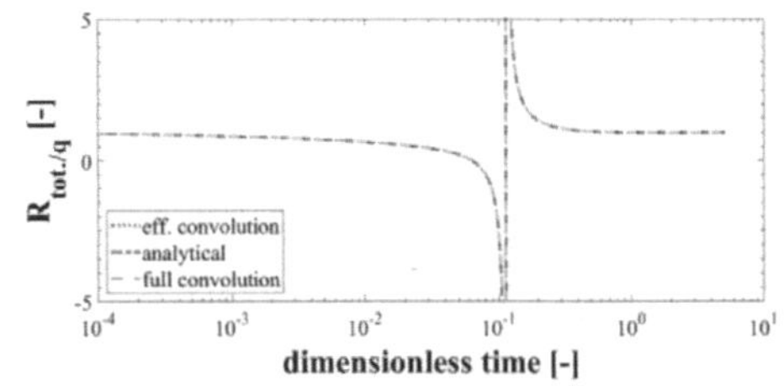

Figure 6. Calculated ratio $R_{tot./q}$: a) accelerated flow, b) decelerated flow, c) flow reversal

The appearance of negative values is related to the fact that the liquid flow direction changes more rapidly near to the pipe wall (Fig. 7). A zero value of the average velocity occurred in the analysed flow for time $\hat{t} = 0.11$. Just after this time, this ratio reaches the value of $+\infty$, before falling to a value of unity (no effect of unsteady flow).

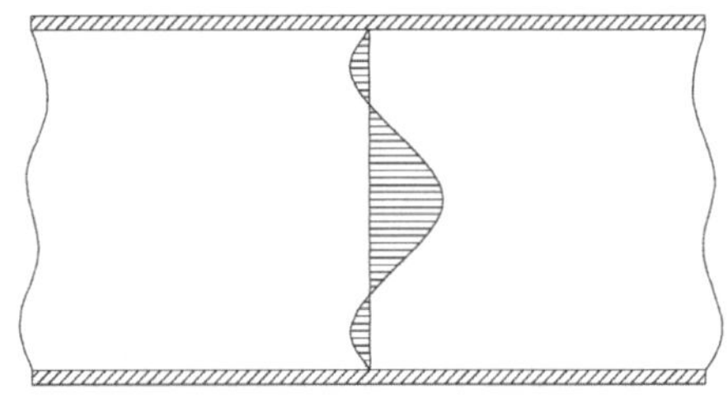

Figure 7. Back flow near the wall

4. Summary

The results presented in this paper clearly show that both the improved Zielke solution (35) and the more efficient solution of the convolution integral (36) describe very well the unsteady laminar flow resulting from acceleration, deceleration, or reversal of liquid in a pipe. The graphs show no significant differences between the analytical and numerical results. Subtle differences between the efficient and the classical solution of the convolution integral are not shown herein but were noted in an extensive analysis of relative errors carried out during the preparation of this paper.

Historical references [1-3] collected and analysed for the purpose of this paper allow the statement that the solution known in the literature relating to accelerated laminar pipe flows, the authorship of which was originally transcribed to Szymański in 1932, had been derived already sixty years earlier in 1871 by Roiti and later on in 1882 by Gromeka. Therefore, referring to these solutions, we should include the names of all authors who have derived them apparently independently: Roiti-Gromeka-Szymański.

While working on these solutions, we can see a heat-conduction analogy. Since the issues of heat conduction have been dealt with since the beginning of the nineteenth century, one can suppose that the same final formulas could have been derived by someone even earlier than 1871 for some special case of non-stationary heat flow - perhaps in a hollow cylinder.

The universal solutions presented in Section 2.3 may be used in any inclined straight pipe with accelerated flow in any direction.

References

[1] Roiti A 1871 Sul movimento dei liquidi *Annali della Scuola Normale Superiore di Pisa – Classe di Scienze* **1** (in Italian) 193

[2] Gromeka I S 1882 On a theory of the motion of fluids in narrow cylindrical tubes *Uch. Zap. Kazan. Inst.* **112** (41) (in Russian)

[3] Szymański P 1932 Quelques solutions exactes des équations de l'hydrodynamique du fluide visqueux dans le cas d'un tube cylindrique *J. Math. Pures et Appliquées* **11(9)** (in French) 67

[4] Gerbes W 1951 Zur instationären, laminaren Strömung einer inkompressiblen, zähen Flüssigkeit in kreiszylindrischen Rohren *Zeitschrift für angewandte Physik* **3(7)** (in German) 267

[5] Poisson S D 1823 Mémoire sur la distribution de la chaleur dans les corps solides *Journal de l'Ecole Polytechnique* **12** (in French) 1

[6] Poisson S D 1831 Mémoire sur les équations générales de l'équilibre et du mouvement des corps solides élastiques et des fluides *Journal de l'Ecole Polytechnique* **13** (in French) 1

[7] Poisson S D 1835 Théorie Mathématique de la Chaleur, Paris: Bachelier (in French)

[8] Betti E 1868 Alcune determinazioni delle temperature variabili di un cilindro, Pisa, *Tip. Nistri* (in Italian)

[9] Betti E 1868 Sopra la determinazione delle temperature nei corpi solidi ed omogenei, Firenze, *Stamperia Reale* (in Italian)

[10] Fassò C A 1956 Avviamento del moto di una corrente liquida in un tubo di sezione costante: Influenza delle resistenze *Reniconti Instituto Lombardo – Academia di Scienze e Lettere* **90** (in Italian) 305

[11] Ghidaoui M S and A A Kolyshkin 2002 A Quasi-steady approach to the instability of time dependent flows in pipes *Journal of Fluid Mechanics* **465** 301

[12] Szymański P 1930 Sur l'écoulement non permanent du fluide visqueux dans le tuyau *Compte Rendu du III Congrès International de Mécanique Appliquée* Stockholm (in French) 249

[13] Chambré P L et al. 1978 Reversal of laminar flow in a circular pipe *Nuclear Engineering Design* **47** 239

[14] Vardy A E and Brown J M B 2010 Evaluation of unsteady wall shear stress by Zielke's method *J. Hydraul. Eng.* **136** 453

[15] Urbanowicz K 2015 Simple modelling of unsteady friction factor *Proc. BHR Group - 12th International Conference on Pressure Surges* Dublin 113

[16] Reynolds O 1883 An experimental investigation of the circumstances which determine whether the motion of water shall be direct or sinuous, and of the law of resistance in parallel channels *Philosophical Transactions of the Royal Society of London* **174** 935

[17] Iguchi M et al 2010 Effect of initial constant acceleration on the transition to turbulence in transient circular pipe flow *Journal of Fluids Engineering* **132**

[18] Vardy A E and Brown J M B 2011 Laminar pipe flow with time-dependent viscosity *Journal of Hydroinformatics* **13** (4) 729

[19] Muzychka Y S and Yovanovich M M 2006 Compact models for transient conduction or viscous transport in non-circular geometries with a uniform source *International Journal of Thermal Sciences* **45** 1091

[20] Muzychka Y S and Yovanovich M M 2010 Unsteady viscous flows and Stokes's first problem *International Journal of Thermal Sciences* **49** 820
[21] Telionis D 2012 Unsteady viscous flows *Springer Series in Computational Physics*
[22] Lam F 2015 On exact solutions of the Navier-Stokes equations for uni-directional flows *arXiv:1503.05149v3 [physics.flu-dyn]*
[23] Chaudhury R A et al. 2015 Length and time for development of laminar flow in tubes following a step increase of volume flux *Exp Fluids* **56** (22)
[24] Pontrelli G 2000 Blood flow through a circular pipe with an impulsive pressure gradient *Mathematical Models and Methods in Applied Sciences* **10** (2) 187
[25] Drazin P G and Riley N 2006 The Navier-Stokes equations: A classification of flows and exact solutions *London Mathematical Society Lecture Note Series* **334** *Cambridge University Press*
[26] Constantinescu V N 2012 Laminar viscous flow (Mechanical Engineering Series) *Springer*
[27] Urbanowicz K and Tijsseling A S 2015 Work and life of Piotr Szymański *Proc. BHR Group - 12th International Conference on Pressure Surges* Dublin 311
[28] Loitsyanskii L G 1966 Mechanics of Liquids and Gases (first English edition) *Oxford: Pergamon Press*
[29] Martin C S and Wiggert D C 2013 Discussion of "Evaluation of unsteady wall shear stress by Zielke's method" by Alan E. Vardy and Jim M. B. Brown. *ASCE Journal of Hydraulic Engineering* **139** 562
[30] Gromeka I S 1952 Collected Works, *USSR Academy of Sciences*, Moscow (in Russian) 149
[31] http://books.e-heritage.ru/book/10079943
[32] Çengel Y A and Cimbala J M 2006 Fluid Mechanics: Fundamentals and Applications. *Boston: McGraw-Hill Higher Education*
[33] Urbanowicz K 2012 New approximation of unsteady friction weighting functions *Proc. of the 11th International Conference on Pressure Surges* Lisbon 477
[34] Zielke W 1966 Frequency–dependent friction in transient pipe flow *University of Michigan* (Phd thesis)
[35] Zielke W 1968 Frequency–dependent friction in transient pipe flow *Trans. ASME, J. Basic Engng* **90** 109
[36] Brown F T 1962 A The transient response of fluid lines *Trans. ASME, J. Basic Engng* **84** 547
[37] D'Souza A F and Oldenburger R 1964 Dynamic response of fluid lines *Trans. ASME, J. Basic Engng* **86** (3) pp. 589-598
[38] Zarzycki Z et al. 2011 Improved method for simulating transients of turbulent pipe flow *Journal of Theoretical and Applied Mechanics* **49** (1) 135
[39] Urbanowicz K and Zarzycki Z 2012 New efficient approximation of weighting functions for simulations of unsteady friction losses in liquid pipe flow *Journal of Theoretical and Applied Mechanics* **50** (2) 487
[40] Urbanowicz K et al. 2012 Universal weighting function in modeling transient cavitating pipe flow *Journal of Theoretical and Applied Mechanics* **50** (4) 889
[41] Urbanowicz K and Zarzycki Z 2015 Improved lumping friction model for liquid pipe flow *Journal of Theoretical and Applied Mechanics* **53** (2) 295
[42] Schohl G A 1993 Improved approximate method for simulating frequency–dependent friction in transient laminar flow *Trans. ASME, Journ. of Fluids Eng.* **115** 420

Journal of Physics: Conference Series **760** (2016) 012037 doi:10.1088/1742-6596/760/1/012037

Modelling water hammer in viscoelastic pipelines: short brief

K Urbanowicz, M Firkowski and Z Zarzycki
West Pomeranian University of Technology Szczecin, Faculty of Mechanical Engineering and Mechatronics, Al. Piastów 19, 70-310 Szczecin, Poland

E-mail: kamil.urbanowicz@zut.edu.pl, mateusz.firkowski@zut.edu.pl, zbigniew.zarzycki@zut.edu.pl

Abstract. The model of water hammer in viscoelastic pipelines is analyzed. An appropriate mathematical model of water hammer in polymer pipelines is presented. An additional term has been added to continuity equation to describe the retarded deformation of the pipe wall. The mechanical behavior of viscoelastic material is described by generalized Kelvin-Voigt model. The comparison of numerical simulation and experimental data from well known papers is presented. Short discussion about obtained results are given.

1. Introduction

Hydraulic transient analysis is important during the design of water supply systems. One of the most dangerous phenomena for pipeline systems is water hammer. Analysis of water hammer allows to choose an appropriate parameters of system like pipe material, wall thickness or surge protection devices. For many years, researches used only the classic water hammer model. The main assumption in this model is linear elastic behavior of pipe walls [1–3]. This theory is good for modelling systems build mainly from steel or brass.

In last 40 years, researchers started investigate a new water hammer model for pipes made from a new generation polymeric materials (see e.g. [7, 10, 11]). High resistant properties (mechanical, chemical, temperature) of polymers, like polyethylene or polyvinyl chloride, caused increasing popularity of this type of pipelines. These materials have viscoelastic behavior which influence on the pressure response of a pipe system.

Viscoelastic rheological behavior of polymers has been studied by many authors [4–6]. This behavior is characterized by a sum of instantaneous elastic strain and retarded strain. It makes a mechanical damping of the pressure wave. There are two main approaches to simulate the viscoelastic behavior of plastic pipe-walls.

The first approach assumes frequency–dependent wave speed. The viscoealstic behavior is modelled by a frequency–dependent creep function to calculate wave speed, instead of a constant elastic modulus [7–10].

The second, more popular, approach assumes a retarded viscoelastic term in continuity equation. The precursory paper were written by Gally et al. [11] and Rieutford and Blanchard [12]. Gally at al. presented the mathematical model and comparison of theoretical and experimental results. Covas et al. in [13] analyzed the dynamic effects of pipe wall viscoelasticity on hydraulic transients. This effects have been observed in data collected on the two experimental setups. In two–part paper [14, 15], Covas with others widely described the viscoelasticity effects in polymer pipelines. The theoretical and experimental results were compared. The creep–function of the pipe material was experimentally determined. Bergant et al. [16,17] investigated parameters that may significantly affect water hammer wave attenuation, shape and timing. The authors considered five possible mechanism that affect pressure waveforms include unsteady friction, cavitation, fluid–structure interaction, viscoelastic behavior of the pipe–wall material, leakages and blockages. In [18], Soares et al. investigated the analysis of hydraulic transients in polyvinyl chloride pipes. The creep function of the PVC pipes was determined by using an inverse transient method. Different characterization of creep function of PVC in comparison with HDPE pipes was observed. Keramat et al. [19] considered water hammer model in viscoelastic pipes with Poisson's ratio as a function of time. K. Weinerowska–Bords discussed in [20] the main problems connected with applying the water hammer model in viscoelastic pipes. The aspects of estimation a group of parameters like wave speed, retardation time and creep compliance was analyzed.

The structure of the work is as follows. After Introduction, Section 2 presents mathematical model of water hammer in elastic pipes. Next, Section 3 contains efficient solution of convolution integral. Section 4 is devoted to present mathematical model of water hammer in viscoelastic pipelines. In Section 5, numerical simulation are conducted. In conclusions, the summary of obtained results is given.

2. Water hammer in elastic pipe – fundamental equations

Unsteady flow of liquid in elastic pipes (made of brass, steel etc.) is often represented by a two 1D hyperbolic partial differential equations. The momentum and continuity equations are [21–27]:

$$\rho\frac{\partial v}{\partial t}+\frac{\partial p}{\partial x}+\frac{2}{R}\tau=0, \tag{1}$$

$$\frac{\partial p}{\partial t}+\rho c^2\frac{\partial v}{\partial x}=0, \tag{2}$$

where t – time $[s]$, x – distance along the pipe axis $[m]$, $v=v(x,t)$ – average value of velocity in cross–section of pipe $[m/s]$, $p=p(x,t)$ – average value of pressure in cross–section of pipe $[Pa]$, τ – shear stress at pipe wall $[Pa]$, ρ – density of liquid (constant) $[kg/m^3]$, R – inside radius of pipe $[m]$, c – pressure wave speed $[m/s]$.

The pressure wave speed in this model is calculated from following equation:

$$c=\sqrt{\frac{1}{\rho\left(\frac{1}{K}+\frac{\alpha D}{E_0 e}\right)}}, \tag{3}$$

where: K – is bulk modulus of elasticity $[Pa]$, E_0 – Young modulus $[Pa]$, D – inner pipe diameter [m], e – thickness of pipe wall [m], α – parameter reporting support condition of pipe $[-]$.

In unsteady flow in pipe, the instantaneous wall shear stress τ may be regarded as the sum of two components [28–33]:

$$\tau=\underbrace{\frac{\rho v|v|}{8}\lambda}_{\tau_q}+\underbrace{\frac{2\mu}{R}\int_0^t w(t-u)\frac{\partial v}{\partial t}(u)du}_{\tau_u}, \tag{4}$$

where λ – Darcy–Weisbach friction factor $[-]$, $w(t)$ – weighting function $[-]$, μ – dynamic viscosity $\left[\frac{kg}{m\cdot s}\right]$, u – time, used in convolution integral $[s]$.

The first component in Equation (4) τ_q presents the quasi–steady wall shear stress and the second τ_u is the additional contribution due to unsteadiness. Equation (4) relates the wall shear stress to the instantaneous average velocity and to the weighted past velocity changes. Among methods, which enable to resolve system of above equations, particular attention should be paid to the method of characteristics (MOC), which perfectly interprets the essence of natural phenomena of transient flow, and at the same time is characterized by fast convergence, easiness to take into account various boundary conditions and the high accuracy of calculation results. With its help one can easily solve a system of partial differential equations of quasi-linear hyperbolic type (1)–(4).

3. Efficient solution of convolution integral

Zielke presented the weighting functions for laminar flow [28]:

$$w(\hat{t}) = \begin{cases} \sum_{i=1}^{6} m_i \hat{t}^{(i-2)/2} & \text{for} \quad \hat{t} \le 0.02 \\ \sum_{i=1}^{5} e^{-n_i \hat{t}} & \text{for} \quad t > 0.02 \end{cases} \tag{5}$$

where $\hat{t} = \nu \cdot t/R^2$ – dimensionless time, and $m_1 = 0.282095$, $m_2 = -1.25$, $m_3 = 1.057855$, $m_4 = 0.9375$, $m_5 = 0.396696$, $m_6 = -0.351563$, $n_1 = 26.3744$, $n_2 = 70.8493$, $n_3 = 135.0198$, $n_4 = 218.9216$, $n_5 = 322.5544$.

Numerical calculation of time–dependent component of the wall shear stress τ_u (second component in equation (4)) can be conducted using the first–order differential approximation [26, 28]:

$$\tau_u = \frac{2\mu}{R} \sum_{j=1}^{k-1} (v_{i,j+1} - v_{i,j})\, w\left((k-j)\Delta t - \frac{\Delta t}{2}\right) = \frac{2\mu}{R} \sum_{j=1}^{k-1} (v_{i,k-j+1} - v_{i,k-j})\, w\left(j\Delta t - \frac{\Delta t}{2}\right), \tag{6}$$

where Δt – time step in MOC $[s]$.

Recently Vardy–Brown [34] rightly pointed out an error in this methodology. To avoid it they suggest to calculate the τ_u component using following equation:

$$\tau_u = \frac{2\mu}{R} \left(\sum_{j=1}^{n-1} \frac{(v_{i,n-j+1} - v_{i,n-j})}{\Delta \hat{t}} \int_{(j-1)\Delta \hat{t}}^{j\Delta \hat{t}} w\left(\hat{t}\right) d\hat{t} \right) = \frac{2\mu}{R} \mathrm{SUM}_c, \tag{7}$$

where $\Delta \hat{t}$ – dimensionless time step $[-]$.

Determination of the wall shear stress using above formulas ((6) or (7)) is very time consuming. Trikha [35] was the first who developed an effective method of solving the integral convolution. Next Kagawa et al. [36] and Schohl [37] have improved this method. But as all this solution fails to model correctly the wall shear stress, a new efficient solution were presented recently [38] that is fully consistent with corrected by Vardy–Brown (7) classic inefficient solution:

$$\tau_u(t+\Delta t) = \frac{2\mu}{R} \sum_{i=1}^{j} \left[y_{i(t)} A_i + \eta B_i \left[v_{(t+\Delta t)} - v_{(t)} \right] + [1-\theta]\, C_1 \left[v_{(t)} - v_{(t-\Delta t)} \right] \right] = \frac{2\mu}{R} \mathrm{SUM}_e, \tag{8}$$

where

$$A_i = e^{-n_i \Delta \hat{t}}, \quad B_i = \frac{m_i}{\Delta t n_i} \left[1 - A_i \right], \quad C_i = A_i B_i.$$

In above equation n_i and m_i are constants of efective weighting function and θ is a correction factor [38].

Journal of Physics: Conference Series **760** (2016) 012037 doi:10.1088/1742-6596/760/1/012037

The new effective method (8) presented above requires that the weighting function have the form of finite sum of exponential expressions:

$$w\left(\hat{t}\right) = \sum_{i=1}^{k} m_i e^{-n_i \hat{t}} \tag{9}$$

The needed number of exponential terms that make up the final form have been discussed recently in some works [33, 38]. And as it was pointed there for a case of water hammer two terms is just enough for proper simulation.

As they value depends on the dimensionless time step used in simulation – a proper weighting function terms n_i and m_i used in completed simulations will be presented with all other constants in section 5.

4. Water hammer in viscoelastic pipe - fundamental equations

Taking into account viscoelastic behavior of the pipe wall one can derive the continuity equation similarly as for the linear elastic pipe. The most important details of derivation are presented in the Appendix A.

Polymer pipeline do not respond according to Hook law when subjected to a certain instantaneous stress σ_0, one notice an immediate–elastic response and a retarded–viscous response. Thus strain can be decomposed into a sum of instantaneous–elastic strain ε_e and a retarded strain ε_r (see e.g. [11, 14, 16, 18]):

$$\varepsilon(t) = \varepsilon_e + \varepsilon_r(t). \tag{10}$$

The final form of continuity equation is very similar to the one used in elastic pipeline:

$$\frac{\partial p}{\partial t} + \rho c^2 \left(\frac{\partial v}{\partial x} + 2\frac{d\varepsilon_r}{dt} \right) = 0. \tag{11}$$

For small strains, a combination of stresses that act independently in a system result in strains that can be added linearly. Thus, the total strain generated by a continuous application a stress $\sigma(t)$ is [5, 13–15]:

$$\varepsilon(t) = J_0\sigma(t) + \int_0^t \sigma(t-u)\frac{\partial J}{\partial t}(u)du. \tag{12}$$

where J_0 – the instantaneous creep–compliance, $J(u)$ – the creep function at u time. For all linear–elastic materials, creep compliance J_0 is assumed equal to the inverse modulus of elasticity $J_0 = 1/E_0$.

For final solution we additionally assume: homogeneous and isotropic pipe material; linear viscoelastic behavior for small strains and constant Poisson's ratio ν (mechanical behavior is only dependent on a creep–function). Then the total circumferential strain $\varepsilon = (D(t) - D_0)/D_0$ can be described by:

$$\varepsilon(t) = \frac{\alpha_0 D_0}{2e_0}\left[p(t) - p_0\right] J_0 + \int_0^t \frac{\alpha(t-u)D(t-u)}{2e(t-u)}\left[p(t-u) - p_0\right]\frac{\partial J}{\partial t}(u)du. \tag{13}$$

where $\sigma = \alpha\Delta pD/2e$ – circumferential stress $[Pa]$, $p(t)$ – pressure at time t, p_0 – initial pressure, $D(t)$ and D_0 – inner diameters at time t and 0 respectively, $e(t)$ and e_0 – the wall thicknesses at time t and 0 respectively, $\alpha(t)$ and α_0 – the pipe–wall constraint coefficient at time t and 0 respectively, $J(t)$ – the creep–compliance function at time t.

The creep function which describes viscoelastic behaviour of the pipe material should be determined experimentally for polymer pipelines in independent mechanical tests. Using generalized Kelvin–Voigt model in this work (Fig. 1) viscoelastic solid are defined with creep function of the pipe wall in form [5] of finite sum:

$$J(t) = J_0 + \sum_{k=1}^{n} J_k \left(1 - e^{-t/T_k}\right), \tag{14}$$

where J_k – the creep–compliance of the spring of the Kelvin–Voigt k–element defined by $J_k = 1/E_k\ [Pa^{-1}]$, E_k – the modulus of elasticity of the spring of k–element, T_k – the retardation time of the dashpot of k–element ($T_k = \eta_k/E_k = \eta_k J_k$, η_k – the viscosity of the dashpots of k–element) $[s]$.

The parameters J_k and T_k of the viscoelastic mechanical model are adjusted to the creep–compliance experimental data.

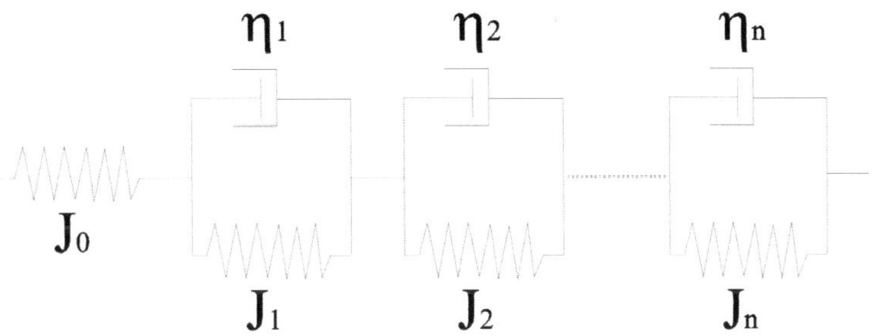

Figure 1. Generalized Kelvin–Voigt Model (viscoelastic solid)

The set of differential equations (1) and (11) is solved by the Method of Characteristics (MOC). With the help of this method this system of equations can be transformed to pair of ordinary differential equations:

$$\pm\frac{1}{\rho c}\frac{dp}{dt} + \frac{dv}{dt} + \frac{2}{\rho R}\tau + 2c\frac{\partial\varepsilon}{\partial t} = 0 \quad \text{if} \quad \frac{dx}{dt} = \pm c. \tag{15}$$

Using for above equation finite differences method on characteristics grid (Fig. 2) one obtain:

$$\begin{aligned} \tfrac{1}{c\rho}(p_D - p_R) + (v_D - v_R) + \tfrac{2\Delta t}{\rho R}\tau_R + 2\Delta t c\tfrac{\partial\varepsilon_D}{\partial t} &= 0 \\ -\tfrac{1}{c\rho}(p_D - p_S) + (v_D - v_S) + \tfrac{2\Delta t}{\rho R}\tau_S - 2\Delta t c\tfrac{\partial\varepsilon_D}{\partial t} &= 0 \end{aligned} \tag{16}$$

In above equations (16)

$$\frac{\partial\varepsilon_D}{\partial t} = p_D F - \sum_{k=1}^{n}\left[p_0 B_k + p_E(A_k - B_k) + C_k\varepsilon_{E,k}\right] = p_D F - V_E, \tag{17}$$

where constants:

$$F = \tfrac{\alpha D}{2e}\textstyle\sum_{k=1}^{n}\tfrac{J_k}{\Delta t}\left(1 - e^{-\frac{\Delta t}{T_k}}\right), \quad A_k = \tfrac{\alpha D}{2e}\tfrac{J_k}{\Delta t}\left(1 - e^{-\frac{\Delta t}{T_k}}\right), \quad B_k = \tfrac{\alpha D}{2e}\tfrac{J_k}{\Delta t}e^{-\frac{\Delta t}{T_k}}, \quad C_k = \frac{e^{-\frac{\Delta t}{T_k}}}{T_k}. \tag{18}$$

Lets now all known terms calculated from previous numerical time step write as:

$$\begin{aligned} C_R &= v_R + \tfrac{p_R}{\rho c} - \tfrac{2\Delta t}{\rho R}\tau_R + 2\Delta t c V_E, \\ C_S &= v_S - \tfrac{p_S}{\rho c} - \tfrac{2\Delta t}{\rho R}\tau_S - 2\Delta t c V_E. \end{aligned} \tag{19}$$

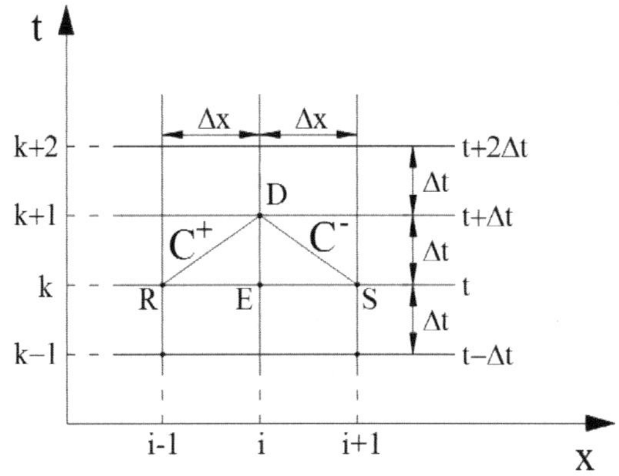

Figure 2. Rectangular grid of characteristics

Finally one can reach the numerical solution in form:

$$p_D = \frac{C_R - C_S}{\left(\frac{2}{\rho c} + 4cF\Delta t\right)} \quad \text{and} \quad v_D = \frac{C_S + C_R}{2}. \tag{20}$$

In numerical procedure after determining these two values there is still need to determine the momentary retarded strain $\varepsilon_{D,k}$ which is used in the equation (17):

$$\varepsilon_{D,k} = T_k \left[p_D \left(B_k e^{\frac{\Delta t}{T_k}} - A_k \right) + p_E \left(A_k - B_k \right) + p_0 B_k \left(1 - e^{\frac{\Delta t}{T_k}} \right) + C_k \varepsilon_{E,k} \right]. \tag{21}$$

5. Simulation example

To verify results of simulation using elastic water hammer model and a modified viscoelastic one, they results have been compared to experimental results. The experiments have been carried out by D. Covas in London on a PE pipe–rig at Imperial College [14, 45]. The experimental reservoir-pipeline-valve system comprises a 271.7 $[m]$ long pipe of 50.6 $[mm]$ internal diameter and 6.3 $[mm]$ wall thickness.

The following parameters where kept constant for purpose of all further considerations: time of operation of quick shut–off of the globe valve $T_f = 0.13$ $[s]$; observed mean pressure wave speed $c = 395$ $[m/s]$; water temperature $T_w = 20$ $[C]$. Pipeline material properties: Young modulus $E_0 = 1.43$ $[GPa]$ and Poisson ratio (given by the manufacturer) $\nu = 0.46$. The water density and viscosity have been determined from characteristics for known temperature ($\rho = 998.2$ $[kg/m^3]$ and $\nu = 1 \cdot 10^{-6}$ $[m^2/s]$]). The creep functions were an experimental functions [15] characterized by following creep and retardation time coefficients: ($J_0 = 0.7 \cdot 10^{-9}$, $J_1 = 0.1394 \cdot 10^{-9}$, $J_2 = 0.0062 \cdot 10^{-9}$, $J_3 = 0.1148 \cdot 10^{-9}$, $J_4 = 0.3425 \cdot 10^{-9}$, $J_5 = 0.0928 \cdot 10^{-9}$) $[Pa^{-1}]$ and ($T_0 = 0$, $T_1 = 0.05$, $T_2 = 0.5$, $T_3 = 1.5$, $T_4 = 5$, $T_5 = 10$) $[s]$.

In all numerical simulations the pipeline was divided for 32 equally long elements ($f = 32$ $[-]$, $\Delta x = 8.5$ $[m]$). With that assumption a constant time step were calculated $\Delta t = 0.0215$ $[s]$ ($\Delta\hat{t} = 3.36 \cdot 10^{-5}$ $[-]$). For this case the estimated coefficient of simplified two term weighting function were: $m_1 = 4.481$, $m_2 = 33.6906$, $n_1 = 747.536$, $n_2 = 3419.3$ (for Re=25300 turbulent case. Note that to get this coeficient a scalling procedure using an extended Vardy-Brown coefficient [39] and a universal weighting function assumption were made [26], also characteristic polymer pipe roughness size $k_s = 1.5 \cdot 10^{-6}$ were assumed) and $m_1 = 4.4413$, $m_2 = 33.6391$, $n_1 = 72.591$, $n_2 = 2744.3$ (for Re=1400 laminar case).

Knowing Poisson ratio one can determine the pipe-wall constraint coefficient α. As there is at least four different equations:

$$
\begin{array}{llll}
a) & \alpha & = & \frac{2e}{D}\left(1+\nu_p\right)+\frac{D}{D+e}\left(1-\nu_p^2\right)=1.07\ [14] \\
b) & \alpha & = & 1+\left(\frac{e}{D}\right)^2+\nu_p\frac{2e}{D}-\nu_p^2\left(1-\frac{e}{D}\right)^2=0.97\ [11] \\
c) & \alpha & = & \left(1-\nu_p^2\right)+0.75\nu_p\left(1+\nu_p\right)\frac{2e}{D}=0.91\ [40] \\
d) & \alpha & = & \left(1-\nu_p^2\right)+0.5\nu_p\left(1+\nu_p\right)\frac{2e}{D}=0.87\ [41]
\end{array}
\qquad (22)
$$

that can be used to determine α value for thick pipes that are anchored along its length. In the simulations carried out below examined will be not only presented models (elastic and viscoelastic) but also the impact of this parameter on the results (using constant $c = 395 m/s$ – which is equivalent with assuming $\alpha \cdot J_0 = const$).

The graphical representations of transient states are presented below on Fig. 3 and Fig. 4.

A) Laminar case ($Re_0 = 1400\ [-]$, $v_0 = 0.028\ [m/s]$, $p_R = 4.738 \cdot 10^5 [Pa]$)

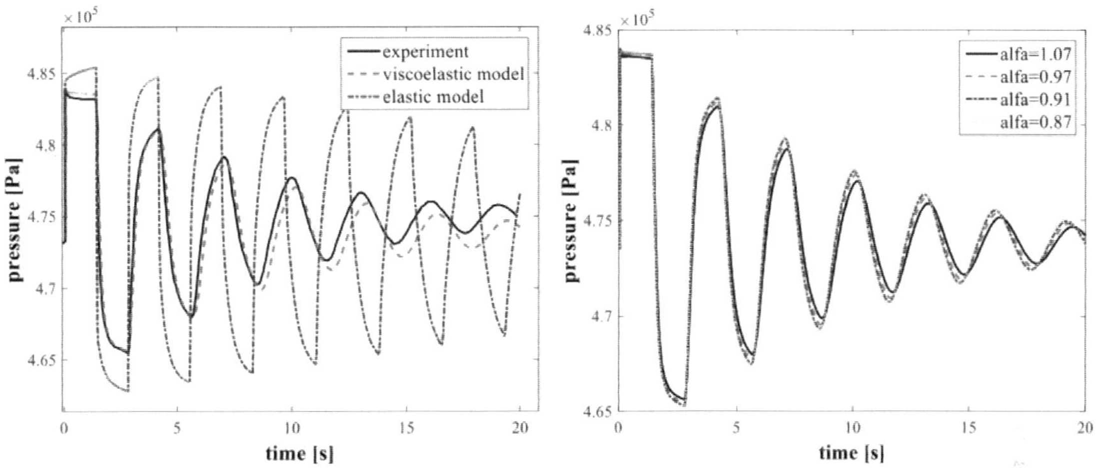

Figure 3. Pressure variation at quick closing globe valve laminar case a) comparison of different models, b) the impact of alpha coefficient.

B) Turbulent case ($Re_0 = 25300\ [-]$, $v_0 = 0.5\ [m/s]$, $p_R = 4.454 \cdot 10^5 [Pa]$)

From simulations presented in Fig. 3 and Fig. 4, we can formulate the following conclusions:

(i) The linear elastic model (with unsteady friction factor) is not an appropriate model to simulate water hammer in pressurized pipes made from polymers. Even, the maximum pressure on first amplitude simulated by this model was overestimated.

(ii) The pipe–wall constraint coefficient α has a slight influence on obtained results. The lower values of α: the higher max. pressures (on first and following pressures amplitudes) and the faster following amplitudes occurs (see Fig. 5).

The differences between numerical simulations of water hammer in viscoelastic pipes and experimental data can be explained by:

(i) Simulations do not take into account the slope of the pipe and local resistance in the elbows.

(ii) An experimental creep function was used, which haved a selected J_0 coefficient, because there are problems in determining the creep function for small time $t < 0.1\ [s]$ [15].

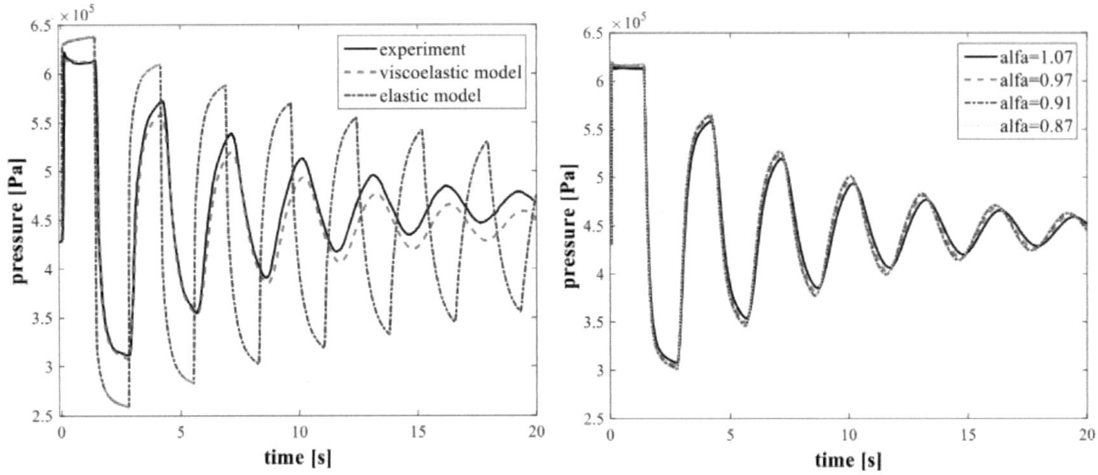

Figure 4. Pressure variation at quick closing globe valve turbulent case a) comparison of different models, b) the impact of alpha coefficient.

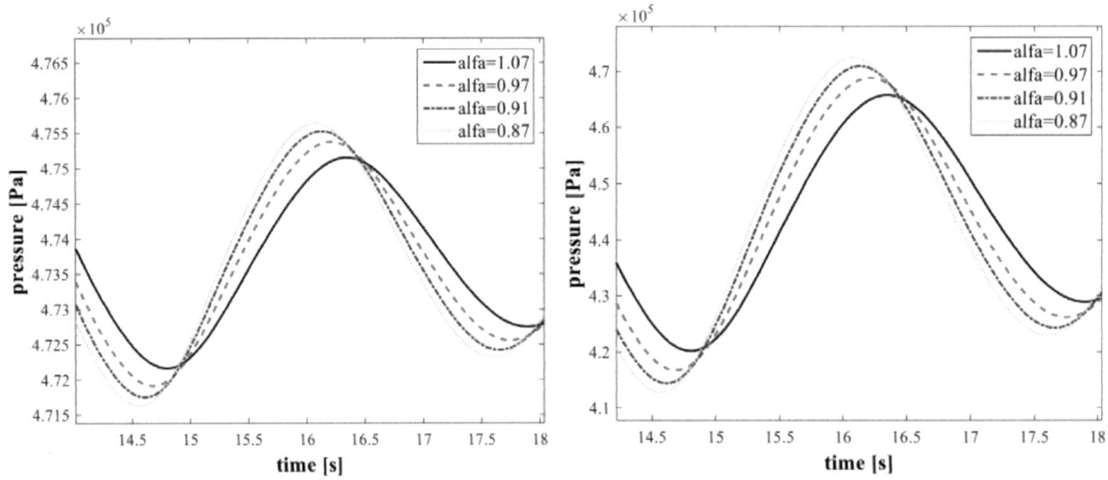

Figure 5. The role of α coefficient a) 6–th amplitude of laminar case, b) 6–th amplitude of turbulent case.

(iii) During simulations a constant pressure on the reservoir were assumed. But the analysis of experimental pressure runs shows that this pressure after the valve closure linearly increases (unfortunately the authors of experiment did not explain what is the reason, despite showing this increas on Fig. 10 on page 65 of their most cited paper [15]).

6. Conclusions

The paper presents a numerical model used to calculate the transient flow (caused by rapid closing of the valve) in the polymeric pipes. From presented exemplary simulation tests and obtained from them results the following conclusions can be formulate:

a) There is a great need to conduct new experimental research in a simple horizontal reservoir-pipe-valve system, in which the pipes are made of a polymeric material. As it is still lack of the accurate results of analysis of water hammer in polymeric pressurized pipes. All known studies [11,13–15] were mostly carried out in more or less inclined pipes. In them occurred an additional local resistance (in elbows or in curvature of the pipes caused by coiling them onto a cylinder) that caused problem with collected experimental data. There is a major problem with separation of their influence and their proper numerical modelling. This is particularly visible from presented in this paper exemplary results.

b) Used by many authors procedure adjusting J coefficients (commonly referred to calibration [15, 42, 43]) that role is to minimize the simulation errors and fit simulation results to experimental one is in the view of the authors of this study unacceptable. Indeed, such a calibration can be performed for existing systems, but the priority should be to find an analytical solution by means of which, it will be possible to predict the unsteady behavior of the liquid at any time in polymeric pipes. To this end, new material research are necessary needed to properly describe the mechanical behavior of the polymeric materials used for pressure pipes.

c) There is a need for additional numerical simulations to determine the significant of the unsteady friction in these type of flows.

d) The real observed velocity pressure wave speed is during transient flow variable and can be calculated from following equation [8, 9, 43, 44]:

$$c = \sqrt{\frac{1}{\rho\left(\frac{1}{K} + \frac{\alpha D}{E(s)e}\right)}},$$

That's the reason why one should consider the attempt to modify the classical numerical solution based on method of characteristics built on a solid rectangular grid.

Appendix A

General deriveration of continuity equation in viscoelastic pipe after Covas [45]:

$$\frac{\partial}{\partial t}(\rho A) + \frac{\partial}{\partial x}(\rho A v) = 0 \qquad \text{(Appendix .1)}$$

Calculating derivatives:

$$A\frac{d\rho}{dt} + \rho\frac{dA}{dt} + \rho A\frac{\partial v}{\partial x} = 0 \qquad \text{(Appendix .2)}$$

Assuming that:

$$\frac{d\rho}{dt} = \frac{\rho}{K}dp \quad \text{then} \quad \frac{d\rho}{dt} = \frac{\rho}{K}\frac{dp}{dt} \quad \text{also} \quad \frac{dA}{dt} = \frac{\pi D^2}{2}\frac{d\varepsilon}{dt} = 2A\frac{d\varepsilon}{dt} \qquad \text{(Appendix .3)}$$

After decomposition of circumferential strain $\varepsilon = \varepsilon_e + \varepsilon_r$:

$$\frac{d\varepsilon}{dt} = \frac{d\varepsilon_e}{dt} + \frac{d\varepsilon_r}{dt} \qquad \text{(Appendix .4)}$$

The instantaneous linear–elastic strain ε_e:

$$\varepsilon_e = \frac{\alpha\sigma_e}{E_0} = \frac{\alpha p D}{2E_0 e} \quad \text{and} \quad \frac{d\varepsilon_e}{dt} = \frac{\alpha}{2E_0 e}\left(p\frac{dD}{dt} + D\frac{dp}{dt}\right) = \frac{\alpha D}{2E_0 e}\left(\frac{p}{D}\frac{dD}{dt} + \frac{dp}{dt}\right), \qquad \text{(Appendix .5)}$$

where σ_e – elastic component of hoop stress, $E_0 = 1/J_0$ –instantaneous modulus of elasticity. Accordingly, the:

$$\frac{d\varepsilon_e}{dt} = \frac{1}{D}\frac{dp}{dt} \qquad \text{(Appendix .6)}$$

One can write:

$$\frac{d\varepsilon_e}{dt} = \frac{\alpha D}{2E_0 e} p \frac{d\varepsilon_e}{dt} + \frac{\alpha D}{2E_0 e} \frac{dp}{dt} \quad \text{(Appendix .7)}$$

whereof:

$$\frac{d\varepsilon_e}{dt} = \frac{dp/dt}{\left(\frac{2E_0 e}{\alpha D} - p\right)} \quad \text{as} \quad \frac{2E_0 e}{\alpha D} \gg p \quad \frac{d\varepsilon_e}{dt} = \frac{dp/dt}{\left(\frac{2E_0 e}{\alpha D}\right)} \quad \text{(Appendix .8)}$$

Introducing equation (Appendix .8) to dA/dt from equation (Appendix .3) having in mind assumption (Appendix .4):

$$\frac{dA}{dt} = \frac{\alpha D A}{E_0 e} \frac{dp}{dt} + 2A \frac{d\varepsilon_r}{dt} \quad \text{(Appendix .9)}$$

Now introducing this equation (Appendix .9) and $d\rho/dt$ from equation (Appendix .3) to equation (Appendix .2) and ordering:

$$A\left(\rho\left[\frac{1}{K} + \frac{\alpha D}{E_0 e}\right]\right)\frac{dp}{dt} + 2A\rho\frac{d\varepsilon_r}{dt} + A\rho\frac{\partial v}{\partial x} = 0 \quad \text{(Appendix .10)}$$

Finally introducing in above equation c^2 (equation (3)) and then dividing all terms by $A\rho$ one have the solution:

$$\frac{1}{\rho c^2}\frac{\partial p}{\partial t} + \frac{\partial v}{\partial x} + 2\frac{d\varepsilon_r}{dt} = 0 \quad \text{(Appendix .11)}$$

References

[1] Chaudhry M H 1987 *Applied Hydraulic Transients* (Litton Educational Publishing: Van Nostrand Reinhold Co., New York)

[2] Almeida A B and Koelle E 1992 *Fluid transients in Pipe Network* (Computational Mechanics Publications, London)

[3] Wylie E B and Streeter V L 1993 *Fluid Transients in Systems*, (Prentice Hall: Englewood Cliffs, New York)

[4] Ferry J D 1970 *Viscoelastic Properites of Polymers* (John Wiley & Sons, New York)

[5] Aklonis J J, MacKnight W J and Shen M 1972 *Intoduction to Polymer Viscoelasticity* (John Wiley & Sons, New York)

[6] Riande E, Diaz–Calleja R, Prolongo M G, Masegosa R M and Salom C 2000 *Polymer Viscoelasticity: Stress and Strain in Practise* (Mercel Dekker, New York)

[7] Meissner E and Franke G 1977 Influence of pipe material on the dampening of water hammer *Proc. of the 17th Congress of the Int. Associacion for Hydraulic Research*

[8] Rieutford E 1982 Transient response of fluid viscoelastic lines *J. Fluid Engng.* **104** 335

[9] Franke G and Seyler F 1983 Computation of unsteady pipe flow with respect to viscoelastic material properties *J. Hydraul. Res.* **21** 345

[10] Suo L and Wylie E B 1990 Complex wave speed and hydraulic transients in viscoelastic pipes *J. Fluid. Engng.* **112** 496

[11] Gally M, Guney M and Rieutford E 1979 An investigation of pressure transients in viscoelastic pipes *J. Fluid Engng, Trans. ASME* **101** 495

[12] Rieutford E and Blanchard A 1979 Pulsating Viscoelastic Pipe Flow – Water–Hammer *J. Hydraulic res.* **17** 217

[13] Covas D, Stoianov I, Ramos H, Graham N, Maksimovic C and Butler D 2004 Water hammer in pressrized polyethylene pipes: conceptual model and expreimantal analysis *Urban Water J.* **1**177

[14] Covas D, Stoianov I, Mano J F, Ramos H, Graham N and Maksimovic C 2004 The dynamic effect of pipe–wall viscoelasticity in hydraulic transients. Part I – experimantal analysis and creep characterization *J. Hydraul. Res.* **42** 516

[15] Covas D, Stoianov I, Mano J F, Ramos H, Graham N and Maksimovic C 2005 The dynamic effect of pipe–wall viscoelasticity in hydraulic transients. Part II – model development, calibration and verification, *J. Hydraul. Res.* **43** 56

[16] Bergant A, Tijsseling A S, Vitkovsky J P, Covas D, Simpson A and Lambert M 2008 Parameters affecting water–hammer wave attenuation, shape and timing–Part 1: Mathematical tools *J. Hydraul. Res.* **46** 373

[17] Bergant A, Tijsseling A S, Vitkovsky J P, Covas D, Simpson A and Lambert M 2008 Parameters affecting water–hammer wave attenuation, shape and timing–Part 2: Case studies *J. Hydraul. Res.* **46** 382

[18] Soares A K, Covas D and Reis L F R 2008 Analysis of PVC pipe-wall viscoelasticity during water hammer *J. Hydraul. Eng,* **134**, 1389
[19] Keramat A, Kolahi A G and Ahmadi A 2013 Whaterhammer modelling of viscoelastic pipes with a time–dependent Poisson's ratio, *J. Fluid. Struct.* **43** 164
[20] Weinerowska–Bords K 2006 Viscoelastic model of waterhammer in single pipeline – problems and questions *Arch. Hydro-Eng. Environ. Mech.* **53** 331
[21] Wylie E B and Victor L S 1978 *Fluid transients* (New York: McGraw-Hill International Book Co)
[22] Szymkiewicz R and Mitosek M 2013 Alternative Convolution Approach to Friction in Unsteady Pipe Flow *J. Fluids Eng* **136(1)** 011202
[23] Zarzycki Z 2000 On weighting function for wall shear stress during unsteady turbulent pipe flow *Proc. of 8th Int. Conf. on Pressure Surges* BHR Group The Hague 529
[24] Zarzycki Z and Urbanowicz K 2006 Modelling of transient flow during water hammer considering cavitation in pressure pipes *Chemical and Process Engineering* **27** 3 915 (in polish)
[25] Urbanowicz K and Zarzycki Z 2012 New efficient approximation of weighting functions for simulations of unsteady friction losses in liquid pipe flow *Journal of Theoretical and Applied Mechanics* **50** (2) 487.
[26] Urbanowicz K et al. 2012 Universal weighting function in modeling transient cavitating pipe flow *Journal of Theoretical and Applied Mechanics* **50** (4) 889
[27] Zarzycki Z. et al. 2011 Improved method for simulating transients of turbulent pipe flow *Journal of Theoretical and Applied Mechanics* **49** 1 135
[28] Zielke W 1968 FrequencyDependent Friction in Transient Pipe Flow *Journal of Basic Engineering ASME* **90** 1 109
[29] Zarzycki Z 1997 Hydraulic resistance of unsteady turbulent liquid flow in pipes *Proc. of 3rd Int. Conf. on Water Pipeline Systems* The Hague BHR Group 163
[30] Vardy A E and Brown J M B 2003 Transient turbulent friction in smooth pipe flows *J. Sound Vibration* 259 Issue 5 1011
[31] Vardy A E and Brown J M B 2004 Transient turbulent friction in fully rough pipe flows *J. Sound Vibration* **270** 233
[32] Urbanowicz K 2012 New Approximation of unsteady friction weighting functions *Proc. of the 11th Int. Conf. on Pressure Surges* Lisbon Portugal October 24–26 477
[33] Urbanowicz K and Zarzycki Z 2015 Improved lumping friction model for liquid pipe flow *Journal of Theoretical and Applied Mechanics* **53** (2) 295
[34] Vardy A E and Brown J M B 2010 Evaluation of unsteady wall shear stress by Zielkes method *J. Hydraul. Eng.* **136** 453
[35] Trikha A K 1975 An efficient method for simulating frequency–dependent friction in transient liquid flow *Journ. of Fluids Eng. Trans. ASME* March 97
[36] Kagawa T, Lee I, Kitagawa A and Takenaka T 1983 High speed and accurate computing method of frequency–dependent friction in laminar pipe flow for characteristics method *Trans. Jpn. Soc. Mech. Eng.* A **49** (447) 2638
[37] Schohl G A 1993 Improved approximate method for simulating frequency – dependent friction in transient laminar flow *Journ. of Fluids Eng. Trans. ASME* **115** September 420
[38] Urbanowicz K 2015 Simple modelling of unsteady friction factor *Proc. BHR Group - 12th Int. Conf. on Pressure Surges* Dublin 113
[39] Vardy A E and Brown J M B 2007 Approximation of Turbulent Wall Shear Stresses in Highly Transient Pipe Flows *J. Hydraulic Eng.* **133** (11) 1219
[40] Tijsseling A S 2007 Water hammer with fluid–structure interaction in thick–walled pipes *Comput. Struct. 85* (11–14) 844
[41] Rachid, F B F and Stuckenbruck S 1990 Transients in liquid and structure in viscoelastic pipes *Proc., 6th Int. Conf. on Pressure Surges*, BHR Group, 69
[42] Keramat A, Tijsseling A S and Ahmadi A 2010 Investigation of transient cavitating flow in viscoelastic pipes *25th IAHR Symposium on Hydraulic Machinery and Systems* IOP Conf. Series: Earth and Environmental Science 12
[43] Gong et al 2016 Determination of the creep function of viscoelastic pipelines using system resonant frequencies with hydraulic transient analysis *J. Hydraul. Eng.* 04016023
[44] Covas, D et al 2005 Impulse response method for solving hydraulic transients in viscoelastic pipes *Proc. 31st IAHR Congress*, 1 Seoul, Korea,11
[45] Covas D 2003 *Inverse transient analysis for leak detection and calibration of water pipe systems modelling special dynamic effects* Phd thesis Imperial College of Science, Technology and Medicine.

Symmetry analysis and invariant solutions of the multipoint infinite systems describing turbulence

M Wacławczyk[1] and M Oberlack[2]

1 Institute of Geophysics, Faculty of Physics, University of Warsaw, Pasteura 7, 02-093 Warsaw, Poland

2 Chair of Fluid Dynamics, Department of Mechanical Engineering, TU Darmstadt, Otto-Berndt Str. 2, 64287 Darmstadt, Germany

E-mail: marta.waclawczyk@igf.fuw.edu.pl

Abstract. The present work concerns multipoint description of turbulence in terms of the probability density functions (pdf's) and the characteristic Hopf functional. Lie symmetries of infinite systems of equations for the pdf's and the Hopf functional equation are discussed. Based on symmetries, invariant solutions for turbulence statistics are calculated.

1. Introduction

With respect to turbulence research three complete statistical descriptions of turbulence, treated as a stochastic field, are known, namely the infnite hierarchy of the multi-point correlation equations (so-called Friedmann-Keller (FK) hierarchy), the infnite hierarchy of the multipoint probability density functions (pdf's) equations (Lundgren-Monin-Novikov (LMN) equations, [1]) and finally the Hopf functional approach [2]. The two latter approaches will be discussed below.

The n-point velocity pdf $f_n(\boldsymbol{x}_{(1)}, \boldsymbol{v}_{(1)}, \boldsymbol{x}_{(2)}, \boldsymbol{v}_{(2)}, \ldots, \boldsymbol{x}_{(n)}, \boldsymbol{v}_{(n)}, t)$ contains information about all statistics up to n-point statistics of infinite order which can be calculated from the pdf by integration over the sample space variables, for example

$$\langle U_{i_{(1)}}(\boldsymbol{x}_{(1)}, t) U_{i_{(2)}}(\boldsymbol{x}_{(2)}, t) \cdot \cdots \cdot U_{i_{(n)}}(\boldsymbol{x}_{(n)}, t)\rangle = \int v_{i_{(1)}} v_{i_{(2)}} \cdot \cdots \cdot v_{i_{(n)}} f_n \mathrm{d}\boldsymbol{v}_{(1)} \ldots \mathrm{d}\boldsymbol{v}_{(n)}.$$

The infinite hierarchy of equations for multipoint pdf's was derived in [1]. E. Hopf introduced another very general approach to the description of turbulence. He considered the case where the number of points in pdf goes to infinity, so that the probability density function becomes a probability density functional and one deals with a continuous set of sample space variables $\boldsymbol{v}(\boldsymbol{x})$. It is more convenient to consider a functional Fourier transform of the probability density functional called the characteristic functional $\Phi([\boldsymbol{v}(\boldsymbol{x})], t)$. The approach of Hopf was later generalised in Ref. [3] where the full space-time functional formalism was introduced and the functional $\Phi([\boldsymbol{v}(\boldsymbol{x}, t)])$ was defined as

$$\Phi([\boldsymbol{v}(\boldsymbol{x}, t)]) = \left\langle \exp\left(i \iint \boldsymbol{U}(\boldsymbol{x}, t) \cdot \boldsymbol{y}(\boldsymbol{x}, t) \mathrm{d}t \mathrm{d}\boldsymbol{x} \right) \right\rangle \tag{1}$$

Journal of Physics: Conference Series **760** (2016) 012038 doi:10.1088/1742-6596/760/1/012038

where $\boldsymbol{y}(\boldsymbol{x},t)$ is an arbitrary vector field which vanishes at spatial infinity. With this definition moments of velocity can be calculated as the functional derivatives of Φ at the origin [3]

$$\left.\frac{\delta^n\Phi}{\delta y_{i_{(1)}}(\boldsymbol{x}_{(1)},t_1)\cdots\delta y_{i_{(n)}}(\boldsymbol{x}_{(n)},t_n)}\right|_{\boldsymbol{y}=0}=i^n\langle U_{i_{(1)}}(\boldsymbol{x}_{(1)},t_1)U_{i_{(2)}}(\boldsymbol{x}_{(2)},t_2)\cdot\dots\cdot U_{i_{(n)}}(\boldsymbol{x}_{(n)},t_n)\rangle \quad (2)$$

where the functional derivative $\delta/\delta y_i(\boldsymbol{x})$ can be understood as a generalisation of a gradient, for the case of infinitely many variables, e.g. we have

$$\frac{\partial}{\partial s_i}\sum_{j=-\infty}^{\infty}f_j s_j=\sum_{j=-\infty}^{\infty}f_j\delta_{ij}=f_i$$

and, for the functional derivative

$$\frac{\delta}{\delta s(\boldsymbol{x})}\int f(\boldsymbol{x}')s(\boldsymbol{x}')\mathrm{d}\boldsymbol{x}'=\int f(\boldsymbol{x}')\delta(\boldsymbol{x}'-\boldsymbol{x})\mathrm{d}\boldsymbol{x}'=f(\boldsymbol{x}).$$

Evolution equation for the characteristic functional was derived in Ref. [3]. It is only one equation (not a hierarchy) which embodies the statistical properties of the fluid flow in a very concise form.

The objectives of the present work are to discuss the classical and new statistical Lie symmetries that were first found for the FK hierarchy [4] and are also present in the LMN hierarchy [5] and the Hopf equation. Based on symmetries, invariant solutions of the considered systems can be derived [6]. Such solutions and their realizability will be addressed within the present work. In particular, a possible route to derive invariant solutions for the characteristic functional of turbulence will be outlined.

The Lie symmetry transformation is such transformation of the independent and dependent variables, which does not change the functional form of a considered equation [6], $F(\boldsymbol{x},\boldsymbol{y}(\boldsymbol{x}),\underset{1}{\boldsymbol{y}}(\boldsymbol{x}),\underset{2}{\boldsymbol{y}}(\boldsymbol{x}),\dots)=0$, where $\underset{1}{\boldsymbol{y}}(\boldsymbol{x})$ denotes any first-order derivative of $\boldsymbol{y}$, $\underset{2}{\boldsymbol{y}}(\boldsymbol{x})$ any second-order derivative etc. The transformed variables

$$\boldsymbol{x}^*=\boldsymbol{g}(\boldsymbol{y}(\boldsymbol{x}),\epsilon),\quad \boldsymbol{y}^*=\mathbf{h}(\boldsymbol{x},\boldsymbol{y}(\boldsymbol{x}),\epsilon)$$

are functions of $\boldsymbol{x}$, $\boldsymbol{y}$ and a group parameter ϵ. The functional form of a considered equation does not change when written in the new variables, i.e. we have

$$F(\boldsymbol{x}^*,\boldsymbol{y}^*,\underset{1}{\boldsymbol{y}^*},\underset{2}{\boldsymbol{y}^*},\dots)=0.$$

The transformations can also be written in infinitesimal forms after the Taylor series expansion about ϵ:

$$\boldsymbol{x}^*=\boldsymbol{x}+\boldsymbol{\xi}(\boldsymbol{x},\boldsymbol{y})\epsilon+\mathcal{O}(\epsilon^2),\quad \boldsymbol{y}^*=\boldsymbol{y}+\boldsymbol{\eta}(\boldsymbol{x},\boldsymbol{y})\epsilon+\mathcal{O}(\epsilon^2). \quad (3)$$

It follows from the Lie first theorem that knowing the infinitesimal forms ξ and η uniquely determines the global form of the group transformation $\boldsymbol{x}^*$ and $\boldsymbol{y}^*$. With the use of infinitesimals invariant solutions of the considered equation may be derived [6]. In fluid mechanics these solutions often represent attractors of the instantaneous fluctuating solutions of the Navier-Stokes equations, i.e. the scaling laws for turbulence statistics. From the Lie symmetry analysis of the LMN hierarchy it followed that the new symmetries are connected with intermittent laminar/turbulent flows [5]. The outcome of the symmetry analysis are the invariant solutions for turbulence statistics and new possibilities to improve turbulence closures, such that invariance under the whole set of symmetries is accounted for.

2. Symmetries of the LMN hierarchy

Symmetries of the LMN hierarchy were investigated in Ref. [5]. Therein, it was shown that the hierarchy is invariant under the classical symmetries of the Navier-Stokes, equations, in particular, time and space translations $t^* = t + a_t$ and $\boldsymbol{x}_i^* = \boldsymbol{x}_i + \boldsymbol{a}_x$, Galilean invariance $t^* = t$, $\boldsymbol{x}_i^* = \boldsymbol{x}_i + \boldsymbol{v}_0\, t$, $\boldsymbol{v}_i^* = \boldsymbol{v}_i + \boldsymbol{v}_0$ and, for non-zero viscosity, one scaling group

$$t^* = \mathrm{e}^{2a_2} t,\ \ \boldsymbol{x}_{(l)} = \mathrm{e}^{a_2}\boldsymbol{x}_{(l)},\ \ \boldsymbol{v}^*_{(l)} = \mathrm{e}^{-a_2}\boldsymbol{v}_{(l)}\ f_n^* = \mathrm{e}^{-3na_2} f_n,\ f^*_{n+1} = \mathrm{e}^{-3(n+1)a_2} f_{n+1}. \tag{4}$$

Moreover, it was shown in [5] that two additional symmetries of the LMN hierarchy exist which transform a pdf of a turbulent signal into the pdf of an intermittent laminar-turbulent flow. These symmetries are equivalent to the additional symmetries derived first for the multipoint velocity correlations in [4]. Written for the the one-point pdf these symmetries are

$$f_1^*(\boldsymbol{v};\boldsymbol{x},t) \;=\; \delta(\boldsymbol{v}) + \mathrm{e}^{a_s}(f_n - \delta(\boldsymbol{v})) \tag{5}$$

and

$$f_1^*(\boldsymbol{v};\boldsymbol{x},t) = f_1(\boldsymbol{v};\boldsymbol{x},t) + \psi(\boldsymbol{v}),\ \ \text{where} \int \psi(\boldsymbol{v})\mathrm{d}\boldsymbol{v} = 0. \tag{6}$$

The transformation (5) forms a semigroup, as the parameter a_s is not arbitrary but is restricted to $a_s \geq 0$.

With the use of new symmetries series of invariant solutions for turbulence statistics can be derived from the characteristic equation [4]. A turbulent/laminar solution for the mean velocity in the plane Poiseuille channel flow was derived in [5]. For this purpose it was necessary to use the classical scaling symmetry (4), the new statistical scaling (5) and translation (6) written for a channel flow (in a modified form due to the presence of boundaries). For the mean velocity these transformations have the following form

$$x_2^* = \mathrm{e}^{k_2}x_2,\ \ \langle U\rangle^* = \mathrm{e}^{a_s - k_2}\langle U\rangle + C_1(1 - {x_2}^2/H^2)$$

where x_2 is the wall-normal coordinate, or, in the infinitesimal forms (3)

$$\xi_{x_2} = k_2 x_2,\quad \eta_{\langle U\rangle} = (a_s - k_2)\langle U\rangle + C_1(1 - {x_2}^2/H^2). \tag{7}$$

Invariant solution for velocity can be found from the solution of the characteristic equation [5]

$$\frac{\mathrm{d}\langle U\rangle}{\eta_{\langle U\rangle}} = \frac{\mathrm{d}x_2}{\xi_{x_2}},\ \Rightarrow\ \frac{\mathrm{d}\langle U\rangle}{(a_s - k_2)\langle U\rangle + C_1(1 - {x_2}^2/H^2)} = \frac{\mathrm{d}x_2}{k_2 x_2}. \tag{8}$$

The system for $a_s = k_2$ has the following solution

$$\langle U\rangle = \frac{C_1}{k_2}\ln(x_2) + \frac{C_1}{2k_2}\left(1 - \frac{x_2^2}{H^2}\right) + \mathcal{C} \tag{9}$$

where, C_1, k_2 and $\mathcal{C}$ are constants and $a_s = k_2 \geq 0$. This solution is a sum of the turbulent (logarithmic) and laminar velocity profiles and follows from a mathematical analysis of the governing equations. In particular, the logarithmic solution for the turbulent part is obtained without the use of the mixing-length assumption and without specification of the mixing-length.

Within the present work the realizability of the solution (9) will be confirmed. Let us consider a flow in a channel with *Re* close to its critical value where the laminar-turbulent transition takes place. In the channel both laminar and turbulent flow is possible with a certain probability. A realisable solution of an intermittent laminar-turbulent flow is given by

$$\langle U\rangle = \gamma u_\tau \frac{1}{\kappa}\ln(x_2) + (1-\gamma)U_L\left(1-\frac{x_2^2}{H^2}\right) + C' \tag{10}$$

where where u_τ is the friction velocity and H- the channel half-width, κ is the von Karman constant, U_L is the centerline velocity of a laminar profile in a channel and γ is the intermittency factor. The value $\gamma = 1$ corresponds to a fully turbulent and $\gamma = 0$ to a fully laminar profile, hence for (10) to be realizable, γ must be contained within the bounds $0 \le \gamma \le 1$.

From the comparison between (9) and (10) it follows that

$$\frac{C_1}{k_2} = \gamma u_\tau \frac{1}{\kappa},\quad \frac{C_1}{2k_2} = (1-\gamma)U_L. \tag{11}$$

The following relations between the driving pressure gradient $\nabla\langle P\rangle$ and u_τ and U_L result from the averaged Navier-Stokes equations for the laminar and turbulent flow, respectively

$$U_L = -\frac{1}{2\nu\rho}\nabla\langle P\rangle H^2,\quad u_\tau = \sqrt{-\frac{H}{\rho}\nabla\langle P\rangle}. \tag{12}$$

We can also define the Reynolds number, based on the pressure gradient as $Re = \sqrt{-H\nabla\langle P\rangle/\rho}H/\nu$. When the flow is turbulent this corresponds to $Re = Re_\tau = u_\tau H/\nu$. It then follows from (12) that $u_\tau/U_L = 2/Re$. Comparing (11) with (12) we obtain

$$\gamma = \frac{\kappa Re}{1+\kappa Re} = \frac{1}{1+(\kappa Re)^{-1}}. \tag{13}$$

Such value of γ is, correctly, contained within the bounds $0 \le \gamma \le 1$, independently of the value of Re, moreover it is a monotonic function which increases with Re. Hence, the realizability of the solution (9) which follows from the Lie group analysis can be confirmed. It is difficult to comment on a possible predictive property of (13) (in terms of the dependence of γ on Re) as the laminar-turbulent transition depends on various factors (e.g. the level of external disturbances) which are not included in the analysis. Still, it is interesting that Eq. (13) could be obtained as a result which follows from the Lie-group invariant solution (9). For $Re = 150$ where low-Re turbulent flows are observed Eq. (13) predicts $\gamma \approx 0.98$, which is a reasonable result. However, for very low Re, γ obtained from Eq. (13) is overpredicted. Possibly, Eq. (13) could be valid within a certain range of Re.

3. Symmetries of the Hopf equation

In this section, symmetries of the Hopf functional formulation for turbulence will be discussed. The evolution of the space-time functional, as defined in Eq. (1) is governed by the following equation [3]

$$\iint \eta_k(\boldsymbol{x},t)\left[\frac{\partial}{\partial t}\frac{\delta\Phi}{\delta y_k(\boldsymbol{x},t)} - i\frac{\partial}{\partial x_l}\frac{\delta^2\Phi}{\delta y_k(\boldsymbol{x},t)\delta y_l(\boldsymbol{x},t)} - \nu\frac{\partial^2}{\partial x_l\partial x_l}\frac{\delta\Phi}{\delta y_k(\boldsymbol{x},t)} + \frac{\partial\Pi}{\partial x_k}\right]\mathrm{d}t\mathrm{d}\boldsymbol{x} = 0, \tag{14}$$

where Π is the pressure functional and $\eta_k(\boldsymbol{x},t)$ is a testing field. If η_k satisfies the condition $\partial\eta_k/\partial x_k = 0$, then, the pressure functional term can be eliminated by integrating (14) by parts with $\int \partial\eta_k/\partial x_k \Pi \mathrm{d}\boldsymbol{x} = 0$. The continuity condition for the incompressible flow in the functional formulation reads

$$\frac{\partial}{\partial x_k}\frac{\delta\Phi}{\delta y_k(\boldsymbol{x},t)} = 0. \tag{15}$$

Moreover, Φ must satisfy the relations

$$\Phi(0) = 1, \quad \overline{\Phi([y])} = \Phi([-y]), \quad |\Phi| \leq 1, \tag{16}$$

where $\bar{\cdot}$ denotes the complex conjugate.

The Hopf functional equation (14) is invariant under the following transformation of variables

$$\Phi^* = \Phi, \quad \boldsymbol{x}^* = \mathrm{e}^{k_2}\boldsymbol{x}, \quad t^* = \mathrm{e}^{2k_2}t, \quad y_i^* \mathrm{d}\boldsymbol{x}^* \mathrm{d}t^* = \mathrm{e}^{k_2} y_i \mathrm{d}\boldsymbol{x}\mathrm{d}t, \quad \boldsymbol{y}^* = \mathrm{e}^{-4k_2}\boldsymbol{y}, \tag{17}$$

where we note that the functional derivative $\delta/\delta y_k(\boldsymbol{x}, t)$ is sometimes denoted by $\partial/(\partial y_k(\boldsymbol{x}, t)\mathrm{d}\boldsymbol{x}\mathrm{d}t)$ which indicates that its dimension is $1/([y][L][T])$. Hence,

$$\frac{\delta}{\delta y_k^*(\boldsymbol{x}, t)} = \mathrm{e}^{-k_2} \frac{\delta}{\delta y_k(\boldsymbol{x}, t)}.$$

The Galilean invariance reads $t^* = t$, $\boldsymbol{x}^* = \boldsymbol{x} + \boldsymbol{U}_0 t$ and $\boldsymbol{U}^* = \boldsymbol{U} + \boldsymbol{U}_0$. With this, the functional Φ transforms as follows

$$\Phi^* = \left\langle \mathrm{e}^{i\iint \boldsymbol{U}^*(\boldsymbol{x},t)\cdot \boldsymbol{y}^*(\boldsymbol{x},t)\mathrm{d}\boldsymbol{x}^*\mathrm{d}t^*} \right\rangle = \left\langle \mathrm{e}^{i\iint \boldsymbol{U}(\boldsymbol{x},t)\cdot \boldsymbol{y}(\boldsymbol{x},t)\mathrm{d}\boldsymbol{x}\mathrm{d}t} \right\rangle \mathrm{e}^{i\int \boldsymbol{y}(\boldsymbol{x},t)\cdot \boldsymbol{U}_0 \mathrm{d}\boldsymbol{x}\mathrm{d}t} = \Phi C([\boldsymbol{y}(\boldsymbol{x}, t)], \tag{18}$$

where the functional $C([\boldsymbol{y}(\boldsymbol{x}, t)]) = \exp(i\iint \boldsymbol{y}(\boldsymbol{x}, t)\cdot \boldsymbol{U}_0 \mathrm{d}\boldsymbol{x}\mathrm{d}t)$ satisfies the condition $C(0) = 1$, hence, also $\Phi^*(0) = 1$ as required in Eq. (16).

The n-th derivative of the transformed functional Φ^* at $\boldsymbol{y} = 0$ gives

$$\left.\frac{\delta^n \Phi^*}{\delta y_{i_{(0)}}(\boldsymbol{x}, t)\cdots \delta y_{i_{(n-1)}}(\boldsymbol{x}, t)}\right|_{\boldsymbol{y}=0} = i^n (U_{i_{(0)}}(\boldsymbol{x}, t) + U_{0i_{(0)}})\cdots (U_{i_{(n-1)}}(\boldsymbol{x}, t) + U_{0i_{(n-1)}}) \tag{19}$$

as expected for the Galilean invariance. In the Galilean transformation, the space derivatives in equation (14) transform as $\partial/\partial x_i^* = \partial/\partial x_i$ and the integral over time and space $\int \mathrm{d}\boldsymbol{x}^*\mathrm{d}t^* = \int \mathrm{d}\boldsymbol{x}\mathrm{d}t$. As the variables $\boldsymbol{y}^* = \boldsymbol{y}$, also the functional derivative remains unchanged $\delta/\delta(y_i(\boldsymbol{x}))^* = \delta/\delta y_i(\boldsymbol{x})$. The derivative $\partial/\partial t$ can be presented as

$$\frac{\partial}{\partial t} = \frac{\partial t^*}{\partial t}\frac{\partial}{\partial t^*} + \frac{\partial x_i^*}{\partial t}\frac{\partial}{\partial x_i^*} = \frac{\partial}{\partial t^*} + U_{0k}\frac{\partial}{\partial x_k}. \tag{20}$$

The transformed functional equation (14) reads

$$\iint \eta_k(\boldsymbol{x}, t)\left[\frac{\partial}{\partial t^*}\frac{\delta C([\boldsymbol{y}(\boldsymbol{x}, t)])\Phi}{\delta y_k(\boldsymbol{x}, t)} - i\frac{\partial}{\partial x_l}\frac{\delta^2 C([\boldsymbol{y}(\boldsymbol{x}, t)])\Phi}{\delta y_k(\boldsymbol{x}, t)\delta y_l(\boldsymbol{x}, t)} - \nu\frac{\partial^2}{\partial x_l \partial x_l}\frac{\delta C([\boldsymbol{y}(\boldsymbol{x}, t)])\Phi}{\delta y_k(\boldsymbol{x}, t)}\right]\mathrm{d}t\mathrm{d}\boldsymbol{x} = 0. \tag{21}$$

With (20) the first term in bracket in (21) is

$$\frac{\partial}{\partial t^*}\frac{\delta C\Phi}{\delta y_k(\boldsymbol{x}, t)} = \left(\frac{\partial}{\partial t} - U_{0l}\frac{\partial}{\partial x_l}\right)\left(C\frac{\delta\Phi}{\delta y_k(\boldsymbol{x}, t)} + \Phi\frac{\delta C}{\delta y_k(\boldsymbol{x}, t)}\right) = C\left(\frac{\partial}{\partial t} - U_{0l}\frac{\partial}{\partial x_l}\right)\frac{\delta\Phi}{\delta y_k(\boldsymbol{x}, t)} \tag{22}$$

where the second equality follows from the fact that $\delta C/\delta y_k = iU_{0k}C$ does not depend explicitly on $\boldsymbol{x}$ or t.

The functional derivative of Φ^* in (21) reads $C\delta\Phi/\delta y_k(\boldsymbol{x}) + \Phi\delta C/\delta y_k$, and we note that Laplacian ∇_x^2 of the second term is zero as this term is not a function of $\boldsymbol{x}$. Hence, the last RHS term of equation (21) inside the integral is $C\nu\nabla_x^2\delta\Phi/\delta y_k$. Further, the second functional derivative of Φ^* reads

$$C\frac{\delta^2\Phi}{\delta y_k(\boldsymbol{x}, t)\delta y_l(\boldsymbol{x}, t)} + \frac{\delta\Phi}{\delta y_k(\boldsymbol{x}, t)}\frac{\delta C}{\delta y_l(\boldsymbol{x}, t)} + \frac{\delta\Phi}{\delta y_l(\boldsymbol{x}, t)}\frac{\delta C}{\delta y_k(\boldsymbol{x}, t)} + \Phi\frac{\delta^2 C}{\delta y_k(\boldsymbol{x}, t)\delta y_l(\boldsymbol{x}, t)}. \tag{23}$$

Again, the derivative $\partial/\partial x_l$ of the last term is zero, as it does not depend on $\boldsymbol{x}$. In addition we also have

$$\frac{\partial}{\partial x_l}\left[\frac{\delta\Phi}{\delta y_l(\boldsymbol{x},t)}\frac{\delta C}{\delta y_k(\boldsymbol{x},t)}\right]=\frac{\delta C}{\delta y_k(\boldsymbol{x},t)}\frac{\partial}{\partial x_l}\frac{\delta\Phi}{\delta y_l(\boldsymbol{x},t)}=0, \tag{24}$$

where the first equality follows from the fact that the derivative of C does not depend explicitly on $\boldsymbol{x}$ and the second from the continuity condition (15).

The second term in Eq.. (23), introduced into Eq. (21), leads to

$$-\frac{\delta C}{\delta y_l(\boldsymbol{x})}\left[i\frac{\partial}{\partial x_l}\frac{\delta\Phi}{\delta y_k(\boldsymbol{x},t)}\right]=U_{0l}C\left[\frac{\partial}{\partial x_l}\frac{\delta\Phi}{\delta y_k(\boldsymbol{x},t)}\right], \tag{25}$$

which cancels with the last term in (22). This finally proves the Galilean invariance of Eq. (14).

An analogue of the new statistical scaling and translation symmetries (found for the multipoint correlations in [4] and discussed in [5] for pdf's, see also Eqs. (5,6)) can also be derived for the Hopf functional. Mathematically, they follow from the linearity of the considered functional equation (14) and the normalisation condition (16), and have the folowing form [7]

$$\Phi^*=1+\mathrm{e}^{k_s}(\Phi-1)\quad\text{and}\quad\Phi^*=\Phi+\Psi([\boldsymbol{y}(\boldsymbol{x})]), \tag{26}$$

where Ψ is a functional which satisfies Eq. (14) such that $\Psi(0)=0$. Substituting (26) into (14) the invariance of (14) under these transformations is easily confirmed.

With the classical scaling (17) and the new symmetries (26) we can attempt to derive invariant solutions for the space-time Hopf functional. For this purpose we can use a procedure analogous to the one used in [8, 9]. We first present the functional in a form of Taylor-series expansion [2]

$$\Phi=1+C_1+C_2+\cdots \tag{27}$$

where

$$C_n=\int K_{i_{(1)}\cdots i_{(n)}}(\boldsymbol{x}_{(1)},\ldots,\boldsymbol{x}_{(n)},t_1,\ldots,t_n)y_{i_{(1)}}(\boldsymbol{x}_{(1)},t_1)\cdots y_{i_{(n)}}(\boldsymbol{x}_{(n)},t_n)\mathrm{d}\boldsymbol{x}_{(1)}\cdots\mathrm{d}\boldsymbol{x}_{(n)}\mathrm{d}t_1\cdots\mathrm{d}t_n \tag{28}$$

with functions $K_{i_{(1)}\cdots i_{(n)}}(\boldsymbol{x}_{(1)},\ldots,\boldsymbol{x}_{(n)},t_1,\ldots,t_n)=i^n/n!\langle U_{i_{(1)}}(\boldsymbol{x}_{(1)},t_1)\cdots U_{i_{(n)}}(\boldsymbol{x}_{(n)},t_n)\rangle$ which are related to the multipoint, multitime moments of velocity. The three symmetries, (17) and the new symmetries (26), written in the infinitesimal forms lead to the following characteristic system of equations

$$\frac{\mathrm{d}t}{2k_2t}=\frac{\mathrm{d}x_i}{k_2x_i}=\frac{\mathrm{d}(y_i\mathrm{d}\boldsymbol{x}\mathrm{d}t)}{k_2(y_i\mathrm{d}\boldsymbol{x}\mathrm{d}t)}=\frac{\mathrm{d}\Phi}{a_s(\Phi-1)+\Psi},\quad i=1,2,3, \tag{29}$$

which should hold for each point $\boldsymbol{x}$ and each time t.

The purpose of the present study is to outline a procedure of deriving invariant solutions for the characteristic functional. Hence, one particular invariant solution will be derived, and discussed although it seems that many more possibilities exist. We will consider two different points $\boldsymbol{x}_{(1)}$ and $\boldsymbol{x}_{(2)}$ and two different times t_1 and t_2, and note that the characteristic system (29) can also be rewritten as

$$\frac{\mathrm{d}t_1}{\mathrm{d}s}=2k_2t_1,\quad\frac{\mathrm{d}t_2}{\mathrm{d}s}=2k_2t_2,\quad\frac{\mathrm{d}x_{(1)i}}{\mathrm{d}s}=k_2x_{(1)i},\quad i=1,2,3,\quad\frac{\mathrm{d}\Phi}{\mathrm{d}s}=a_s(\Phi-1)+\Psi \tag{30}$$

$$\frac{\mathrm{d}(y_i(\boldsymbol{x}_{(1)},t_1)\mathrm{d}\boldsymbol{x}_{(1)}\mathrm{d}t_1)}{\mathrm{d}s}=k_2y_i(\boldsymbol{x}_{(1)},t_1)\mathrm{d}\boldsymbol{x}_{(1)}\mathrm{d}t_1,$$

$$\frac{\mathrm{d}(y_i(\boldsymbol{x}_{(2)},t_1)\mathrm{d}\boldsymbol{x}_{(2)}\mathrm{d}t_2)}{\mathrm{d}s}=k_2y_i(\boldsymbol{x}_{(2)},t_2)\mathrm{d}\boldsymbol{x}_{(2)}\mathrm{d}t_2.$$

With further, mathematical transformations Eqs. (30) can be reformulated e.g. to the following form

$$\frac{\mathrm{d}t_1}{2k_2t_1}=\frac{\mathrm{d}t_2}{2k_2t_2}=\frac{\mathrm{d}(t_1-t_2)}{2k_2(t_1-t_2)}=\frac{\mathrm{d}(x_{(1)i}-x_{(2)i})}{k_2(x_{(1)i}-x_{(2)i})}=\frac{\mathrm{d}(y_j(\boldsymbol{x}_{(1)},t_1)\mathrm{d}\boldsymbol{x}_{(1)}\mathrm{d}t_1)}{k_2(y_j(\boldsymbol{x}_{(1)},t_1)\mathrm{d}\boldsymbol{x}_{(1)}\mathrm{d}t_1)}= \tag{31}$$
$$=\frac{\mathrm{d}(y_k(\boldsymbol{x}_{(1)},t_1)y_l(\boldsymbol{x}_{(2)},t_2)\mathrm{d}\boldsymbol{x}_{(1)}\mathrm{d}t_1\mathrm{d}\boldsymbol{x}_{(2)}\mathrm{d}t_2)}{2k_2(y_k(\boldsymbol{x}_{(1)},t_1)y_l(\boldsymbol{x}_{(2)},t_2)\mathrm{d}\boldsymbol{x}_{(1)}\mathrm{d}t_1\mathrm{d}\boldsymbol{x}_{(2)}\mathrm{d}t_2)}=\dots=\frac{\mathrm{d}\Phi}{a_s(\Phi-1)+\Psi},\quad i,j,k,l=1,2,3$$

The first five integration constants, obtained from (31) are

$$a_1=\frac{t_1}{t_2},\quad a_2=\frac{t_1-t_2}{t_1},\quad a_3=\frac{(x_{(1)1}-x_{(2)1})^2}{t_1-t_2},\quad a_4=\frac{(x_{(1)2}-x_{(2)2})^2}{t_1-t_2},\quad a_5=\frac{(x_{(1)3}-x_{(2)3})^2}{t_1-t_2}.$$

A solution, which follows from the comparison of the first and the sixth term in (31) is

$$A_6(\boldsymbol{x}_{(1)},\boldsymbol{x}_{(2)},t_1,t_2)=\frac{1}{t_1}F_{ij}(a_1,a_2,a_3,a_4,a_5)y_i(\boldsymbol{x}_{(1)},t_1)y_j(\boldsymbol{x}_{(2)},t_2)\mathrm{d}\boldsymbol{x}_{(1)}\mathrm{d}t_1\mathrm{d}\boldsymbol{x}_{(2)}\mathrm{d}t_2, \tag{32}$$

for each $\boldsymbol{x}_{(1)}$,$\boldsymbol{x}_{(2)}$,t_1,t_2. Integrating (32) we obtain another constant

$$C_2=\iint\frac{1}{t_1}F_{ij}(a_1,a_2,a_3,a_4,a_5)y_i(\boldsymbol{x}_{(1)},t_1)y_j(\boldsymbol{x}_{(2)},t_2)\mathrm{d}\boldsymbol{x}_{(1)}\mathrm{d}t_1\mathrm{d}\boldsymbol{x}_{(2)}\mathrm{d}t_2. \tag{33}$$

We next consider the last term in (31) and assume $a_s=0$ and $\Psi=0$. We will present the solution Φ as a sum of constants $\Phi=1+C_1+C_2+\dots$, as in the Taylor-series expansion (27). For the homogeneous isotropic turbulence with zero mean velocity we have $C_1=0$, the constant C_2 is given in Eq. (33) and it could also be possible to derive other constants, i.e. C_3, C_4 by an analogous procedure, considering larger number of points $\boldsymbol{x}_{(1)}$, $\boldsymbol{x}_{(2)}$, $\boldsymbol{x}_{(3)}$, $\boldsymbol{x}_{(4)}$ etc. and larger number of times.

The second functional derivative of Φ at the origin can provide a relation for the decay law of the second-order turbulence statistics

$$\left.\frac{\delta^2\Phi}{\delta y_i(\boldsymbol{x}_{(1)},t_1)\delta y_j(\boldsymbol{x}_{(2)},t_2)}\right|_{\boldsymbol{y}=0}=i^2\langle U_i(\boldsymbol{x}_{(1)},t_1)U_j(\boldsymbol{x}_{(2)},t_2)\rangle=\frac{1}{t_1}F_{ij}(a_1,a_2,a_3,a_4,a_5). \tag{34}$$

For example, if $F_{ij}=const$ for a possible choice $\boldsymbol{x}_{(1)}=\boldsymbol{x}_{(2)}$ and $t_1=t_2$, we obtain from (34) the power-law decay of the kinetic energy in the homogeneous, isotropic turbulence

$$\langle U_i(\boldsymbol{x}_{(1)},t_1)U_i(\boldsymbol{x}_{(1)},t_1)\rangle\sim\frac{1}{t_1}.$$

In the literature, the decay of kinetic energy $k\sim t^{-m}$ is reported [10]. Although there is a discussion on the value of m, several studies suggest that m approaches the value 1 as the initial *Re* number increases [11].

4. Conclusions

The present work concerns the Lie symmetry analysis of infinite systems describing the statistics of turbulence, the LMN hierarchy for pdf's and the Hopf functional equation. First, the symmetries of the LMN hierarchy, as derived in [5] were addressed, together with an invariant solution for the mean velocity in a laminar-turbulent channel flow. The new contribution of the present work is a discussion of the realizability of this solution. It was shown that the intermittency factor, following from the solution is correctly contained within the bounds $0 \leq \gamma \leq 1$.

Next, the Hopf functional in the spatio-temporal formulation introduced in Ref. [3] was discussed. It was shown that the evolution equation for the functional is invariant under the transformations which follow from the Navier-Stokes equations, in particular, the Galilean invariance was shown. Next, a possible procedure to derive invariant solutions for the Hopf functional was proposed, which is another new contribution of the present work. Studying other invariant solutions and their physical consequences is a perspective for a further study.

Another interesting issue is connected with the numerical study of the multipoint equations. A numerical approach for the multipoint pdf's has recently been devised in [12]. A new method of numerical approximation of nonlinear functionals and functional differential equations has been put forward in [13]. Both novel numerical approaches provide a new perspective for further, numerical study of the multipoint approaches to turbulence.

Acknowledgments
The financial support of the National Science Centre, Poland (project No. 2014/15/B/ST8/00180) is acknowledged.

References

[1] Lundgren T. S. 1967 *Phys. Fluids* **10** 969
[2] Hopf E. 1952 *J. Rational Mech. Anal* **1** 87
[3] Lewis R. M., Kraichnan R. 1962 *Comm. Pure Appl. Math.* **15** 397
[4] Oberlack M., Rosteck A. 2010 *Disc. Cont. Dyn. Sys. Ser. S.* **3** 451
[5] Wacławczyk M., Staffolani N., Oberlack M., Rosteck A., Wilczek M., Friedrich R. 2014 *Phys. Rev. E* **90** 013022
[6] Bluman G., Cheviakov A., Anco S. 2009 *Applications of Symmetry Methods to Partial Differential Equations* (New York: Springer)
[7] Janocha D. D., Wacławczyk M., Oberlack M. 2015 *Symmetry* **7**, 1536
[8] Oberlack M., Wacławczyk M. 2006 *Arch. Mech.* **54** 072901
[9] Wacławczyk M., Oberlack M. 2013 *J. Math. Phys.* **58** 597
[10] Pope S. B. 2000 *Turbulent flows* (Cambridge: Cambridge University Press)
[11] Burattini P., Lavoie P., Agrawal A., Djenidi L., Antonia R. A. 2006 *Phys. Rev. E* **73** 066304
[12] Cho H., Venturi D., Karniadakis G. E. 2016 *J. Comp. Phys.* **305** 817
[13] Venturi D. 2016 arXiv:1604.05250

Numerical investigations of flow structure in gas turbine shroud gap

F Wasilczuk[1,2], P Flaszyński[1] and P Doerffer[1]

[1]Institute of Fluid-Flow Machinery, Polish Academy of Sciences, Fiszera 14, 80-952 Gdańsk, Poland
[2]Interdepartmental PhD Studies at the Gdansk University of Technology Narutowicza 11/12, 80–233 Gdańsk, Poland

Abstract. The structure of the flow in the labyrinth sealing of an axial gas turbine was investigated by means of numerical simulations. Additionally, the flow structure for two- and three-dimensional axisymmetric models was compared. The porous disc as a model for the pressure drop relevant to the obtained in the cascade was proposed and tested. Several flow structure features existing in the sealing cavities are investigated: vortical structure and separation bubble on the rib and the correlation between the pressure drop and the clearance size. The carried out investigations indicate that the innovation aimed at decreasing the leakage flow through implementation of the flow control devices is possible. Furthermore the comparison between 2D and 3D models shows good agreement, thus application of less demanding 2D model introduces negligible differences. It is shown that the proposed porous disc model applied to mimic pressure drop in cascade can be effectively used for rotor blade sealing simulations.

1. Introduction

The leakage between the shroud and the casing is one of the most important sources of losses in shrouded cascades of gas turbines. It influences on the loss coefficient increase. Firstly it reduces the working potential, by decreasing the mass flow rate in blade passages [1]. Secondly, there is a difference of velocity distribution downstream of the rotor blade between the fluid passing through the gap and the blade passage. This results in mixing in highly non-uniform flow zone and entropy production [1,2,3]. Moreover, the vortices created in shear layer can affect the efficiency of the subsequent stator [2,3] and they can have negative influence on thermal conditions on following nozzle guide vane. The leakage losses can be limited by reduction of the amount of the mass flow in the gap between the shroud and the casing.

One of the most popular methods to decrease the gap flow between the shroud and the casing is a labyrinth seal. It is made of ribs mounted on the shroud, creating cavities along the gap. The kinetic energy fluid is dissipated in those cavities influencing on mass flow reduction in the gap.

The ultimate aim of the research, INNOLOT-COOPERNIK is the investigation of the labyrinth seal modifications that will limit the mass flow through the seal, decreasing losses caused by the leakage. There is a lot of studies aimed at leakage loss reduction by analysing of the 3-dimensional flow structure and application flow control devices to match the fluid velocity at the gap and in the mainstream, to limit the mixing [4,5,6,7]. This paper presents results of flow structure investigations in the labyrinth seal as the first step of the further modifications and flow control methods application.

2. Numerical model description

2.1. Geometry

The results presented in the paper are obtained for the geometry created as the simplification of the configuration for a large-scale, low-speed gas turbine rotor stage, investigated in European project AITEB-2 [8]. The main objective of the numerical investigations it to study the effect of applied model simplifications on the flow structure. An important result of the leakage arises from the passage flow interaction with leakage flow downstream of the sealing. Simulations for full cascade configuration should include passage flow and sealing. However, if the investigations are focused on the sealing configuration only, the cascade is simplified or modelled. The results in the paper are carried out for the model where blade row is substituted with porous disc which enables to model the effect of pressure drop in cascade. The porous disc model is described in the next chapter. Some changes were also applied to the geometry of the labyrinth seal, which is located on the disc. Firstly, the last of the three seal ribs was neglected in the model. Secondly, the roundings were simplified by corners, which significantly reduce the complexity of the mesh, but not influence on these basic investigations. The sketch of the cross section model with the most important dimensions is shown in Figure 1.

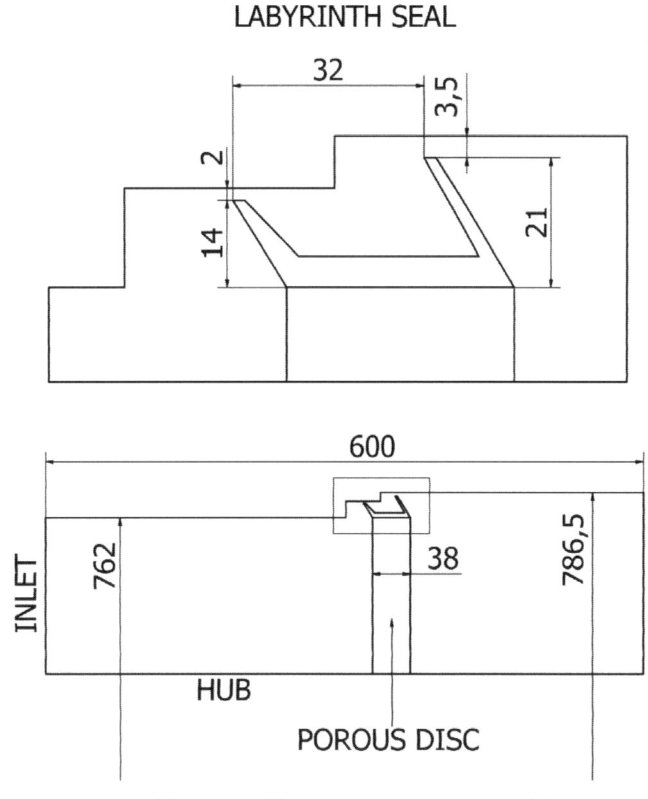

Figure 1. The geometry of the model.

2.2. Boundary Conditions

The boundary conditions were set according to the operating conditions of test rig [8]. –At the inlet, total temperature is 293K, the turbulence viscosity is 0.0001. Five cases for different inlet total pressures were analysed (Table 1). The outlet static pressure was set for all cases to 98325 Pa. The speed of rotation at labyrinth walls was assigned 359 RPM, corresponding to the rotational velocity at the test section.

Table 1. Boundary conditions.

Case No.	Inlet [Pa]	Outlet [Pa]	Pressure ratio (π)
1.	99419	98325	1.01
2.	101419	98325	1.03
3.	103419	98325	1.05
4.	105419	98325	1.07
5.	107419	98325	1.09

2.3. Numerical parameters

Two structured meshes were used for the numerical simulation: 2D and 3D. 2-dimensional, axisymmetric mesh contains ~70000 elements. 3-dimensional consists of 1.3 million cells with the 2^0 angle of extrusion. The calculations were carried out in Fine/Turbo Numeca as steady ones by means of Spalart-Allmaras turbulence model [9]. The solver uses multigrid technique with coarse grid initialization. In order to improve convergence Merkle low-speed preconditioning and CFL number of 1 was used. The perfect gas model is used.

3. Porous disc

Simulating the whole blade row including all 3 dimensional features is time and resource consuming. Thus, due to the investigations effectiveness of wide range of various seal configurations, the simplified model was applied. The analysis of the flow over rotating, impermeable disc [10,11,12] showed formation of the vortex upstream of the disc that changed the flow conditions in a gap. Thus in the numerical investigation the labyrinth seal cannot be just attached to the impermeable disc. There is a need to simulate the main features of the blade with a less complex model. There are two methods that can be utilized. First is to take data of flow features upstream and downstream of a blade row from the numerical simulation or experimental measurements, and apply them as additional outlet and inlet in the area, where blades are supposed to be (Figure. 2) [13]. However, this approach requires accurate data to set as boundary conditions. Moreover, with the flow becomes constrained with extra boundary conditions, and the impact of the seal flow on the main flow is difficult to observe.

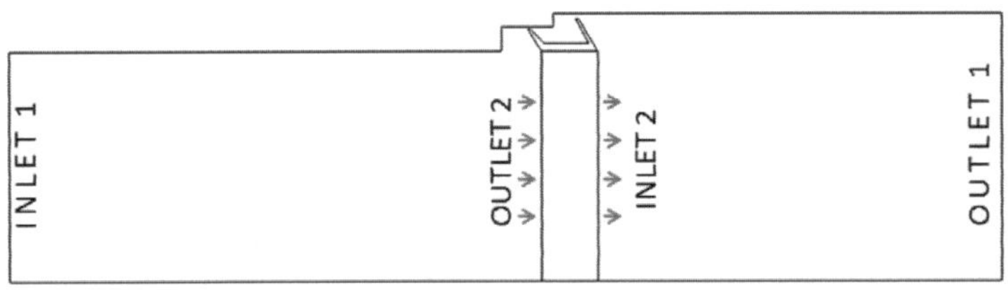

Figure 2. Sketch of the model with additional inlet and outlet instead of the blade row.

Another considered approach is to model the pressure drop in the blade row by means of the porous disc. Based on appropriately chosen porosity parameters, one can achieve a pressure drop similar to the one in a blade row, keeping the same the mean axial velocity downstream of cascade. The main drawback of this solution is that it is only two-dimensional, all of the complex 3-D flow structures are not modelled, and one cannot study the interaction of the leakage flow with the main flow. However, since the ultimate goal of the investigation is to study the features of the flow within the seal, this drawback is not much affecting the main objective. Thus, the porous disc approach was chosen for this study.

Fine/Turbo allows various porous media formulation, but for this case the Darcy law was used to determine the pressure drop on the disc (1).

$$\vec{\nabla} p = -\frac{\mu}{\kappa}\vec{v} \qquad (1)$$

where:

$\vec{\nabla} p$ – pressure drop

μ – viscosity

$\vec{v}$ – velocity

κ – intrinsic permeability given by (2)

$$\kappa = \frac{\varepsilon^3 \left(\frac{6}{A}\right)^2}{180(1-\varepsilon)^2} \tag{2}$$

where A is the specific surface area and ε is the porosity. [14]

In the specific case of blades row simulation, the entire pressure drop will occur on the porous disc, independently of its parameters and its value will depend on the boundary conditions. However, the porosity parameters will significantly affect velocity of the flow. Using the relation between pressure drop and velocity it is impossible to estimate both porosity and specific surface area. For that reason, porosity was constrained to 0.9, and from that the estimate specific surface area was calculated. Then it was tuned using numerical calculations to mimic the velocity from the experiment [8]. Figure 3 shows the pressure distribution in the model, with visible pressure drop in the porous zone, proving that the model is working as intended.

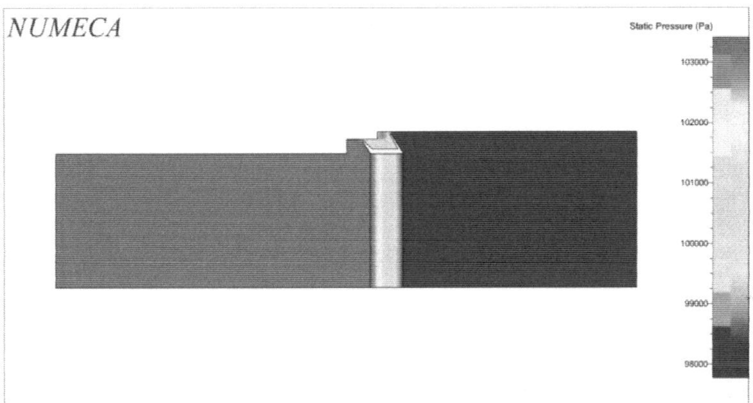

Figure 3. Pressure contour in the meridional view.

4. Results

The static pressure distribution in the whole computational domain is shown in the figure 3. Application of the porous disc model leads to the uniform pressure upstream and downstream of the disc as assumed at inlet and outlet plane. The same pressure decrease occurs inside the modelled cascade (disc) domain and in the labyrinth seal.

As mentioned above, the presented results are considered as the reference flow structure investigations, which is planned to be modified and controlled in the next step by means of flow control methods. Due to that, selected flow features are analysed below.

4.1. Pressure drop in the gaps between the ribs and the casing

In the figure 4, the pressure drop in each gap between the rib and the casing is shown. The pressure drop above each rib is proportional to the pressure ratio in the cascade (figure 5) and the gap height (distance between rib and shroud). In the investigated case, the clearance is different above both ribs (2 mm for the first gap and 3.5 mm for the second), what influences on the pressure drop differences. It is several times bigger in the first gap than in the second one. The first rib is responsible for approximately 80% of the pressure drop, while the second one for the remaining 20%.

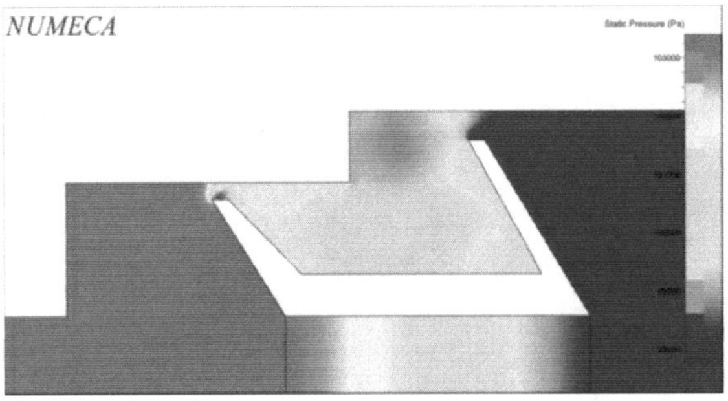

Figure 4. The pressure distribution in the labyrinth seal

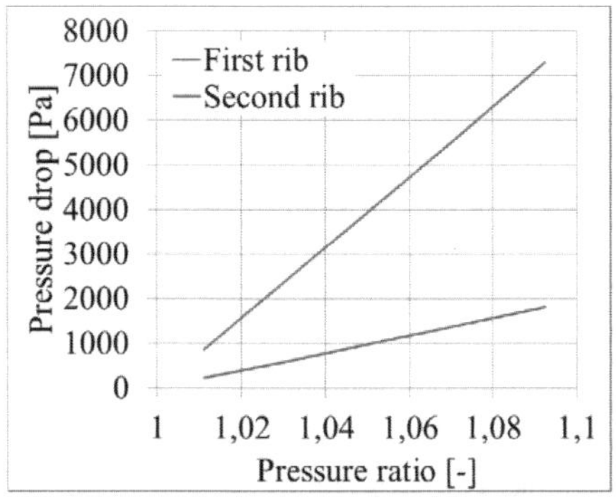

Figure 5. Pressure drop in the gap between the rib and the casing as a function of pressure ratio.

4.2. Acceleration and separation in the contractions

The clearance between the first rib and the casing influences on the flow acceleration. As the effect of the local acceleration, the jet is generated which interacts with the second rib. Such interaction is dependent on the second rib location, its shape and sealing geometry (figure 6). The maximum velocity of the jet as the function of pressure ratio is shown in figure 7. Moreover, the separation is induced at the leading edge of the rib, which reduces the effective area of the gap. This phenomenon can be also noticed in the flow over rotating disc with sharp edges [11, 12]. Similar situation occurs on the second rib, although the gap is higher, so the jet velocity is lower. Anyway, one has to emphasize that this flow feature is strongly dependant on the inflow velocity, flow direction and the geometry details of the rib (inclination, sharp or rounded edge). [12, 15]

Figure 6. Streamlines and absolute Mach number contour.

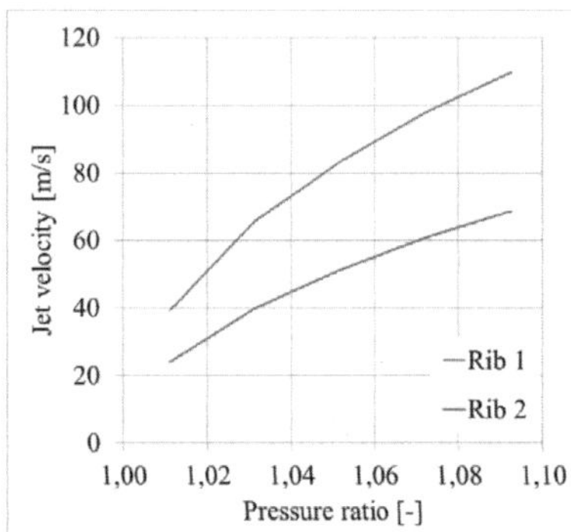

Figure 7. Maximum velocity of the jet.

4.3. Pressure rise on the surface of the second rib.

The pressure rise can be observed on the surface of the second rib at the upstream side, next to the tip of the rib (figure 8). It is caused by the jet generated by the contraction above the first rib influencing on the created stagnation zone on the second rib (figure 8). The pressure distribution on the second rib and the location of the maximum value (stagnation point) is dependent on the pressure ratio in cascade (figure 9). This feature is very important if flow control method based on the local pressure difference is considered to apply.

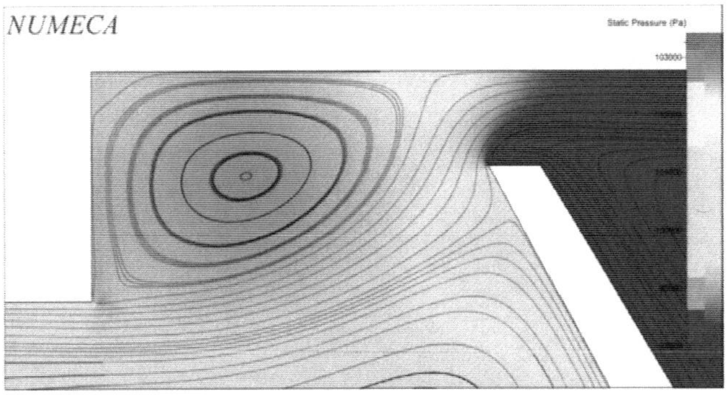

Figure 8. Streamlines and static pressure contour.

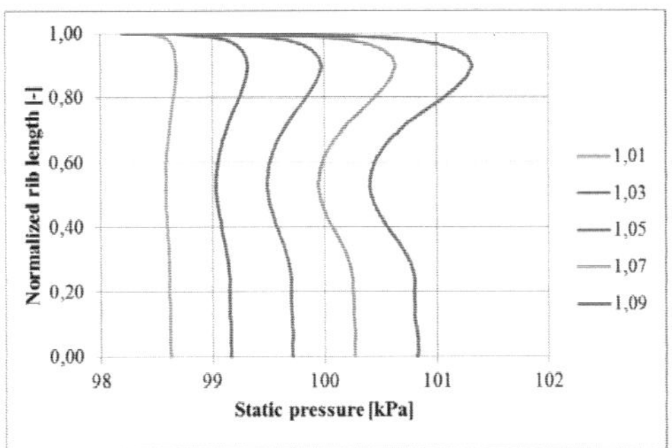

Figure 9. Static pressure on the wall of the second rib.

4.4. Vortex structure

Vortex structure of the flow upstream and inside the labyrinth seal was analyzed. The flow downstream of the sealing in real cascade is strongly affected by interaction with the passage flow, so its analysis downstream of the blade row on the simplified model could not lead to any meaningful conclusions. However, the flow structure inside the cavity is weakly affected by the outer (passage) flow and it is similar for different pressure ratio in all analyzed cases.

Upstream of the seal, there is one large vortex created by low velocity, which located in the corner of a casing (figure 6). The size and shape of this vortex is highly dependent on the casing geometry.

The large vortex that is powered by the jet generated above the first rib is created in the cavity between the two ribs. Due to the viscosity, the circumferential velocity of the shroud drives the fluid in this direction. For that reason the vortex has significant tangential velocity, unlike the two others, which is shown in figure 10.

The last major vortex is located in the corner of the casing, upstream of the second rib. Similarly to the main vortex in the cavity, it is driven by the jet and its intensity (strength) is dependent on kinetic energy transfer between the jet and the corner vortex. However, because it is separated from the rotating seal by the high velocity jet, its tangential velocity component is not significant. In the center of this vortex, there is a region of low pressure (figure 6).

It is worthwhile to note, that various labyrinth seals differ from each other significantly. The configuration depends on the stage geometry: rotor, upstream and downstream nozzle guide vanes and casing. Finally, the vortex structure, its intensity and location in the sealing zone is reliant on the investigated geometry.

Journal of Physics: Conference Series **760** (2016) 012039 doi:10.1088/1742-6596/760/1/012039

Figure 10. Streamlines and circumferential velocity contour

4.5. 2D and 3D model comparison

The velocity components distribution at four traverses (figure 11) for 2D and 3D models were compared, to compare the effect of the model simplification. The example of the velocity components comparison at traverse 1, which is located in the gap above the first rib, is shown on figures 12, 13 and 14. The curves are marked by different colors, the 3D model with brighter line and dots while 2D model with darker solid line. In case of all traverses, for all velocity components, results obtained by means of two-dimensional and three-dimensional model are very similar, showing always the same trend in spite of little local differences Thus, using substantially less computationally demanding 2D axisymmetrical model for further studies is justified and does not lead to significant differences in comparison with 3D model.

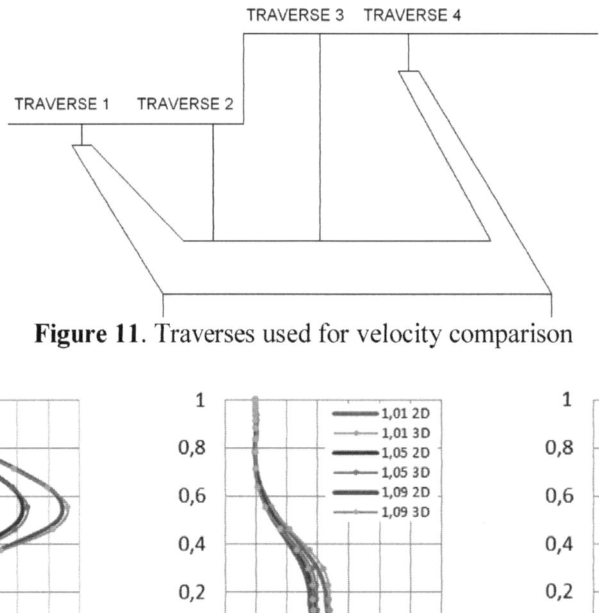

Figure 11. Traverses used for velocity comparison

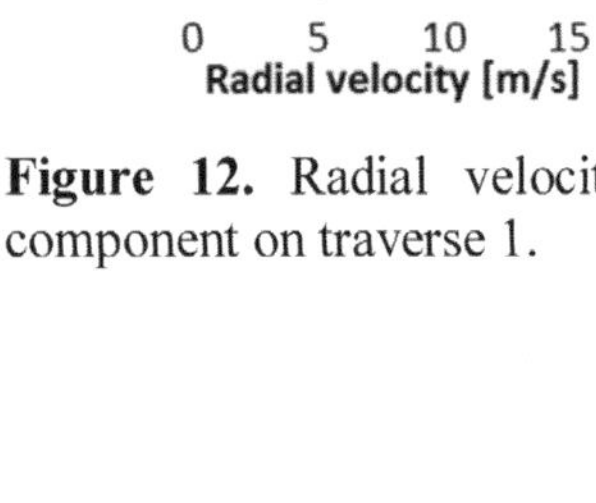

Figure 12. Radial velocity component on traverse 1.

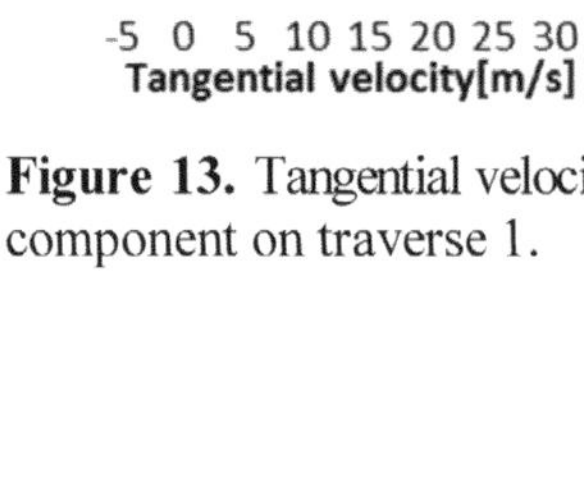

Figure 13. Tangential velocity component on traverse 1.

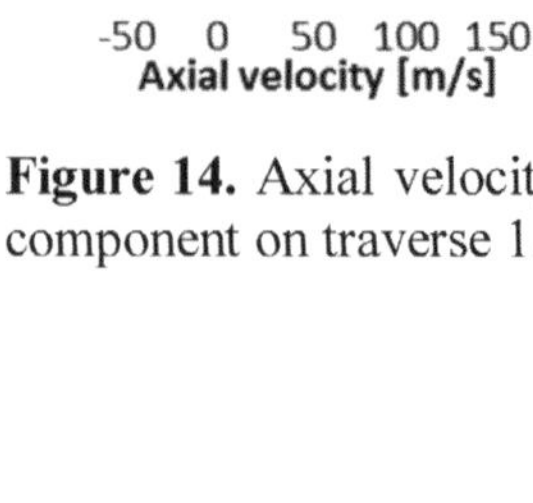

Figure 14. Axial velocity component on traverse 1.

5. Summary, conclusions and future work

The flow features of the rotor blade labyrinth sealing were investigated. It was discovered that the vortical structure relies heavily on the geometry and its location and intensity is dependent on the pressure ratio and rotational speed of cavity walls. Moreover the jet created downstream of the contraction above the first rib is a source of the pressure rise at the upstream side of the second rib. This feature is specific for the investigated labyrinth sealing, but it can be harnessed for flow control application. Finally, the comparison of 2D and 3D models shows the same trends with little local differences.

Several important conclusions, valid for the future implementation of flow control devices aimed at decreasing the gap flow, can be drawn from this study. Firstly, for the purpose of preliminary investigations of flow control devices much less computationally demanding 2D axisymmetrical model can be used. Secondly, most of the flow features are highly dependent on the sealing cavities geometry. Thus the geometry modification or introducing a flow control device is likely to change the flow structure significantly and possibly reduce the leakage flow. For this particular case, the pressure rise on the upstream side of the second rib surface is promising, as the potential location the flow control device based on local pressure difference.

Future studies will be focused on various air jet flow control devices, which will be optimized and implemented on ribbed configuration.

Acknowledgments

This work was supported by the Coopernik project "Cooperative Research for Next Generation High Efficiency LP Turbine" INNOLOT/I/11/NCBR/2014. This research was supported by CI TASK and by PL-Grid Infrastructure in part.

References

[1] Denton J D 1993 *J. Turbomach* **115(4)**, 621–656
[2] Rosic B, Denton J D and Pullan G 2005 *J. Turbomach* **128(4)**, 699–707
[3] Lampart P, 2006 *Task Quarterly* **10(2)**, 139-175
[4] Wallis A M, Denton J D and Demargne A A J 2000 *J. Turbomach* **123(2)**, 334-341
[5] Porreca L, Kalfas A I and Abhari R S 2008 *J. Turbomach* **130(3)**
[6] Rosic B, Denton J D, Curtis E M and Peterson A T 2008 *J. Turbomach* **130(4)**
[7] Rosic B, Denton J D 2008 J. Turbomach **130(2)**
[8] Lehmann K, Thomas R, Hodson H and Stefanis V 2009 *ASME Paper No. GT2009-59531*, 515-525
[9] Spalart, P R and Allmaras S R 1992 *AIAA Paper* **92**
[10] Wasilczuk F, Flaszyński P and Doerffer P 2015 *Task Quarterly* **19(2)**, 89-100
[11] Puzyrewski R and Jakubek A 1989 *Prace Instytutu Maszyn Przepływowych*, **89**
[12] Puzyrewski R and Jakubek A.1991 *Prace Instytutu Maszyn Przepływowych*, **93**
[13] Frączek D, Wróblewski W and Chmielniak T 2015 *Proc. of 11th European Conf. On Turbomach. Fluid Dyn. & Thermodynamics ETC11*
[14] Numeca International 2014 *Theoretical Manual* Brussels
[15] Zimmerman H, Kammerer A and Wolff K H 1994 *ASME Paper No. 94-GT-131*,

Impact of numerical method on a side jets formation in a round jet

K Wawrzak[1] **and A Boguslawski**[1]

[1] Czestochowa University of Technology, Faculty of Mechanical Engineering and Computer Science, Al. Armii Krajowej 21, 42-201 Czestochowa, Poland

E-mail:wawrzak@imc.pcz.czest.pl

Abstract. Numerical analysis of a formation of side jets in an externally modulated round jet is presented. The research is performed applying Large Eddy Simulation method and high-order codes based on Cartesian and cylindrical coordinates. The main attention is paid to an impact of numerical approach on the formation of the side jets, their number and localisation. The results obtained suggest that on Cartesian meshes the number and directions of the side jets are dependent on the distribution of the mesh nodes.

1. Introduction

Contemporary Computational Fluid Dynamics (CFD) offers a wide range of numerical methods, which are capable to predict complex phenomena occurring in fluid flows with a very high accuracy. From the point of view of data validation, knowledge on how a numerical approach influences the physics of the flows and how to exclude such an artificial impact seems to be crucial issues. The present work focuses on Large Eddy Simulation (LES) of an externally excited round jet. A particular attention is devoted to the impact of the numerical approach on the physics of a flow and a possible formation of the so – called side – jets . The side – jets were observed experimentally in a globally unstable and externally modulated round jet, where very strong vortical structures appeared in the near field of the jet as streams released from the jet inclined at some angle to the jet axis [1,2]. Monkewitz et al. [2] claimed that the primary mechanism of the side – jets formation was an azimuthal instability of the vortical structures, that caused a deformation of vortices. However, a complete mechanism driving this phenomenon is not fully understood so far and needs further analyses. The phenomenon of the side – jets formation was observed in a recent numerical investigation [3] devoted to the global instability in a variable density round jet. The LES predictions [3], using Cartesian meshes, showed four well visible side – jets oriented along the mesh lines. Whereas, experimental studies [2] demonstrated that the number of side – jets was between two and six and they were randomly oriented and approximately equally spaced in the azimuthal direction. According to [3] the mechanism leading to the formation of side – jet phenomenon was captured accurately and only the localization of the side – jets was forced by the Cartesian mesh since periodic variation in the azimuthal direction is unavoidable. As concluded in [3] the only difference between LES predictions and a real flow was the mechanism introducing the perturbations, namely random (real situation) versus determined by Cartesian meshes. The discrepancies between numerical and experimental observations were a motivation to perform detailed LES analysis consisting in a comparison of the results coming from two different numerical codes, one based on Cartesian meshes and the second on the cylindrical coordinate system.

2. Numerical methods

As mentioned above the LES predictions were performed employing two numerical codes differing in an application of the coordinate system namely Cartesian and cylindrical. The main difference between these numerical approaches is a mesh resolution in the azimuthal direction. The figures 1 and 2 show inlet planes with the nodes distribution for both numerical meshes. It is seen that the application of the cylindrical mesh ensures a uniform mesh resolution in the azimuthal direction.

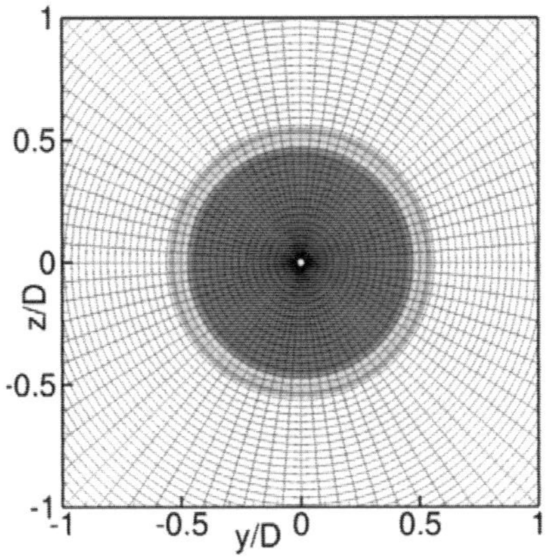

Figure 1. View of the mesh at the inlet plane in the case of the cylindrical coordinates.

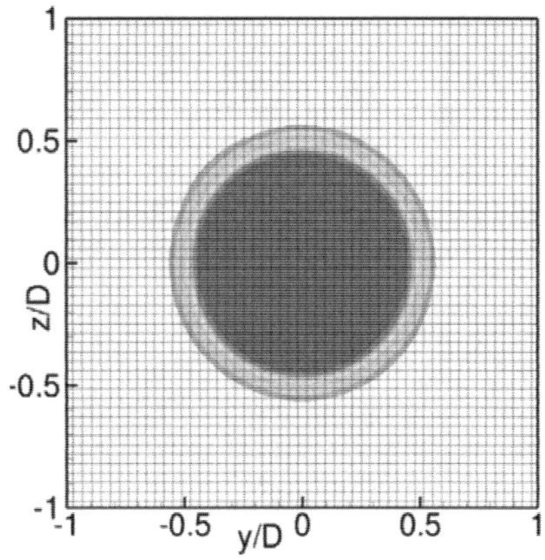

Figure 2. View of the mesh at the inlet plane in the case of the Cartesian coordinates.

2.1. Cartesian coordinate system.

The LES calculations in the case of the Cartesian coordinate system were carried out with the use of high-order solution algorithm based on the projection method for pressure-velocity coupling and a half-staggered mesh arrangement described in details in [4].The time integration was performed by a predictor-corrector (Adams-Bashfort/Adams-Moulton) method and the spatial discretisation was based on 6th order compact difference scheme. The code used in this study was previously verified in isothermal, non-isothermal, exited and reacting jets [3,5,6,7].

2.2. Cylindrical coordinate system.

The application of the cylindrical coordinate system for the Navier–Stokes equations introduces the singularity at the axis which is due to presence of terms involving the term $1/r$, where r – radial distance. Moreover, a definition of the computational domain as $(0,2\pi)\times(0,\hat{R})$, whereboundary of the domain, forces to specify the boundary conditions at $r=0$, even if there is no physical boundary at the axis [8]. In order to avoid these problems the solution algorithm was formulated applying a coordinate transformation from $(0,2\pi)\times(0,\hat{R})$ to $(0,\pi)\times(-\hat{R},\hat{R})$ and the $r=0$ was omitted. The time integration was performed using a low storage Runge-Kutta method and the spatial discretisation was based on compact difference scheme for the radial and axial directions and the Fourier spectral method for the azimuthal one.

3. Boundary conditions and computational domain

The investigation presented in the paper is focused on an externally modulated jet. The computational domain is a simple rectangular box $14D\times14D\times18D$ in the case of Cartesian coordinates and the cylinder $7D\times18D$ in the case of the cylindrical coordinate system as shown in the figures 3 and 4, where $D=2R$ is a jet diameter . In the case of the Cartesian coordinate system the mesh consists of 180 nodes in axial direction, whereas in normal and spanwise directions the number of mesh points is equal to 128. The cylindrical mesh contains $64\times64\times180$ nodes in radial, azimuthal and axial directions, respectively. The inlet boundary condition is specified in terms of instantaneous velocity

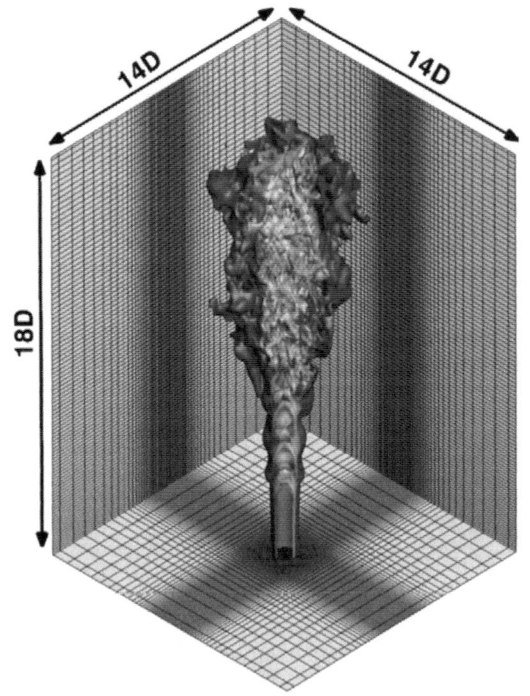

Figure 3. The computational domain with the mesh at the inlet and side boundaries in the case of the Cartesian mesh.

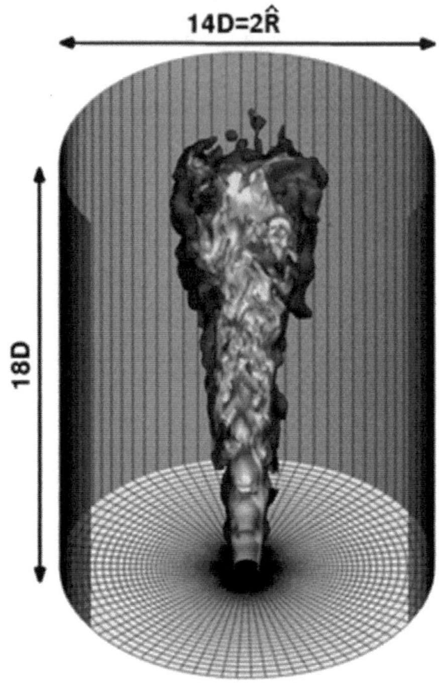

Figure 4. The computational domain with the mesh at the inlet and side boundaries in the case of the cylindrical mesh.

profile as $u(\mathbf{x},t) = U_{mean}(\mathbf{x}) + u_{turb}(\mathbf{x},t) + u_{forcing}(\mathbf{x},t)$. The velocity fluctuations $u_{turb}(\mathbf{x},t)$ are computed using digital filtering method proposed in [9] in the case of Cartesian code and according to algorithm proposed in [10] for the calculations on cylindrical meshes. Both methods for the fluctuation velocity guarantee the proper turbulent velocity field. The mean velocity profile is described by the hyperbolic tangent function as follows:

$$U_{mean}(\mathbf{x}) = \frac{U_0 + U_c}{2} - \frac{U_0 - U_c}{2} \tanh\left(\frac{1}{4}\frac{R}{\theta}\left(\frac{r(\mathbf{x})}{R} - \frac{R}{r(\mathbf{x})}\right)\right)$$

where: U_0 - jet centerline velocity, U_c - co-flow velocity, R -radius of the jet, θ -momentum thickness. The co-flow $U_c = 0.01U_0$ is added to compensate lack of natural suction and entrainment from the lateral walls. The external excitation is modeled by adding the axial forcing $u_{forcing}(\mathbf{x},t)$ to the streamwise velocity. The axial forcing is defined as:

$$u_{forcing}(\mathbf{x},t) = A\sin\left(2\pi St\frac{U_0}{D}t\right)$$

where: A - amplitude of the excitation, $St = fD/U_0$ - the Strouhal number of the excitation. The axial velocity is equal to co-flow velocity while the rest of velocity components are set equal to zero at the side boundaries. The pressure is computed by the Neumann conditions $\nabla p \cdot \mathbf{n} = 0$ at the inlet and side boundaries. At the outlet plane velocity components are computed using the convective boundary condition $\partial U_i / \partial t + \partial U_{conv} / \partial \mathbf{n}$, where U_{conv} is a convection velocity which is computed at every time step as mean velocity at the outlet plane. The pressure is assumed constant at the outlet plane.

4. Results

The LES calculations were performed at the Reynolds number $\mathrm{Re} = U_0 D / \nu = 3000$. The jet with the boundary layer characterized by the parameter $R/\theta = 20$ is considered. Numerical analysis apart from an influence of the coordinate system is also focused on impact of inlet parameters on the formation of side–jets. The turbulence intensity at the nozzle exit varied in the range $Ti = 0.01\%U_0 - 1\%U_0$, the amplitude of the excitation is equal to $A = 1\%U_0, 5\%U_0$ and the Strouhal numbers are varying in the range $St = 0.4 - 0.6$. In figures 5,6,7,8 and 9 the selected computations obtained in Cartesian coordinate system are presented. The results obtained for the cylindrical coordinate system are presented in figure 11 and 12.

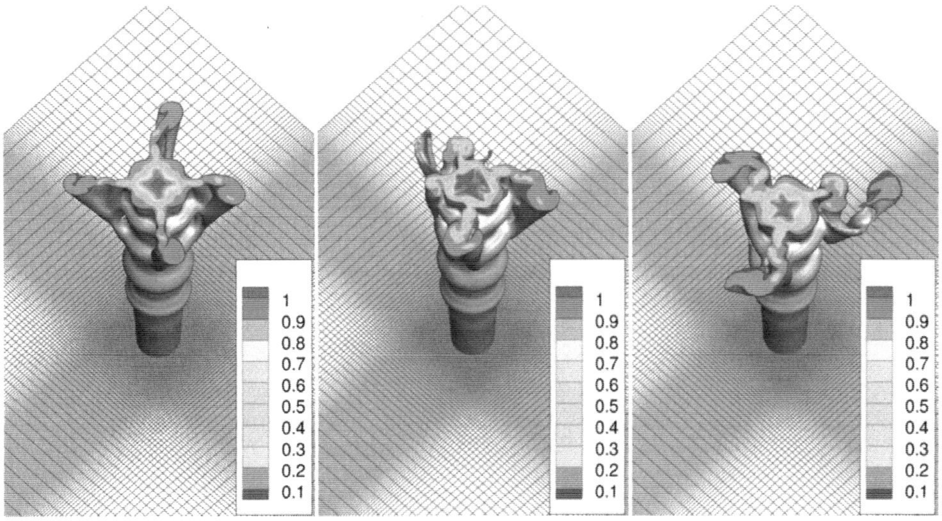

Figure 5. Iso – surfaces of the axial velocity with contours of axial velocity at cross – sections perpendicular to the jet for $St = 0.4$, $A = 1\%U_0$, and turbulence intensities equal to $Ti = 0.01\%U_0$ (left), $Ti = 0.1\%U_0$ (centre), $Ti = 1\%U_0$ (right).

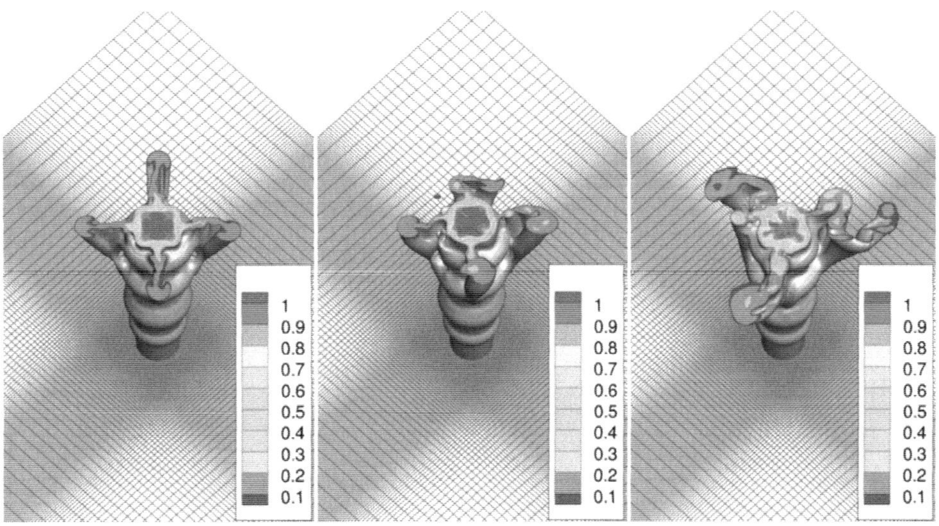

Figure 6. Iso – surfaces of the axial velocity with contours of axial velocity at cross – section perpendicular to the jet for $St = 0.4$, $A = 5\%U_0$, and turbulence intensities equal to $Ti = 0.01\%U_0$ (left), $Ti = 0.1\%U_0$ (centre), $Ti = 1\%U_0$ (right).

Figures 5 and 6 show iso–surfaces of the axial velocity with the cross–sections perpendicular to the jet for Strouhal number equal to $St = 0.4$ and for all turbulence intensities and amplitudes considered. The formation of the side – jets are very well visible in the case of the lowest turbulence intensity ($Ti = 0.01\%U_0$) at the nozzle exit. In this case four side–jets equally spaced in azimuthal direction and located along diagonal of the computational domain are seen (left pictures in figures 5 and 6). An increase of the turbulence intensity at the inlet plane affects the localization and number of side–jets significantly. Only for the amplitude equal to $A = 5\%U_0$ and $Ti = 0.1\%U_0$ (figure 6 centre) results reveal a similar image to the lowest perturbation (cf left pictures in figures 5 and 6). For the same turbulence intensity ($Ti = 0.1\%U_0$) but lower amplitude only two side – jets are observed (figure 5 centre). The results obtained from LES calculations with the $Ti = 1\%U_0$ assumed at the nozzle exit, presented in right subfigures in figures 5 and 6, are characterized by insignificant impact of the amplitude of excitation on the side – jets formation. In this case three side – jets nearly equally

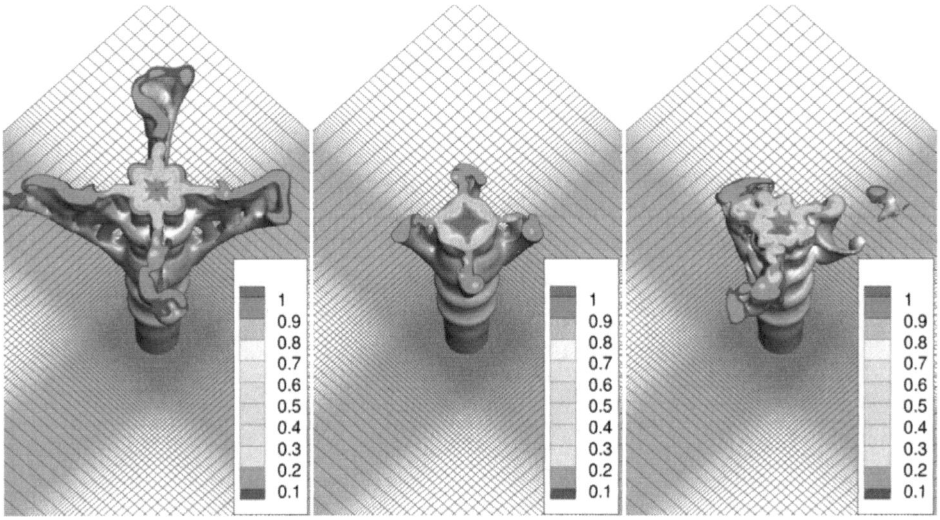

Figure 7. Iso – surfaces of the axial velocity with contours of axial velocity at cross – sections perpendicular to the jet for $St = 0.5$, $A = 1\%U_0$, and turbulence intensities equal to $Ti = 0.01\%U_0$ (left), $Ti = 0.1\%U_0$ (centre), $Ti = 1\%U_0$ (right).

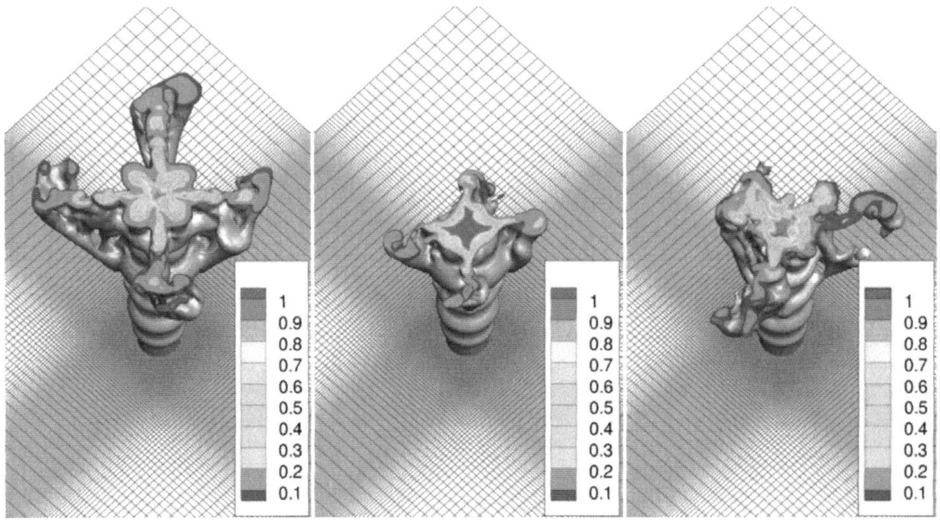

Figure 8. Iso – surfaces of the axial velocity with contours of axial velocity at cross – sections perpendicular to the jet for $St = 0.5$, $A = 5\%U_0$, and turbulence intensities equal to $Ti = 0.01\%U_0$ (left), $Ti = 0.1\%U_0$ (centre), $Ti = 1\%U_0$ (right).

distributed in the azimuthal direction are obtained. However, even in this case strong dependence on Cartesian mesh is seen. There are one side–jet located along the diagonal of computational domain and two located along mesh orientation. The subsequent LES predictions, for the Strouhal number equal to $St = 0.5$, are presented in the figures 7-8. The assumption of the higher Strouhal number leads to a stronger dependence of the side – jets formation phenomenon on the mesh resolution. In this case four side – jets which cover diagonal of computational domain appear for the turbulence intensities equal to $Ti = 0.01\%U_0$ and $Ti = 0.1\%U_0$. However, in this case, for the lowest turbulence intensity, side–jets are much more pronounced than for the $St = 0.4$. They expand in the radial direction at the angle to the jet axis larger than for $St = 0.4$ and cover a larger region of the computational domain. The biggest differences are visible in the case of the highest perturbation at the inlet. In this case for both amplitudes of excitations only one side – jet is predicted. The last part of the LES calculations with Cartesian coordinate system are concentrated on the excitation of the jet characterized by $St = 0.6$.

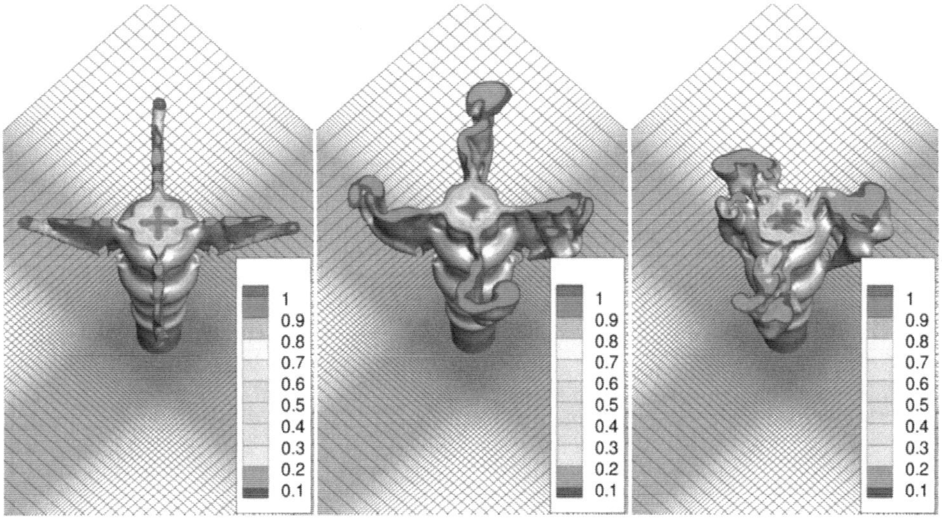

Figure 9. Iso – surfaces of the axial velocity with contours of axial velocity at cross – sections perpendicular to the jet for $St = 0.6$, $A = 1\%U_0$, and turbulence intensities equal to $Ti = 0.01\%U_0$ (left), $Ti = 0.1\%U_0$ (centre), $Ti = 1\%U_0$ (right).

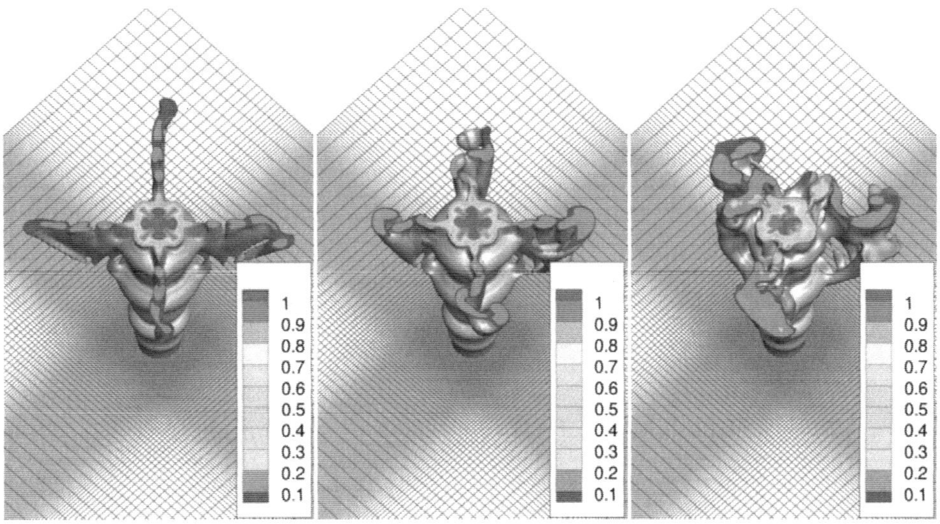

Figure 10. Iso – surfaces of the axial velocity with contours of axial velocity at cross – sections perpendicular to the jet for $St = 0.6$, $A = 5\%U_0$, and turbulence intensities equal to $Ti = 0.01\%U_0$ (left), $Ti = 0.1\%U_0$ (centre), $Ti = 1\%U_0$ (right).

The results obtained from these calculations are shown in the figures 9 and 10. The iso – surfaces and contours reveal the axial velocity field analogical to the lower Strouhal number. For the cases with $Ti = 0.01\%U_0$ and $Ti = 0.1\%U_0$ the round jet flow is characterized by occurrence of four side – jets located at the diagonal of the computational domain again. Only the results with $Ti = 1\%U_0$ show weaker dependence of the formation of side – jets on the grid points distribution.
In order to verify the results obtained with the Cartesian coordinates the LES predictions were also performed for cylindrical meshes. In this case, the jet was stimulated with Strouhal number $St = 0.5$ chosen based on the previous analysis concerning the Cartesian mesh resolution whereas the rest of inlet parameters remained unchanged. Figures 11 and 12 present the results of LES calculations for the cylindrical coordinate system. The application of the cylindrical coordinates demonstrates significantly different results. In the case of the lowest turbulence intensity at the nozzle exit the formation of six side – jets is visible.

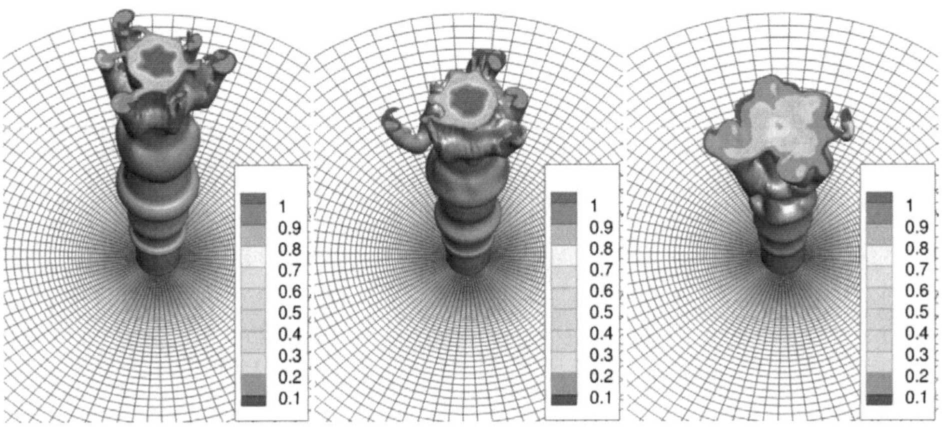

Figure 11. Iso – surfaces of the axial velocity with contours of axial velocity at cross – sections perpendicular to the jet for $St = 0.5$, $A = 1\%U_0$, and turbulence intensities equal to $Ti = 0.01\%U_0$ (left), $Ti = 0.1\%U_0$ (centre), $Ti = 1\%U_0$ (right).

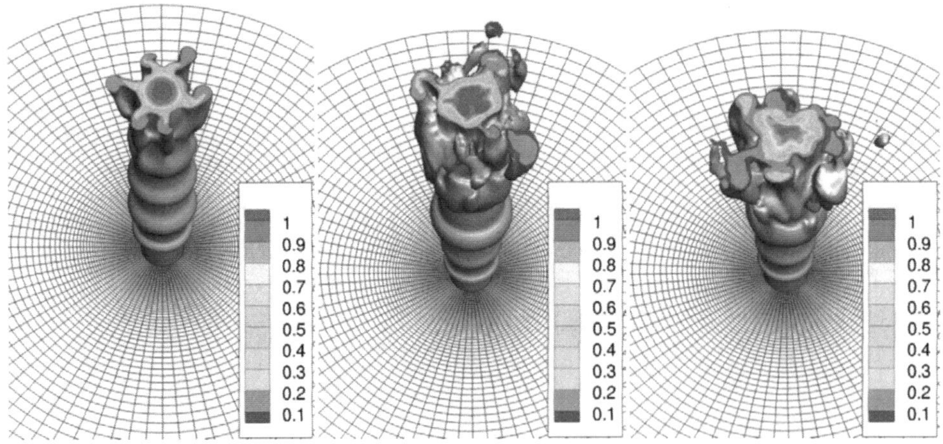

Figure 12. Iso – surfaces of the axial velocity with contours of axial velocity at cross – sections perpendicular to the jet for $St = 0.5$, $A = 5\%U_0$, and turbulence intensities equal to $Ti = 0.01\%U_0$ (left), $Ti = 0.1\%U_0$ (centre), $Ti = 1\%U_0$ (right).

In the case with $Ti = 0.1\%U_0$ only one side – jet can be observed while in the case of the highest turbulence intensity no side – jet can be singled out. Moreover, it should be stressed that the side – jets are inclined at a smaller angle to the jet axis, compared to the results obtained for the Cartesian meshes. This seems to be caused by a decreased mesh resolution along the radial direction. It is also worth noting that in the case of the Cartesian mesh the side – jets remained at the same location and were persistent during the simulation. By contrast, in the case of the cylindrical coordinates the side – jet was rather an instantaneous phenomenon changing the orientation. This observation indicates strong dependence of this phenomenon on the mesh resolution. However, the mechanism leading to the formation of side – jets seems to be predicted correctly with both numerical approaches.

5. Conclusions

The investigation presented in the paper was focused on the influence of numerical approach on the formation of side – jets in the externally modulated round jet. The analysis showed strong dependence of the localization and number of side – jets on the mesh resolution. The LES predictions performed with the Cartesian meshes demonstrated formation of four side – jets, in most of the cases, which cover diagonal of the computational domain. This was especially well visible in the case of a very low turbulence intensity at the nozzle exit. Above-mentioned dependence was significantly less important with increasing the perturbation at the inlet. The application of the cylindrical coordinate system excluded the mesh influence and revealed random number and localization of the side – jets. However, the side – jets predicted on the cylindrical meshes were less inclined to the jet axis. Analysis concentrated on influence of inlet parameter on side – jets formation did not reveal significant impact of the amplitude and Strouhal number.

Acknowledgements

This work was supported by and statutory funds BS/PB-1-103-3010/11/P. PL-Grid infrastructure was used to carry out the computations.

References

[1] Raghu S, Lehmann B and Monkewitz P A 1990 *Proc. 3rd European Turbulence Conference* 221-6
[2] Monkewitz P A, Lehmann B, Barsikow B and Bechert D W 1989 *Phys. Fluids* A **1** 446-8
[3] Boguslawski A, Tyliszczak A and Wawrzak K 2016 *Phys. Fluids* **28(2)** 025108
[4] Tyliszczak A 2014 *J. Comp. Phys.* **276** 438-67
[5] Wawrzak K Boguslawski A and Tyliszczak A 2015 *Flow Turbul. Combust.* **95** 437-59
[6] Tyliszczak A and Geurts BJ 2014 *Flow Turbul. Combust.* **93** 221-47
[7] Rosiak A and Tyliszczak A 2016 *Int. J. Hydrogen Energ.* **41** 9705-17
[8] Constantinescu G S and Lele S K 2002 *J. Comp. Phys.* **183** 165-86
[9] Klein M, Sadiki J and Janicka J 2003 *J. Comp. Phys.* **186** 662-5
[10] Kempf A, Klein M and Janicka J 2005 *Flow Turbul. Combust.* **74** 67-84

Journal of Physics: Conference Series **760** (2016) 012041 doi:10.1088/1742-6596/760/1/012041

On the unsteady Reynolds thermal transpiration law

P Ziółkowski [1,2] **and J Badur** [1,3]
[1] Energy Conversion Department, The Szewalski Institute of Fluid-Flow Machinery PAS-ci, Fiszera 14, 80-231 Gdańsk, Poland

E-mail: [2]`pziolkowski@imp.gda.pl`, [3]`jb@imp.gda.pl`

Abstract. This paper presents the phenomenon of unsteady Reynolds thermal transpiration flow. The possible constitutive equations in the transpiration shell-like layer were studied analytically and numerically. There has been also examined experimental case of helium flow from cold to hot reservoir in nanopipe.

1. Introduction

Since 1879, when Reynolds has discovered the phenomena of thermal transpiration, this issue is still theoretically analyzed. Thermal transpiration can be defined as the macroscopic movement induced in a rarefied gas by a wall temperature gradient. The thermal mobility force, responsible for thermal transpiration, can be classified as a typical Newtonian "*vis impressa*" – a force which can be able to accelerate any massive substances along the wall [2]. In the literature there are still developed phenomenological models of the slip-regime for high Knudsen numbers [3,4], which can be also fitted to thermal transpiration for the case of gas in highly rarefied conditions, namely, in the free molecular regime. Hence, there is a need to prepare numerical implementation of Reynolds model of thermal transpiration at the contemporary experiment for example conducted by Rojas-Cárdenas et al [5].

In this paper, basing on experience in modelling of various surface phenomena [2-4,6-10] and starting from the first concepts by Reynolds [1], we have developed a consistent theory of thermal transpiration, that is able to describe thermal mobility both in rare and dense gases moving on an arbitrary solid surfaces. In particular, we discussed possible constitutive equations in the transpiration shell-like layer. The proposed model has been implemented and numerically assured by simulation of unstationary thermal transpiration benchmark experiment made by Rojas-Cárdenas et al [5].

1.1. Motivation

The *thermal transpiration* is not any pure academic problem and it has been recently used to construct the micro-electro-mechanical systems (MEMS) and nano-electro-mechanical systems (NEMS). The increased pumping effect could be obtained by a steeper gradient of temperature, extending tube length or making stages of pumping. Other micro-devices, which are based on both thermal transpiration and adherence slipping, called accommodation pumping, have been developed in nanotechnology [4].

Additionally, both recent developments in miniaturization and nanotechnology, as well as fullness unrecognized phenomena's occurring on a fluid-solid inter-phase within micro- and nano-scale compel us to reformulate conventional solid and fluid modelling. In nearby dividing surface, new inter-atomic forces make transport phenomena behave differently from their macroscopic counterparts. The most important ones are surface tension, surface curvature and surface twist that possess quite different elastic properties [2-4]. Yet other one is generalized slip described by the Duhem, Navier, and du Buat numbers [7-10].

Owing to fast progress of nano-fluidic technology the nature of boundary conditions of fluid at solid surfaces has been revisited for the recent ten years. The huge surface to volume ratio turned our attention for fundamental understanding of nature of "surface-type driving forces" and the variety of different "jump" phenomena like slipping, rolling, spin-slipping, surface mobility, surface friction, thermo-porosis, etc [2-4]. In a very tiny channel there is no place for typical bulk phenomena, since whole flow is dominated by surface processes [6-11]. Moreover, these surface processes, in general, lead to some enhancement of transport [mass, momentum, thermal energy, electric current] in comparison with the bulk classical transport. Thus, quite surprising phenomena, unobserved in the bulk, appears in a surface layer and if a cross-section of a channel or pore becomes relatively small. Mentioned enhancement appearance could be easily visible and measurable [6-7].

However in this paper, we focused on this surprising phenomenon that the thermal transpiration is in theoretical point of view. Our main assumption is that this phenomenon is completely governed by a contact interface between a hot surface and colder gas. This is a surface migration of gas induced in a pipe with a gradient of temperature along its surface. This surface migration goes from colder to hotter places of surface. Thus, from mathematical modeling viewpoint, the thermal transpiration should be understood as some kind of "boundary conditions".

The notion of "boundary condition" traditionally comes from well understood vocabulary of continuous fluid mechanics but numerous examples of the literature shows, that it could be misleading sometimes. In such case we wanted to change a paradigmatic look on "boundary condition" by assuming that within a boundary layer there are some additional, physically distinguished and measurable quantities which are necessary to create new class of "constitutive equations". It practically means that we need to define the constitutive equations not only in the fluid bulk but also on the fluid-solid surface. Since, in practice, "boundary conditions" are seen as auxiliary mathematical-type constrains that allow solving partial-differential equation in a bulk, in the paper, we shell resign from exposing of "boundary conditions" notion and instead of it we will introduce a notion of a "boundary layer" or, equivalently, "interface layer" with boundary forces.

From this viewpoint, the boundary conditions related to the bulk equation of fluid motion [linear momentum balance], becomes "the surface layer equation of motion". In a solid-fluid surface it is necessary to make up the separate equation of motion. Besides of the classical bulk, wall stress also appears quit new quantities like: surface friction force, surface mobility force, surface flux of momentum, etc. [3,5].

A main hypothesis of our paper is that the thermal transpiration is an example of surface mobility force which undergoes an unclassical procedure of constitutive modeling. We postulate that the status of surface constitutive equations should be equal to the status of constitutive equations in the fluid bulk. As has been already mentioned the thermal mobility force can be classified as a typical Newtonian "*vis impressa*" – a force which can be able to accelerate any massive substances. The phenomenological models of thermal transpirations existing in the literature are based on different approaches to kinetic theory of gases [12], which are usually restricted to a very small calculation domain. The more consist one, proposed here, can be also fitted to slip-regime for high Knudsen numbers and, can be more correctly extended the so-called "second order models" [13,14].

2. Model description

In this paper, we have assumed that the continual motion of gas in a nano-channel is governed simultaneously by two kinds of equations of motion [8], mainly, the bulk equation of motion:

$$\partial_t(\rho\,\mathbf{v}) + \mathrm{div}(\rho\,\mathbf{v}\otimes\mathbf{v} + \mathbf{p}) = \rho\,\mathbf{b} \tag{1}$$

and the surface layer equation of motion:

$$\partial_t(\rho_s\mathbf{v}_s) + \mathrm{div}_s(\rho_s\mathbf{v}_s\otimes\mathbf{v}_{s\parallel}) + \mathrm{div}_s(\mathbf{p}_s) - \varpi\,\mathbf{n} + \mathbf{p}\mathbf{n} + \mathbf{f}_{\partial V} = \rho_s\mathbf{b}_s \tag{2}$$

In above $\rho\mathbf{v}$ and $\rho_s\mathbf{v}_s$ are the bulk and the surface momentum density vectors, ρ and ρ_s are the gas density in the bulk and on the boundary, respectively. Next, having an original Reynolds' notations in mind, $\mathbf{p} = p_{ij}\mathbf{e}_i \otimes \mathbf{e}_j = \mathbf{p}^{\mathrm{T}}$ and $\mathbf{p}_s = \mathbf{p}_s^{\mathrm{T}}$ are the bulk and the surface flux of momentum, $\mathbf{n}$ is the unit normal vector on the boundary surface, $\mathbf{f}_{\partial V}$ – is the boundary force, $\mathbf{b}$ and $\mathbf{b}_s$ are the body forces in the bulk and on the boundary. Additionally, ϖ is the Stokes normal surface pressure, due to assumption that in the boundary layer a usual spherical pressure tensor changes into an ellipsoidal pressure tensor. Surface divergence is defined as a right contraction of surface gradient: $\mathrm{grad}_s(\cdot) = \mathrm{grad}(\cdot)\mathbf{I}_s$. Tangent to surface component of the slip velocity is $\mathbf{v}_{s\|} = \mathbf{v}_s\mathbf{I}_s$ where the surface Gibbs identity is defined to be: $\mathbf{I}_s = \mathbf{I} - \mathbf{n}\otimes\mathbf{n}$ and generally $\mathbf{v}_s$ - slip velocity is equal $\mathbf{v}_{s|} = \mathbf{v} - \mathbf{v}_{wall}$. Hence Poiseuille flow can be modeled by equation (1).

In the case, when we can omit whole dynamics of a surface layer, and the surface flux of momentum is represented only by the Young-Laplace surface tension: $\mathbf{p}_s = \gamma\,\mathbf{I}_s$, the surface equation of motion (2) reduces to the Stokes boundary condition [2]:

$$\mathrm{div}_s(\gamma\,\mathbf{I}_s) - \varpi\mathbf{n} + \mathbf{f}_{\partial V} + \mathbf{p}\,\mathbf{n} = 0 \tag{3}$$

Let us note that this boundary condition acts independently from any constitutive definition of Stokes pressure tensor – it can be defined traditionally to be: $\mathbf{p} = p\mathbf{I} - 2\mu\,\mathbf{d} + 2/3\mu\,\mathrm{I}_{\mathbf{d}}\mathbf{I}$ or it can also consist a transpiration part or a second order part. Hitherto, we denote the rate of deformation as: $\mathbf{d} = 1/2(\mathrm{grad}\,\mathbf{v} + \mathrm{grad}^{\mathrm{T}}\mathbf{v})$; $\mathrm{I}_{\mathbf{d}} = \mathrm{tr}(\mathbf{d})$.

2.1. The boundary forces

The boundary force, responsible for a so-called generalized slip can be splinted on two contradictive components: surface friction and surface mobility: $\mathbf{f}_{\partial V} = \mathbf{f}_r + \mathbf{f}_m$. Both forces are the subject of constitutive modeling. For instance, the friction force possesses three known contributions: Duhem, Navier and du Buat [4]:

$$\mathbf{f}_r = \nu_0(p-\varpi)\mathbf{e}_f + \nu(\mathbf{v}-\mathbf{v}_{wall}) + \nu_2(\mathbf{v}-\mathbf{v}_{wall})\mathbf{e}_f \tag{4}$$

where slip versor is defined as: $\mathbf{e}_f = (\mathbf{v}-\mathbf{v}_{wall})/|\,\mathbf{v}-\mathbf{v}_{wall}\,|$ and ν_0; ν; ν_2 are three friction coefficients [Duhem, Navier, du Buat, respectively] which are simultaneously under influence of the material of fluid and solid surface. However the surface mobility forces, which act usually against a surface friction, describe surface contributions from a numerous transpiration phenomena [2]:

$$\mathbf{f}_m = -c_{mp}\mathrm{grad}_s(p-\varpi) - c_{m\theta}\mathrm{grad}_s(\theta) - c_{mN}\mathrm{grad}_s(N) - .... \tag{5}$$

which depends linearly on surface gradient of surface pressure, surface temperature and surface concentration. More information is in section about mobility force reconstruction. In particular, the main goal of the paper is to discuss only Reynolds' contributions to definition of surface mobility modeling: $-c_{m\theta}\mathrm{grad}_s\theta$.

2.2. Thermal transpiration discovery

Osborne Reynolds is the one author, who entirely devoted to the discovery of thermal transpiration [1]. He asserts that a primary reason of thermal transpiration in the bulk motion is not the second gradient of temperature along axes of a capillary, but the axial gradient of acceleration that acts close to the wall surface it makes the enhancement of the normal Navier slip.

Journal of Physics: Conference Series **760** (2016) 012041 doi:10.1088/1742-6596/760/1/012041

Let us now return, to the paper [1], mainly concerning with the experimental discovery of the thermal transpiration and its description by a Maxwellian-like kinetic theory of transpirated gases.

Reynolds unified three phenomena: the Graham *pressure transpiration*, the Graham *concentration transpiration* [15] and, just discovered by Reynolds, the *thermal transpiration* became into a one generalized model of transpiration phenomena.

Thermal transpiration was originally discovered in thin (5.3-6.3 mm) porous plates made of meerschaum, stucco or graphite. Reynolds has observed that there is the difference in rates at which different gases transpire through minuscule pore channels for the same initial pressure and the same difference of temperature on two sides of plate. Simultaneously, the changing of initial pressure on both sides of the plate was observed. After some time (equilibrium time) the difference of pressure between the hot and cold sides becomes constant and no more migration of gas is observed. It means that the difference of temperature on the two sides of porous plate might cause gas, without any initial difference of pressure to pass through the plate from the cold to the hot side.

For instance, the permanent (equilibrium) difference of pressure equal to an inch of mercury and the higher pressure on the hotter side is achieved in the atmospheric pressure, a difference 160 °F (from 52 to 212 °F) with hydrogen on both sides of the stucco plate of thickness 6.3 mm. Moreover, Reynolds has found that a wall gradient of temperature leads to different mass flow of different gases. This effect can be potentially used as a separation mechanism for gas mixtures [5].

Summarizing the results of his experiments, Reynolds has deduced eight first phenomenological laws of the thermal transpiration. Nowadays, the most known of them is, the "Law IV" , which says that in the equilibrium, for those cases when rarefaction has reached the point at which the difference of pressure is nearly proportional to the density, the ratio of hot and cold gas pressures is in square roots to the absolute temperature:

$$\frac{p_H - p_C}{p_H} = \frac{\sqrt{\tau_H} - \sqrt{\tau_C}}{\sqrt{\tau_H}} \tag{6}$$

Let us note, that in both – meerschaum and stucco plates of 6.3 mm thick, it is tendency of the pressure difference $\delta p = p_H - p_C$, for hydrogen, air and carbon acid [see[1]; original Figures 4,5,6], as the initial pressure rises, to a maximum value. The indices H (hot), C (cold) have appeared in Reynolds' paper only just in further part of theoretical considerations. Note that denotation of the absolute temperature (τ) coming from Rankine's proposal.

2.3. Thermal mobility force reconstruction

In order to show how Reynolds made his model of thermal transpiration one should follows in details of Reynolds construction of kinetic theory of rarefied gases – in the bulk and on a surface. Original Reynolds' line of reasoning is based on his own previous achievements concerning forces which must result when "heat" is communicated from a solid surface to a gas. Reynolds' model of special could define of tangential and normal forces, especially when small spaces might be allowed for the gas molecules to pass. Moreover, Reynolds was able to show that it is a "complete theory of transpiration phenomena" which describes not only thermal transpiration but any other transpiration phenomena discovered by Graham [15], for instance a phenomena called "impulsion".

Reynolds' denotations - ϖ, θ, N - means the normal pressure, temperature and mole concentration of a gaseous component. Hence, we could a unified model of pressure, temperature and concentration transpirations on the boundary layer, by adding particular contributions described by equations: [see [1]: original eqs.:(110)(112)(117)], respectively. Now, reducing eq. (2) only to so-called Stokes' boundary layer equation $\mathbf{f}_{\partial V} - \varpi\mathbf{n} + \mathbf{p}\mathbf{n} = 0$, the Reynolds model of slip-mobility layer can be reconstructed shortly to be:

$$-c_{m\varpi}\text{grad}_s\varpi - c_{m\theta}\text{grad}_s\theta - c_{mN}\text{grad}_s N + \nu(\mathbf{v} - \mathbf{v}_{wall}) + (p - \varpi)\mathbf{n} - 2\mu\mathbf{d}\mathbf{n} + 2\beta\mathbf{d}_{(2)}\mathbf{n} = 0 \tag{7}$$

Realizing the Graham phenomena, the appearing of pressure transpiration gradient is explicit in the model. The pressure mobility coefficient $c_{m\varpi}$ is different for any gases in the same material of capillary tube. The definition of the thermal mobility coefficient $c_{m\theta}$ where Reynolds has introduced four accommodation coefficients f: mainly f_1, f_2, f_3, f_4 - two first depend, separately, on the properties of solid surface material and a kind of gas comes across in this model. Next, we are going to describe two processes of interaction of a gas with a boundary. The diffusion transpiration of Graham is defined by the concentration mobility coefficient c_{mN}. About β and $\mathbf{d}_{(2)}$ next section presents.

2.4. Consistency of thermal transpiration with a 3D model of momentum transport

It should be added that the definition of the Cauchy momentum flux on the boundary - $\mathbf{p}$ contains Reynolds' proposition extends a definition of $\mathbf{p}$ taking the diade of acceleration of deformation $\mathbf{d}_{(2)}$. Reynolds, further applied a coefficient β responsible for accelerative contributions, it should be find on the base of a previous researches of "dimensional properties of mater".

The consistency of mathematical model of fluid in a bulk and within the layer boundary was the reason for writing the paper [16]. Here, Reynolds proposes a new definition of the Cauchy momentum flux, where it appears an *acceleration of deformation diade* - $\mathbf{d}_{(2)}$, defined as a symmetric tensor:

$$\mathbf{d}_{(2)} = 1/2\,(\mathrm{grad}\,\mathbf{a} + \mathrm{grad}^{\mathrm{T}}\mathbf{a}) \neq \dot{\mathbf{d}}\ . \tag{8}$$

Thus, the Stokes form of the momentum flux is extended to be:

$$\mathbf{p} = p\mathbf{I} - 2\mu\,\mathbf{d} + \frac{2}{3}\mu\,\mathrm{I}_{\mathbf{d}}\mathbf{I} + 2\beta\mathbf{d}_{(2)} \tag{9}$$

where, the Reynolds coefficient $\beta = r/p$ is inversely proportional to pressure. Hence, substituting the above constitutive equation to the momentum balance $\rho\,\dot{\mathbf{v}} + \mathrm{div}\,\mathbf{p} = \rho\,\mathbf{b}$, and omitting $\mathrm{div}(\mathrm{grad}^{\mathrm{T}}\mathbf{a}) \approx 0$, we obtain after Reynolds [1, eq.(12)]:

$$\rho\,\mathbf{a} + \beta\,\mathrm{lap}\mathbf{a} + \mathrm{grad}p - \mu\,\mathrm{lap}\mathbf{v} - \frac{1}{3}\mu\,\mathrm{grad}\mathrm{I}_{\mathbf{d}} = \rho\,\mathbf{b}\ . \tag{10}$$

The part with β coefficient can be further omitted in the bulk domain of fluid flow, but on the boundary it represents very important - an accelerative - contribution to the slip conditions.

2.5. Law of mass flow rate in a circular tube

Finally, Reynolds has combined Poiseuille pressure driven flow with transpiration pressure driven flow showed in eq (10) by adding the transpiration mobility force (5). To find a formula describing an average velocity in a long pore on a diameter a, it was necessary to assume that characteristic mean-free path of the gas molecules, say l, is the same range as the geometrical parameter a. Then the mass flow rate is [1, eq. 101]:

$$\dot{m} = \rho(\pi a^2)v_z^{aver} = -\rho(\pi a^2)a\sqrt{\frac{2p}{\rho}}\,\{\,\mathcal{F}_1(a/l)\frac{1}{p}\frac{dp}{dz} - \mathcal{F}_2(a/l)\frac{1}{\theta}\frac{d\theta}{dz}\,\}\ . \tag{11}$$

If $\dot{m} = 0$ then the net motion vanish and some constant proportion between $dp/d\theta$ is obtained. The time needed to reach $\delta\,p = p_H - p_C = const$ is easy to be measured. Appearing above function $\mathcal{F}_1$, $\mathcal{F}_2$ are not only the function of (a/l) but also of four accommodation coefficients f_1, f_2, f_3, f_4.

2.6. Simplified Reynolds' model of a boundary layer with the thermal mobility

Despite the fact that investigation on the thermal transpiration has been started long time ago [see: 13,17] many questions remain up to now unsolved. This is connected with a special feature of the thermal transpiration, which is very sensitive to a kind of solid material and a kind of rarified gas. In this paper, thermal transpiration mobility force has been modeled as:

$$\mathbf{f}_m = -c_{m\theta}\mathrm{grad}_s\theta \,. \tag{12}$$

Contrary, mass flux proposed by Reynolds is, in general, sensitive, both, on the properties of gas and solid surface and depends on two *thermal momentum accommodation coefficients* f_3, f_4. However, having a complete experimental data, for instance works of Rojas-Cárdenas at al. [5,18,19], Daggerbart [17], Reese et al. [12], Ewart et al. [20], Pitakarnnop et al. [21], one tray to develop a consistent closure to the $c_{m\theta}$ coefficient within the framework of continuum modeling.

3. Analysis

It should be mentioned that the above cited law (eq 6) had validity only at a zero-flow $\dot{m} = 0$ final equilibrium state, far from transitional stage of gas displacement, especially at a point where $\dot{m} = \max$. Additionally Daggerhart [17] proposed semi-empirical formula that goes over $p_H / p_C = (\theta_H / \theta_C)^{1/2}$ and predict how the thermo-molecular pressure ratio p_H / p_C at the final equilibrium zero-flow state varies in function of the gas rarefaction, the gas physical properties, and the applied temperature difference to the tube. Many various papers are discussed and verified by Reese, Zheng, and Lockerby [12] as well as by Storvick, Park and Loyalka [13] however a new benchmark experiment has appears in 2011 made by Rojas-Cárdenas et al. [5] and was extended in another works [18,19].

3.1. Geometry and boundary conditions

Rojas-Cárdenas et al [5] have applied an original method for thermal transpiration induced mass flow rate measurements conducted via measuring *in situ* the pressure evolution in time at both ends of the tube using two high-speed response pressure gauges. In the long circular cross-section glass (borosilicate) microtube D=2R=490 μm; L=3.053 cm is connected to two reservoirs (Fig.1): cold (no 1, environmental temperature) and hot (no 2, $\theta_2 = 80^o C$, heated by an internal heater) (Fig.1) with volume ratio of the two reservoirs $V_H / V_C = 0.81$. Before experiment starts, the pressure inside the both reservoirs was regulated by means the vacuum system and stay between 13.3 and 1330 Pa for helium. For numerical simulations we have chose $V_1 = V_C = 14.85$ cm^3.

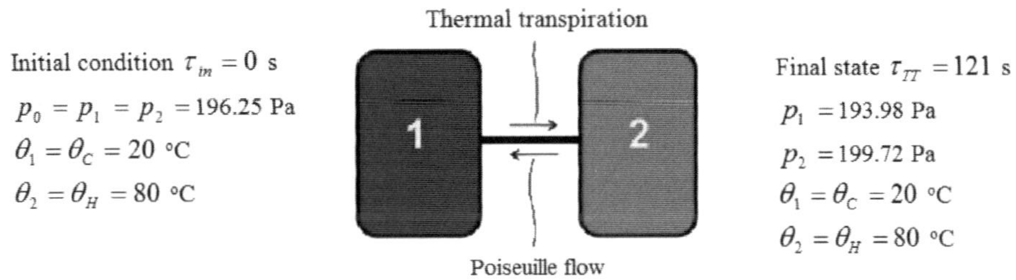

Figure 1. Scheme of helium flow from cold (1) to hot (2) reservoir Rojas-Cárdenas et al [5].

After implementation of mobility force (12) into a commercial code, the Rojas-Cárdenas et al benchmark experiment [5] has been preformed for a basic data corresponding to helium. For numerical simulation we have taken the following data: initial pressure $p_0 = p_{in} = 196.25$ [Pa], cold temperature $\theta_C = 20$ [°C], hot temperature $\theta_H = 80$ [°C] (with linear distribution along the micro-pipe), referential viscosity $\mu = 1.96x10^{-5}$ [Pas], individual gas constant $\widetilde{R} = 2077$ [J/kgK], the ratio of specific heats $\kappa = 1.667$.

Basing on our previous considerations [3,4,7,8] and having in mind proposed in the literature value of thermal accommodation coefficient [12,18,19], we have made a calibration of $c_{m\theta}$ value. Standing on Rojas-Cárdenas et al data [5] and Ewart et al considerations [20], this coefficient is defined as follows:

$$c_{m\theta} = \frac{3}{4}\frac{\mu\nu}{\rho\theta} = \frac{3}{4}\frac{\mu^2}{\rho\theta l_s} \quad . \tag{13}$$

The main unknown was the Navier slip viscosity ν that can be defined by slip length also [2-4]. We assume that numerical value of $c_{m\theta}$ has been calculated using the definition the Helmholtz-Piotrowski slip length: $l_s = \mu/\nu$, and helium-glass slip length $l_s = 0.00016$ cm. In eq (13) the temperature along the micro-pipe was taken according to linear distribution.

4. Results and discussion

Opening a disconnecting valve the flow induced by thermal transpiration makes pressure in both reservoirs equal and flow from cold to hot places derived only by the wall mobility force (2), that depends on $c_{m\theta}$ value between glass and helium and the wall temperature gradient. The pressure in the cold decrease and at the hot palace increases (Fig. 2a). Now, thermal transpiration forces achieve maximal value and mass flux attain pick in a characteristic time τ_R (Fig. 2b).

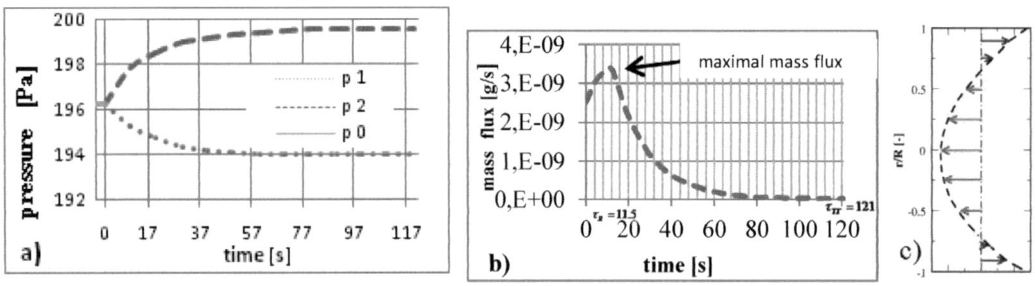

Figure 2. Results of unsteady thermal transpiration modeling of the Rojas-Cárdenas et al benchmark experiment [5] with the thermal transpiration mobility force (eq. 2): a) plot of pressure change in (2) and (1) reservoir; b) the plot of mass flux c) stationary velocity profile.

After that we included the Poiseuille flow governed by difference of pressure (Fig. 2a), that finally leads, after time τ_{TT} to equilibrate state when the pressure difference arrives to its maximum and the resulting mass flux is zero. Our results agreed with the experiment [5]. It is easy to numerically prove that at the final equilibrium zero-flow state (after 121 seconds) is connection of Poiseuille flow in the centre of channel and counter thermal transpiration flow on the surface (Fig. 2c).

5. Conclusions

Reassuming, the numerical implementation of Reynolds model of thermal transpiration and its usefulness for description of an experiment by Rojas-Cárdenas et al [5] have been performed. Finally, Poiseuille flow in the centre of channel and counter thermal transpiration flow on the surface has been achieved.

References

[1] Reynolds O 1879 On certain dimensional properties of matter in the gaseous state *Phil. Trans. Royal Society London* **170** 727–845.
[2] Badur J, Karcz M and Lemański M 2011 On the mass and momentum transport in the Navier-Stokes slip layer *Microfluid Nanofluid* **11** 439-49
[3] Ziółkowski P and Badur J 2014 Navier number and transition to turbulence *Journal of Physics*: Conference Series **530** 012035
[4] Badur J, Ziółkowski PJ and Ziółkowski P 2015 On the angular velocity slip in nano-flows *Microfluid Nanofluid* **19** 191-8
[5] Rojas-Cárdenas M, Graur I, Perrier P and Méolans JG 2011 Thermal transpiration flow: A circular cross-section microtube submitted to a temperature gradient *Phys. Fluids* **25** 031702
[6] Badur J, Karcz M, Lemański M and Nastałek L 2011 Foundations of the Navier-Stokes boundary conditions in fluid mechanics *Transactions of IFFM* **123** 3-55
[7] Badur J, Ziółkowski P and Ziółkowski PJ 2014 Do we need the Navier number? *5th International Conference on "Heat Transfer and Fluid Flow in Microscale"* Marsylia: Campus St. Charles, France, 1-6.
[8] Badur J and Ziółkowski P 2013 Further remarks on the surface vis impressa caused by a fluid-solid contact *in Proceedings of the 12th Joint European Thermodynamics Conference, JETC 2013.* Brescia, Italy: Eds. M. Pilotelli and G.P. Beretta 581-586.
[9] Ziółkowski P and Badur J 2014 On the Boussinesq eddy viscosity koncept based on the Navier and du Buat number *XIV* Aplied Mechanics *Scientic Session PTMTiS*, Book of Abstracts (J. Sawicki Ed.) 87-88, Bydgoszcz 20 November 2014.
[10] Ziółkowski P, Badur J and Ziółkowski PJ 2014 Do we need Navier number? – Further remarks and comparison with another dimensionless numbers *PhD Interdisciplinary Journal* **01** 259-69
[11] Karniadakis GE, Beskok A and Aluru N 2005 *Microflows and Nanoflows: Fundamentals and Simulation* (New York: Springer)
[12] Reese JM, Zheng Y and Lockerby DA 2007 Computing the near-wall region in gas micro- and nanofluidics: critical Knudsen layer phenomena *J. Computational and Theoretical Nanoscience* **4**(4) 807-13
[13] Storvick TS, Park HS and Loyalka SK 1978 Thermal transpiration: a comparison of experiment and theory *J. Vac. Sci. Technol.* A **15**(6) 1844–52
[14] Roşca AV and Pop I 2013 Flow and heat transfer over a vertical permeable stretching/shrinking sheet with a second order slip *Int. J. of Heat and Mass Transfer* **60** 355–64
[15] Graham T 1846;1849 On the motion of gases *Phil. Trans. Royal Soc. London,* **131** 573-632; **137** 349-362
[16] Reynolds O 1883 On the equation of motion and the boundary conditions for viscous fluid *British Association Sec. A Papers* II **46** 132-137
[17] Daggerhart JA 1969 Correction of pressure data for thermal transpiration effects in pneumatic lines *NASA Rep.* CR-1489 Washington D.C. December
[18] Rojas-Cárdenas M, Graur I, Perrier P and Méolans JG 2013 Time-dependent experimental analysis of a thermal transpiration rarefied gas flow *Phys. Fluids* **25** 072001
[19] Rojas-Cárdenas M, Graur I, Perrier P and Méolans JG 2015 A new method to measure the thermal slip coefficient *Int. J. of Heat and Mass Transfer* **88** 766–74
[20] Ewart T, Perrier P, Graur I and Méolans JG 2007 Tangential momentum accomodation in microtube *Microfluid Nanofluid* **3** 689-695
[21] Pitakarnnop J, Varoutis S, Valougergis D, Geoffroy S, Baldas L and Colin S 2010 A novel experimental setup for gas microflows *Microfluid Nanofluid* **8** 57-72